COMPACT MOSFET MODELS FOR VLSI DESIGN

COMPACT MOSFET MODELS FOR VLSI DESIGN

A.B. Bhattacharyya

Jaypee Institute of Information Technology University
India

TK
7874.75
B52
2009
web

IEEE PRESS

IEEE Communications Society, Sponsor

John Wiley & Sons (Asia) Pte Ltd

Other Wiley Editorial Offices

John Wiley & Sons, Ltd, The Atrium, Southern Gate, Chichester, West Sussex, PO19 8SQ, UK

John Wiley & Sons Inc., 111 River Street, Hoboken, NJ 07030, USA

Jossey-Bass, 989 Market Street, San Francisco, CA 94103-1741, USA

Wiley-VCH Verlag GmbH, Boschstrasse 12, D-69469 Weinheim, Germany

John Wiley & Sons Australia Ltd, 42 McDougall Street, Milton, Queensland 4064, Australia

John Wiley & Sons Canada Ltd, 5353 Dundas Street West, Suite 400, Toronto, ONT, M9B 6H8, Canada

Wiley also publishes its books in a variety of electronic formats. Some content that appears in print may not be available in electronic books.

Library of Congress Cataloging-in-Publication Data
Bhattacharyya, A. B. (Amalendu Bhushan)
    Compact MOSFET models for VLSI design / A.B. Bhattacharyya.
      p. cm.
    Includes bibliographical references and index.
    ISBN 978-0-470-82342-2 (cloth)
1. Integrated circuits–Very large scale integration–Design and construction. 2. Metal oxide semiconductor field-effect transistors–Design and construction. I. Title.
    TK7874.75.B52 2009
    621.39'5–dc22

                                                                          2008045585

ISBN 978-0-470-82342-2 (HB)

Typeset in 10/12pt Times by Thomson Digital, Noida, India.
Printed and bound in Singapore by Markono Print Media Pte Ltd, Singapore.
This book is printed on acid-free paper responsibly manufactured from sustainable forestry in which at least two trees are planted for each one used for paper production.

Dedication

To my parents to whom I promised in my childhood that I would write a book.

To my wife Pranati, daughters Usree and Urmi, and family members Arun, Isha and Bela, for their selfless sacrifice and support, which allowed me to honor the most sacred of all obligations—the promise to my parents.

Dedication

Contents

Preface

The microelectronics industry has witnessed an explosive progress in the capability to integrate components on a silicon chip with CMOS as the predominant technology. The present day System-on-Chip (SOC) contains billions of devices with the Metal–Oxide–Semiconductor (MOS) transistor forming the basic building block. The successful design of such a complex integrated system requires extensive computer simulation where an accurate and faithful mathematical description of the MOS transistor is an essential prerequisite. MOS characteristics, therefore, are described by a set of mathematical equations producing a *compact model* for the device. These models are ported into a SPICE circuit simulator for faithful representation of MOSFET characteristics. Models provide the communicating interface between design and manufacturing.

As the complexity and application domains of silicon chips continue to enlarge, the challenge to satisfy apparently conflicting conditions of precision and simplicity is demanding and formidable. The litmus test of a compact model describing the MOS transistor lies in its capability to describe the device characteristics in a way that it is computationally efficient and reliable to handle simulation of ever increasing complex circuits of an electronic system ensuring cost effectiveness and fast turn around time. Further, the model needs to be scalable, predictive for the generations of technology nodes to follow, and capable of addressing statistical process variations. With such multidimensional requirements, different approaches have emerged which can be broadly categorized as *Engineering models, Physical models, and Application Specific hybrid models*.

Historically, development of the threshold voltage, V_T, based compact MOSFET model was pioneered by the University of California, Berkeley (UCB) and placed in the public domain. MOSIS (MOS Implementation Service) provides to the customers parameters for UCB models for a specified technology. The pedagogical approach, interfacing design with technology, has consequently been exclusively through V_T based models. However, there have been developments in recent years which have led to a paradigm shift. Apart from the threshold voltage based method, new approaches using inversion charge and surface potential as state variables have emerged, providing a new perspective in the field of compact models. PSP is already recognized as an alternative standard for the compact MOSFET model.

The measured characteristics reflecting various phenomena related to the device is the ultimate physical reality as far as users are concerned. The *perceived* model representations for the same phenomena, however, may be altogether different depending on the framework on which the model is structured. In the VLSI community, technologists use the model parameters to project the *most* optimistic limit of the MOSFET performance, making parameter extraction as important as model development. Designers, on the other hand, judge the model parameters from the perspective of technological details revealed, and their compliance to simulation requirements such as convergence, etc.

Scaling of feature size of MOSFETs is associated with ever increasing number of model parameters. Thus, VLSI designers are confronted with a complex situation which forces them to handle a large number of parameters for a given model. Further, they need to make a judicious choice from a multiplicity of competing models. The formation of the Compact Model Council (CMC), more of an industry response, has been a natural and inevitable development for providing benchmarks and standards for compact models. As there is very little possibility that a single model will be able to meet the requirements of a wide range of application domains, the diversity of models has to be accepted as a reality. Thus, VLSI designers need to be better informed and broadly educated with *inclusive* coverage of various compact MOSFET models, as against the prevailing practice of training on a specific or an *exclusive* model. However, the emerging pattern in the field of the compact MOSFET model rarely gets reflected in VLSI education in undergraduate and postgraduate courses. On the other hand, cutting edge insights of new perceptions can stimulate the students and model users to a deeper understanding of the models. This will lead to a class of educated model consumers capable of exploiting the models judiciously.

The objective of the author is to create awareness regarding the different approaches prevalent in compact MOSFET modeling in a single text at an introductory level. The present text is the outcome of a project initiated by the author over the last six years in organizing broad-based courses on compact MOSFET models supporting VLSI design at both the undergraduate and postgraduate levels. Selected topics from the material given in this text have also formed the basis for customized modular courses on VLSI design for industry professionals. The author received encouragement from both the academic and industry communities for preparing a text that highlights the core concepts of different models, which can prepare the foundation for further model-specific specialization.

The text has been organized in such a way that it is self-contained. It summarizes in Chapter 1 the basic equations of semiconductor physics necessary for understanding phenomena related to MOS transistors. Chapter 2 deals with an ideal MOS capacitor structure, where the basic concepts and equations that are required to understand various compact models are developed. In Chapter 3 non-ideal effects, which are an integral part of real world practical MOS structures, are considered as add-on phenomena or perturbations over the ideal structure. Keeping in view the fact that the ITRS projected scaled MOSFET performance targets require intervention of non-classical structures, the basic MOS capacitor platform has been discussed with potential gate and channel engineering alternatives requiring the use of an Si–Ge heterostructure and undoped silicon as the substrate, and the use of a high-k dielectric as a replacement of the silicon–dioxide gate dielectric. Chapter 4 presents the formulation of four types of compact models, namely, V_T based models, charge based models, and two surface-potential based models. Chapter 5 deals with the effects of scaling on MOS transistor characteristics, and discusses how such phenomena are integrated into the various compact models. Chapter 6 presents dynamic MOSFET models in quasistatic and non-quasistatic conditions outlining approaches of the BSIM, EKV, HiSIM, and PSP models. The noise model has also been presented briefly. Quantum mechanical effects in nanoscale-MOSFETs have been outlined in Chapter 7. The final chapter is on emerging non-classical structures which hold the prospect to replace mainstream conventional bulk CMOS structures. The basic equations for a double gate MOSFET, which is projected to spearhead nanoscale CMOS, are derived.

There is modularity in the organization of the chapters which enables the reader to be selective. The field trials carried out by the author for undergraduate and postgraduate courses are as follows: at both undergraduate and postgraduate levels the materials which are generic are discussed as core material. For undergraduates, the course was not for beginners but for those who opted for the course on compact MOSFET models as an elective for specializing in VLSI design. The core material prepares the platform on which any particular model can be launched. At the undergraduate level, the author experimented with threshold voltage based models for classroom teaching, using the EKV approach for assignments. PSP/HiSIM was referred only qualitatively. The non-classical structures, quantum effects, and non-quasistatic models were not part of curriculum. At the postgraduate level, threshold voltage based models formed the classroom teaching material, and PSP/HiSIM constituted the assignment activity. EKV was referred qualitatively. The non-classical MOSFET structure, non-quasistatic effect, quantum phenomena and noise form the core material at the postgraduate level. Modularity in the presentation of topics, built in the text, allows flexibility to the instructor to mix and match the content with the requirement of the course. Proficiency in the use of a Computer Algebra System (CAS) is helpful, as compact model equations can be simulated. Such activity can form an important component of assignments.

As more than one category of compact models have been presented in the text, it has been impractical to impose a common list of symbols globally. This would have required modifications to all compact model equations with symbols that differ from those given in the corresponding model documentation. As readers are expected to turn to the official documentation of a specific compact model for details, such a change in symbols would create unnecessary confusion. Hence, a set of generic symbols, which relate to core material on MOS physics independent of model framework specifics, has been identified. The symbols of specific models, by and large, have been kept intact, and have been defined locally wherever used.

The book discusses model equations for highlighting the conceptual framework, and selects them based on suitability from a pedagogical viewpoint. Therefore, equations for the latest version have not necessarily been considered to be the most suitable for introducing a model. At times, an earlier version happens to be a more convenient entry point to the model. The selection of model equations in the text has to be viewed from this perspective. The framework of each class of model being different, the emphasis of the author has been to get the readers initiated to the core concepts that form the basis of an approach, rather than on evolutionary improvements in a given approach. The text is neither a handbook, nor a replacement to the official documentation available for each category of models. Further, as the models are regularly upgraded, the model equations used in the book do not necessarily reflect the current status of a model. For some models information was sketchy and restricted. At many places, the text has used equations suitably truncated to serve the limited objective of pedagogy. Therefore, the readers are again advised to refer to the official documentation for the selected model.

In the presentation of models the author has taken the position to be non-judgmental on the merits, suitability, or otherwise, of a given model, leaving the reader to draw his/her own conclusions. The extent of coverage of a given model, at times, has been dictated by the availability of materials rather than assignment of any planned weightage on the coverage.

The text will be supported by a solution manual for instructors, which will be available from the publisher's website for the book. The simulation plots given in the text have mostly been implemented using MATLAB. The source code for these simulations shall also be made available. These important and useful supplements can be found at the following URL: http://www.wiley.com/go/bhattacharyya.

It is believed that the broad pedagogical presentation of various models will bring them out of the captive domain of the few select industries using them for design and product development. Increased awareness regarding the various compact models available will hopefully lead to wider participation in benchmarking the performance and evaluation of the models. The author believes that such widespread participation from universities, through courses on compact models, will culminate in foundries providing model parameters for *all* categories of models. This will allow VLSI designers to make optimal use of the possibilities provided by the diversity of compact models.

A. B. Bhattacharyya
Noida, Uttar Pradesh
India

Acknowledgements

I express my gratitude for the grace and blessings I received from my mentors, academic and spiritual, but for whom I could not have overcome many insurmountable roadblocks which came on the way since the project of writing a book on compact models for VLSI design was launched at the University of Goa, about five years ago. Prof. B. S. Sonde and Prof. P. R. Sarode provided the initial support. The author initiated an academic program where the compact MOSFET model, *albeit* in a small way, formed the nucleus of the concept of bringing different categories of MOSFET models as a part of the curriculum. At a later stage, MOSFET compact model related topics formed the course content at the Jaypee Institute of Information Technology University (JIITU), NOIDA (India), both at the undergraduate and the postgraduate level. The author records his appreciation for Gajanan Dessai, for useful interaction as a student, having participated in model related courses delivered by the author both at Goa University and at JIITU. His involvement later transcended beyond the classroom assignments. His contribution in Chapter 7 is specially acknowledged. He was also associated with the implementation of simulations for the text and the solution manual. The following students of JIITU made supporting contributions for the solution manual: Umang Sah, Parul Gupta, Ankur Srivastava, Yashshri Negi, Kirminder Singh, Saurabh Chaturvedi, Nishant Chandra, and Apoorva Yati. No less has been the influence of my doctoral, postgraduate, and undergraduate students whose enthusiasm has been responsible for my remaining abreast with the current state of developments in microelectronics. Prof. M. S. Tyagi provided useful editorial suggestions for Chapter 1. The timely help from Dr Manoj Saxena, University of Delhi, for his contributions and review on MOSFET noise, is thankfully acknowledged. Rohit Sharma assisted the author in checking the manuscript of a few chapters. Usree Bhattacharya contributed significantly in arranging reference materials which was immensely useful. Urmi Battu offered her bit in helping me with the illustrations.

The author is grateful to Prof. C. Hu, University of California, Berkeley, Prof. G. Gildenblat, Arizona State University, and Prof. C. Enz, CSEM and EPFL, Switzerland, for granting permission to use equations as necessary for the BSIM, PSP, and EKV models which they have respectively pioneered. Supporting material received from Prof. M. Miura-Mattausch on HiSIM is also appreciated. The permission from Prof. D. Vasilevska of Arizona State University, and Prof. S. Banerjee of University of Texas, Austin, for using SCHRED and UTQUANT simulators is gratefully acknowledged. Prof. H. Wallinga, of the University of Twente, The Netherlands, favored the author with his lecture notes on the Memelink–Wallinga model. Prof. Ed Kinnen, the University of Rochester (NY), and Prof. R. Bowman, presently with the Rochester Institute of Technology, Rochester (NY), are remembered for suggesting the need of a textbook relating MOSFET models with VLSI design, when the author was with them on a sabbatical. Prof. Kinnen's practical suggestions regarding how not to get lost in the expanding MOS universe, and his abiding interest in the project, have been extremely refreshing.

The works of the authors of various compact modeling groups have been referred to frequently, as their models have been the subject of discussion. The book, however, has been no less influenced by the work of Prof. C. T. Sah, Prof. J. R. Brews, and Prof. Y. Tsividis. The author is apologetic about the fact that many deserving works might not have been referred to.

The author records his deep appreciation for the sustained support of Prof. Suneet Tuli, Indian Institute of Technology, Delhi, and Prof. D. N. Singh, University of Punjab, Chandigarh, on making available supporting technical literature. The constructive suggestions of unknown reviewers were responsible for the inclusion of some enriching content not originally planned in the book.

It has been a pleasure to work with James Murphy and Roger Bullen, of John Wiley & Sons, Singapore, who pleasantly guided the author to successful compilation of the book.

I take this opportunity to express my sincere thanks to Dr Y. Meduri and Prof. J. P. Gupta of JIITU, who provided me the support and encouragement to complete the project. It was a rewarding experience to interact with Prof. Sanjay Goel, Head, Computer Science Department, JIITU, Noida, in the planning of innovative learning centric and activity based approach for teaching compact models.

The author is conscious of omissions and the scope for improvement of the book. Suggestions are welcome from readers.

Last but not the least, I have no adequate words to express my gratitude to my colleague Harkesh Singh Dagar, but for whom perhaps the completion of the book would have ever remained so near yet so far. Even though not belonging to the area of VLSI design, the project benefited enormously from his varied knowledge and experience spanning multiple science and engineering disciplines. His knowledge of digital typography also proved indispensable in the preparation of the manuscript. Stimulating and objective discussions with him, free from subjective biases, led to a better appreciation of the limitations of the existing approaches to compact modeling.

Further, he undertook the painful task of carrying out the review of the entire text and the solution manual. His efforts have resulted in the ironing out of inconsistencies in the content, and simplified the presentation. He set up editorial discipline in the absence of which, in spite of all technical content, the book would not have been anywhere near the level attained. His uncanny ability to uncover errors and inconsistencies has hopefully eliminated most blemishes at the pre-production stage. The author owns responsibility for any errors or omissions which remain. I remain indebted to him forever for his selfless effort. I take this opportunity to thank his family, and record my deep appreciation for their self-effacing support and sacrifice, while Harkesh got involved with this project.

This is practice, but remember first to set forth the theory.

Leonardo da Vinci

John Wheeler was an advocate of doing more with less. That is how scientific modeling should be done, rather than heaping on a bunch of parameters to match phenomena—a good model accomplishes more, explains more than what it contains.

Todd Rowland
Academic Director
NKS Summer School
Wolfram Research, Inc.

List of Symbols

Physical Constants

$\epsilon_0 = 8.854 \times 10^{-14}\,\mathrm{F\,cm^{-1}}$	Permittivity of free space
$\epsilon_{Si} = 11.7\epsilon_0$	Permittivity of Si
$\epsilon_{ox} = 3.97\epsilon_0$	Permittivity of SiO$_2$
$q = 1.6 \times 10^{-19}\,\mathrm{C}$	Electron charge
$k = 1.38 \times 10^{-23}\,\mathrm{JK^{-1}}$	Boltzmann's constant
$h = 6.63 \times 10^{-34}\,\mathrm{J\,sec}$	Planck's constant
$m = 9.1 \times 10^{-31}\,\mathrm{kg}$	Free electron mass

Select Parameter Values for Silicon

$m_n^* = 0.33$ (relative to unity)	Effective mass of electron
$m_p^* = 0.55$ (relative to unity)	Effective mass of hole

Values at a temperature of 300 K

$n_i = 1.4 \times 10^{10}\,\mathrm{cm^{-3}}$	Intrinsic carrier concentration
$\mu_n = 1400\,\mathrm{cm^2\,V^{-1}\,s^{-1}}$	Electron mobility in bulk silicon
$\mu_p = 500\,\mathrm{cm^2\,V^{-1}\,s^{-1}}$	Hole mobility in bulk silicon
$v_{sat,n} = 8.5 \times 10^6\,\mathrm{cm\,s^{-1}}$	Saturation velocity for electrons
$v_{sat,p} = 5.0 \times 10^6\,\mathrm{cm\,s^{-1}}$	Saturation velocity for holes
$F_{c,n} = 2.5 \times 10^3\,\mathrm{V\,cm^{-1}}$	Critical electric field for electron saturation velocity
$F_{c,p} = 2.5 \times 10^3\,\mathrm{V\,cm^{-1}}$	Critical electric field for hole saturation velocity

Dimensions

L_G	Mask defined gate length
$L = L_G - 2L_D$	Channel length
L_D	Length of lateral diffusion from source/drain under the gate
L_e	Corrected channel length in saturation
$L(S, D)$	Length of source, drain diffusion region in the direction of channel
L_b	Debye length
ΔL	Length of pinched-off/high field region near drain
L_n	Electron diffusion length
L_p	Hole diffusion length
$\bar{l}$	Mean free path
l	Characteristic length (context dependent)
$W = W_g - 2DW$	Effective channel width

W_d	Width of depletion region
W_{ds}	Depletion width at source
W_{dd}	Depletion region at source
W_{dc}	Channel depletion width
W_{dm}	Maximum width of depletion region at threshold
W_{dps}	Depletion width in polysilicon
W_i	Inversion layer thickness
W_I	Implant depth
W_E	Equivalent depletion layer depth source/drain
X_j	Source/drain junction depth
T_{ox}	Gate oxide thickness
T_{fox}	Field oxide thickness
a_0	Lattice constant of silicon
x	Direction perpendicular to channel
y	Direction parallel to channel

Doping and Carrier Concentrations

N_B	Bulk impurity doping volume concentration
N_D	Donor impurity doping volume concentration
N_A	Acceptor impurity doping volume concentration
n_{n0}	Electron concentration in n-type Si at thermal equilibrium
p_{n0}	Hole concentration in n-type Si at thermal equilibrium
N_P	Polysilicon doping concentration
N_c	Effective density of states in conduction band (cm^{-3})
N_v	Effective density of states in valence band (cm^{-3})
n_n	Non-equilibrium free electron density in n-type Si
p_p	Non-equilibrium free hole density in p-type Si
n_p	Non-equilibrium free electron density in p-type Si
p_n	Non-equilibrium free hole density in n-type Si
n_i	Instrinsic carrier concentration
n_{ie}	Effective instrinsic carrier concentration
n^+	Highly doped n-region
N_I	Implant layer doping volume concentration
N_E	Equivalent transformed doping volume concentration for inhomogeneous substrate

Energy

E	Energy
E_c	Lower edge of the conduction energy band
E_{cb}	Lower edge of the conduction band of the bulk (substrate) material
E_{cg}	Lower edge of the conduction band of the gate material
E_v	Upper edge of the valence energy band
E_f	Fermi energy level
E_g	Energy band gap (eV)
ΔE_g	Energy band gap narrowing
χ_{Si}	Electron affinity for Si (eV)
E_i	Intrinsic Fermi level
E_d	Donor energy level
E_a	Acceptor energy level

E_{fg}	Fermi energy level of gate material
E_{fn}	Quasi-Fermi energy level for electrons
E_{fp}	Quasi-Fermi energy level for holes

Voltages and Currents

ϕ_g	Potential corresponding to work function of gate material
ϕ_{Si}	Potential corresponding to work function of Si
ϕ	Electrostatic potential
ϕ_F	Fermi potential
ϕ_f	Bulk potential; the potential difference between Fermi and intrinsic energy level
ϕ_t	Thermal voltage $(= kT/q)$
ϕ_s	Surface potential with reference to bulk
ϕ_{ss}	Surface potential with reference to source
ϕ_{sS}	Surface potential at source end of the channel
ϕ_{sD}	Surface potential at drain end of the channel
Φ_b	Conduction band offset between Si–SiO$_2$
ϕ_{bi}	Built-in potential barrier across junction
ϕ_{gs}	Potential corresponding to work function difference between gate and substrate
ϕ_n	Quasi-Fermi potential of electrons
ϕ_p	Potential drop across polysilicon depletion layer
ϕ_{sm}	Surface potential at midpoint of the channel $(L/2)$
ϕ_0	Surface potential at onset of strong inversion
$\Delta\phi_0$	Drain induced barrier lowering w.r.t. source
ϕ_i	Potential at high–low/low–high transition point in inhomogeneous substrate
V_{GB}	Gate voltage w.r.t. bulk
V_{DB}	Drain voltage w.r.t. bulk
V_{SB}	Source voltage w.r.t. bulk
V_{ox}	Voltage drop across gate oxide
V_{DS}	Drain voltage w.r.t. source
V_{GS}	Gate voltage w.r.t. source
V_{CB}	Channel potential w.r.t. bulk
V_{DG}	Drain to gate voltage
V_{CS}	Channel potential w.r.t. source
V_{fb}	Flat band voltage
V_T	Threshold voltage
V_{T0}	Threshold voltage without substrate bias
V_{TB}	Threshold voltage w.r.t. bulk
V_{TS}	Threshold voltage w.r.t. source
V_{MI}	Gate voltage corresponding to $\phi_s = 2\phi_F$
V_{WI}	Gate voltage corresponding to $\phi_s = \phi_F$
V_{SI}	Gate voltage corresponding to $\phi_s = (2\phi_F + m\phi_t)$
V_{DSsat}	Drain current saturation voltage
V_P	Pinch-off voltage
I_{DSsat}	Saturation drain current
I_{DS}	Drain to source current
I	Current

I_{diff}	Diffusion current
I_{drift}	Drift current
I_{DB}	Drain to bulk current
I_{sub}	Substrate current due to hot electrons
I_G	Gate current
I_{GS}	Gate to source tunneling current
J	Current density
$J_{n,diff}$	Diffusion current density due to electrons
$J_{p,diff}$	Diffusion current density due to holes
$J_{n,drift}$	Drift current density due to electrons
$J_{p,drift}$	Drift current density due to holes
TC	Tunneling coefficient
S	Supply function

Charges

Q'_b	Bulk charge density per unit area
Q'_{bm}	Maximum bulk charge density per unit area
Q'_n	Inversion layer charge density per unit area
Q'_g	Gate charge density per unit area
Q'_{sc}	Surface space charge density per unit area
Q'_f	Fixed oxide charge density per unit area in SiO_2
Q'_m	Mobile charge density per unit area in SiO_2
Q'_{it}	Interface trapped charge density per unit area
Q'_{ot}	Oxide trapped charge density per unit area
Q_s	Total inversion charge assigned to source
Q_d	Total inversion charge assigned to drain
ρ	Volume space charge density per unit volume

Capacitances

C'_{ox}	Oxide capacitance per unit area
C'_{sc}	Space charge capacitance per unit area
C'_b	Depletion layer capacitance per unit area
C'_n	Inversion layer capacitance per unit area
C'_{eq}	Equivalent gate–substrate capacitance per unit area
C'_{min}	Minimum gate–bulk capacitance per unit area
C'_{dmin}	Minimum depletion capacitance per unit area
C'_{fb}	Flat band capacitance per unit area
C_{diff}	Diffusion capacitance
C'_j	Junction capacitance per unit area
C'_{j0}	Zero bias junction capacitance per unit area
C_{gs}	Gate–source intrinsic capacitance
C_{gd}	Gate–drain intrinsic capacitance
C_{gb}	Gate–bulk intrinsic capacitance
C_{bs}	Bulk–source intrinsic capacitance
C_{bd}	Bulk–drain intrinsic capacitance
$C_{ox} = C'_{ox}WL$	Total oxide capacitance
C_{ov}	Total gate to source/drain capacitance

1

Semiconductor Physics Review for MOSFET Modeling

1.1 Introduction

The idea of controlling electric current through the field effect on a semiconductor surface dates back to as early as 1930. It was proposed in a patent claim by Lilienfeld which forms the conceptual framework of present day MOSFET operation [1]. From the account of the evolution of the MOS transistor, as outlined by Sah [2], it appears that MOS technology did not emerge primarily as a consequence of a focussed pursuit to realize Lilienfeld's field effect concept, but as a by-product of a series of revolutionary developments in the field of solid-state electronics. These developments include the invention of point contact and junction bipolar transistors by Bardeen [3], and Shockley [4], innovations in planar technology and selective diffusion through windows etched in oxide as demonstrated by Hoerni [5], integration of passive and active components by Kilby [6], to name only a few landmark milestones. These technological capabilities enabled the realization of the Metal–Oxide–Semiconductor Field Effect Transistor (MOSFET) structure by Kahng and Atalla [7–9], which was considered suitable for replacement of prevailing bipolar integrated transistor logic circuits. The complementary PMOS–NMOS inverter structure by Wanlass and Sah [10] provided a defining innovation which heralded the CMOS digital era. While the realization of MOSFET on silicon took about as long as thirty years, taking the initial patent as the reference in the time scale, the history of the *subsequent* development and scaling phase driven by Moore's law has been breathtaking [14]. The acronym MOS, originally signifying a Metal–Oxide–Semiconductor structure, has now become synonymous with Microsystem-On-Silicon [12].

After the initial metal gate form, the *polysilicon gate* MOS structure has been central to continuous MOS technology upgradation [11]. Its essential constituents are shown in Figure 1.1 and illustrated below.

(i) A *single crystal* silicon serves as the substrate at the surface of which the field effect transistor action is manifested. At the interface of the silicon substrate and the gate dielectric layer, a channel of mobile carriers is formed by the application of a gate voltage which

Compact MOSFET Models for VLSI Design A.B. Bhattacharyya
© 2009 John Wiley & Sons (Asia) Pte Ltd

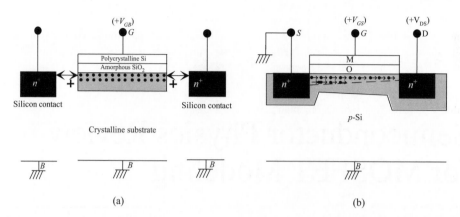

Figure 1.1 Components of the field effect transistor structure: (a) MOS capacitor and contacts; (b) composite MOSFET device

controls the conductivity of the channel. For the device to be reproducible and reliable, it is obligatory that the substrate material be a single crystal where *atoms are arranged in an orderly array.*

(ii) A silicon-dioxide layer is integrated with the substrate at the top of the silicon surface which insulates the silicon substrate from the gate electrode. The insulating film is structurally *amorphous* where *no perceivable long-range crystalline order exists.*

(iii) Over the insulating dielectric a *polysilicon* film is deposited where *a relatively small segment of the film has an ordered crystalline order.* The poly-layer at the top serves as the gate electrode, and is heavily doped to make its conductivity approach that of a metal. Though high conductivity polycrystalline silicon replaces the original metal electrode on top of the gate oxide, it continues to be labelled as a metal as a historic continuity of the nomenclature of MOS for the device structure.

(iv) Finally, *ohmic* contacts are established with the polysilicon layer at the top and the silicon substrate at the bottom, for which aluminum has been the most commonly used material. The stack of the above three layers, with the top and bottom electrodes designated as *Gate* and *Bulk* terminals, respectively, provide the platform for *Field Effect* characteristics.

(v) The two terminal MOS capacitor described above, provides the platform to observe the *Field Effect* phenomenon at the Si–SiO$_2$ interface in the form of a gate controlled space charge constituted by mobile carriers and fixed bulk charges. The mobile carriers induced at the silicon substrate under the gate at the interface are connected to the external world through *conducting silicon probes* of the *same conductivity* as that of the mobile carriers in the channel. The external voltage applied at the two end terminals results in a current which is controlled by the gate field. The basic *four terminal* MOSFET structure shown in Figure 1.1 is formed by *integrating* the diffusion contacts with the MOS capacitor on a common substrate. The capacitor under the gate forms the *intrinsic* part of the MOS transistor, whereas the contact terminals form the *extrinsic* part of the device.

MOS technology has undergone unprecedented evolution to keep pace with projection on *scaling* as envisaged by Gordon Moore [14]. Unlike the pre-birth history of the MOS transistor, the future has been made almost predictive. The International Technology Roadmap for

Semiconductor (ITRS) lays down projections for a comprehensive development of MOS technology which is indicative of the trends, tasks, challenges, and opportunities in the years ahead in terms of process integration, system development, materials, device architecture, bottlenecks, etc. The document is dynamically updated. Table 1.1 is an illustrative projection [13].

Table 1.1 ITRS 2003 roadmap for high-performance logic

Year	2003	2006	2009	2012	2015	2018
Technology node (nm)	100	70	50	35	25	18
MPU gate length (nm)	45	28	20	14	10	7
V_{dd} (V)	1.2	1.1	1.0	0.9	0.8	0.7
T_{ox} (Å)	13	10	8	7	6	5
Clock frequency (MHz)	2 976	6 783	12 369	20 065	33 403	53 207
Maximum power (W)	149	180	210	240	270	300
Static power (W/μm)	4.0×10^{-7}	6.1×10^{-7}	7.7×10^{-7}	9.9×10^{-7}	2.6×10^{-6}	3.9×10^{-6}
Nominal gate delay (ps)	30.24	18.92	12.06	7.47	4.45	2.81
Power delay product (J/μm)	1.4×10^{-15}	9.7×10^{-15}	7.7×10^{-16}	4.8×10^{-16}	3.0×10^{-16}	1.7×10^{-16}
Transistors per chip ($\times 10^6$)	153	307	614	1227	2454	4908

Though the basic MOS structure remains the core of development forecast, the primitive Metal–Oxide–Semiconductor has evolved into a significant complex structure which resembles a stack of multiple layers of varying properties and increasing number of interfaces as shown in Figure 1.2 due to the requirements of gate, channel, and interface engineering.

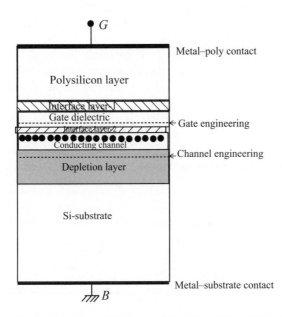

Figure 1.2 Stacked layers and interfaces in the evolving MOS structure

The channel engineering has resulted in substrate *inhomogeneity, stratification, and heterogeneity* of material near the Si–SiO$_2$ interface as shown by dashed lines. The gate engineering has lead to stacking of dielectric layers to avoid tunneling through the thin gate insulator. The *transitions* at the interfaces are not abrupt, introducing effectively transition layers with graded material composition. Thus, the device characteristics and modeling are closely linked with the properties of the materials forming the layers. With miniaturization, the dimensions correspond to only a few atomic layers, where quantum effects start governing the device physics.

The inputs required for MOSFET modeling are: material properties, device structure and dimensions, process details, parameters related to carrier statistics, mobility, band gap narrowing, carrier recombination, generation and multiplication, quantization and tunnelling effects, etc. The present chapter provides a brief summary of the above aspects, more in the form of statements, for the purpose of ready reference without going to details which are available in a number of texts on the subject [17–22].

1.2 Crystal Planes

The crystal structure of silicon (Si) is *diamond cubic* type, where each atom makes covalent bonds with four neighboring atoms to form a tetrahedral structure. In a MOSFET, the device characteristics depend on the crystal plane on which the device is fabricated. In a crystal structure, the orientation of planes is defined by Miller indices, which are expressed by a set of integer numbers. The procedure to obtain Miller indices of a crystal plane is explained through an illustration shown in Figure 1.3, where x, y, and z define the directions of the primitive vectors of the lattice.

- **Step1:** Find the intercept of the plane under consideration in terms of the *axial units* along the axes. In Figure 1.3 the intercepts are A, B, and C where $A = 3$, $B = 2$, and $C = 2$ units.
- **Step2:** Take reciprocals of the intercepts which, for the present case, are: $1/A$, $1/B$, and $1/C$, i.e. $1/3$, $1/2$, and $1/2$ respectively.
- **Step3:** Get the common denominator D of the intercepts A, B, and C. Multiplying the reciprocal numbers by D we get the sets of numbers defining the Miller indices (h, k, l) where $h = D/A$, $k = D/B$, and $l = D/C$, i.e. $(2, 3, 3)$ for the plane under consideration.

In case a plane is parallel to an axis, the intercept of the plane with the above axis is taken as infinity and the reciprocal of the intercept becomes $1/\infty = 0$.

Figure 1.3 shows the (100) plane in a cubic crystal. The preferred orientation of the plane of a silicon crystal for fabrication of conventional MOS devices is (100) with SiO$_2$ as the dielectric

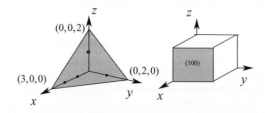

Figure 1.3   Crystal plane representation in a lattice

due to process related advantages. The orientations (110) and (111) are being examined for an optimum performance with the alternative high-k dielectrics, oxynitride, metal oxides, etc., emerging as possible candidates replacing silicon-dioxide as the gate insulator [16].

1.3 Band Theory of Semiconductors

Classical physics which considers current carrying carriers as particles is inadequate to explain many characteristics of semiconductors as well as the behavior of scaled MOS transistors. In a quantum mechanical description the mobile carriers are viewed as matter waves represented mathematically by wave functions $\Psi(x)$, assuming x as the direction of propagation of the particle or the wave associated with it. The motion of a particle, therefore, is described by a wave equation, which is named after Schrödinger. The wave function $\Psi(x)$, mentioned above is the solution of the wave equation which has an important attribute that only a complex wave function is able to satisfy Schrödinger's equation. This leads to the interpretation that it is the product of the wave function $\Psi(x)$ and its complex conjugate $\psi(x)^*$, that has a physical meaning which provides an estimate of its position in terms of the probability of finding the particle in the given volume element. Due to the wave nature of the particle, a particle with a mass m moving with a velocity v, and energy E, is associated with a wavelength given by:

$$\lambda = \frac{h}{p} \tag{1.1}$$

where ($p = mv$) represents the momentum of the particle. The frequency v is given by $v = \frac{E}{h}$ and the angular frequency $\omega = \frac{E}{\hbar}$ where $\hbar = \frac{h}{2\pi}$.

Assuming that the electron motion is confined only to the x-direction, the appropriate time independent Schrödinger equation is given by:

$$\frac{\hbar^2}{2m} \frac{d^2 \Psi(x)}{dx^2} + [E - U(x)]\Psi(x) = 0 \tag{1.2}$$

where $U(x)$ is the potential energy associated with the electron at the position, x, E is the total electron energy, and m is the mass.

For a free particle, $U(x) = 0$. Substituting for the kinetic energy of the electron in terms of wavelength ($E = (h/\lambda)^2/2m$), and generalizing the wave equation with respect to direction of propagation, we can write Equation (1.2) in the following forms:

$$\nabla^2 \Psi + \left(\frac{2m}{\hbar^2} E \right) \Psi = 0 \tag{1.3}$$

$$\nabla^2 \Psi + \mathbf{k}^2 \Psi = 0 \tag{1.4}$$

where $\mathbf{k}$ is a vector representing the direction of propagation of the wave. It has a magnitude $|\mathbf{k}| = \frac{2\pi}{\lambda}$. We briefly describe below some important applications of the wave properties of electrons in semiconductors. Electrons in a semiconductor crystal can be broadly viewed belonging to two categories [21, 26].

(i) **Tightly bound electrons:** These electrons in a crystal belong to the inner energy levels of atoms where they are tightly bound to the nucleus. Such an electron can be considered to be confined within a potential well, as in an isolated atom, for which we assume that the well has width a and has an infinite height. The wave equation, Equation (1.2), can be written in the following form by substituting $k_1 = \sqrt{2m(E - U)}/\hbar$.

$$\frac{d^2 \Psi(x)}{dx^2} + k_1^2 \Psi(x) = 0 \tag{1.5}$$

The solution of Equation (1.5) is given as

$$\Psi = A \exp ik_1 x + B \exp -ik_1 x \tag{1.6}$$

The following boundary conditions are appropriate for the situation under consideration:

$$x \geq a, \ x \leq 0, \ U = \infty, \ \Psi(x) = 0 \tag{1.7}$$

$$0 < x < a, \ U(x) = 0 \tag{1.8}$$

At the two extreme ends of the potential well, $x = 0$ and $x = a$, $\Psi(0) = \Psi(a) = 0$ from the consideration of the wave function continuity. It follows from Equation (1.6) that $(B = -A)$, and:

$$\Psi = A \left(\exp i(k_1 x) - \exp -i(k_1 x)\right) = A_0 \sin(k_1 x) \tag{1.9}$$

Considering that, $A_0 \neq 0$, and from the boundary condition that for $x = a$, $\Psi(a) = 0$, we have $\sin(k_1 x) = 0$. It follows that $k_n = \frac{n\pi}{a}$ with $n = 1, 2, \ldots$, etc., and Equation (1.9) gives the solution in the form:

$$\Psi_n = A_n \sin \frac{n\pi}{a} x \tag{1.10}$$

The eigenvalues of energy are given by [26]:

$$E_n = \frac{h^2}{8ma^2} n^2 = \left(\frac{\pi^2 \hbar^2}{2ma^2}\right) n^2 \tag{1.11}$$

In a semiconductor crystal with a density of, say, N atoms there will be an energy level corresponding to each atom and therefore there is n-fold *degeneracy*. As for the transport of carriers, tightly bound electrons are not of prime interest, we shall discuss in more detail the behavior of so called weakly bound electrons which participate in the conduction of the current in a semiconductor.

(ii) **Weakly bound electrons:** Weakly bound electrons correspond to valence electrons, and due to the wave nature of electrons, they move through the array of the crystalline periodic potential. The motion of electrons in such an environment is described by the *time independent* Schrödinger wave equation which is solved assuming *single electron* approximation, where it is implied that the ith electron moves in an average field of all the core atoms and other electrons except the selected one. Assuming further that the electron motion is confined to the x direction, Equation (1.2) is required to be solved keeping in view the fact that $U(x)$ is the potential energy which has a periodicity defined by the lattice described by $U(x) = U(x + a_0)$, with a_0 denoting the lattice constant defining the periodicity of the potential barrier. The solution of the Schrödinger amplitude (Equation (1.2)) in the presence of periodicity of the potential barrier through which the electrons are moving is given by the *Bloch function*:

$$\Psi(x) = u(x) \exp(ik_x) \tag{1.12}$$

where k_x is the wave vector component in the x-direction, and $u(x)$ is the modulation function of the amplitude of the plane wave of wavelength $\lambda = 2\pi/k_x$. The amplitude function $u(x)$ has the property that it has the same periodicity as that of the potential energy $U(x)$, and its form is determined by the nature of the potential defining $U(x)$.

Figure 1.4 represents the Kronig–Penney model, which is an *idealized* one dimensional representation of the *periodic* coulomb potential well in a crystal structure. It has two distinct regions.

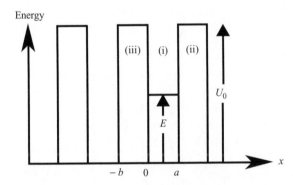

Figure 1.4   The Kronig–Penney one dimensional model of the crystal periodic potential

- **Region 1:** $(0 \leq x \leq a)$. This is identified with the region of location of the atom constituting the crystal which has a width a and $U = 0$. Ψ_1 is the corresponding wave function.
- **Region 2:** $(-b \leq x \leq 0)$. This presents an energy barrier U_0 to the electron of region 1 of width b. Ψ_2 is the corresponding wave function.

The length $a_0 = (a + b)$ defines the spatial periodicity of the potential well in the x-direction. The energy E of the electron is assumed to be less than the barrier energy U_0.

The wave functions Ψ_1 and Ψ_2, the solutions of Equation (1.2) for regions 1 and 2 respectively, are given by:

$$\Psi_1 = A_1 \exp(i\alpha x) + B_1 \exp -(i\alpha x) \tag{1.13}$$

$$\Psi_2 = A_2 \exp(i\beta x) + B_2 \exp -(i\beta x) \tag{1.14}$$

where $\alpha = (2mE/\hbar^2)^{1/2}$ and $\beta = (2m(U_0 - E)/\hbar^2)^{1/2}$.

The unknown constants A_1, B_1, A_2, and B_2 are determined from four equations obtained using the boundary conditions that the *function* $u_x = \Psi \exp -(ik_x x)$ *and its derivative* du_x/dx are (i) continuous at $x = 0$, and (ii) due to the spatial periodicity $(a + b)$, they have the same values at $x = a$ and at $x = -b$.

From the system of equations we get the following condition:

$$\cos k(a+b) = \frac{\beta^2 - \alpha^2}{2\alpha\beta} \sin(\alpha a) \sinh(\beta b) + \cosh(\beta b) \cos(\alpha a) \qquad (1.15)$$

Equation (1.15), where the RHS is a function of the electron energy E, relates it to the wave vector k through the LHS, and can be solved graphically, providing the E–k relationship and energy band representation of a semiconductor.

The RHS in Equation (1.15) is an oscillatory function. The electron energy value E must be restricted within the limits $+1$ and -1 to satisfy the LHS. The range of E for which the value of $f(E)$ lies within the above limits forms the energy band, and that outside the maximum value form the forbidden band. Figure 1.5 shows the allowed and forbidden energy bands in the E–k plane as derived from the Kronig–Penney model.

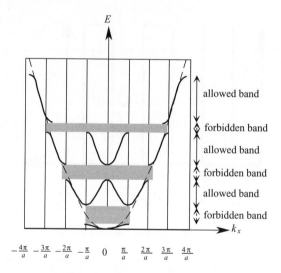

Figure 1.5   Allowed and forbidden energy bands in the E–k plane

We shall provide a qualitative summary of some of the important features of the E–k diagram.

- The E–k diagram for a free electron, given by the relation $E(k) = (\hbar^2/2m)k^2$, is shown by a dashed line, which is a parabola. For a solid, on the other hand, at values of the wave vector $k_x = \frac{n\pi}{a}$, where n is an integer, there is a discontinuity in the E–k diagram, indicating the onset of a forbidden band or energy gap.
- Within the energy band, the electron energy E is an increasing function of the wave vector k. Near the bottom and top of an energy band the E-k curve can be represented by a parabola

of the type represented by the equation $E = E_{k_0} + A(k_x - k_0)^2$, where E_{k_0} is the value of the electron energy at the band edge corresponding to $k_x = k_0$. The constant $A = \mathrm{d}^2 E/\mathrm{d}k^2$ is positive at the bottom of the band and negative at the top, and passes through a point of inflexion where $\mathrm{d}^2 E/\mathrm{d}x^2 = 0$. The double derivative of E w.r.t. k is related to the effective mass of the electron, which will be discussed in a later sub-section.

- We can rewrite Equation (1.12) in the following form, multiplying and dividing the RHS by $\exp i(\frac{2\pi n x}{a})$, which gives:

$$\Psi(x) = u^*(x) \exp k_x^* \tag{1.16}$$

where $u^*(x) = u(x) \exp -i(\frac{2\pi n x}{a})$, and $k_x^* = (k_x + \frac{2\pi n}{a})$. It may be noted that $u^*(x)$ is a product of two components, each of which has the same periodicity. Consequently, $u^*(x)$ also retains the same periodicity as that of the components, and k_x and k_x^* are equivalent. It follows that Equation (1.16) can as well be considered a solution of the Schrödinger equation. The entire range of the electron energy can, therefore, be embedded within the *first Brillion zone, $-\frac{\pi}{a} \le k_x \le \frac{\pi}{a}$.*

1.3.1 Energy–Momentum Relation and Energy Band Diagram

The motion of an electron in a semiconductor in response to an externally applied field F_e is different from that in free space due to the presence of an internal lattice field F_i. Using the classical Newton's law of motion we can write the following expression:

$$F_e + F_i = m\frac{\mathrm{d}v}{\mathrm{d}t}$$

where m is the electron mass in free space, and v is the velocity.

We shall show that the quantity m can be modified to be replaced by m^*, called an *effective mass*, so that the electron motion in a solid can be described by:

$$F_e = m^*\frac{\mathrm{d}v}{\mathrm{d}t} \tag{1.17}$$

From a quantum mechanical viewpoint, the velocity of an electron is given by the group velocity v_g of the wave packet around the wave number k and an angular frequency ω given by:

$$v_g = \frac{\mathrm{d}\omega}{\mathrm{d}k} = \frac{1}{\hbar}\frac{\mathrm{d}E}{\mathrm{d}k} \tag{1.18}$$

where we have made use of the well-known equation:

$$E = \hbar\omega \tag{1.19}$$

The acceleration a, generated on the electron by the external electric field, is obtained from Equation (1.18) by differentiating it with w.r.t. t, which gives:

$$a = \frac{dv_g}{dt} = \frac{1}{\hbar}\frac{d}{dk}\frac{dE}{dt} \tag{1.20}$$

The term dE on the RHS is related to the external field F_e through the following equation:

$$dE = F_e dx = F_e v_g dt \tag{1.21}$$

which gives:

$$\frac{dE}{dt} = F_e v_g \tag{1.22}$$

Combining Equations (1.17), (1.18), and (1.20) we get the expression relating force and acceleration:

$$a = \frac{1}{\hbar^2} F_e \frac{dv_g}{dk} \tag{1.23}$$

Substituting for v_g in Equation (1.23) we get:

$$m^* = \frac{F_e}{a} = \frac{1}{\hbar^2}\frac{1}{\frac{d^2E}{dk^2}} \tag{1.24}$$

From the E–k relationship we can get the curvature d^2E/dk^2, and hence, from Equation (1.24), the effective mass can be estimated. At the bottom of the conduction band, d^2E/dk^2 is positive, which makes the effective mass positive as is the case for normal conduction electrons. At the top of the band, however, $d^2E/dk^2 < 0$ makes the effective mass negative indicating that in response to the applied field the motion of the electron will behave as if its charge is positive. This corresponds to *holes* at the top of the valence band, which can be visualized as a particle with a positive charge and a positive mass.

For current transport in a solid, it is the energy states at the top and bottom of the energy band that matter, as these states correspond to electrons and holes participating in current flow. Assuming a cubic crystal, and that the point of extremum is located at $k_x = k_0$, the energy (E)-wave vector (or momentum) (k) relationship can be represented by:

$$E(k) = E(k - k_0) + \frac{\hbar^2}{2m^*}(k_x^2 + k_y^2 + k_z^2) \tag{1.25}$$

where $m^* = \hbar^2/(d^2E/dk^2)$. This relationship has been obtained using a Taylor series expansion and retaining terms up to the second derivative only. Due to symmetry, the function $E(k)$ contains only even terms. If the energy is measured with respect to the bottom of the conduction band at k_0, then the above equation is given by:

$$E(k) = \frac{\hbar^2}{2m^*}(k_x^2 + k_y^2 + k_z^2)$$ (1.26)

This is similar to the E–k relation of a free particle, $E = \hbar^2 k^2 / 2m$, establishing the point that the motion of an electron in a crystal can be considered to be that of a free electron, replacing the free electron mass by an effective mass.

In general, the effective mass depends on crystal orientation. The components of effective mass can be represented by:

$$m_{ij}^* = \frac{\hbar^2}{\frac{d^2 E}{dk_i dk_j}}$$ (1.27)

When the components of the effective mass in the three directions of k-space are the same, i.e., it is isotropic, the constant energy surfaces are spherical. In general, however, the iso-energy surface will be determined by the anisotropic nature of the effective mass.

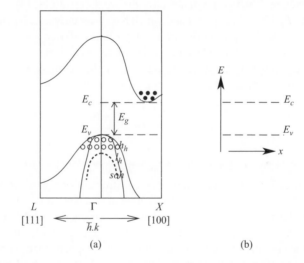

Figure 1.6   (a) *Primary*, and (b) *simplified* silicon energy band diagrams

Figure 1.6(a) shows the energy–momentum E–k diagram for silicon and Figure 1.6(b) shows the simplified spatial, E–x, representation of the energy band diagram. Figure 1.7 shows the iso-energy surfaces or valleys for silicon at the conduction band minima.

We summarize below some facts related to band structure of silicon which shall later be referred to in the text.

- Single crystal silicon has a diamond cubic crystal structure with a lattice constant $a_0 = 0.543$ nm.
- Silicon is an indirect band gap semiconductor where the lowest and highest band edges are identified as the main conduction and valence bands. As seen from Figure 1.6, the bottom

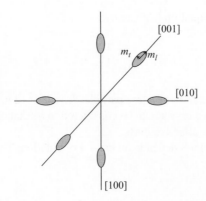

Figure 1.7   Iso-energy valleys for the silicon conduction band near the energy minima

of the conduction band E_c, and top of the valence band E_v, are displaced in the k-axis. The valence band is located at the centre of the Brilloin zone at $k = 0$, which is labeled as point Γ. The valence band has two branches which converge at the top of the band, where it is said to be degenerate. The two branches have different curvatures. $(E_c - E_v)$ defines the energy band gap E_g [25]. The upper band which has lower curvature has heavy holes (h_h), and the lower band with higher curvature has light holes l_h occupying them respectively. There is a third band at $k = 0$ displaced downwards corresponding to split-off holes (*soh*).

- The energy band gap is temperature dependent due to the thermal coefficient of the lattice constant. The temperature dependence is empirically expressed by [17]:

$$E_g(T) = E_g(0) - \frac{AT^2}{B + T} \tag{1.28}$$

where $E_g(0) = 1.17\,\text{eV}$, $A = 4.3 \times 10^{-4}$, and $B = 636$.

- A simplified energy band diagram can now be constructed for Figure 1.6(b) retaining only the band edges E_c and E_v, as the use of effective mass takes care of the curvature of the energy–momentum diagram. It may be noted that the term energy band diagram, henceforth, will mean the representation of *spatial* dependence of electron/hole energy in the crystal. An electron at the bottom of the conduction band E_c corresponds to the conduction electron which is at rest, and has zero kinetic energy. Similarly, a hole positioned at the valence band energy level E_v implies a hole at rest.

- The energy band minima are displaced by an amount of approximately $0.2(\frac{2\pi}{a})$ from the Brilloin zone boundary. The point X represents the first Brilloin zone boundary in the direction [100]. The point L denotes the zone boundary in the direction [111]. There are six conduction band minima corresponding to six equivalent directions, [100], [$\bar{1}$00], [010], [0$\bar{1}$0], [001], and [00$\bar{1}$] which are shown in Figure 1.7. Silicon, therefore, is known as a multi-valley semiconductor with six valleys.

- In the conduction band, silicon has three components of effective mass: (i) one component is $m_1^* = m_l^*$, where m_l^* is designated as the longitudinal mass defined by the major axis of the ellipsoid, and (ii) the other two components are $m_2^* = m_3^* = m_t^*$, which are defined by the minor axes of the ellipsoid.

- For the conduction band electrons in silicon $m_l^*/m = 0.98$, $m_t/m = 0.19$, and the anisotropy coefficient $m_l/m_t = 5.1$.
- The valence band has three branches all having a maximum value at $k = 0$. Two branches are degenerate as they have the same energy values at $k = 0$. The upper and lower branches correspond respectively to heavy and light holes, m_{ph}^* and m_{pl}^* respectively, which is evident from the curvature of the E–k diagram, the upper one having a lower curvature. The values of heavy and light holes are given by: $m_{ph}/m = 0.49$ and $m_{pl}/m = 0.16$. The third branch, originating due to spin-orbit interaction, is displaced from the valence band edge by 0.035 eV. The value of the split-off hole effective mass is $m_{psoh}/m = 0.245$.
- It is general practice to use an average value of mass m_c^* which is given by:

$$\frac{1}{m_c^*} = \frac{1}{3}\left[\frac{1}{m_1^*} + \frac{1}{m_2^*} + \frac{1}{m_3^*}\right] \tag{1.29}$$

For an ellipsoidal iso-energetic surface Equation (1.29) reduces to:

$$\frac{1}{m^*} = \frac{1}{3}\left[\frac{1}{m_l^*} + \frac{2}{m_t^*}\right] \tag{1.30}$$

- The extrinsic n- and p-type silicon is formed due to doping of pentavalent and trivalent impurities which act as donors of electrons to the conduction band, and acceptors of electrons from the valence band, respectively. The donors and acceptors create energy levels in the band gap which are positioned close to the conduction and valence bands, respectively. For an n-type silicon, at room temperature, the donor impurities are ionized with a positive charge as the pentavalent impurity donates the fifth valence electron of the outermost orbit to the conduction band. Similarly, for p-type silicon, trivalent impurities are ionized with negative charge as they capture (accept) electrons from the valence band, thereby creating holes (vacancies) in the valence band. The silicon remains electrically neutral satisfying the following equations in a thermal equilibrium condition assuming all the dopants to be ionized:

–for n-type

$$n_{n0} = N_D^* + p_{n0}$$

where n_{n0} = majority carrier concentration, p_{n0} = minority carrier concentration, and N_D^* = donor concentration, and

–for p-type

$$p_{p0} = N_A^* + n_{p0}$$

where p_{p0} = majority hole concentration, n_{p0} = minority electron concentration, and N_A^* = acceptor ion concentration. Figure 1.8 shows the energy band diagram of p- and n-type silicon with acceptor and donor energy level positions E_a and E_d respectively.

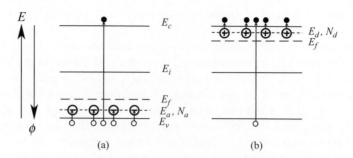

Figure 1.8 Extrinsic *p*- and *n*-type silicon

1.4 Carrier Statistics

In a semiconductor two types of free carriers, electrons and holes, are responsible for current conduction. The concentrations of electrons (n_0) and holes (p_0) in thermodynamical equilibrium are obtained from the product of the following:

- Density of energy levels (states) in the conduction and valence bands.
- The probability of occupation of states by electrons and holes.

Figure 1.9 shows the graphical representation of the calculation of carrier concentration in a semiconductor taking intrinsic and *p*-type semiconductors as an illustration.

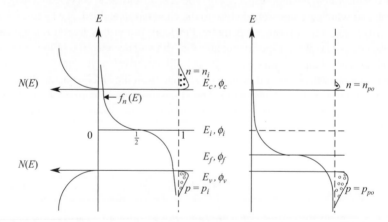

Figure 1.9 Electron and hole concentrations in intrinsic and *p*-type silicon

1.4.1 Density of Energy States

The density of energy states in the conduction and valence bands implies the number of allowed energy levels dN in a small energy interval dE centred around E, and is given by $g(E) = \frac{dN}{dE}$. As the density of states is normally required near the edges of the conduction bands where the energy momentum relation is parabolic, the density of states in the conduction and valence bands can be shown to be given by [18]:

$$g_n(E) = 4\pi \left(\frac{2m_n^*}{h^2} \right)^{\frac{3}{2}} \sqrt{E - E_c} \tag{1.31}$$

$$g_p(E) = 4\pi \left(\frac{2m_p^*}{h^2} \right)^{\frac{3}{2}} \sqrt{E_v - E} \tag{1.32}$$

For an ellipsoidal iso-energy surface the anisotropic nature of the mass can be accounted for with the following expression:

$$g_n(E) = 4\pi (m_1^* m_2^* m_3^*)^{\frac{1}{2}} \left(\frac{2}{h^2} \right)^{3/2} (E - E_c)^{1/2} \tag{1.33}$$

The above equation is normally represented by an equivalent or effective *density of state mass*, m_d^*, given by:

$$m_d^* = (m_1^* \, m_2^* \, m_3^*)^{\frac{1}{3}} \tag{1.34}$$

1.4.2 Probability of Occupation of Energy States

The probability of occupation of an energy level by an electron is represented by the Fermi function $f_n(E)$ which is given below:

$$f_n(E) = \frac{1}{1 + \exp \frac{(E - E_f)}{kT}} \tag{1.35}$$

The probability of an energy state being occupied by a hole is given by:

$$f_p(E) = [1 - f_n(E)] = \frac{1}{1 + \exp \frac{(E_f - E)}{kT}} \tag{1.36}$$

For $(E - E_f) \geq 3kT$, $\exp((E - E_f)/kT) \gg 1$, the Fermi function reduces to an exponentially decaying Maxwellian distribution.

1.4.3 Carrier Concentration for Non-degenerate Silicon

For the normal doping range, the Fermi energy level position in the band gap lies a few kT away from the conduction and valence band edges in the band gap. For such levels of doping the semiconductor is called *non-degenerate*. The probabilities of occupation are given by:

$$f_n(E) = \exp\left(-\frac{(E - E_f)}{kT}\right) \tag{1.37}$$

$$f_p(E) = \exp\left(-\frac{(E_f - E)}{kT}\right) \tag{1.38}$$

With the knowledge of density of states in the bands and the probabilities of occupation, we can estimate the thermal equilibrium electron and hole concentrations as given below:

$$n_0 = \int_{E_c}^{\infty} g_n(E) f_n(E)\, dE \tag{1.39}$$

$$p_0 = \int_{-\infty}^{E_v} g_p(E) f_p(E)\, dE \tag{1.40}$$

Substituting for the expressions $g_n(E)$ and $f_n(E)$, from Equations (1.31) and (1.37), and evaluating the integral on the RHS of Equation (1.39), we get the expression for the equilibrium electron concentration in the conduction band. Similarly, substituting for $g_p(E)$ and $f_p(E)$ from Equations (1.32) and (1.38), and evaluating the integral on the RHS of Equation (1.40), we shall get the thermal equilibrium hole concentration in the valence band. Thus, the final expressions for n_0 and p_0 are:

$$n_0 = N_c \exp\left(-\frac{(E_c - E_f)}{kT}\right) \tag{1.41}$$

$$p_0 = N_v \exp\left(-\frac{(E_f - E_V)}{kT}\right) \tag{1.42}$$

where the expressions for N_c and N_v are given by the following equation:

$$N_{(c,v)} = 2\left[\frac{2\pi m_{(n,p)}^* kT}{h^2}\right]^{\frac{3}{2}} \tag{1.43}$$

$N_{(c,v)}$, known as the *effective density of states*, has the physical significance that its product with the probability of occupation at the band edges gives the carrier concentrations. For silicon, at room temperature $N_c = 2.8 \times 10^{19} \text{cm}^{-3}$ and $N_v = 1.04 \times 10^{19}\text{cm}^{-3}$.

From the value of the density of states we can conclude that if the number of electrons or holes contributed by donors and acceptors are much less than N_c or N_v, then there are few carriers compared to the number of energy levels, and they can be considered as classical particles obeying Boltzmann's statistics.

Equations (1.41) and (1.42) give electron and hole concentration, respectively in the conduction and valence bands under thermodynamical equilibrium conditions, in terms of the effective density of states in the conduction and valence bands and the difference between the Fermi levels and band edges.

It is often required to express electron and hole concentrations in terms of intrinsic carrier concentration, n_i, which is obtained from Equations (1.41) and (1.42) for n_0 and p_0 respectively as shown below:

$$n_0\, p_0 = n_i^2 = N_c N_v \exp\left(-\frac{E_g}{kT}\right) \tag{1.44}$$

where $E_g = E_c - E_v$.

It is noted from the above equation that the intrinsic carrier concentration in silicon has an exponential dependence on the energy band gap of the material as well as temperature. At room temperature, for silicon, $n_i = 1.4 \times 10^{10}\text{cm}^{-3}$. *The relation is valid as long as the concentration is non-degenerate and the relations are obtained using Boltzmann's statistics.*

1.4.4 Degenerate Silicon and Band Gap Narrowing

When the Fermi energy level gets close to the band edges, the Maxwell–Boltzmann statistics no longer hold good. The Fermi functions given in Equations (1.35) and (1.36) are to be used, which give the following expressions for electron and hole concentrations respectively:

$$\frac{n_0}{N_c} = F_{\frac{1}{2}}(\eta_f) = \frac{2}{\sqrt{\pi}} \int_0^\infty \frac{\eta^{\frac{1}{2}}\,d\eta}{1 + \exp(\eta - \eta_f)} \tag{1.45}$$

$$\frac{p_0}{N_v} = F_{\frac{1}{2}}(\varepsilon_f) = \frac{2}{\sqrt{\pi}} \int_0^\infty \frac{\varepsilon^{\frac{1}{2}}\,d\varepsilon}{1 + \exp(\varepsilon - \varepsilon_f)} \tag{1.46}$$

where $\eta_f = (E_f - E_c)/kT$ and $\varepsilon_f = (E_v - E_f)/kT$. The error in the calculation of electron concentration by the use of Boltzmann statistics instead of the Fermi–Dirac approach is expressed by:

$$\delta_n = 1 - \frac{F_{\frac{1}{2}}(\eta_f)}{\exp(\eta_f)} \tag{1.47}$$

As doping concentration is increased the inter-dopant spacing starts decreasing. For concentrations at which the separation between the atoms becomes comparable to the first Bohr orbit r_B of the impurity atom ($r_B \approx N_B^{-1/3}$), the energy levels of the impurities start splitting up. They form a band of energy which merge with the band edge of silicon and, thereby, effectively reduce the band gap [21]. This phenomenon is known as *Band Gap Narrowing* (BGN), which has been established both theoretically and experimentally [27, 29, 30]. The empirical relation by Slotboom giving the bandgap narrowing ΔE_g as a function of doping concentration [27] is given below:

$$\Delta E_g = q V_1 \left[\ln \frac{N}{N_0} + \sqrt{\left(\ln \frac{N}{N_0}\right)^2 + C} \right] \tag{1.48}$$

In Equation (1.48), for silicon, the appropriate values are: $N = (N_A + N_D)$ is the total impurity concentration, $V_1 = 9.0\text{ mV}$, $N_0 = 10^{17}\text{cm}^{-3}$ and $C = 0.5$. With BGN, an *effective intrinsic carrier concentration*, n_{ie}, related to doping concentration is proposed as [27, 28]:

$$n_{ie} = n_i \exp \left[\frac{4.5 \, \text{mV}}{\phi_t} \left(F_N \sqrt{F_N^2 + 0.5} \right) \right] \tag{1.49}$$

$$F_N = \ln \left(\frac{N_D + N_A}{N_R} \right) \tag{1.50}$$

where $N_R = 10^{17} \text{cm}^{-3}$.

The expression for the doping and temperature dependence of n_{ie} is given by [27]:

$$n_{ie}^2 = n_{i0}^2(T) \exp \left(\frac{\Delta E_{g0}}{kT} \right) \tag{1.51}$$

where the symbol '0' indicates room temperature values, and the bandgap narrowing is distributed equally at both the conduction and valence band edges.

1.4.5 Fermi Level Position

For an intrinsic silicon, the condition $n_0 = p_0 = n_i$ is the manifestation of the *condition of charge neutrality* in the semiconductor. This condition provides the equation from which the unknown parameter E_f can be obtained. Combining Equations (1.44), (1.41), and (1.42), the expression for the Fermi energy level, E_i, for an intrinsic silicon is obtained as given below:

$$E_f = E_i = \frac{E_c + E_v}{2} + \frac{kT}{2} \ln \frac{N_v}{N_c} \tag{1.52}$$

We can obtain the position for the Fermi energy level for an extrinsic silicon from the condition of charge balance using a similar approach.

For *n*-type silicon (representing n_0 as n_{n0} to emphasize *n*-type silicon):

$$n_{n0} = N_D^* + p_{n0} \approx N_D \tag{1.53}$$

Equation (1.53) shows that the majority carrier electron concentration may be considered to be the same as the donor concentration as the minority carrier concentration is a few orders lower than the majority carrier concentration, and hence, can be neglected.

Using Equation (1.41) for $n_0 = N_D$ we get:

$$E_f = E_c + kT \ln \frac{N_D}{N_c} \tag{1.54}$$

For p-type silicon using the charge balance condition and Equation (1.42), we can similarly write:

$$p_{p0} = N_A^* + n_{p0} \approx N_A^* \tag{1.55}$$

$$E_f = E_v + kT \ln \frac{Nv}{N_A} \tag{1.56}$$

Combining Equations (1.52), (1.41), and (1.42), the desired relations for electron and hole concentrations, n_{n0} and p_{p0}, in terms of intrinsic carrier concentration and the displacement of the Fermi level E_f from the intrinsic position E_i are obtained for the non-degenerate doping concentrations:

$$n_{n0} = n_i \exp \frac{(E_f - E_i)}{kT} \tag{1.57}$$

$$p_{p0} = n_i \exp \frac{(E_i - E_f)}{kT} \tag{1.58}$$

Some important observations worth mentioning are as follows:

- It is seen that the donor type impurity doping in silicon, which increases the electron concentration in the conduction band, increases the separation $(E_f - E_i)$. It implies that an increase in n-type doping shifts the position of the Fermi level closer to the conduction band. A similar conclusion holds as well for p-type doping; increased acceptor doping shifts the equilibrium Fermi level position E_f towards the valence band edge E_V.
- When the Fermi energy level reaches the conduction band edge E_C or the valence band edge E_V, due to heavy doping, the Boltzmann statistics do not hold good and Fermi–Dirac statistics have to be used.
- From Equations (1.41) and (1.42), we can write:

$$\phi_f = \phi_c - \phi_t \ln \left(\frac{n_0}{N_c} \right) = \phi_c - \underline{\chi}_n \tag{1.59}$$

$$\phi_f = \phi_v + \phi_t \ln \left(\frac{p_0}{N_v} \right) = \phi_v + \underline{\chi}_p \tag{1.60}$$

where $\underline{\chi}_{(n,p)} \infty$ are chemical potentials corresponding to electrons and holes. It is seen that the Fermi potential has two components, namely, (i) electrostatic potential, $\phi_{(c,v)}$, corresponding to band edge energy, and (ii) chemical potential, $\underline{\chi}_{(n,p)}$, corresponding to electron and hole concentrations. Therefore, the Fermi potential is also designated as *electrochemical potential*. In a semiconductor, a gradient of electrostatic potential causes a drift current, and that of the chemical potential leads to a diffusion component of the current. This leads to an important conclusion that *in equilibrium condition* when the net current flow is zero, grad $\phi_f = 0$, and equivalently, *the Fermi potential ϕ_f gets aligned horizontally.*

1.4.6 Quasi-Fermi Level

We have so far discussed the equilibrium situation for which the Fermi distribution is defined. For a non-equilibrium situation, due to factors such as external excitation, injection, etc., the carrier concentrations deviate from the equilibrium value. The important property of the Fermi level relating it uniquely to the equilibrium carrier concentrations can be extended for the non-equilibrium condition as well.

For a non-equilibrium carrier concentration, a parameter known as the quasi-Fermi level, or Imref (*Fermi spelt backwards*) is introduced, which relates non-equilibrium electron and hole densities with its position. Following Equations (1.57) and (1.58), for the non-equilibrium condition, we can write:

$$n = n_i \exp \frac{E_{fn} - E_i}{kT} \tag{1.61}$$

$$p = n_i \exp \frac{E_i - E_{fp}}{kT} \tag{1.62}$$

where E_{fn} and E_{fp} are the quasi-Fermi energy levels for electrons and holes respectively. Multiplying Equations (1.61) and (1.62), we get:

$$np = n_i^2 \exp \frac{(E_{fn} - E_{fp})}{kT} \tag{1.63}$$

It is seen that the product of the majority and minority carrier concentrations in the non-equilibrium case does not equal n_i^2, as happens for the equilibrium condition. The non-equilibrium carrier concentrations can be expressed in terms of the quasi-Fermi potentials with the following substitutions:

$$\frac{np}{n_0 p_0} = \frac{n_i^2 \exp \frac{q}{kT}(\phi_p - \phi_n)}{n_i^2} = \exp \frac{q}{kT}(\phi_p - \phi_n) \tag{1.64}$$

where $\phi_n = -E_{fn}/q$ and $\phi_p = -E_{fp}/q$.

It may be noted that unlike the equilibrium case, where a single Fermi level characterizes the majority and minority carrier concentrations, in the non-equilibrium condition the Fermi level is split. Figure 1.10 shows the Fermi level positions for equilibrium and

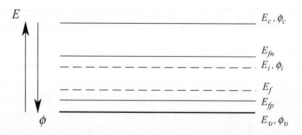

Figure 1.10   Quasi-Fermi level positions for non-equilibrium electron and hole concentrations

non-equilibrium conditions when electron and hole concentrations are in excess of the equilibrium values for a p-type semiconductor. E_f stands for the Fermi level for the equilibrium condition.

It may be seen that in a non-equilibrium situation $\Delta n = \Delta p$, from consideration of the charge neutrality. For a p-type semiconductor, $(p_{p0} + \Delta p)$ is the non-equilibrium majority carrier concentration and $(\Delta p / p_{p0} \ll 1)$, assuming low-level perturbation. Hence, the Fermi level position is disturbed marginally for the majority of carrier holes. E_{fp} moves towards the valence band marginally w.r.t. the equilibrium Fermi level position E_f. On the other hand, for the minority carriers, the non-equilibrium concentration is $(n_p = n_{p0} + \Delta n)$ where Δn is large compared to the equilibrium concentration n_{p0}. In order to reflect the increase in electron concentration, the quasi-Fermi potential moves towards the conduction band edge significantly.

1.5 Carrier Generation and Recombination

In a semiconductor, at thermal equilibrium, the generation rate of electrons and holes equals their recombination rate and $p_0 n_0 = n_i^2$. When the equilibrium condition is disturbed due to factors such as injection or extraction, etc., $pn \neq n_i^2$ where p and n are non-equilibrium carrier concentrations. If the carriers are in excess of the equilibrium value, through a recombination process of excess carriers, the system tries to reach the equilibrium value. For a carrier concentration less than the equilibrium value, a generation mechanism is in place to restore the equilibrium value. Silicon being an indirect band gap semiconductor, direct recombination or generation of electron and hole pairs are considered very unlikely. Shockley, Read, and Hall proposed a model where the recombination/generation of electrons and holes takes place through intermediate recombination centers located midway between the energy band gap. The intermediate recombination centers in the band gap are created by defects in the material. Below, the physical mechanism involved in electron–hole recombination through the participation of intermediate recombination centers, represented by a single level, is briefly outlined [31, 33].

1.5.1 Shockley–Read–Hall (SRH) Model

This model assumes the presence of deep level recombination centers, having a concentration of N_r, near the centre of the band gap. It identifies the following capture and emission related processes.

1. An electron is captured by the recombination center which makes it negatively charged.
2. The negatively charged center captures a hole. Steps 1 and 2 complete the electron–hole *recombination* process.
3. The negatively charged center after completion of step 1 emits the electron back to the conduction band and becomes neutral.
4. The neutral level, after emission of an electron to the conduction band in step 3, captures an electron from the valence band, or equivalently can be considered to have emitted a hole, to complete the electron–hole generation process.

If U^* denotes the net recombination rate of electron-hole pairs $(cm^{-3}\,s^{-1})$ in the steady state, it can be shown that:

$$U^* = \frac{pn - n_i^2}{\frac{1}{\sigma_p v_{th} N_r}(n + n_1) + \frac{1}{\sigma_n v_{th} N_r}(p + p_1)} \tag{1.65}$$

or:

$$U^* = \frac{d\Delta n}{dt} = \frac{d\Delta p}{dt} = \frac{pn - n_i^2}{\tau_{p0}(n + n_1) + \tau_{n0}(p + p_1)} \tag{1.66}$$

where $\tau_{p0} = 1/c_p N_r$ and $\tau_{n0} = 1/c_n N_r$ are the carrier lifetime and have the dimension (sec). $c_p = \sigma_p v_{th}$ and $c_n = \sigma_n v_{th}$ are the capture probabilities $(cm^3\,s^{-1})$, σ_n and σ_p are respectively the capture cross-sections of electrons and holes (cm^2), and $v_{th} = \sqrt{3kT/m^*}$ is the thermal velocity of the carriers $(cm\,s^{-1})$. n_1 and p_1 are the values of free electron and hole concentrations (cm^{-3}), respectively, that would result when the Fermi level lies at the level of the recombination centers. Making simplifying approximation that $c_n = c_p$, $\tau_{n0} = \tau_{p0} = \tau$, and assuming that recombination centers are located near the center of the energy gap when $n_1 = p_1$ equals n_i, Equation (1.66) is simplified to:

$$U^* = \frac{1}{\tau}\left(\frac{pn - n_i^2}{p + n + 2n_i}\right) \tag{1.67}$$

We shall consider below two cases:

Case (i): $pn > n_i^2$, a condition encountered due to excess of carriers over the equilibrium value. Here, U^* is positive and a *net recombination* or loss of excess carriers is denoted. We consider a p-type substrate for illustration and a low injection level where $\Delta p \ll p_0$. The term $(pn - n_i^2) = p(n - n_i^2/(p_0 + \Delta p)) = p(n - n_0)$. For the denominator, $p \gg n_i \gg n$. Equation (1.67) reduces to:

$$U^* = \frac{n - n_0}{\tau} \tag{1.68}$$

where τ is called the *lifetime* of the minority carrier electrons.

Case (ii): $pn < n_i^2$, a condition encountered in a reverse biased depletion region ($n = p = 0$). Thus, U^* is negative and signifies generation of carriers to restore equilibrium. Equation (1.67) now takes the form:

$$U^* = -\frac{n_i}{2\tau} \tag{1.69}$$

The recombination through deep level traps provides a physical explanation of the dependence of lifetime of carriers on processing conditions. It may be mentioned that the SRH recombination mechanism does not account for lifetime at high impurity doping. The Auger recombination mechanism, which requires a large number of free carrier concentrations, is believed to be a more appropriate model at high doping concentrations [32, 66].

1.6 Carrier Scattering

Considering the charge carriers in a semiconductor as classical particles, we can apply the laws of the kinetic theory of gases to understand the scattering mechanism that occurs in the semiconductor. The mobile electrons or holes in a semiconductor move randomly within the solid with fixed ionized donor or acceptor atoms. Due to random movement of free carriers the net displacement is zero in a given direction, and the random motion of charged carriers does not constitute a net current. As the thermal energy of the carriers is manifested in the form of kinetic energy, considering the three degrees of freedom, the average random thermal velocity is given by:

$$\frac{1}{2}m^* v_{th}^2 = \frac{3}{2}kT \tag{1.70}$$

where m^* stands for the effective mass of electrons or holes. The motion of the carriers is random due to their collision with the scattering center.

For an electric field F applied on the semiconductor sample, the average drift velocity in the direction of electric field is:

$$v_{dr} = \frac{1}{2}\left(0 + \frac{qF}{m_n^*}\tau_c\right) = \frac{1}{2}\left(\frac{qF}{m_n^*}\tau_c\right) = \mu_n F \tag{1.71}$$

where τ_c is related to the *mean interval* between successive collisions, also known as the *relaxation time*, when the carriers lose the energy acquired from the electric field and relax back to the initial state.

Equation (1.71) shows that the *average* drift velocity of electrons is proportional to the electric field F and the proportionality constant, called μ_n, is a measure of the mobility of electrons:

$$\mu_n = \left(\frac{1}{2}\frac{q}{m_n^*}\right)\tau_c \tag{1.72}$$

The value of τ_c is determined by the scattering mechanism in the semiconductor, and therefore, is dependent on doping concentration, temperature, and even electric field which may influence the scattering processes. Physically, the *scattering implies a change in velocity and direction of the carrier* after the collision takes place. The quantity τ_c can be related to the mean free path distance, $\bar{l}$, and electron velocity v through a simple equation:

$$\tau_c = \frac{\bar{l}}{v} \tag{1.73}$$

The mean free time equals the relaxation time when the scattering is isotropic.

From Equations (1.72) and (1.73) we note that the mobility of electrons directly depends on the mean free path between collision $\bar{l}$, and inversely on the electron effective mass m_n^*. If the carriers in the semiconductor are holes, their mobility μ_p, would be less compared to that of the electrons because $m_p^* > m_n^*$. *For silicon the ratio $m_p^*/m_n^* = 2.9$.* The relaxation time is determined by the nature of the collision center from which charge carriers are scattered. In a semiconductor there are several independent mechanisms of scattering amongst which the two predominant ones are: (i) scattering by lattice vibrations, and (ii) scattering by ionized impurity atoms. The basic characteristics of these mechanisms are summarized below.

1.6.1 Lattice Vibration Scattering

The atoms in a crystal lattice are in a state of vibration due to thermal energy. The vibrational energies are quantized like photons and called phonons. These vibrations cause displacement of atoms from their normal positions resulting in perturbation at the band edges in the energy band diagram which act as scattering centers for electrons and holes. Phonons are of two types, *acoustical* and *optical*, depending on their frequency of vibration. When the impurity concentrations are not high, the lattice scattering due to acoustic phonon is the predominant scattering mechanism.

$$\frac{1}{\tau_L} = \frac{v_{th}}{\bar{l}} \sim \frac{\sqrt{\frac{T}{m^*}}}{\frac{1}{T(m^*)^2}} \sim m^{*\frac{3}{2}} T^{\frac{3}{2}} \tag{1.74}$$

Using Equation (1.72), we can write:

$$\mu_L \sim (\frac{q}{m^*})\tau_L \sim m^{*-\frac{5}{2}} T^{-\frac{3}{2}} \tag{1.75}$$

where τ_L is the relaxation time corresponding to lattice scattering. In general, acoustic and optical scattering have different temperature dependencies. In MINIMOS, the temperature dependence of both acoustic and optical phonon related lattice mobility, μ_L, is represented by a single relation, with parameters η and μ_{L0} as given below [44]:

$$\mu = \mu_{L0} \left(\frac{T}{T_0}\right)^{-\eta} \tag{1.76}$$

Typical values of lattice mobility at reference room temperature ($T_0 = 300$ K), μ_{L0}, and η, are:

For electrons: $\mu_{L0} = 1200$ to 1500 cm^2 V^{-1}s^{-1}
For holes: $\mu_{L0} = 400$ to 600 cm^2 V^{-1}s^{-1}
For both electrons and holes: $\eta = 2.0$ to 2.7

1.6.2 Ionized Impurity Scattering

The ionized impurity atoms act as scattering centers in a semiconductor as they cause deviation in the trajectory of the moving carriers through coulomb interaction between the field of the ionic charge and transporting mobile charged carriers. Brooks–Herring [40], Conwell–Weisskopf [41] models, etc., are a few amongst a number of physical models of impurity scattering. We shall present a functional outline of the Conwell–Weisskopf formula to highlight the dependence of impurity scattering related mobility on doping concentration N_I, temperature T, and charge of the scattering atoms.

$$\mu_I = \frac{q\,\vec{\tau}}{m_{n*}} \sim \frac{T^{\frac{3}{2}}}{N_I Z^2 (m_n^*)^{\frac{1}{2}}} \tag{1.77}$$

where a multiplying logarithmic term has been ignored as it does not influence the dependence on impurity concentration and temperature in a significant way. A modified form of the Brooks–Herring expression, obtained by Li [35, 36] which has been used in MOSFET simulators is given as [69]:

$$\mu_I = \frac{C_1}{N_t G_b} \left(\frac{T}{T_0}\right)^{\frac{3}{2}} \tag{1.78}$$

$$G_b = \ln(b+1) - \frac{b}{b+1} \tag{1.79}$$

$$b = \frac{C_2}{N_t} \left(\frac{T}{T_0}\right)^2 \tag{1.80}$$

where C_1, C_2 are constants, and N_t represents the total impurity concentration.

1.6.3 Carrier–Carrier Scattering

This mode of scattering takes place due to collision amongst mobile carriers, particularly at high carrier density, when the electron population becomes degenerate. Normally, this mechanism is not considered to be significant in MOSFET operation. In our discussion, we shall ignore its contribution to the scattering process.

1.6.4 Multiple-Scattering: Mathiessen's Rule

When there are several scattering mechanisms in action which are not inter-dependent, the resultant mobility is given by Mathiessen's formula, $\mu^{-1} = \sum \mu_i^{-1}$, where $i = L$ (lattice), I (impurity), C (carrier), ..., etc. The resultant mobility expressions obtained from physical modeling of individual scattering processes have involved expressions which calls for numerical computation, which is not suitable for the development of analytical expressions for device characteristics. In such cases, one of the most common approaches has been to develop suitable simplified expressions fitting experimental or simulation data and apply it for device modeling. The Caughey–Thomas model [42], a fitting formula for experimental data involving

doping and temperature dependence of mobility over a useful range of practical interest, is
reproduced below:

$$\mu_{LI} = \mu_{mn} + \frac{(\mu_L - \mu_{mn})}{1 + \left(\frac{N_t}{N_R}\right)^\alpha} \tag{1.81}$$

$$\mu_{mn} = \mu_{mn0} \left(\frac{T}{T_0}\right)^{-\eta_2} \tag{1.82}$$

$$\mu_L = \mu_{L0} \left(\frac{T}{T_0}\right)^{-\eta_1} \tag{1.83}$$

$$N_R = N_{R0} \left(\frac{T}{T_0}\right)^{\eta_3} \tag{1.84}$$

$$\alpha = \alpha_0 \left(\frac{T}{T_0}\right)^{-\eta_4} \tag{1.85}$$

where μ_{mn0} and μ_{L0} are the minimum and maximum mobilities at the highest doping and
lowest concentrations respectively. N_R is a reference, N_t is the total impurity concentration,
and α represents the gradient of the μ_{LI}–N_t curve around $N_t = N_R$. The typical values of
the parameters used in MINIMOS4 as Caughey–Thomas coefficients for electrons and holes
are [69]:

For electrons
$\mu_{mn0} = 80 \text{ cm}^2 \text{ V}^{-1}\text{s}^{-1}$, $\mu_{L0} = 1430 \text{ cm}^2 \text{ V}^{-1}\text{s}^{-1}$, $N_{R0} = 1.21 \times 10^{17} \text{ cm}^{-3}$,
$\alpha_0 = 0.72$, $\eta_1 = 2.0$, $\eta_2 = 0.45$, $\eta_3 = 3.2$, $\eta_4 = 0.065$
For holes
$\mu_{mn0} = 45 \text{ cm}^2 \text{ V}^{-1}\text{s}^{-1}$, $\mu_{L0} = 460 \text{ cm}^2 \text{ V}^{-1}\text{s}^{-1}$, $N_{R0} = 2.23 \times 10^{17} \text{ cm}^{-3}$,
$\alpha_0 = 0.72$, $\eta_1 = 2.18$, $\eta_2 = 0.45$, $\eta_3 = 3.2$, $\eta_4 = 0.065$

There are other mobility models, such as that of Arora [38], which also provide doping and
temperature dependence of the mobility. Klaassen [39] provides a more comprehensive and
unified majority and minority carrier model for silicon including carrier–carrier scattering, etc.
It may be mentioned, however, that the above give the low-field mobility value.

1.6.4.1 Surface Scattering

In MOSFET operation, carriers are confined in a small thickness near the silicon–oxide interface, or equivalently, at the surface of the silicon substrate on which the oxide is grown.
Due to the roughness at the interface there is an additional scattering of carriers, known in the
literature as the *surface scattering*, which are not experienced when carriers transport in the
bulk. More appropriately, the surface scattering is essentially an interface roughness scattering
phenomenon. In MOSFETs, surface scattering has significant importance as the carriers which
constitute current, move near the Si–SiO$_2$ interface. It is obvious that the scattering model must
take into consideration the proximity of charge carriers to the interface plane, whose influence

should diminish if the carriers are positioned away from it. The field due to the gate potential has a strong influence on the inversion layer. A physical approach where the diffused surface scattering imposes a boundary condition on the carrier distribution involves a parameter, p_F, named after Fuchs [45], where $0 \leq p_F \leq 1$, in which the maximum value unity corresponds to an ideal scatter-free surface. From a simulation viewpoint, where a wide range of operating and process conditions are to be kept in view, Selberherr empirically builds a factor in the surface scattering model which provides a smooth transition from surface to bulk scattering [44].

1.6.5 Drift Velocity–Electric Field Characteristics

In response to an electric field, carriers in a semiconductor move with an average drift velocity along the direction of the force by the field. This velocity is a function of the field itself. Figure 1.11 shows typical v_{dr}–F characteristics which show that it is nonlinear, reflecting different physical mechanisms taking place as the field in increased.

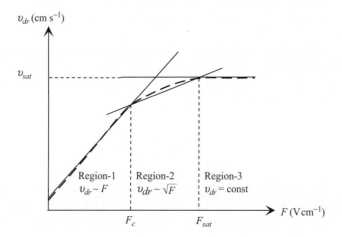

Figure 1.11 Drift velocity–electric field characteristics in silicon

As drift velocity is the product of electric field and mobility, the slope of the v_{dr}–F curve represents the mobility. The curve indicates that *mobility decreases with increasing electric field*, transitioning from a linear dependence of velocity on electric field (Region-1) to a limiting field-independent saturation velocity (Region-3).

The v_{dr}–F characteristics can be broadly divided into three regions for the purpose of physical interpretation [47]:

1. Region-1 (low-field region): $(F < 10^3 \text{ V cm}^{-1})$; $v_{dr} \sim F \rightarrow \mu = \text{const.}$
2. Region-2 (intermediate-field region): $(10^3 \text{ V cm}^{-1} < F < 10^4 \text{ V cm}^{-1})$; $v_{dr} \sim F^{\frac{1}{2}}$.
3. Region-3 (high-field region): $(F > 10^4 \text{ V cm}^{-1})$; $v_{dr} = \text{const} = v_{sat}$.

The drift velocity–electric field curve has the following characteristics:

- At low field ($F < 10^3$ V cm^{-1}), the drift velocity is linearly related to the applied electric field, which defines Region-1. In this region, the mobility, μ, which is given by the slope of the v_{dr}–F curve, dv_{dr}/dF is constant. This is also called the *low-field region* where Ohm's law holds good. The thermal velocity of an electron at room temperature T is given by $v_{th} = \sqrt{3kT/m}$ ($\sim 10^7$ cm s^{-1}), which is dependent on ambient temperature. At low field, the drift velocity is negligible compared to the thermal velocity of the carriers, and the low field mobility value is determined by lattice and coulomb scattering.
- As the field is increased, the v_{dr}–F curve deviates from linear behavior. The carriers acquire energy from the electric field and this energy is added to the thermal energy corresponding to the ambient temperature ($\frac{3}{2}kT$), which changes the energy distribution of electrons. The electron energy, expressed in terms of thermal velocity, thus corresponds to a higher temperature than that of the lattice. The electric field causing deviation from the linear v_{dr}–F region is attributed to the so-called *hot carrier*. The hot carriers scatter or collide with acoustic phonons and lead to the following:

  (i)   heat the crystal and raise its temperature;
  (ii) generate new types of high energy optical phonons.

In the intermediate values of field strength (Region-2), with the onset of the hot carrier effect, the energy distribution of carriers is not dependent on crystal temperature, but on electric field. This leads to field dependent mobility relation. For $F > F_c$, the mobility–field relation can be put in the form:

$$\mu = \mu_0 \left(\frac{F_c}{F} \right)^{\frac{1}{2}}$$

where F_c is designated as the *critical field* from where deviation from Ohm's law takes place, and μ_0 is the low field mobility value. F_c has been found to increase with doping concentration ($\sim \sqrt{N}$) and temperature. It has an inverse relationship with the mobility value of the material ($\sim \mu^{-1}$).

For $F > F_c$ and $F = F_{sat} \approx (4\text{–}5)F_c$, the optical phonon scattering dominates and the drift velocity does not increase with the applied field, attaining a maximum saturation value of v_{sat}.

At the saturation field F_{sat}, the carriers gain enough energy from the electric field required to excite optical phonons. When the carriers attain saturation velocity, v_{sat}, the entire energy acquired from the field by the carriers is dissipated by scattering and generation of optical phonons.

- A first order estimate of the magnitude of the saturation velocity may be obtained as below, based on simplified assumptions:

$$\frac{1}{2}m_n^* v_d^2 = E_{op} \tag{1.86}$$

where E_{op} is the energy required to generate an optical phonon.

Equating E_{op} to the thermal velocity of the 'cold carriers' $(=\frac{3}{2}kT)$, one gets:

$$v_d \approx \left(\frac{3kT}{m_n^*}\right)^{\frac{1}{2}} \approx v_{th} \approx 10^7 \text{ cm s}^{-1} \tag{1.87}$$

The velocity–field relation discussed above is very important for modeling of the MOSFET drain current as mobility is a key parameter. To make MOSFET modeling analytically amenable, the $v_{dr}-F$ non-linear relation has been approximated in various forms which will be discussed in Chapter 5.

The temperature dependence of carrier saturation velocities for electrons and holes are reported as:

$$v_{sat,n} = v_{sato,n} \left(\frac{T}{T_0}\right)^{-0.87} \tag{1.88}$$

$$v_{sat,p} = v_{sato,p} \left(\frac{T}{T_0}\right)^{-0.87} \tag{1.89}$$

From the above, it is seen that the saturation velocity decreases with temperature.

1.7 Contacts and Interfaces

In a semiconductor device, metals are commonly used to establish contacts with different regions of the device. A metal and a semiconductor lead to two types of contacts [70]: (a) Schottky, and (b) Ohmic. Figure 1.12 shows the energy band diagram of a metal and silicon without physical contact.

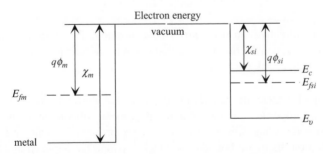

Figure 1.12   Electron affinity and work function in metals and semiconductors

The barrier formation at a metal–semiconductor contact may be understood from the following considerations.

- The potential energy of electrons within a metal or semiconductor is less than that of free electrons in vacuum, as energy has to be supplied to extract them out of the material (metal/semiconductor).

- χ, designated as *electron affinity*, is defined as the amount of minimum energy required for the conduction electron at the lower edge E_c of the conduction band to be emitted to vacuum with zero kinetic energy. It may be considered as an energy barrier holding the electrons within the material. χ_m and χ_{Si} are the electron affinities respectively for metal and silicon. Their values, expressed in electron-volt, are characteristics of the material, and can vary over a wide range.

- Some of the electrons can yet overcome the energy barrier due to their energy distribution, which can be defined by Maxwell–Boltzmann's statistics. These electrons get emitted into the vacuum constituting the thermionic emission current density $J \sim \exp(-\frac{\chi}{kT})$. Hence, $\chi = \chi_0 + (E_c - E_f)$ can be defined as the thermodynamical energy required to be given to the electron for emission into the vacuum *referred to the Fermi energy level*. In the literature, it is commonly expressed in terms of potential energy, and is called the *work function* of the material. When the metal and the semiconductor are brought in physical contact, an equilibrium condition is established when the net current flow is zero and $J_{m \to s}$ equals $J_{m \to s}$ as a result. A potential barrier is formed at this contact whose height is given by $\phi_b = \phi_m - \phi_{Si}$. In a semiconductor, as the position of the Fermi level depends on doping concentration, the work function depends on the doping concentration, and is thus a variable quantity. It is a desirable requirement that contact resistance be small. This is realized through a heavily doped silicon interface, such as $m-n^+-n$ or $m-p^+-p$, so that the current transport between metal and heavily doped semiconductor at the contact is implemented not through thermionic emission, as in the Schottky mechanism, but by tunneling through a very thin potential barrier formed at the interface.

  From a modeling viewpoint, the boundary condition at ohmic contacts assumes that thermal equilibrium relations hold good. Considering commonly employed heavy doping technique, the condition can be written as:

$$n_0 p_0 = n_{ie}^2 = n_i^2 \exp \frac{\Delta E_g}{kT} \tag{1.90}$$

where n_{ie} is the effective intrinsic carrier concentration. The intrinsic concentration has been taken to be n_{ie}, considering the fact that at ohmic contact it is an accepted practice to heavily dope the material where contact is taken.

In a MOS transistor there are various interfaces at the physical transition between materials with different properties, such as silicon–SiO$_2$, metal–silicon, and dielectric–dielectric in stacked gate structures, and silicon–silicon when the properties change due to material engineering. In addition, there are boundaries which separate the depleted and neutral silicon region. The boundary condition at a dielectric–Si interface, say, at $x = x_0$, assuming that the displacement vector directed along the x-direction is given by the continuity of the normal component of the displacement vector follows Gauss' law:

$$\epsilon_D \left. \frac{dF}{dx} \right|_{x=x_0} - \epsilon_{Si} \left. \frac{dF}{dx} \right|_{x=x_0} = Q_i \tag{1.91}$$

where ϵ_D and ϵ_{Si} are, respectively, the permittivity of the dielectric material and silicon, and Q_i is the charge at the interface per unit area. The above condition holds at a $p-n$ homojunction as well, where the dielectric constant remains unchanged on both sides of the junction. Normally, the charge at the interface is considered to be negligible.

1.8 Strained Silicon

Strained silicon (s-Si) is becoming increasingly important in MOS technology in the pursuit of a high mobility silicon substrate. This is realized through engineering the silicon surface layers. The basic principle is to first create a *defect free* Si–Ge layer of composition represented by $Si_{(1-x)}Ge_x$, where x denotes the germanium mole fraction in the Si–Ge relaxed layer which acts as a substrate for the film. It may be mentioned that such a layer itself can be developed on a graded Si–Ge layer acting as a buffer between the base silicon substrate and the relaxed Si–Ge layer. The incorporation of Ge atoms, which have a larger size compared to the silicon atoms, in silicon crystal results in an increase in the magnitude of the lattice constant in the alloy compared to that in unalloyed silicon: $a(x)Si_{1-x}Ge_x = a_s > a_0(Si)$. Now, if a silicon film is grown or deposited on top of this Si–Ge composite layer, the lattice constant of the deposited silicon film, a_f, adopts the spacing of the host layer a_s. The atomic spacing of the deposited silicon layer is now higher than the lattice constant of silicon, thereby, developing *biaxial* strains in the film along the surface of the silicon. In the plane normal to the surface a *compressive* strain is developed. This means, the interatomic spacing in the direction normal to surface is smaller than that along the plane. The material parameters, such as band gap, electron affinity, mobility, band offsets, etc., are dependent on x, the Ge mole fraction. Figure 1.13 shows a representative strained silicon film on a relaxed Si–Ge layer on a silicon substrate.

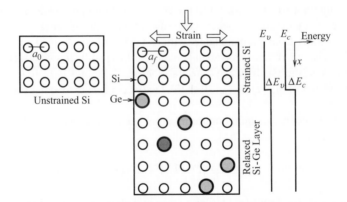

Figure 1.13   Strained silicon

The process of creating layers with differing material properties is known as *hetero-epitaxy*. There is a critical thickness within the limit of which a high quality *dislocation free* layer can be grown over a *host* substrate. Let us assume that the substrate has a lattice constant, a_s, over which a layer is grown of a material having a different lattice constant, a_x, resulting in a lattice mismatch of $\Delta a = (a_x - a_s)$. The quantity $f = \Delta a / \bar{a}$, where $\bar{a}$ is an average value of lattice constant, is known as the lattice mismatch parameter. For a small amount of strain, the strain energy adjusts the lattice constant from a_s to a_f. The film is known as *pseudomorphic*.

With the increase in film thickness, however, the strain energy forms dislocations. The critical thickness, t_c, for which a dislocation free hetero-epitaxial film can be grown, is shown to be approximately given by $t_c = \frac{0.55}{x} \ln(10t_c)$, wherein t_c is expressed in nm [48].

Ge mole concentration dependent values of important parameters of $Si_{(1-x)}Ge_x$ layer are given in Table 1.2 [29, 52, 53, 55].

Table 1.2 Si–SiGe energy-band parameters and equations

Parameter	Equation
Bandgap (eV)	$E_{g,\text{SiGe}}(x) = 1.155 - 0.43x + 0.00206x^2$ $(0 < x < 0.85)$
CB Si–SiGe offsets (eV)	$\Delta E_c(x) = 0.63x \quad (0 < x < 0.4)$
VB Si–SiGe offsets (eV)	$\Delta E_v(x) = (0.74 - 0.53x)x \quad (0 < x < 0.4)$
CB effective density of states (cm^{-3})	$N_{c,\text{SiGe}}(x) = [1.04 + 2.80(1 - x)]10^{19}$
VB effective density of states (cm^{-3})	$N_{v,\text{SiGe}}(x) = [0.60 + 1.04(1 - x)]10^{19}$
Band gap narrowing (eV)	$\Delta E_g = K_1 \left[\left(\ln \frac{N}{K_2} + \sqrt{\left(\ln \frac{N}{K_2} \right)^2 + 0.5} \right) \right]$ $K_1 = 9.0,\ K_2 = 10^{17} \text{cm}^{-3}; (0 < x \le 0.3)$
Permittivity (F cm^{-1})	$\epsilon_{\text{SiGe}}(x) = (11.8 + 4.2x)\epsilon_0$

1.8.1 Mobility Enhancement

The strain in silicon has been found to affect the E–k diagram which, in turn, affects the effective mass, and consequently, the mobility. (i) For a $Si_{1-x}Ge_x$ film grown on (100)-silicon, six conduction minima are split: four into lower and two into upper valleys. (ii) The valency band degeneracy is modified; the heavy hole edge moves into the energy gap and the light hole edge moves away from the band edge, with respect to the unstrained silicon position, by ≈ 77 meV. This change in the valence band characteristics results in two effects: (i) reduction in the inter-valley scattering between carriers of heavy hole and light hole bands, and (ii) decrease in the values of m_{ph} and m_{pl}. Both the factors contribute to enhancement of the hole mobility.

The mobility dependence on temperature, T, doping concentration, N, and Ge concentration in a Si–Ge layer, x, is give below [51]:

$$\mu(x, T, N) = \mu_A(T, N)(1 + 4.31x - 2.28x^2) \tag{1.92}$$

where μ_A represents the proposed doping and temperature dependent low field mobility in Arora's model [38], and N, the total (sum of) all ionized impurity atoms. The proposed expression for μ_A is:

$$\mu_A(T, N) = \mu_{1p} \left(\frac{T}{300} \right)^{\alpha_p} + \frac{\mu_{2p} \left(\frac{T}{300} \right)^{\beta_p}}{1 + \frac{N}{N_{Rp}} \left(\frac{T}{300} \right)^{\gamma_p}} \tag{1.93}$$

The values of the parameters used in Equation (1.93) are: $\mu_{1p} = 54.3$ cm^2 V^{-1}s^{-1}, $\mu_{2p} = 407.0$ cm^2 V^{-1}s^{-1}, $\alpha_p = -0.57$, $\beta_p = -2.23$, $\gamma_p = 2.546$, $N_{Rp} = 2.67.10^{17}$ cm^{-3}.

1.9 Basic Semiconductor Equations

The basic semiconductor equations required for MOSFET device modeling are:

(i) Poisson's and current continuity equations which follow from Maxwell's equations.
(ii) Carrier transport expressions which are derived from Boltzmann transport equation.

These basic equations are outlined below.

1.9.1 Poisson's Equation

Poisson's equation provides a relation between the volume space charge density and electric field. The one-dimensional Maxwell's equation, considering an isotropic and homogeneous medium, is given by:

$$\frac{dF}{dx} = \frac{\rho(x)}{\epsilon_{Si}} \tag{1.94}$$

where $\rho(x)$ is the volume space charge density at a given position x. Using the relation, $F(x) = -\frac{d\phi}{dx}$, we can relate the potential with the space charge density. Thus:

$$\frac{d^2\phi}{dx^2} = -\frac{\rho(x)}{\epsilon_{Si}} \tag{1.95}$$

The space charge density in a semiconductor (silicon) is constituted by the following charges:

p: holes which are mobile and have positive charge;
N_D^+: ionized donor atoms which are fixed and carry positive charge;
n: electrons which are mobile and have negative charge;
N_A^+: ionized acceptor atoms which are fixed and carry negative charge.

Thus, the net space charge density per unit volume is given by:

$$\rho = q(p + N_D^* - n - N_A^*) \tag{1.96}$$

For complete ionization, $N_D^* = N_D$, $N_A^* = N_A$. Substituting $n = n_i \exp(\phi/\phi_t)$, $p = n_i \exp(-\phi/\phi_t)$, $\Phi = \phi/\phi_t$, and combining Equations (1.95) and (1.96), we get:

$$\frac{d^2\Phi}{dx^2} = \frac{1}{L_{bi}^2}\left(\sinh \Phi - \frac{N_D - N_A}{2n_i}\right) \tag{1.97}$$

where $L_{bi} = (\epsilon_{Si}\phi_t/2qn_i)^{1/2}$ stands for the Debye length for an intrinsic semiconductor. Furthermore, where trapped charges are to be considered, $\pm q\Sigma n_{tr}$ is to be included in the volume charge density term on the RHS of Equation (1.96), where n_{tr} is the density of trapped charges in the localized levels in the energy band [26]. It is a common practice to neglect the mobile carriers in solving the Poisson's equation for the depletion approximation.

1.9.2 Current Density Equation

In a semiconductor there are two transport mechanisms which cause the current flow:

(i) Diffusion, which takes place due to the concentration gradient of carriers.
(ii) Drift, which is caused by the force that acts on the carriers due to the presence of an electric field or a potential gradient.

Further, the total current $\vec{J}$ is constituted of two components, namely, $\vec{J_n}$ and $\vec{J_p}$, due to electrons and holes respectively. Thus, the electron and hole current equations can be written as:

$$\vec{J_n} = qn\mu_n \vec{F} + qD_n \, \text{grad}(n) \tag{1.98}$$

$$\vec{J_p} = qp\mu_p \vec{F} - qD_p \, \text{grad}(p) \tag{1.99}$$

The diffusion constant $D_{n,p}$ and mobility $\mu_{n,p}$ are related through the following Einstein's relation:

$$D_{(n,p)} = \phi_t \mu_{(n,p)} \tag{1.100}$$

The current density Equations (1.98) and (1.99) can be expressed in an alternative compact and useful form in terms of the quasi-Fermi potentials ϕ_n and ϕ_p. Assuming the current flow to be in the positive y-direction, we can obtain expressions for the gradients of the quasi-Fermi potentials for electrons and holes respectively, using Equations (1.61) and (1.62).

For electrons:

$$\text{grad } \phi_n = -\vec{F} - \phi_t \frac{1}{n} \frac{dn}{dy} \tag{1.101}$$

and, for holes:

$$\text{grad } \phi_p = -\vec{F} + \phi_t \frac{1}{p} \frac{dp}{dy} \tag{1.102}$$

From Equations (1.101) and (1.102), it is seen that the gradients of the quasi-Fermi potential for electrons and holes take care of the electric field which causes the drift current, and the concentration gradient which causes the diffusion current. Thus, the use of the quasi-Fermi potential in the formulation of the general transport of the carriers leads to the inclusion of both drift and diffusion mechanisms of current in a compact manner. This follows by combining Equations (1.101) and (1.102) for electron and hole concentration gradients, with Equations (1.98) and (1.99) for electron and hole current densities respectively. Following the above approach, the complete expressions for electron and hole current densities are:

$$\vec{J_n} = -q\,n\,\mu_n\,\text{grad}\,\phi_n \qquad (1.103)$$

$$\vec{J_p} = -q\,p\,\mu_p\,\text{grad}\,\phi_p \qquad (1.104)$$

1.9.3 Current Continuity Equation

The current continuity equation follows from the principle of conservation of charge, which can be obtained from Maxwell's second and third equations as shown below:

$$\text{curl}(\vec{H}) = \frac{\partial \vec{D}}{\partial t} + \vec{J} \qquad (1.105)$$

where $\vec{D} = \epsilon \vec{F}$ represents the displacement vector, and $\vec{J}$ is the total conduction current density, being the sum of electron and hole currents, $(\vec{J_n} + \vec{J_p})$, in a semiconductor. Taking the divergence on both sides of the Equation (1.105), we note that the LHS becomes zero, by definition from vector algebra. The term $\partial \vec{D}/\partial t$ after the divergence operation equals $\partial q(p - n)/\partial t$ as, $\text{div}\,\vec{D} = \rho = q(p - n)$, following Poisson's equation. Separating now the terms for electrons and holes from Equation (1.105), we arrive at the following sets of current continuity equations, neglecting any generation–recombination process:

$$\frac{\partial n}{\partial t} = \frac{1}{q}\,\text{div}(\vec{J_n}) \qquad (1.106)$$

$$\frac{\partial p}{\partial t} = -\frac{1}{q}\,\text{div}(\vec{J_p}) \qquad (1.107)$$

The current continuity equations can be generalized further considering the fact that apart from divergence of current, generation and recombination of carriers can also contribute to the change of carrier concentration in the volume under consideration. We then have:

$$\frac{\partial n}{\partial t} = \frac{1}{q}\,\text{div}(\vec{J_n}) + (g - r) \qquad (1.108)$$

$$\frac{\partial p}{\partial t} = -\frac{1}{q}\,\text{div}(\vec{J_p}) + (g - r) \qquad (1.109)$$

where g and r are represent the generation and recombination rates of carriers respectively. We shall designate $U^* = (r - g)$ as the net rate of recombination. For a steady state condition, the time dependent term in Equations (1.108) and (1.109) is zero which gives:

$$\frac{1}{q}\,\text{div}(\vec{J_n}) = U^* \qquad (1.110)$$

$$-\frac{1}{q}\,\text{div}(\vec{J_p}) = U^* \qquad (1.111)$$

In MOSFET current modeling the recombination–generation terms are normally neglected. Thus, the term U^* in Equations (1.110) and (1.111) can be set equal to zero.

In stand-alone device simulators, which provide accurate device characteristics, the semiconductor equations are solved *numerically* with inputs related to device fabrication, material parameters, device dimension, etc., and using appropriate boundary conditions. PISCES, MEDICI, ATLAS, MINIMOS, etc., are some popular device simulators which are used for the purpose of device design optimization and a deeper insight of physical processes. The use of such numerical solution based device simulators, however, is unsuitable and impractical for use in *circuit simulators*, which involves handling a large number of devices constituting the circuit. Therefore, analytic, accurate, and simple expressions are required to describe MOSFET characteristics for VLSI circuit design for use in circuit simulators such as SPICE.[1] A set of such expressions is called the *Compact MOSFET Model*, and can only be obtained by adopting suitable approximations and appropriate empirical fittings of parameters.

1.10 Compact MOSFET Models

More precisely, MOSFET models need to be compact and accurate to meet the demanding requirements of present day circuit designs on silicon. Due to proliferation of MOSFET models which were developed to meet a wide range of applications, the need to benchmark and standardize models started gaining ground. This resulted in the formation of the Compact Model Council (CMC) [75]. The IEEE Standard Working Draft defines a compact MOSFET model as that which is characterized by a collection of mathematical equations whose parameters are used as an input to a SPICE-like circuit simulator [71, 72]. The model, through *mathematical equations* representing the perceived physical mechanism, is expected to be able to reproduce the device characteristics for different device dimensions, range of temperature, and process variations, etc. Additionally, the description must hold good under a variety of operating conditions. The model parameters provide the vital interface between the design community, the model users or *consumers*, and the device manufacturers or *foundries* as they are popularly known, who print the design on silicon. The communication between designers and technologists through model parameters assumes added importance as the two communities are more often located geographically far apart.

The models, to be acceptable to a diverse category of users, must meet application specific requirements. Depending on application domains, electronics circuits can be classified into various categories: digital, analog, RF (radio frequency), memory, low-voltage, high-speed, and mixed analog–digital type. For the models to meet the specific need of such a wide variety of applications, along with the compulsion to keep pace with new technology nodes, makes model development a challenging task. For instance, in digital design an accurate drain current prediction and charge conservation are key requirements. For analog design, on the other hand, satisfying more stringent conditions such as continuity of current and its derivative in the entire region of operation are important. Similarly, for high frequency RF applications, noise modeling, distortion, etc., are issues of concern. In addition, non-quasistatic conditions set in, which are to be incorporated in the modeling strategy for the development of high frequency equivalent circuits.

---

[1] SPICE: **S**imulation **P**rogram with **I**ntegrated **C**ircuit **E**mphasis.

In the initial phase, MOSFET modeling had the sole emphasis on a precise description of the device behavior. However, with the enlarging sphere of application and growing circuit complexity, there is a distinct paradigm shift on emphasis. Computational overheads and model parameter extraction become additional overriding considerations. With the growing complexity of phenomena in scaled devices, the parameters become prone to *empirical fitting*, less physically meaningful, and large in number.

In this background, a number of approaches, each with an individual characteristic framework, emerged both in public and proprietary domains, the account of which has been detailed by C.T. Sah [74]. In this text we shall restrict the scope of discussion to a representative few model categories, that are currently available in the public domain, and have undergone a sustained phase of development. These models can be broadly classified under three categories [74, 77] as follows.

1. **Threshold Voltage, V_T, Based Models** [80, 81].
   Berkeley's LEVEL 1, LEVEL 2, LEVEL 3, BSIM1, BSIM2, BSIM3, and BSIM4[2] models, along with the Philips–MM9 model, belong to this category. Inspite of MM9 having been developed with a rich industry perspective, the Berkeley models played a pioneering role in the history of MOSFET modeling. The threshold voltage based model introduces a major simplifying step over the computationally intensive but benchmark Pao–Sah model [79]. It is a piece-wise approach with respect to the description of MOSFET behavior in weak and strong inversion demarcated by threshold voltage. The versions of the Berkeley models mentioned above represent three generations of modeling [92]. BSIM1 represents a second generation approach which involves device dimensions in the model equations and parameters, but introduced many fitting parameters. BSIM3 represents a third generation approach where both computational robustness and physical basis form the guiding philosophy. The ever increasing number of model parameters is turning out to be a major concern for this approach, yet it remains a *de facto* industry standard.

2. **Charge, Q_n, Based Models** [82, 83].
   In this approach the drain currrent is formulated in terms of the inversion charge density at the source and drain ends of the channel. This approach has been particularly very effective where low-power analog design is involved. It is physics-based, with a minimum requirement of empirical fitting, and has a relatively easy parameter extraction procedure [87, 88]. EPFL–EKV,[3] ACM,[4] and Berkeley's BSIM5 belong to this catagory.

3. **Surface Potential, ϕ_s, Based Models** [84–86].
   Philips MM11, HiSIM,[5] and PSP belong to this category [89–91]. PSP is built on the features of the Philips MOS model, MM11, and the Pennsylvania State University's SP model. In this approach the drain currrent is formulated in terms of surface potential at the source and drain ends of the channel. The computational overhead originally associated with surface potential calculation is significantly overcome by efficient analytical (PSP) or numerical algorithms (HiSIM). Further, this method does not adopt a regional modeling approach. It considers drift–diffusion current transport, which provides a single-piece expression for the current,

---

[2]BSIM: **B**erkeley **S**hort Channel **I**GFET **M**odel.
[3]EKV: **E**nz–**K**rummenacher–**V**ittoz.
[4]ACM: **A**dvanced **C**ompact **M**odel.
[5]HiSIM : **Hi**roshima University **S**TARC **I**GFET **M**odel.

with continuity of its derivatives, over the entire operating regions. Being physics-based it has acclaimed advantages in RF circuit simulation, and responds well to the phenomena modeling due to scaling.

1.11 The *p–n* Junction Diode

In a MOS transistor, junction diodes are formed when source and drain terminals are added to the basic MOS capacitor structure to have electrical contact with the inversion layer formed under the gate. These diodes are, therefore, external to the field effect action taking place under the gate. Source and drain terminals form junctions with respect to the bulk terminal. For normal MOSFET operation, source–bulk and drain–bulk junctions are reverse biased. Under certain conditions, though, the junction gets forward biased but it does not belong to the normal mode of operation.

The junction is defined as the transition point around which the semiconductor conductivity type changes. With specific reference to a MOS transistor with a *p*-substrate in the source–drain junctions, the donor concentration decreases from surface to bulk vertically downward in the *x*-direction, and at the junction:

$$N(X_j) = N_A \tag{1.112}$$

where X_j is the junction depth measured from the surface, and N_A is the background acceptor doping concentration. It is worth noting that impurities diffuse within the semiconductor isotropically. The profile details are normally available from Process Simulators such as SUPREM. If the transition of doping is abrupt around the junction, the junction is known as a *step-junction*. For gradual transition of dopant concentration, the junction is said to be a *graded-junction*.

The aspects that receive attention in modeling the junction characteristics are: (i) electric field and potential variation, (ii) transition region width, (iii) depletion capacitance, and (iv) leakage current, etc [56, 67, 68]. This section summarizes important expressions related to the above aspects which figure in MOSFET modeling, the details of which are available in standard texts introducing solid state devices [17, 18].

1.11.1 Transition Region

The transition region formed across the metallurgical junction is depleted of mobile charge due to diffusion of the majority of carriers away from the region and, therefore, consists of space charge due to fixed ionized impurity atoms. The Poisson's equation, which is required to be solved for obtaining the spatial dependence of electric field and potential in the transition region, normally assumes *depletion approximation* neglecting the presence of majority carriers. Figure 1.14 depicts the transition region features in a reverse biased *p–n* junction diode.

Important expressions related to the transition region under equilibrium are given below.

- Equilibrium condition

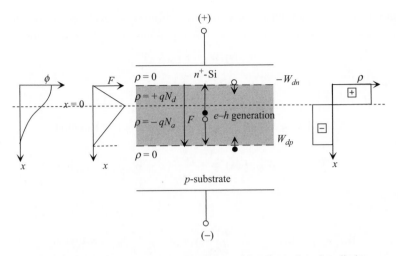

Figure 1.14 Transition region in a reverse biased *p–n* junction diode

Poisson's equation:

$$\frac{\mathrm{d}^2\phi}{\mathrm{d}x^2} = -\frac{\rho(x)}{\epsilon_{Si}} \tag{1.113}$$

For *n*-region ($-W_{dn} < x < 0$):

$$\rho(x) = \text{const.} = qN_D \tag{1.114}$$

Boundary condition at $x = -W_{dn}$:

$$\text{Electric field,} \ \frac{\mathrm{d}\phi(x)}{\mathrm{d}x} = 0 \tag{1.115}$$

$$\text{Potential,} \ \phi = \phi_n \tag{1.116}$$

For *p*-region ($0 < x < W_{dp}$):

$$\rho(x) = \text{const.} = -qN_A \tag{1.117}$$

Boundary condition at $x = W_{dp}$:

$$\text{Electric field,} \ \frac{\mathrm{d}\phi(x)}{\mathrm{d}x} = 0 \tag{1.118}$$

$$\text{Potential,} \ \phi = \phi_p \tag{1.119}$$

Interface boundary condition at $x = 0$, due to continuity of displacement vector:

$$\epsilon_{Si} N_D W_{dn} = \epsilon_{Si} N_A W_{dp} \tag{1.120}$$

Electric field:

$$-W_{dn} < x < 0; F(x) = \frac{qN_D}{\epsilon_{Si}}(x + W_{dn}) \quad : n\text{-region} \tag{1.121}$$

$$0 < x < W_{dp}; F(x) = -\frac{qN_A}{\epsilon_{Si}}(x - W_{dp}) \quad : p\text{-region} \tag{1.122}$$

Potential:

$$-W_{dn} < x < 0; \phi(x) = -\frac{qN_D}{2\,\epsilon_{Si}}(x + W_{dn})^2 + \phi_n \quad : n\text{-region} \tag{1.123}$$

$$0 < x < W_{dp}; \phi(x) = \frac{qN_A}{2\,\epsilon_{Si}}(x - W_{dp})^2 + \phi_p \quad : p\text{-region} \tag{1.124}$$

$$\phi_{bi} = \phi_n - \phi_p = \text{barrier potential} = \phi_t \ln\left(\frac{N_A N_D}{n_i^2}\right) \tag{1.125}$$

$$W_{d0} = \text{zero-bias depletion width} = \sqrt{\frac{2\epsilon_{Si}\phi_{bi}}{q}}\,\sqrt{\left(\frac{1}{N_A} + \frac{1}{N_D}\right)} \tag{1.126}$$

For a one-sided abrupt junction, $N_D \gg N_A$:

$$W_{do} \approx W_{dp} = \sqrt{\frac{2\epsilon_{Si}\phi_{bi}}{qN_A}} = \sqrt{2\epsilon_{Si}\mu_p\rho_p\phi_{bi}} \tag{1.127}$$

where ρ_p is the resistivity of the p-type substrate.

- Non-equilibrium condition
  The equilibrium values of electric field, potential, and depletion width, all become modified when an external voltage, V, is applied across the junction diode. The potential barrier is now modified as:

$$\phi_b(V) = (\phi_{bi} - V) \tag{1.128}$$

For forward bias, $V = V_F$ is positive, and for reverse bias, $V = -V_R$ is negative.

1.11.2 Recombination/Generation Current

- **Recombination in the Quasi-Neutral Bulk Region**
  The *ideal* forward biased junction diode current is due to recombination of the injected carriers in the neutral bulk region, where it is assumed that *the quasi-Fermi potential is*

constant in the space charge region. For a low injection level the forward diode current is given by:

$$I_F = I_0 \left[\exp \left(\frac{\phi_n - \phi_p}{\phi_t} \right) - 1 \right] = I_0 \exp \left[\exp \left(\frac{V_F}{\phi_t} \right) - 1 \right] \qquad (1.129)$$

where the applied forward voltage V_F equals the quasi-Fermi potential difference, $(\phi_n - \phi_p)$, at the edges of the junction transition region.

- **Recombination in the Space Charge Region**

  The current–voltage characteristics of a practical junction diode is significantly affected by the carrier recombination in the space charge in the forward bias condition, and by the carrier generation in the space charge region in the reverse bias condition. In the operating region, where this dominates over recombination in the quasi-neutral region, the current voltage relation can be expressed as:

$$I_F \approx \exp \frac{V_F}{n \phi_t} \qquad (1.130)$$

  where n is the diode ideality (in reality, non-ideality!) factor. In many models, n is used as a fitting parameter to match the experimental I–V characteristics to the diode model [92].

- **Generation Current in the Transition Region**

  When a p–n junction diode is reversed biased, the carrier concentration in the depletion region is below the equilibrium value. For such a condition, generation of electron–hole pairs takes place within the depletion region. The generated carriers encounter an electric field in the depletion region which is directed from the n- to the p-region. The holes are directed towards the p-region, and electrons to the n-region, by the drift mechanism. They together constitute the space charge generation current, adding to the ideal reverse saturation current which originates from the diffusion of minority carriers from quasi-neutral bulk regions. The space charge generation current is bias dependent, since the volume of the depletion region, where generation of carriers take place, increases with the increase in the width of the depletion region. On the other hand, the diffusion related current from the bulk is saturated and has a constant value. In some models, the reverse leakage current is assumed to be constant, even though the current in a diode in the reverse bias condition has a dependence on bias.

  In the reverse biased non-equilibrium condition it can be shown that, assuming that generation centers are located in the center of the band gap, the generation rate is given by $n_i/2\tau$, and the generation current density by:

$$J_G = \frac{q \, n_i \, W_{dp}}{2\tau} \qquad (1.131)$$

For a one-sided step junction:

$$W_{dp} \propto |V_R|^{\frac{1}{2}} \qquad (1.132)$$

which gives:

$$J_G \propto |V_R|^{\frac{1}{2}} \qquad (1.133)$$

where V_R is the applied reverse bias.

1.11.3 Carrier Multiplication

We shall illustrate the carrier multiplication within the transition region of a reverse biased p–n junction due to a phenomenon called impact ionization. The minority carriers are injected in the depletion region from the neutral region by the *extraction process*, as the carrier concentration at the depletion edge falls below the equilibrium value in a reverse biased junction. This causes a concentration gradient which makes the minority carriers flow from the bulk towards the depletion region. The minority carriers injected into the depletion region gain energy from the high electric field present in this region, and some of them collide with the bound electrons in the valence band and thus create electron–hole pairs. These electron–hole pairs, in a subsequent collision, further multiply the number of carriers. The carrier multiplication results in a sharp increase in the reverse current. This is known as the *avalanche breakdown* of the junction.

For impact ionization to take place, there is a need of a minimum threshold energy to be acquired by the carriers to enable the creation of electron–hole pairs by breaking of covalent bonds. The critical field required depends on the mean free path of the carriers, on traversal of which the carriers acquire the necessary energy to break bonds. The mean free path is related to impurity doping. The higher the doping, the lower is the mean free path. Hence, the critical field required to provide ionization energy increases with doping concentration as the required energy needs to be acquired over a short distance of travel. We shall consider a simple case of a one-sided n^+–p junction to obtain an analytical expression for the avalanche breakdown condition. Figure 1.15 outlines the approach to estimate the carrier multiplication within the reverse biased depletion region.

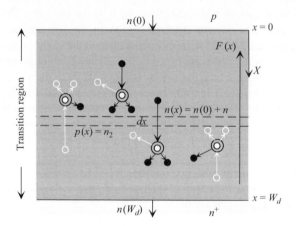

Figure 1.15   Carrier multiplication in a reverse biased p–n junction

The width of the depletion region, W_d, is assumed to be entirely confined to the lowly doped p-type substrate region conforming to the one-sided abrupt junction approximation for an assumed n^+–p junction. The field is directed from the n- to the p-region within the depletion. Due to the one-sided junction approximation, we consider only the electron injection from the lowly doped p-type substrate into the depletion region.

We shall now define certain terms which are required for the formulation of carrier multiplication and avalanche breakdown voltage of a p–n junction. We define $\alpha_n(F)$ and $\alpha_p(F)$ as the ionization coefficients for electrons and holes respectively, which give the rate of electron–hole pair generation per unit length along the direction of the electric field, while carriers transverse through the depletion region.

The ionization coefficient dependence on electric field for silicon is given by the following empirical relation [59, 61]:

$$\alpha_{(n,p)} = A_{(n,p)} \exp \frac{-B_{(n,p)}}{F} \tag{1.134}$$

where the coefficients $A_{(n,p)}$, $B_{(n,p)}$ are obtained experimentally, and F is the electric field along the current flow direction. Physically, $A_{(n,p)}$ means the value of the ionization coefficient at very high field, whereas $B_{(n,p)}$ represents the critical field at which the ionization coefficient reduces to $\frac{1}{e} A_{(n,p)}$.

There have been several other models for the ionization coefficient to account for field and temperature dependence [60, 62] which can be described by a general expression of the form [69]:

$$\alpha_{n,p} = \left(A_{1(n,p)} + A_{2(n,p)} F\right) \exp - \left(\frac{B_{(n,p)}}{F}\right)^{\beta} \tag{1.135}$$

MINIMOS3 uses the following values of the parameters A and B, and they are taken as independent of temperature.

For electrons: $A_{1(n)} = 7.03 \times 10^5 \, \text{cm}^{-1}$, $A_{2(n)} = 0 \, \text{V}^{-1}$, $B_{(n)} = 1.23 \times 10^6$, $\beta = 1.0$, and for holes: $A_{1(p)} = 1.58 \times 10^6 \, \text{cm}^{-1}$, $A_{2(p)} = 0 \, \text{V}^{-1}$, $B_{(p)} = 2.04 \times 10^6$, $\beta = 1.0$.

The electron hole multiplication factor M is defined as the ratio of the concentration of electrons/holes at the exit to that of the entry boundary planes of the depletion layer. Assuming $\alpha_n = \alpha_p = \alpha$, the electron multiplication factor M is given by:

$$M = \frac{n(W_d)}{n(0)} = \frac{1}{\left(1 - \int_0^{W_d} \alpha(F) \, dx\right)} \tag{1.136}$$

It is seen from Equation (1.136) that for the reverse bias voltage V_{BR} at which the electric field is such that the integrand $\int_0^{W_d} \alpha(F) \, dx \rightarrow 1$, $M_n \rightarrow \infty$ which leads to uncontrolled carrier multiplication, and the junction breakdown takes place.

1.11.4 SPICE Model Parameters for a p–n Junction Diode

In a MOS transistor, source and drain diffused layers form junction diodes with the substrate. The model parameters that characterize the junction diode for SPICE-like circuit simulators are as follows.

Process Parameters

$XJ(x_j)$ junction depth
$RSH(R_{sh})$ source–drain diffusion sheet resistance ($\Omega/\square$)

Electrical Parameters   The capital letters denote the SPICE names and the parameters are described by the terms in parentheses.

CJ (C_{j0})	zero-bias capacitance per unit area for the bottom junction
MJ (mj)	aerial junction grading coefficient
PB (ϕ_0)	potential barrier of the junction required for computation of the capacitor at a given bias
$CJSW$ (C_{jsw0})	sidewall capacitance per unit length
$MJSW$ (m_{jsw0})	average grading coefficient of the sidewall junction
AS	area of source
AD	area of drain
PS	periphery length for source
PD	periphery length for drain
JS (j_0)	source/drain reverse saturation current density

PS and PD are used for calculation of sidewall capacitance and current of source and drain respectively.

Similarly, AS and AD multiplied with C_{j0} gives the total bottom capacitance at $V_{BS} = 0$.

1.12 Tunneling Through Potential Barrier

The tunneling phenomenon is playing an increasingly important role in modern MOSFET devices as their dimensions are shrinking. We have already seen in the development of the band theory of semiconductors that electrons and holes can be described by Schrödinger's wave equation. Under certain conditions a wave particle can be shown to have a passage through a potential barrier quantum mechanically, which is not feasible from a classical viewpoint. In view of its importance we shall briefly describe the essential steps involved in the development of the theory of tunneling through a potential barrier. The details are available in Levi [22].

We shall illustrate tunneling through a potential barrier using the diagram shown in Figure 1.16 where a square potential barrier is assumed separating two regions through which propagation of the particle is being considered. We have, therefore, three regions to consider for the motion of the particle from Region-1 to Region-3 through Region-2, having a potential barrier of height U_0 and width d. The motion is assumed to be in the direction x.

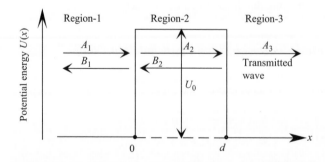

Figure 1.16   Tunneling of particles through a potential barrier

Schrödinger's equations corresponding to the three regions are as follows.

- Region-1 ($x < 0$; $U(x) = 0$).   The wave function solution, Ψ_1, is an oscillatory function which has forward (propagating), and backward (reflected) components:

$$\frac{d^2\Psi_1}{dx^2} + k_1^2\Psi_1 = 0 \qquad (1.137)$$

where:

$$k_1 = \frac{1}{\hbar}\sqrt{2mE} \qquad (1.138)$$

- Region-2 ($0 < x < d$; $E < U_0$).   The wave function solution, Ψ_2, is a monotonically decaying/increasing function:

$$\frac{d^2\Psi_2}{dx^2} + k_2^2\Psi_2 = 0 \qquad (1.139)$$

$$k_2 = \frac{1}{\hbar}\sqrt{2m(E - U_0)} = ik \qquad (1.140)$$

$$k = \frac{1}{\hbar}\sqrt{2m(U_0 - E)} \qquad (1.141)$$

where the wave number, k_2, is imaginary for particle energy E less than the barrier height U_0, and k is a real number as $U_0 > E$.
- Region-3 ($x > d$; $U(x) = 0$).   The wave function solution, Ψ_3, is an oscillatory function with a forward (transmitted) wave component, and no backward wave component:

$$\frac{d^2\Psi_3}{dx^2} + k_3^2\Psi_3 = 0 \qquad (1.142)$$

where:

$$k_3 = k_1 \qquad (1.143)$$

The solutions of Equations (1.137), (1.139), and (1.142) are:

$$\Psi_1 = A_1 \exp i(k_1 x) + B_1 \exp -i(k_1 x) \qquad (1.144)$$

$$\Psi_2 = A_2 \exp i(k_2 x) + B_2 \exp -i(k_2 x) \qquad (1.145)$$

$$\Psi_3 = A_3 \exp i(k_3 x) \qquad (1.146)$$

As the propagating particle wave does not encounter any barrier in Region-3, the second term in Equation (1.146) representing a backward wave must be equated to zero making $B_3 = 0$.

In Equation (1.144), the first term on the RHS, $A_1 \exp i(k_1 x)$, stands for the wave propagation representing the particle in the positive x-direction with an amplitude A_1, and the second term $B_1 \exp -i(k_1 x)$ is the wave that is reflected from the boundary of the potential barrier at $x = 0$. The coefficients of the exponential terms A_1 and B_1 represent respectively, the amplitudes of the incident and reflected waves which pertain to Region-1. The ratio of the square of the amplitudes of *reflected* and *incident* waves R represents the *reflection coefficient* of the barrier given by: $R = |B_1|^2/|A_1|^2$. Our primary aim is to obtain the amplitude of the propagating wave in Region-3, A_3, in terms of the incident wave amplitude A_1 in Region-1, for which we have to evaluate the amplitude terms associated with the waves in the three regions from boundary conditions. For the boundary conditions, we use the fact that the wave function and its derivative are continuous at the boundaries, as outlined below.

(i) $\Psi_1 |_{x=0} = \Psi_2 |_{x=0}$ (ii) $\left. \frac{d\Psi_1}{dx} \right|_{x=0} = \left. \frac{d\Psi_2}{dx} \right|_{x=0}$

(iii) $\Psi_2 |_{x=d} = \Psi_3 |_{x=d}$ (iv) $\left. \frac{d\Psi_2}{dx} \right|_{x=d} = \left. \frac{d\Psi_3}{dx} \right|_{x=d}$

Carrying out the above steps we get the expression for the tunnelling probability or the transmission coefficient, TC, which is given by:

$$TC = \frac{|A_3|^2}{|A_1|^2} \tag{1.147}$$

The terms A_1 (and A_3) can be obtained from equations related to the boundary conditions by eliminating the coefficients B_1, B_2, and A_2 using the following steps:

(i) Express A_1 in terms of A_2 and B_2.
(ii) Express A_2 and B_2 in terms of A_3.
(iii) Steps (i) and (ii) establish a relation between A_3 and A_1.

A straightforward algebraic manipulation provides the value of A_3/A_1, which happens to be a complex quantity. Therefore, to obtain $|A_3/A_1|^2$, we need to get the complex conjugate A_3^*/A_1^* and use the identity:

$$\frac{A_3}{A_1} \cdot \frac{A_3^*}{A_1^*} = \frac{|A_3|^2}{|A_1|^2} \tag{1.148}$$

By evaluating Equations (1.147) and (1.148), we arrive at the final expression which establishes the functional dependence of the transmission coefficient, TC, or equivalently, the tunnelling probability/penetrability, D. This is obtained in terms of the width and height of the potential barrier and given energy of the particle:

$$TC = D = \frac{4k_1^2 k^2}{[(k_1^2 + k^2)^2 \sinh^2 (kd) + 4k^2 k_1^2]} \tag{1.149}$$

For the condition satisfying $kd \gg 1$ the following approximations hold good:

$$\exp(kd) \gg 1 \tag{1.150}$$

$$\sinh^2(kd) = \left[\frac{\exp(kd) - \exp(-kd)}{2}\right]^2 \approx \frac{1}{4}\exp(2kd) \tag{1.151}$$

Equation (1.149) can then be simplified and be put in the form:

$$D \approx \frac{16k^2 k_1^2}{(k^2 + k_1^2)^2}\exp -(2kd) \tag{1.152}$$

Generally, k and k_1 are of the same order. Hence, the value of the pre-exponential factor in Equation (1.152), $(16k^2 k_1^2/(k^2 + k_1^2)^2 = D_0)$, is approximately equal to unity. We need to keep in view the fact that the exponential term is the more sensitive factor. Thus, we can write:

$$D = D_0 \exp -(2kd) \approx \exp -\frac{2d}{\hbar}\left[\sqrt{2m(U_0 - E)}\right] \tag{1.153}$$

We can physically arrive at the same conclusion through inspection of the wave function in Region-2, where by ignoring the backward component, we have $\Psi_2 = A_2 \exp -(kx)$. This form of equation is not representative of a propagating plane wave, implying non-existence of a particle inside the potential barrier. Yet we have the existence of a finite probability density function at $x = d$ given by $\Psi_2 = A_2 \exp -(kd)$, following the condition of continuity of the wave function at the boundary between Region-2 and Region-3. Thus, the probability of finding the particle at the edge of Region-3 is given by the square of the probability density function $|\Psi|^2 = A_2^2 \exp -\frac{2d}{\hbar}\sqrt{2m(U_0 - E)}$, which is a finite quantity. Approximating A_2 to unity, the expression is identical to Equation (1.153). The important aspects of the tunnelling phenomenon are summarized as under:

- The tunnelling probability has a negative exponential dependence on barrier thickness, implying that a thin barrier increases the transmission or penetration probability.
- The tunnelling probability has a negative exponential square root dependence on the barrier height U_0.
- The transmission probability through the barrier increases for particles with higher energy, as the term $(U_0 - E)$ decreases in the negative exponential term.
- In the tunnelling process through the barrier, the particle energy remains unchanged.
- For carriers in a semiconductor, m represents the effective mass of the particle. For an arbitrary shape of the potential barrier, $U(x)$, the transmission coefficient, TC or D, can be expressed as:

$$TC = D \approx \exp\left(-\frac{2\sqrt{2m}}{\hbar}\right)\int_{x_2}^{x_1}\sqrt{(U(x) - E)}\,dx \tag{1.154}$$

where x_1 and x_2 represent the *turning points* of the barrier.
- Last, but not the least, it is implied that when particles tunnel, there exists an energy state on both sides of the barrier *at the same energy level*.

References

[1] J.E. Lilienfeld, Method and Apparatus for Controlling Electric Currents, *US Patent, 1745175*, Filed October 8, 1926, Issued January 28, 1930.

[2] C.T. Sah, Evolution of the MOS Transistor—From Conception to VLSI, *Proc. IEEE*, **76**(10), 1280–1326, 1988.

[3] J. Bardeen and W.H. Brattain, Three-Electrode Circuit Element Utilizing Semiconductive Materials, *US Patent, 2524035*, Filed June 17, 1948, Issued October 3, 1950.

[4] W. Shockley, Semiconductor Amplifier, *US Patent, 2502488*, Filed September 24, 1948, Issued April 4, 1950.

[5] J.A. Hoerni, Method of Manufacturing Semiconductor Devices, *US Patent, 3025589*, Filed May 1, 1959, Issued March 20, 1962.

[6] J.S.C. Kilby, Miniaturized Electronics Circuits, *US Patent, 3138743*, Filed February 6, 1959, Issued June 23, 1964.

[7] D. Kahng and M.M. Atalla, Silicon–Silicon Dioxide Field Induced Surface Devices, *IRE Solid-State Device Research Conference,* Carnegie Institute of Technology, Pittsburg, PA, 1960.

[8] M.M. Atalla, Semiconductor Triode, *US Patent, 3056888*, Filed August 17, 1960, Issued October 2, 1962.

[9] D. Kahng, Electric Field Controlled Semiconductor Device, *US Patent, 3102230*, Filed May 31, 1960, issued August, 27, 1963.

[10] F.M. Wanlass and C.T. Sah, Nanowatt Logic Using Field-effect Metal–Oxide–Semiconductor Triodes, *IEEE International Solid-State Circuit Conference*, Pennsylvania, pp. 32–33, 1963.

[11] L.L. Vadasz, A.S. Grove, T.A. Rowe and G.E. Moore, Silicon-Gate Technology, *IEEE Spectrum*, **6**(10), 28–35, 1969.

[12] A.B. Bhattacharyya, Microsystem-On-Silicon: A Revolution on the Grains of Sand, *Ann. Ind. Acad. Eng.*, IV, 13–18, 2007.

[13] International Technology Roadmap for Semiconductors (http://public.itrs.net).

[14] G.E. Moore, Cramming More Components on to Integrated Circuits, *Electronics*, **38**(8), 114–117, 1965.

[15] D.A. Buchanan, Scaling the dielectric gate dielectric: Materials, integration, and reliability, *IBM J. Res. Develop.*, **43**(3), 245–263, 1999.

[16] D.A. Buchanan, Beyond Microelectronics: Materials and Technology for Nano-Scale CMOS Devices, *Phys. Stat. Sol. c*, **1**(S2) S155–S162, 2004.

[17] S.M. Sze, *Physics of Semiconductor Devices*, 2nd Edition, John Wiley & Sons, Inc., New York, NY, 1981.

[18] M.S. Tyagi, *Semiconductor Materials and Devices*, John Wiley & Sons, Inc., New York, NY, 1991.

[19] J. Singh, *Modern Physics for Engineers*, John Wiley & Sons, Inc., New York, NY, 1999.

[20] G.E. Pikus, *Osnovi Teorii Poluprovodnikovykh Priborov*, Nauka Publishing House, Moscow, USSR, 1965 (In Russian).

[21] G. Yepifanov, *Physical Principles of Microelectronics*, Mir Publishers, Moscow, USSR, 1974.

[22] A.F.J. Levi, *Applied Quantum Mechanics*, Cambridge University Press, Cambridge, UK, 2003.

[23] S. Selberherr, *Analysis and Simulation of Semiconductor Devices*, Springer-Verlag, Wien, Austria, 1984.

[24] J.W. Mayer and S.S. Lau, *Electronic Materials Science: For Integrated Circuits in Si and GaAs*, Macmillan Publishing Company, New York, NY, 1990.

[25] M.L. Cohen and T.K. Bergstresser, Band Structure and Pseudopotential Form Factors for Fourteen Semiconductors of Diamond and Zinc-Blende Structures, *Phys. Rev.*, **141**, 789–796, 1966.

[26] J.J. Liou, Semiconductor Device Physics and Modelling. Part 1: Overview of Fundamental Theories and Equations, *IEE Proc.*, **139**(6), 646–654, 1992.

[27] J.W. Slotboom and H.C. de Graaff, Measurement of Bandgap Narrowing in Si Bipolar Transistors, *Solid-State Electron.*, **19**, 857–862, 1976.

[28] A.F. Franz and G.A. Franz, BAMBI–A Design Model for Power MOSFETs, *IEEE Trans. Comput. Aided Design Integr. Circ. Syst.*, **4**(3), 177–189, 1985.

[29] S.C. Jain and D.J. Roulston, A Simple Expression For Band Gap Narrowing (BGN) in Heavily Doped Si, Ge, GaAs, and $Ge_x Si_{1-x}$ Strained Silicon Layers, *Solid-State Electron.*, **34**(5), 453–465, 1991.

[30] S.R. Dhariwal and V.N. Ojha, Band Gap Narrowing in Heavily Doped Silicon, *Solid-State Electron.*, **25**(9), 909–911, 1982.

[31] W. Shockley and W.T. Read, Statistics of the Recombination of Holes and Electrons, *Phys. Rev.*, **87**, 835–842, 1952.

[32] J.D. Beck and R. Conradt, Auger-Recombination in Si, *Solid-State Commun.*, **3**, 93–95, 1973.

[33] R.N. Hall, Electron–Hole Recombination in Germanium, *Phys. Rev.*, **87**, 387, 1952.

[34] S. Reggiani, M. Valdinoci, L. Colalongo, M. Rudan, G. Baccarani, A.D. Stricker, F. Illien, N. Felber, W. Fichtner and L. Zullino, Electron and Hole Mobility in Silicon at Large Operating Temperatures–Part 1: Bulk Mobility, *IEEE Trans. Electron Dev.*, **49**(3), 490–499, 2002.

[35] S.S. Li and W.R. Thurber, The Dopant Density and Temperature Dependence of Electron Mobility and Resistivity in *n*-type Silicon, *Solid-State Electron.*, **20**, 609–616, 1977.

[36] S.S. Li, The Dopant Density and Temperature Dependence of Hole Mobility and Resistivity in Boron Doped Silicon, *Solid-State Electron.*, **21**, 1109–1107, 1978.

[37] A.B. Bhattacharyya, M.L. Gupta and V. K. Garg, Temperature Coefficient Calculations for Monolithic Resistors, *Microelectron. Reliability*, **9**(4), 349–354, 1970.

[38] N.D. Arora, J.R. Hauser, and D.J. Roulston, Electron and Hole Mobilities in Silicon as a function of Concentration and Temperature, *IEEE Trans. Electron Dev.*, **ED-2**(2), 292–295, 1982.

[39] D.B.M. Klaassen, A Unified Mobility Model for Device Simulation–Part I: Model Equations and Concentration Dependence, *Solid-State Electron.*, **35**, 953–959, 1992.

[40] H. Brooks, Scattering by Ionized Impurities in Semiconductors, *Phys. Rev.*, vol. 83, 879, 1951.

[41] E. Conwell and V.F. Weisskopf, Theory of Impurity Scattering in Semiconductors, *Phys. Rev.*, **77**, 388–390, 1950.

[42] D.M. Caughey and R.E. Thomas, Carrier Mobilities in Silicon Empirically Related to Doping and Field, *Proc. IEEE*, **52**(12), 2192–2193, 1967.

[43] D.B.M. Klaassen, A Unified Mobility Model for Device Simulation–Part II: Temperature Dependence of Carrier Mobility and Lifetime, *Solid-State Electron.*, **35**(7), 961–967, 1992.

[44] S.Selberherr, W. Hänsch, M. Seavey and J. Slotboom, The Evolution of the MINIMOS Mobility Model, *Solid-State Electron.*, **33**(11), 1425–1436, 1990.

[45] J.R. Watling, L. Yang, M. Boriçi, R.C.W. Wilkins, A. Asenov, J.R. Barker and S. Roy, The Impact of Interface Roughness Scattering and Degeneracy in Relaxed and Strained Si *n*-Channel MOSFETs, *Solid-State Electron.*, **48**(8), 1337–1346, 2004.

[46] S. Selberherr, A. Schütz and H.W. Pötzl, A Two-Dimensional MOS Transistor Analyzer, *IEEE Trans. Electron Dev.*, **ED-27**(8), 1540–1550, 1980.

[47] I.P. Stepanenko, *Ocnovy Theorii Transistorov i Transistornykh Skhem*, 4th Edition, Energiya Publishing House, Moscow, USSR, 1977.

[48] S.C. Jain and M. Willander, Silicon–Germanium Strained Layers and Heterostructures, *Semiconductors and Semimetals*, Vol. 74, Academic Press, Amsterdam, The Netherlands, 2003.

[49] M. Orlowski, CMOS Challenges of Keeping up with Moore's Law, *13th International Conference on Advanced Thermal Processing of Semiconductors–RTP*, 3–21, 2005.

[50] M.M. Rieger and P. Vogl, Electronic-band Parameters in Strained $Si_{1-x}Ge_x$ alloy on $Si_{1-y}Ge_y$ Substrates, *Phys. Rev. B*, **48**(19), 14276–14287, 1993.

[51] C.K. Maiti, L.K. Bera and S. Chattopadhyay, Strained-Si Heterostructure Field Effect Transistors, *Semicond. Sci. Technol.*, **13**, 1225–1246, 1998.

[52] R. People and J.C. Bean, Band Alignment of Coherently Strained Ge_xSi_{1-x}/Si Heterostructures on (001) $Ge_y Si_{1-y}$ Substrates, *Appl. Phys. Lett.*, 48, 538–541, 1986.

[53] K. Chandrasekaran, X. Zhou, S.B. Chiah, W. Shangguan, and G.H. See, Physics-Based Single-Piece Charge Model for Strained-Si MOSFETs, *IEEE Trans. Electron Dev.*, **52**(7), 1555–1562, 2005.

[54] K. Rim, J. Welser, J.L. Hoyt and J.F. Gibbons, Design of Si/SiGe Heterojunction Complementary Metal-Oxide-Semiconductor Transistor, *IEDM Tech. Digest*, 517–520, 1995.

[55] V. Venkataraman, S. Nawal and M.J. Kumar, Compact Analytical Threshold Voltage model of Nanoscale fully Depleted Strained-Si-On-Silicon-Germanium-on-Insulator (SGOI) MOSFETs, *IEEE Trans. Electron Dev.*, **54**(3), 554–562, 2007.

[56] A.J. Scholten, G.D.J. Smit, M. Durand and R. van Langevelde, The Physical Background of JUNCAP2, *IEEE Trans. Electron Dev.*, **53**(9), 2098–2107, 2006.

[57] G.A.M. Hurks, H.C. de Graff, W.J. Kloosterman and P.G. Knuvers, A New Analytical Diode Model Including Tunneling and Avalanche Breakdown, *IEEE Trans. Electron Dev.*, **39**(9), 2090–2098, 1992.

[58] M.A. Green, Intrinsic Concentration, Effective Density of States and Effective Mass of Silicon, *J. Appl. Phys.*, **67**(6), 2944–2954, 1990.

[59] R. Van Overstraeten and H. De Man, Measurement of the Ionization Rates in Diffused Silicon *p–n* Junctions, *Solid-State Electron.*, **13**, 583–608, 1970.

[60] W. Maes, K. De Meyer and R. van Overstaeten, Impact Ionization in Silicon: A Review and Update, *Solid-State Electron.*, **33**(6), 705–718, 1990.

[61] A.G. Chynoweth, Ionization Rates for Electrons and Holes in Silicon, *Phys. Rev.*, **109**, 1537–1540, 1958.

[62] Y. Okuto and C.R. Crowell, Threshold Energy Effect on Avalanche Breakdown Voltage in Semiconductor Junctions, *Solid-State Electron.*, **18**, 161–168, 1975.

[63] G.A.M. Hurkx, D.B.M. Klaassen and M.P.G. Knuvers, A New Recombination Model for Device Simulation Including Tunneling, *IEEE Trans. Electron Dev.*, **39**(2), 331–338, 1992.

[64] A.J. Scholten, G.D.J. Smit, M. Durand, R. van Langevelde, C.J.J. Dachs and D.B.M. Klaassen, A New Compact Model for Junctions in Advanced CMOS Technologies, *IEDM Tech. Digest*, 209–212, 2005.

[65] C.T. Sah, R.N. Noyce and W. Shockley, Carrier Generation and Recombination in *p–n* Junction Characteristics, *Proc. IRE*, **45**(9), 1228–1243, 1957.

[66] M.S. Tyagi and R. van Overstraeten, Minority Carrier Recombination in Heavily Doped Silicon, *Solid-State Electron.*, **26**, 577–597, 1983.

[67] P.U. Calzolari and S. Graffi, A Theoretical Investigation on the Generation current in Silicon *p–n* Junctions Under Reverse Bias, *Solid-State Electron.*, **15**(9), 1003–1011, 1972.

[68] M. Shur, Recombination Current in Forward-Biased *p–n* Junctions, *IEEE Trans. Electron Dev.*, **35**(9), 1564–1565, 1988.

[69] P. Wolbert, *Modeling and Simulation of Semiconductor Devices in TRENDY*, Ph.D. Thesis, University of Twente, Twente, The Netherlands, 1991.

[70] E.H. Rhoderick, Metal Semiconductor Contacts, *IEE Proc.*, **129**(1), 1–14, 1982.

[71] L.W. Nagel and D.O. Pederson, *SPICE (Simulation Program with Integrated Circuit Emphasis)*, Electronics Research Laboratory Memorandum, No. ERL-M382, University of California, Berkeley, CA, 1973.

[72] *Test Procedures for Micro-electronic MOSFET Circuit Simulator for Model Validation*, Micro Electronic MOSFET Task Group IEEE Standard Working Draft, 1–41, 2001.

[73] (http://www.eigroup.org/cmc).

[74] C.T. Sah, A History of MOS Transistor Compact Modeling, *Proc. Tech. WCM 2005*, 347–385, 2005.

[75] B. Brooks, Standardizing Compact Models for IC Simulation, *IEEE Circ. Dev. Mag.*, **15**(4), 10–13, 1999.

[76] J. Mattausch, M. Miyake, T. Yoshida, S. Hazama, D. Navarro, N. Sadachika, T. Ezaki and M. Miura-Mattausch, HiSIM2 Circuit Simulation–Solving the Speed versus Accuracy Crisis, *IEEE Circ. Dev. Mag.*, **22**(5), 30–37, 2006.

[77] X. Zhou, S.B. Chiah, S. Chandrasekaran, G.H. See, W.Z. Shangguan, S.M. Pandey, M. Cheng, S. Chu and L.C. Hsia, A Compact Model for Future Generation Predictive Technology Modeling and Circuit Simulation, 12th International Conference, MIXDES, Krakow, Poland, 22–25, June 2005.

[78] X. Zhou, The Missing Kink to Seamless Simulation (Nanometer CMOS ULSI), *IEEE Circ. Dev. Mag.*, **19**(3), 9–17, 2003.

[79] H.C. Pao and C.T. Sah, Effect of Diffusion Current on Characteristics of Metal–Oxide (Insulator)–Semiconductor Transistors, *Solid-State Electron.*, **9**(10), 927–937, 1966.

[80] D.P. Foty, *MOSFET modeling with SPICE: Principles and Practice*, Prentice Hall, Inc., New Jersey, NJ, 1997.

[81] W. Liu, *MOSFET Models for SPICE Simulation, including BSIM3v3 and BSIM4*, John Wiley & Sons, Inc., New York, NY, 2001.

[82] M. Butcher, C. Lallement and C. Enz, *The EPFL-EKV MOSFET Model*, Version 2.3, Technical Report, EPFL, Lausanne, Switzerland, 1995.

[83] A.I.A. Cunha, M.C. Schneider and C. Galup-Montoro, An MOS Transistor Model for Analog Circuit Design, *IEEE J. Solid-State Circ.*, **33**(10), 1510–1519, 1998.

[84] G. Gildenblat, H. Wang, T.L. Chen, X. Gu and X. Xai, SP: An Advanced Surface Potential-based Compact MOSFET Model, *IEEE J. Solid-State Circ.*, **39**(9), 1394–1406, 2004.

[85] M. Miura-Mattausch, N. Sadachika, D. Navarro, G. Suzuki, Y. Takeda, M. Miyake, T. Warabino, Y. Mizukane, R. Inagaki, T. Ezaki, H.J. Mattausch, T. Ohguro, T. Iizuka, M. Taguchi, S. Kumashiro and S. Miyamoto, HiSIM2: Advanced MOSFET Model Valid for RF Circuit Simulation, *IEEE Trans. Electron Dev.*, **53**(9), 1994–2007, 2006.

[86] G. Gildenblat, X. Li, W. Wu, H. Wang, A. Jha, R. van Langevelde, G.D.J. Smit, A. J. Scholten and B.M. Klaassen, PSP: An Advanced Surface-Potential-Based MOSFET Model for Circuit Simulation, *IEEE Trans. Electron Dev.*, 53(9), 1979–1993, 2006.

[87] EKV (http://legwww.epfl.ch/ekv/).

[88] ACM (http://eel.ufsc.br/ lci/acm/introduction.html).

[89] HiSIM (http://home.hiroshima-u.ac.jp/usdl/HiSIM.html).

[90] PSP (http://pspmodel.asu.edu/).

[91] Phi (http://www.nxp.com/models/mos_models/model11/).

[92] BSIM (http://www-device.eecs.berkeley.edu/~bsim3/).

Problems

1. Calculate the error incurred due to the use of Boltzmann statistics in place of Fermi–Dirac statistics, for an n-type doping concentration of 10^{20} cm^{-3} in silicon, the temperature varying from 273 K to 100 K.

2. Find the minimum conductivity of silicon and compare its value with the intrinsic conductivity. Which is higher, and by what percentage?

3. (a) Find T_E, the exhaustion temperature at which all impurity donor atoms in an n-type silicon are fully ionized.

   (b) Find the temperature at which the silicon sample with the above doping behaves as intrinsic. Assume donor atoms are located 0.015 eV below the conduction band edge and the donor concentration is $N_D = 5 \times 10^{16}$ cm^{-3}.

4. Assume that in silicon, in a given volume, excess electron and hole pairs are injected with concentrations Δn and Δp creating a volume space charge density of $\rho = q(\Delta n - \Delta p)$. With the help of the time dependent continuity equation and neglecting recombination, show that the excess carrier decays with a dielectric relaxation time $\tau_d = \epsilon_{Si}/\sigma$ where σ represents the conductivity of the sample.

5. (a) Show the quasi-Fermi level variation in the quasi-neutral region of a forward biased p–n junction, assuming that in the space charge region the quasi-Fermi level is constant and the minority carrier concentration varies exponentially in the quasi-neutral bulk region.

   (b) Show the quasi-Fermi level positions for electrons and holes in a semiconductor when the carrier concentration is below the equilibrium value ($p \times n < n_i^2$) such as in the depletion region.

6. A p–n junction is formed by diffusion of a p-type impurity on a homogeneously doped n-silicon. Obtain the expression for the diode current considering the contribution of current component of the inhomogeneous layer. Assume that the junction is formed at a distance X_j from the surface.

7. Consider a symmetric p–n junction described by the following equations:

$$N = N_0 \left(\frac{x}{x_0}\right)^m ; \quad x \geq 0$$

$$N = N_0 \left(\frac{x}{x_0}\right)^m ; \quad x \leq 0$$

where N_0 and x_0 are parameterizing quantities for the doping concentration and the depletion width respectively; m is a characteristic of the junction grading coefficient, P, given by $P = \frac{1}{m+2}$. As special cases, $m = 0$ and 1, represent step and linearly graded junctions respectively. Obtain general expressions for the spatial variation of potential and electric field in the transition region for the zero bias condition in terms of the junction gradient coefficient P, assuming depletion approximation.

8. Consider the equilibrium transition region of a p–n junction, and show through the consideration of spatial variation of majority carriers within the transition region, that the Debye length is a measure of the penetration of the majority carriers in the transition region. Prove that the ratio of the transition width to Debye length depends on doping. Assume a one sided step junction.

9. Show the variation of electron mobility against total impurity concentration (range 10^{15} to 10^{19}cm^{-3}) using Equation(1.81) and data for MINIMOS3 given in the text using a CAS. Compare it with the variation using parameter values used by BAMBI as given below.

For electrons:
$\mu_{mn0} = 65$ cm^2 V^{-1}s^{-1}, $\mu_{I0} = 1365$ cm^2 V^{-1}s^{-1}, $N_{R0} = 8.5 \times 10^{16}$ cm^{-3},
$\alpha_0 = 0.72$, $\eta_1 = 0$, $\eta_2 = 0$, $\eta_3 = 0$, $\eta_4 = 0$

For holes:
$\mu_{mn0} = 48$ cm^2 V^{-1}s^{-1}, $\mu_{I0} = 447$ cm^2 V^{-1}s^{-1}, $N_{R0} = 6.3 \times 10^{16}$ cm^{-3},
$\alpha_0 = 0.76$, $\eta_1 = 0$, $\eta_2 = 0$, $\eta_3 = 0$, $\eta_4 = 0$

Obtain a piecewise regional model of the simulation obtained above in the form given below:

$$\mu = C_1 \qquad\qquad \text{for } N < N_1$$
$$\mu = A(B - \log N) \qquad\qquad \text{for } N_1 < N < N_2$$
$$\mu = C_2 \qquad\qquad \text{for } N > N_2$$

Obtain values of C_1, C_2, A, B, N_1, and N_2 for electrons and holes from the simulation.

10. (a) Estimate the amount of band gap narrowing at room temperature in silicon due to impurity level broadening for a donor doping concentration of 10^{19}cm^{-3} and compare it with that at a doping of 10^{17}cm^{-3}.
(b) Draw the energy band diagram of a hetero-junction formed by two dissimilar semiconductors SC_x and SC_y with band gap, work function and electron affinity values, E_{gx}, E_{gy}, ϕ_x, ϕ_y, χ_x, and χ_y respectively. Show the equilibrium energy band diagram of the junction and obtain the expressions for conduction and valence band offsets. Show that the homo-junction band diagram can be shown to be a particular case of the hetetro-junction.

11. Assuming an one-sided step p-n junction for a silicon diode, obtain the condition for the avalanche junction breakdown voltage, $\int_0^{W_d} \alpha \, dx = 1$. Relate the doping concentration of the heavily doped side with the maximum potential (applied voltage added to the barrier potential) in the junction. Assume $\alpha = A \exp\left(\frac{-B}{F}\right)$ where α is the carrier ionization rate per

cm of length and F is the electric field. Plot the junction breakdown voltage as a function of doping concentration varying from 10^{15} to 10^{19}cm^{-3}.

12. Assume that an electron with an energy E is confined in a potential well has to cross a barrier height of 10.0 eV. Calculate the tunneling probabilities for potential barrier thicknesses of $d = 0.5, 1.5, 2.5, 3.5$, and 4.5 Å and an energy barrier of a magnitude $(U - E) = 4$ eV.

13. Draw the energy band diagram of an m–n^+–n ohmic contact in equilibrium, forward and reverse bias conditions and explain thereby the non-rectifying characteristics of the contact.

14. Arrive at the expression given in Equation (1.149) for the electron tunneling probability, TC, evaluating the constants A_1 and A_3.

15. Highlight the differences between BSIM3v3 and JUNCAP (PSP) junction diode models with respect to reverse leakage current.

2

Ideal Metal Oxide Semiconductor Capacitor

2.1 Physical Structure and Energy Band Diagram

A Metal Oxide Semiconductor (MOS) capacitor realized on a silicon substrate, is a two terminal device shown schematically in Figure 2.1(a) where the insulator, conventionally a thermally grown silicon-dioxide (SiO_2) film, is sandwiched between a conducting electrode called the gate (G) at one end, and a silicon substrate connected through an ohmic contact acting as the other electrode (B) [1–3]. In modern MOS technology the gate is formed by a highly doped polysilicon material which is accessed through an ohmic contact. The MOS capacitor shows non-linear $C–V$ characteristics as against a conventional capacitor which is a linear element.

For the purpose of illustration and discussion we shall consider a p-type silicon substrate. The physical concepts developed are applicable for an n-type substrate as well. Figure 2.1(b) shows the energy band diagram of an *ideal* MOS capacitor structure. We use ideality to mean that the structure is *simple*, and some of inherent complexities are not considered. Further, the structure has assumed attributes such as *uniform substrate doping, charge free oxide, and equal work function for the substrate and the polysilicon gate electrode*. Though these conditions do not hold good for real structures, the assumptions help us to focus on the basic physics of the structure to which corrections can be introduced later for the deviation from the *so-called* ideality.

The symbols used in the diagram have the following meanings: ϕ_g = potential corresponding to the work function of the gate material, ϕ_{Si} = potential corresponding to the work function of the silicon substrate, χ_{Si} = electron affinity of silicon (in eV), E_g = band gap energy of the silicon substrate, E_{fg} = Fermi energy level of gate electrode, E_f = Fermi energy level of the p-silicon substrate, and N_B is the doping concentration of the substrate. For a p-type silicon, N_B will be taken to be the acceptor concentration N_A. It may be noted that we emphasize potential representation in the energy band diagram. The energy levels depicted imply a corresponding multiplicative factor for electronic charge (q).

Compact MOSFET Models for VLSI Design   A.B. Bhattacharyya
© 2009 John Wiley & Sons (Asia) Pte Ltd

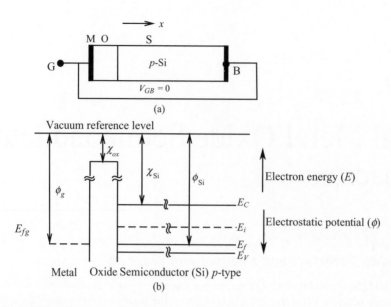

Figure 2.1 (a) Physical structure of a two terminal MOS capacitor with SiO_2 as the insulator and p-type silicon as the substrate. (b) Energy band diagram of a MOS structure in thermodynamical equilibrium; $V_{GB} = 0$ and $\phi_g = \phi_{Si}$

The work function is defined as the amount of work required to extract the electron from the Fermi level of the material (gate/substrate) to the reference level in vacuum, and has units of electron volts.

For an ideal MOS structure, ϕ_g has been assumed to be equal to ϕ_{Si}, i.e. there is no work function difference between the gate and substrate materials. We first consider the case for no voltage applied between the gate and bulk terminals. In this condition, the gate and substrate Fermi levels are aligned. As the work functions of the gate and substrate are the same, there is no net transport of electrons between the gate and substrate terminals in the thermal equilibrium condition. There is no build up of charge in the system. This leads to a *flat-band condition* when there is no space charge in the substrate, and the electric field is zero both in the substrate and the insulator. The energy band being flat, ($\phi(x) =$ constant), implies that the electric field $-\frac{d}{dx}\phi(x) = F(x) = 0$.

2.2 Modes of Operation of MOS Capacitors

The voltage applied at the gate with the substrate as reference, gives rise to an electric field in the insulator which is sustained by the space charge on the two sides of the insulating layer. The space charge region in the semiconductor can be understood from the following consideration. From the condition of charge neutrality or charge balance of the MOS structure, the space charge density in the semiconductor at the Si–SiO$_2$ interface equals that on the gate induced by the applied voltage. The above condition is mathematically expressed by:

$$Q_g + Q_{sc} = 0$$

or:

$$Q_g = -Q_{sc}$$

where Q_g is the total gate charge, and Q_{sc} is total space charge in the substrate at the interface. Thus, for a positive voltage applied to the gate, the conducting gate electrode is charged positive, and the net space charge on the semiconductor adjoining the insulating layer is negative.

For a p-type silicon substrate the space charge region at the interface has the following constituents:

1. Fixed acceptor ions which are negatively charged. These charges are immobile and are called fixed bulk charges.
2. Mobile electrons.
3. Mobile holes.

For a positive gate voltage, the thermally generated electrons in the depletion region are collected at the interface by the normal electric field in the x-direction. This field also pushes away thermally generated holes from the interface. The generation rate $g \approx n_i/\tau$, where τ is the lifetime of the carriers. Therefore, it takes a finite time to form the inversion layer where electron–hole pairs attain an equilibrium condition ($pn - n_i^2 = 0$).

The net space charge volume concentration is given by the sum of the concentrations of negatively charged fixed acceptor ions, mobile electrons, and holes. In the space charge region, $\rho \neq 0$. Below, we will obtain the expression for the volume charge density in terms of electric potential in the space charge region.

2.2.1 Volume Space Charge Density

The electric field terminating on the space charge region of the semiconductor, which is normal to the surface, leads to bending of the energy band in the space charge region. Figure 2.2 shows the energy band diagram of an MOS capacitor on a p-substrate for three normally

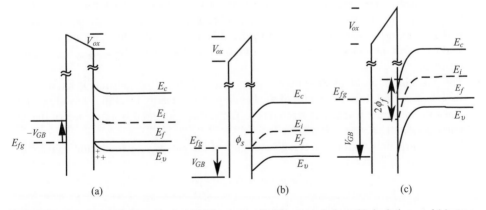

Figure 2.2 Energy band diagram of an MOS structure in (a) accumulation, (b) depletion, and (c) strong inversion

important operating conditions with negative and positive bias on the gate with respect to substrate.

The following features of the energy band diagram describe the formation of the space charge layer components.

- The Fermi level is aligned horizontally in the substrate for all biasing conditions shown, as there is no current flow.
- For negative bias on the gate, the gate is negatively charged and holes accumulate at the interface to maintain the charge balance or neutrality. The hole concentration is more and the electron concentration is less than the equilibrium value, resulting in a net positive space charge at the interface. Figure 2.2(a) shows the corresponding energy band diagram, for the *accumulation mode* of operation.
- For a small positive voltage on the gate, the valence band moves away from the Fermi level to account for negative space charge at the interface. The band bending is such that the charge balance is preserved. The negative space charge region is created due to mobile holes being pushed out. Figure 2.2(b) shows the corresponding energy band diagram. The separation between the valence band edge and the Fermi level near the interface is larger than that in the bulk neutral region, indicating decrease in mobile hole concentration near the interface. The removal of holes exposes acceptor ions which form fixed bulk charges of acceptor ions.
  The conduction band bends closer to the Fermi level, signifying an increase in electron concentration in the space charge region compared to that in the neutral region. For band bending less than the intrinsic level, electron and hole concentrations can be considered to be negligible, to a first approximation, compared to that of fixed charge. In other words, the space charge region may be considered to be depleted of mobile carriers. The operation of a MOS capacitor under such a condition is designated to be in the *depletion* mode.
- As the gate voltage is increased, the band bending increases implying an increase in the surface potential. In the case where the bending is large enough to position the conduction band edge at the Si–SiO$_2$ interface below the intrinsic level by an amount of bulk potential ϕ_f, it will denote that for that region the conductivity has been inverted from p-type to n-type by the applied field. The inverted region is known as the *channel*, whose conductivity is controlled by the gate. It is this property of a MOS capacitor which forms the basis of MOS transistor action. For the range of band bending between ϕ_f and $2\phi_f$ the channel is designated as *weakly inverted*, and for band bending greater than $2\phi_f$ the channel is said to be *strongly inverted*. Figure 2.2(c) represents the energy band diagram in strong inversion.
- The potential ϕ_s at the oxide–silicon interface ($x = 0$) w.r.t. the substrate is known as the *surface potential*. It is a measure of the bending of the energy band in terms of potential when the substrate is taken as a reference. The surface potential is a function of the gate voltage V_{GB}.
- In the neutral bulk region, the charge neutrality of the p-type semiconductor is given by:

$$p_{p0} = N_A^- + n_{p0} \tag{2.1}$$

where p_{p0} and n_{p0} are the equilibrium hole (majority carrier) and electron (minority carrier) concentrations respectively, and N_A^- is the concentration of negatively charged ionized acceptor atoms. The LHS of Equation (2.1) represents the positive charge concentration, and the RHS represents the negative charge concentrations constituted by mobile electrons and fixed acceptor ions.

- The equilibrium majority and minority carrier concentrations, p_{p0} and n_{p0} respectively, can be expressed in terms of the bulk potential ϕ_f following Boltzmann's approximation, which is valid at room temperature and for normal doping:

$$p_{p0} = n_i \exp\left(\frac{\phi_f}{\phi_t}\right) \tag{2.2}$$

$$n_{p0} = n_i \exp\left(-\frac{\phi_f}{\phi_t}\right) \tag{2.3}$$

The band bending modifies the electron and hole concentrations from the equilibrium values. The spatial variation of carrier concentrations depends on the potential variation $\phi(x)$ in the space charge region. The potential at the neutral region of the substrate is considered as zero.

- The potential at any point in the space charge is higher than that of the neutral region for a positive gate voltage. Making use of Boltzmann's approximation, the electron concentration is given by:

$$n_p(x) = n_{p0} \exp\left(\frac{\phi(x)}{\phi_t}\right) \tag{2.4}$$

where $\phi_t = \frac{kT}{q}$ is the thermal voltage, and n_{p0} is the equilibrium electron concentration in the neutral substrate region.

- For the above condition, the hole concentration decreases with increasing potential and is given by:

$$p_p(x) = p_{p0} \exp\left(-\frac{\phi(x)}{\phi_t}\right) \tag{2.5}$$

where p_{p0} is the equilibrium hole concentration in the neutral region. The hole concentration decreases exponentially with increasing potential.

- The net volume space charge density $\rho(x)$ is the sum of mobile and immobile charge components. For a p-type semiconductor assuming complete ionization of acceptor ions, we can write:

$$\rho(x) = q\{p_p(x) - [N_A + n_p(x)]\} \tag{2.6}$$

Substituting for $n_p(x)$ and $p_p(x)$ from Equations (2.4) and (2.5) respectively, Equation (2.6) reduces to:

$$\rho(x) = q\left(-N_A + p_{p0}e^{-\frac{\phi(x)}{\phi_t}} - n_{p0}e^{\frac{\phi(x)}{\phi_t}}\right) \tag{2.7}$$

At the Si–SiO$_2$ interface $x = 0$ and $\phi|_{x=0} = \phi_s$ (surface potential), which gives:

$$\rho(\phi_s) = q\left(-N_A + p_{p0}e^{-\frac{\phi_s}{\phi_t}} - n_{p0}e^{\frac{\phi_s}{\phi_t}}\right) \tag{2.8}$$

The surface potential ϕ_s depends on the applied gate voltage applied across the two terminal MOS capacitor. The volume space charge density is maximum at $x = 0$, decreases with x, and reduces to zero at the edge of the neutral region.

Figure 2.3 shows the plot of surface volume charge density against surface potential using Equation (2.8). In the above equation, the third term represents the electron concentration at the surface, n_s, which equals hole concentration of the substrate (qN_A) in the neutral region at $\phi_s = 2\phi_f$. Thus we conclude that at $\phi_s = 2\phi_f$, the electron volume concentration at the surface equals that of hole volume concentration in the bulk. In other words, due to band bending of $2\phi_f$ at the surface, the conductivity of the surface is inverted, i.e. the surface changes from p- to n-type due to the field effect. The surface potential $\phi_s = 2\phi_f$ is known in the literature as the *surface inversion potential*. The gate voltage corresponding to the surface inversion potential is known as *Threshold Voltage* which will be discussed in detail in later sections.

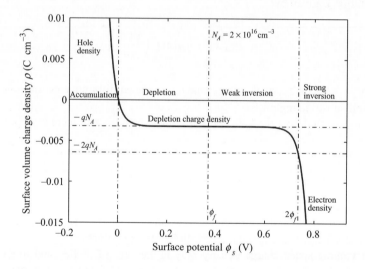

Figure 2.3 Variation of surface volume charge density with surface potential

2.2.2 Surface Space Charge Density

With the application of voltage across the MOS capacitor, a field is created in the oxide which terminates on the space charge created in a thin layer at the interface. The space charge can be expressed through the surface space charge density Q'_{sc}, the charge per unit area. The field in silicon at the Si–SiO$_2$ interface $(x = 0)$ is related to the surface space charge density Q'_{sc}

through Gauss' relation [8]:

$$F_{Si}\big|_{x=0} = F_{Si}(0) = -\frac{Q'_{sc}}{\epsilon_{Si}} \tag{2.9}$$

where ϵ_{Si} is the permittivity of silicon.

The field in the oxide, in turn, is related to the field in silicon at the interface through continuity of the normal displacement vector at the interface. For the silicon–oxide interface, we can write:

$$\epsilon_{ox} F_{ox} = \epsilon_{Si} F_{Si}(0) \tag{2.10}$$

In the oxide, the electric field is constant. As the permittivity of silicon and oxide are different there is a discontinuity of the electric field at the silicon–oxide interface. The potential in silicon and oxide at the interface ($x = 0$), however, is continuous.

The surface space charge density can be obtained from expressions of the volume charge density outlined in the previous section. Poisson's equation relates electric field with volume charge density, and Gauss' theorem relates electric field with surface charge density. Thus, we can obtain expressions for surface charge density in terms of surface potential. The surface potential can be expressed implicitly in terms of gate voltage as described later.

According to Poisson's equation:

$$\frac{d^2\phi(x)}{dx^2} = -\frac{\rho(x)}{\epsilon_{Si}} \tag{2.11}$$

For solving Equation (2.11), boundary conditions appropriate to the structure have to be imposed, taking into account physical considerations of the structure: potential (ϕ) and field (F) equal zero at the boundary of the space charge and neutral region. The neutral substrate is taken as a reference for potential.

A *general* analytical solution of Equation (2.11), giving the spatial distribution of potential $\phi(x)$, not being feasible, the problem has to be solved numerically. We shall, however, confine ourselves to a few specific cases where, using approximations, analytical expressions can be obtained.

Two important quantities, electric field and surface space charge density, related to the understanding of MOS device physics, are described below. The detailed derivation is given in Appendix A.

1. **Electric Field**. $F(x)$ is obtained by integrating Poisson's equation (Equation (2.11)) and using the appropriate boundary conditions mentioned earlier. $F(x)$ is given by:

$$F(x) = \text{sgn}(\phi_s)\frac{\phi_t}{L_b}\left[e^{-\frac{\phi(x)}{\phi_t}} + \frac{\phi(x)}{\phi_t} - 1 + e^{-\frac{2\phi_f}{\phi_t}}\left(e^{\frac{\phi(x)}{\phi_t}} - \frac{\phi(x)}{\phi_t} - 1\right)\right]^{\frac{1}{2}} \tag{2.12}$$

where $L_b = \sqrt{\frac{\epsilon_{Si}\phi_t}{2qN_A}}$ is the Debye length corresponding to the doping concentration N_A.

2. **Surface Space Charge density.** Q'_{sc} is obtained from Equation (2.12) by finding the value of electric field at the surface, and using Gauss' law relating surface density with the electric field as $Q'_{sc} = -\epsilon_{Si} F_{Si}(0)$. At the silicon surface $\phi(0) = \phi_s$. Now Q'_{sc} is given by:

$$Q'_{sc} = -\text{sgn}(\phi_s)Q'_x \left\{ \left(e^{\frac{-\phi_s}{\phi_t}} - 1 \right) + \left[\frac{\phi_s}{\phi_t} \left(1 - e^{\frac{-2\phi_f}{\phi_t}} \right) \right] \right.$$

$$\left. + \left[e^{\frac{-2\phi_f}{\phi_t}} \left(e^{\frac{\phi_s}{\phi_t}} - 1 \right) \right] \right\}^{\frac{1}{2}} \tag{2.13}$$

where $Q'_x = \epsilon_{Si}\phi_t/L_b = \sqrt{2q\epsilon_{Si} N_A \phi_t}$. Equation (2.13) is rewritten as:

$$Q'_{sc} = -Q'_x (A_1 + A_2 + A_3)^{\frac{1}{2}} \tag{2.14}$$

where;

$$A_1 = e^{\frac{-\phi_s}{\phi_t}} - 1 \tag{2.15}$$

$$A_2 = \frac{\phi_s}{\phi_t} \left(1 - e^{\frac{-2\phi_f}{\phi_t}} \right) \tag{2.16}$$

$$A_3 = e^{\frac{-2\phi_f}{\phi_t}} \left(e^{\frac{\phi_s}{\phi_t}} - 1 \right) \tag{2.17}$$

It may be noted that Equation (2.13) has three components A_1, A_2, and A_3 which contribute to the surface space charge density. The physical origin of the individual terms may be attributed to holes, fixed acceptor ions, and electrons respectively.

Figure 2.4 shows the variation of normalized space charge density Q'_{sc}, given by Equation (2.13), as a function of surface potential. We draw the following conclusions from the characteristics of the plot shown in Figure 2.4.

- The slope of the curve in the accumulation region, is obtained from the expression:

$$Q'_{sc} = Q'_x \exp \left(\frac{-\phi_s}{2\phi_t} \right) \tag{2.18}$$

where the proper sign for the charge has been adjusted considering the sign of the surface potential. The slope of space charge density w.r.t. surface potential gives the rate of variation of hole space charge density with potential. The slope is $1/2\phi_t$ and is negative, which means that the hole space charge density decreases with increase of surface potential.
- The term characterizing the dependence of minority carrier electron density with surface potential in Equation (2.13), for $\phi_s > 2\phi_c$, is:

$$Q'_{sc} = -Q'_x \exp \left(\frac{\phi_s - 2\phi_f}{2\phi_t} \right) \tag{2.19}$$

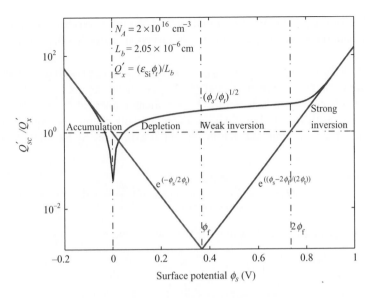

Figure 2.4 Variation of normalized surface space charge density with surface potential

It follows that the Q'_{sc}/Q'_x versus ϕ_s curve, with the y-axis in the log scale, is a straight line. The extrapolated value of the inversion charge component of the space charge due to electrons is Q'_x at $\phi_s = 2\phi_f$. As Q'_n increases with ϕ_s, the slope of the straight line $\ln \left| Q'_{sc}/Q'_x \right|$ versus ϕ_s is positive, and equals $1/2\phi_t$.

- Thus, the slope of the space charge curve versus surface potential representing the variation of surface electron and hole charge density with surface potential, shows that it has a value equal in magnitude but opposite in sign.
- There are three important *surface potential nodes* in Figure 2.4: (i) $\phi_s = 0$, (ii) $\phi_s = \phi_f$, and (iii) $\phi_s = 2\phi_f$. At $\phi_s = \phi_f$, the two slopes drawn asymptotically for the hole accumulation and electron inversion charge meet, where $n = p = n_i$. Lastly, $\phi_s = 2\phi_f$, denotes the conventional inversion or threshold potential. At this surface potential node, the inversion charge density Q'_n turns out to be much smaller in magnitude than the bulk fixed charge density—*an assumption on which the threshold voltage is conventionally defined*. Thus, at threshold potential, we normally assume, $\phi_s = 2\phi_f$, and $Q'_{sc} = Q'_b + Q'_n \approx Q'_b$.
- For $\phi_s > 5\phi_t$, which covers the depletion and inversion region, Equation (2.13) assumes the simplified form given below:

$$Q'_{sc} = -Q'_x \left(\frac{\phi_s}{\phi_t} + e^{\frac{\phi_s - 2\phi_f}{\phi_t}} \right)^{\frac{1}{2}} \tag{2.20}$$

The general expression given by Equation (2.13) for surface space charge density can be approximated to simplified expressions corresponding to specific regions of operation. In the following section we will discuss specific regions in some detail.

2.2.3 Surface Inversion Charge Density

In the previous section we obtained an expression for surface space charge density, which is constituted of electrons, holes, and fixed ion charges. Holes and fixed ion charges together form the bulk charge (Q_b). In MOSFET modeling, it is the inversion charge that constitutes the drain current and, therefore, proper modeling of inversion charge has been an area of special interest.

Though we have an expression for Q'_{sc} in terms of surface potential as given in Equation (2.13), we cannot separate out Q'_n and Q'_b. We can, however, make use of the expression for volume charge density in Equation (2.7) to evaluate numerically the inversion and bulk surface charge densities. Identifying the inversion and bulk charge components from Equation (2.7), and integrating from the neutral region ($\phi = 0$), to the Si–SiO$_2$ interface ($\phi = \phi_s$), we get:

$$Q'_n = -q \int_{\text{neutral region}}^{\text{interface}} n_{p0} \exp\left(\frac{\phi(x)}{\phi_t}\right) dx \qquad (2.21)$$

Making the change of variable, we can write the above equation as:

$$Q'_n = -q n_{p0} \int_0^{\phi_s} e^{\frac{\phi(x)}{\phi_t}} \frac{1}{F(x)} d\phi \qquad (2.22)$$

Similarly for Q'_b, we have:

$$Q'_b = -q N_A \int_0^{\phi_s} \left(1 - e^{\frac{-\phi(x)}{\phi_t}}\right) \frac{1}{F(x)} d\phi \qquad (2.23)$$

where we have used $p_{p0} = N_A$. An accurate estimation of Q'_n and Q'_b requires numerical computation of Equations (2.22) and (2.23) respectively.

For $\phi_s < 0$ the inversion charge can be considered to be zero, and for $\phi_s > 0$ the inversion charge density can be written as:

$$Q'_n = Q'_{sc} - Q'_b \qquad (2.24)$$

In order to evaluate the bulk charge density we make an important approximation, which we shall designate as the *charge sheet approximation* (CSA). In the CSA we assume that the inversion charge forms a sheet of infinitesimal thickness, and the field associated with it does not penetrate in the bulk. Thus, the surface potential is determined by the bulk charge under such approximation, and we can write:

$$Q'_b = -\epsilon_{Si} F_b \Big|_{x=0} \qquad (2.25)$$

where $F_b\big|_{x=0}$ is the surface electric field due to the bulk charge. F_b can be obtained by solving Poisson's equation (Equation (2.11)) in the bulk with $\rho = -q N_A + p_{p0} e^{-\phi_s/\phi_t}$. On substituting the calculated F_b in Equation (2.25), we get:

$$Q'_b = -Q'_x \sqrt{\frac{\phi_s}{\phi_t} - 1 + e^{-\frac{\phi_s}{\phi_t}}} \qquad (2.26)$$

Substituting for Q'_b in Equation (2.24), we get:

$$Q'_n = Q'_{sc} + Q'_x \sqrt{\frac{\phi_s}{\phi_t} - 1 + e^{-\frac{\phi_s}{\phi_t}}} \qquad (2.27)$$

The above expression can be further simplified by noticing that for $\phi_s > \phi_t$, the third term in the root argument, which corresponds to one of the terms for the hole contribution, is insignificant as compared to other terms. Hence, we have:

$$Q'_n = Q'_{sc} + Q'_x \sqrt{\frac{\phi_s}{\phi_t} - 1} \qquad (2.28)$$

For still greater ϕ_s, the '1' in the above equation can also be neglected, and we have:

$$Q'_n = Q'_{sc} + Q'_x \sqrt{\frac{\phi_s}{\phi_t}} \qquad (2.29)$$

The second term in the above equation is nothing but the expression for depletion charge density under depletion approximation, which is charge sheet approximation with bulk charge consisting of only depletion charge (more on which is discussed later).

$$Q'_b = -\sqrt{2q\epsilon_{Si}N_A}\sqrt{\phi_s} \qquad (2.30)$$

Equations (2.27), (2.28), and (2.29) are all valid in strong inversion. Equation (2.27) is used in the MOS11 series of Philips [32], while Equation (2.28) finds usage in HiSIM [31] and SP [30]. Equation (2.29) has been a popular text book material [1] for the Brews charge sheet model using depletion approximation. The features and error involved in the use of the above approximations in the depletion and weak inversion regions have been investigated in He *et al.* [12], McAndrew and Victory [13], and Gummel and Singhal [34]. Equation (2.29) is simple, and is used for the inversion charge density for $\phi_s > 5\phi_t$ in most of the earlier compact models [1].

Using the expression for Q_{sc} given in Equation (2.20), which is valid in weak and strong inversion, and substituting in Equation (2.29), the inversion charge can be written as

$$Q'_n = -Q'_x \left(\frac{\phi_s}{\phi_t} + e^{\frac{\phi_s - 2\phi_f}{\phi_t}} \right)^{1/2} + Q'_x \sqrt{\frac{\phi_s}{\phi_t}} \qquad (2.31)$$

which is valid for $\phi_s > 5\phi_t$.

For $5\phi_t < \phi_s < 2\phi_f$, the region which is identified as weak inversion, the second term in the round brackets of the above equation is small. This enables a Taylor series expansion, which gives:

$$Q'_{nw} = -\frac{Q'_x}{2} \left(\frac{\phi_t}{\phi_s} \right)^{1/2} e^{\frac{\phi_s - 2\phi_f}{\phi_t}} \qquad (2.32)$$

For $\phi_s > 2\phi_f + 3\phi_t$, the exponential term in Equation (2.31), corresponding to inversion charge, dominates over the square root term representing the depletion charge. This sets the condition for strong inversion in terms of surface potential. More generally, the strong inversion condition is defined by the surface potential $\phi_s = 2\phi_f + m\phi_t$, where m is assigned a value between 4 to 6.

$$Q'_{ns} = -Q'_x e^{\frac{\phi_s - 2\phi_f}{2\phi_t}} \qquad (2.33)$$

2.2.4 Moderate Inversion

To get the expression for the inversion charge density in strong inversion an assumption was made that the surface potential is $2\phi_f + m\phi_t$ instead of $2\phi_f$. The range of surface potential for which ϕ_s lies between $2\phi_f$ and $2\phi_f + m\phi_t$ is called the *moderate inversion* region. In this region of operation, simplification of the general Equation (2.31) for the inversion charge density Q'_n is not feasible as we have done for weak and strong inversion. Figure 2.5 shows the plot of approximate equations for inversion charge density in weak and strong inversion, and its correspondence with the general expression for Q'_n.

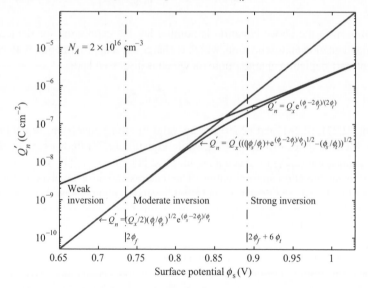

Figure 2.5   Surface inversion charge density versus surface potential with various approximations

From Figure 2.5, we note the deviation of weak and strong inversion approximations from the actual curve in the moderate inversion region. The slope of the inversion charge density with respect to surface potential is different for the weak and strong inversion regions. The slope is $1/\phi_t$ in weak inversion, and is $1/2\phi_t$ in strong inversion.

It may be noted that the upper limit of the surface potential for moderate inversion is usually defined arbitrarily. In the moderate inversion region the changes in the depletion and inversion charge densities are comparable for changes in surface potential.

The moderate inversion region is gaining importance due to scaling of voltage in modern low-power MOS devices, where the need for accurate modeling of this region can no longer be ignored [17, 22, 23, 26]. In the literature, at times, the moderate inversion region is identified with strong inversion. Unless specifically mentioned for moderate inversion, in this text the convention mentioned above will be followed.

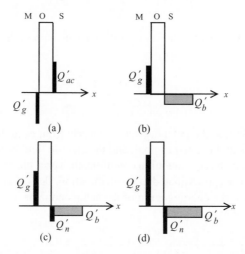

Figure 2.6 Charge balance in a MOS capacitor for (a) accumulation, (b) depletion, (c) weak inversion and (d) strong inversion modes of operation

Figure 2.6 shows the charge balance diagram of the MOS capacitor for (a) accumulation, (b) depletion, (c) weak inversion, and (d) strong inversion modes of operation. Equation (2.18) represents the positive sheet charge density due to holes for (a). Equation (2.30) gives the bulk fixed charge density for (b), and in this region mobile charges are assumed to be absent. Equations (2.32) and (2.33) denote *mobile* inversion electron charge densities for modes of operation in weak and strong inversion regions as shown by (c) and (d) respectively.

2.3 Electric Field and Potential Distributions

2.3.1 Distribution in Oxide

The potential and the electric field distribution in the insulator can be obtained through the use of Poisson's equation. The applied voltage provides the necessary boundary condition. The charge on the terminals may then be obtained from Gauss' theorem which relates charge to the electric field. For the ideal oxide considered, no charge is present within the insulator, i.e. $\rho(x) = 0$. From Poisson's equation:

$$\frac{d^2\phi}{dx^2} = 0 \tag{2.34}$$

$$\phi(x) = C_1 x + C_2 \tag{2.35}$$

Applying the appropriate boundary conditions, the constants of integration are evaluated. At $x = 0$, $\phi(x) = \phi_s$, which implies $C_2 = \phi_s$. Also, at $x = -T_{ox}$, $\phi_x = V_{GB}$ gives

$C_1 = -(V_{GB} - \phi_s)/T_{ox}$. Thus, Equation (2.35) gives:

$$\phi(x) = -\frac{V_{GB} - \phi_s}{T_{ox}}x + \phi_s \tag{2.36}$$

Differentiating both sides w.r.t. x for a given ϕ_s, we can express the electric field variation in the oxide as:

$$F(x) = -\frac{d\phi(x)}{dx} = \frac{V_{GB} - \phi_s}{T_{ox}} \tag{2.37}$$

Equation (2.36) shows that the potential varies linearly from ϕ_s at the Si–SiO$_2$ interface to the applied voltage V_{GB} at the gate terminal, and the electric field is constant in the oxide as illustrated in Figure 2.7. In the accumulation condition, the space charge at the silicon interface being infinitely thin, can be approximated to a delta function. Thus, there is negligible potential across the accumulated space charge region of the semiconductor. The entire applied voltage gets developed across the oxide layer.

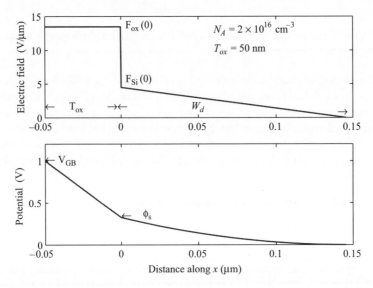

Figure 2.7   Spatial variation of electric field and potential in the oxide and space charge regions in the depletion condition

2.3.2 Distribution in Space Charge Region

The field and potential distribution in the space charge region in silicon are obtained analytically for the uniformly doped substrate assuming depletion condition, where we can ignore mobile charges in the space charge region. With this approximation, the volume charge density $\rho(x)$ can be taken as constant, and for the p-substrate equals $-qN_A$. Poisson's equation for the region is expressed as:

$$\frac{d^2\phi}{dx^2} = \frac{qN_A}{\epsilon_{Si}}$$ (2.38)

Integrating Equation (2.38) w.r.t. x:

$$\frac{d\phi}{dx} = \frac{qN_A}{\epsilon_{Si}}x + C_1$$ (2.39)

C_1 can be evaluated from the boundary condition that at $x = W_d$, $\frac{d\phi}{dx} = 0$, which gives $C_1 = -(qN_A/\epsilon_{Si})W_d$, where W_d is the depletion layer width. Therefore, from Equation (2.39) we can write:

$$F(x) = -\frac{d\phi}{dx} = \frac{qN_A}{\epsilon_{Si}}(W_d - x)$$ (2.40)

Integrating the above equation, and using the boundary condition for potential that at $x = W_d$, $\phi = 0$, we get for the potential distribution:

$$\phi(x) = \frac{qN_A(W_d)^2}{2\epsilon_{Si}}\left(1 - \frac{x}{W_d}\right)^2$$ (2.41)

Figure (2.7) shows the spatial variation of electric field and potential in the oxide and the space charge region. The surface potential ϕ_s can be expressed in terms of depletion width by putting $\phi(x) = \phi_s$ in Equation (2.41), which gives for $x = 0$:

$$\phi_s = \frac{qN_A(W_d)^2}{2\epsilon_{Si}}$$ (2.42)

Equation (2.41) can be written in the form:

$$\phi(x) = \phi_s\left(1 - \frac{x}{W_d}\right)^2$$ (2.43)

Figure 2.8 shows the spatial variation of electric field and potential in the space charge region at the Si–SiO$_2$ interface for a homogeneously doped substrate with different doping concentrations for $\phi_s = 2\phi_f$. The following conclusion may be obtained.

- As the substrate doping increases the inversion surface potential $\phi_s = 2\phi_f$ increases and the depletion depth decreases.
- The electric field at the Si–SiO$_2$ interface increases with doping concentration.
- The area of the triangle in the F–x plane gives the surface potential as shown below:

$$\frac{1}{2}F_{Si}(0)W_d = \frac{1}{2}\frac{qN_A}{\epsilon_{Si}}W_d^2$$ (2.44)

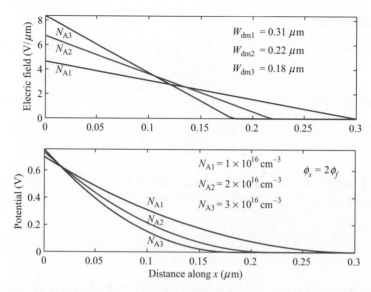

Figure 2.8 Variation of electric field, potential and maximum depletion width in the case of a strong inversion operation for different substrate concentrations

Substituting for W_d^2 from Equation (2.42) in Equation (2.44), we have:

$$\frac{1}{2}F_{Si}(0)W_d = \frac{1}{2}\frac{qN_A}{\epsilon_{Si}}\frac{2\epsilon_{Si}}{qN_A}\phi_s = \phi_s \tag{2.45}$$

• The slope of the line joining $F_{Si}(0)$ and W_d is given by:

$$\frac{F_{Si}(0)}{W_d} = \frac{qN_A}{\epsilon_{Si}} \tag{2.46}$$

From Equation (2.40), it is noted that the slope of the F–x curve is directly related to the doping concentration of the substrate.

A discontinuity in the electric field occurs at the interface because of the difference in the permittivity of Si and SiO$_2$ ($\epsilon_{Si} > \epsilon_{ox}$). It follows from the requirement of Gauss' law that the normal component of the electric displacement must be continuous at the interface. This gives:

$$\epsilon_{ox} F_{ox}(0) = \epsilon_{Si} F_{Si}(0) \tag{2.47}$$

$$F_{ox} \approx 3F_{Si}(0) \tag{2.48}$$

as $\epsilon_{Si} \approx 3\epsilon_{ox}$.

2.4 Potential Balance

As the surface potential is controlled by the gate voltage in a two terminal MOS capacitor, it is useful from the practical viewpoint to have a relation between the gate voltage and the resultant surface potential at the interface. From the potential balance we can write:

$$V_{GB} = \phi_s + V_{ox} \qquad (2.49)$$

where V_{ox} is the potential drop across the oxide which can be written as $V_{ox} = -Q'_{sc}/C'_{ox}$.

$$V_{GB} = \phi_s - \frac{Q'_{sc}}{C'_{ox}} \qquad (2.50)$$

From Equation (2.13), substituting for Q'_{sc} and assigning a negative sign to the charge for a positive surface potential, we get

$$V_{GB} = \phi_s$$

$$+ \frac{Q'_x \left\{ \left(e^{\frac{-\phi_s}{\phi_t}} - 1 \right) + \left[\frac{\phi_s}{\phi_t} \left(1 - e^{\frac{-2\phi_f}{\phi_t}} \right) \right] + \left[e^{\frac{-2\phi_f}{\phi_t}} \left(e^{\frac{\phi_s}{\phi_t}} - 1 \right) \right] \right\}^{\frac{1}{2}}}{C'_{ox}} \qquad (2.51)$$

We consider the operation of a MOS capacitor for V_{GB} such that $\phi_s > 3\phi_t$, because it is in this condition that an MOS transistor is predominantly used. We can then use the simplified expression for Q'_{sc} given in Equation (2.20), and substitute it in Equation (2.50), which gives:

$$V_{GB} = \phi_s + \frac{\sqrt{2q\epsilon_{Si}N_A}}{C'_{ox}} \left(\phi_s + \phi_t e^{\frac{\phi_s - 2\phi_f}{\phi_t}} \right)^{\frac{1}{2}} \qquad (2.52)$$

or:

$$V_{GB} = \phi_s + \gamma \left(\phi_s + \phi_t e^{\frac{\phi_s - 2\phi_f}{\phi_t}} \right)^{\frac{1}{2}} \qquad (2.53)$$

where:

$$\gamma = \frac{\sqrt{2q\epsilon_{Si}N_A}}{C'_{ox}} \qquad (2.54)$$

where γ is an important parameter in MOS operation known as the *body effect* or *substrate bias coefficient*. It has a dimension of $\sqrt{V}$.

We now introduce parameters such as V_{WI}, V_{MI}, and V_{SI}, which are gate voltages corresponding to surface potentials of ϕ_f, $2\phi_f$, and $(2\phi_f + m\phi_t)$ respectively. These parameters indicate the upper boundary limits for weak, moderate, and strong inversion conditions respectively, in terms of applied gate voltage V_{GB}. To obtain the expressions for V_{WI}, V_{MI}, and V_{SI} we substitute the respective values of ϕ_s in Equation (2.53). The expressions for V_{WI}, V_{MI}, and V_{SI} are as follows.

- For $\phi_s = \phi_f$, the term $e^{(\phi_s - 2\phi_f)/\phi_t}$ in Equation (2.53) is very small compared to 1, which gives:

$$V_{WI} = \phi_f + \gamma\sqrt{\phi_f} \qquad (2.55)$$

- For $\phi_s = 2\phi_f$, Equation (2.53) becomes:

$$V_{MI} = 2\phi_f + \gamma\sqrt{2\phi_f + \phi_t} \qquad (2.56)$$

- Similarly, for $\phi_s = 2\phi_f + m\phi_t$, Equation (2.53) becomes:

$$V_{SI} = 2\phi_f + m\phi_t + \gamma\sqrt{2\phi_f + m\phi_t + \phi_t e^m} \qquad (2.57)$$

Figure 2.9 gives the plot of Equation (2.53) specially illustrating the potential balance at $V_{GB} = V_{MI}$ and in strong inversion. It is seen that as the surface potential crosses $2\phi_f$, it becomes a weak function of gate voltage, meaning that it requires a large change in the gate voltage to bring a significant change in surface potential. That is the reason for the conventional approximation that the surface potential remains fixed at $2\phi_f$ once the value is reached at the

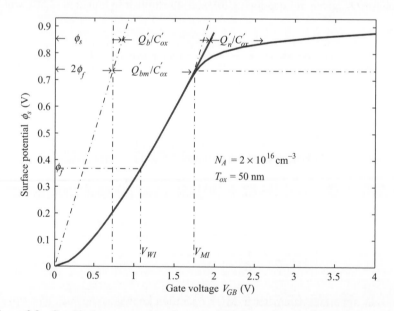

Figure 2.9   Graphical representation of a potential balance diagram for a MOS capacitor

gate voltage $V_{GB} = V_{MI}$. For more precise calculation, the depletion charge corresponding to $\phi_s > 2\phi_f$ ($\phi_s = 2\phi_f + \Delta\phi_s$) is calculated from the formula $Q'_b = -\sqrt{2q\epsilon_{Si}N_A(2\phi_f + \Delta\phi_s)}$ with delta-depletion approximation. It can be shown quantitatively from Equation (2.53) that for $V_G < V_{MI}$, we have $\phi_s \propto V_G$, and for $V_G > V_{MI}$ we have $\phi_s \propto \ln(V_G)$. Also note that that for a gate voltage greater then V_{MI}, most of the excess gate voltage above V_{MI} is dropped across the oxide.

In Equation (2.53) the gate voltage is related implicitly with the surface potential, where gate voltage is expressed as a function of surface potential. It is useful to have an explicit relation which expresses ϕ_s in terms of V_{GB}. Equation (2.53) cannot be inverted to get an explicit analytical expression. But in the depletion/weak inversion condition, however, we can have an approximate explicit relation for ϕ_s in terms of gate voltage.

In the depletion/weak inversion condition, neglecting the inversion charge contribution in Equation (2.53), we have:

$$V_{GB} = \phi_s + \gamma\sqrt{\phi_s} \qquad (2.58)$$

where the $\gamma\sqrt{\phi_s}$ term corresponds to the potential drop across the oxide. Solving Equation (2.58) for ϕ_s, and retaining the physically admissible root, we have:

$$\phi_s = V_{GB} + \frac{\gamma^2}{2}\left[1 - \left(1 + \frac{4V_{GB}}{\gamma^2}\right)^{\frac{1}{2}}\right] \qquad (2.59)$$

which can be also written as:

$$\phi_s = \left(\sqrt{V_{GB} + \frac{\gamma^2}{4}} - \frac{\gamma}{2}\right)^2 \qquad (2.60)$$

2.4.1 An Explicit Relation of ϕ_s with V_{GB}

In Section 2.2.1 we obtained the relationships between the inversion charge density and the surface potential. In order to model the MOS device for circuit simulation, an explicit relation between inversion charge density and terminal voltages is required. If we have an explicit expression for the surface potential in terms of gate voltage, the space charge and therefrom the inversion charge density can be found out analytically in terms of gate voltage from Equation (2.13). Equation (2.53) gives an implicit dependence of surface potential on the gate voltage which needs to be solved numerically. For circuit simulation, in general, such an approach is not computationally efficient. We outline below an intuitive approach, following Langevelde and Klaassen [27], to obtain an approximate but explicit expression showing the dependence of surface potential on the gate voltage which is valid for the weak, moderate, and strong inversion regions of operation of a MOS capacitor.

For the weak inversion region, $\phi_s < 2\phi_f$, and from Equation (2.59) we have:

$$\phi_{s(wi)} \approx V_{GB} + \frac{\gamma^2}{2}\left[1 - \left(1 + \frac{4V_{GB}}{\gamma^2}\right)^{\frac{1}{2}}\right] \tag{2.61}$$

In the strong inversion region, on the other hand, it is noted from a numerical calculation of Equation (2.53), that the surface potential has a weak dependence on the gate voltage. Due to this fact it has been normal practice to assume that in the strong inversion region of operation:

$$\phi_s \approx 2\phi_f \tag{2.62}$$

which is illustrated in Figure 2.9. But with the scaling of terminal voltages in modern MOS devices, it has become necessary to consider a more accurate expression for surface potential in the strong inversion region of operation. One can assume, to a first order approximation, that in strong inversion the relation between the gate voltage and surface potential is given by:

$$V_{GB} = \phi_0 + \gamma\left(\phi_0 + \phi_t e^{\frac{\phi_s - \phi_0}{\phi_t}}\right)^{\frac{1}{2}} \tag{2.63}$$

where $\phi_0 = 2\phi_f$. The above equation is obtained by substituting $2\phi_f$ for ϕ_s in all terms except in the exponential term of Equation (2.53), since it is the latter term that dominantly contributes to an increase in surface potential above $2\phi_f$ in strong inversion. Rearranging Equation (2.63) we get:

$$\phi_{s(si)} = \phi_0 + \phi_t \ln\left\{\frac{1}{\phi_t}\left[\left(\frac{V_{GB} - \phi_0}{\gamma}\right)^2 - \phi_0 + \phi_t\right]\right\} \tag{2.64}$$

The last term in the square brackets, ϕ_t, is added so that $\phi_{s(si)} = \phi_{s(wi)}$ at $\phi_s = 2\phi_f$, i.e. at $V_{GB} = 2\phi_f + \gamma\sqrt{2\phi_f}$. Equation (2.64), however, does not represent the moderate inversion region accurately because in this region ϕ_s is close to $2\phi_f$, and the assumption made that the exponential term dominates is not strictly valid. The accuracy of Equation (2.64) can be further enhanced by substituting an empirical expression ϕ^* replacing the term ϕ_0 in the quadratic term only [27], where:

$$\phi^* = \phi_0 + \frac{\phi_{s(wi)} - \phi_0}{\left[1 + \left(\frac{\phi_{s(wi)} - \phi_0}{4\phi_t}\right)^2\right]^{\frac{1}{2}}} \tag{2.65}$$

Thus, we get:

$$\phi_{s(si)} = \phi_0 + \phi_t \ln\left\{\frac{1}{\phi_t}\left[\left(\frac{V_{GB} - \phi^*}{\gamma}\right)^2 - \phi_0 + \phi_t\right]\right\} \tag{2.66}$$

For a single piece expression of surface potential as a function of gate voltage for all regions of operation, including the moderate inversion region lying between weak and strong, we can use Equation (2.66). We replace ϕ_0 by a smoothing function $f(V_{GB})$ given by:

$$f(V_{GB}) = \phi_0 + \phi_{s(wi)} - \frac{1}{2}\left[(\phi_{s(wi)} - \phi_0)^2 + 4c^2\right] \tag{2.67}$$

where c is a fitting parameter.

The function $f(V_{GB})$ approaches ϕ_0 for strong inversion, and $\phi_{s(wi)}$ for weak inversion. The substitution of $f(V_{GB})$ in Equation (2.66) gives an expression of surface potential as a function of gate voltage which is continuous covering weak, moderate, and strong inversion regions.

$$\phi_s = f(V_{GB}) + \phi_t \ln\left\{\frac{1}{\phi_t}\left[\left(\frac{V_{GB} - \phi^*}{\gamma}\right)^2 - f(V_{GB}) + \phi_t\right]\right\} \tag{2.68}$$

Figure 2.10 gives the plot of surface potential as a function of gate voltage using Equation (2.68). It should be noted that due to an exponential dependence of inversion charge density on surface potential, for meeting the requirement of high precision in modern compact modeling, alternative analytical approximation has been proposed by Chen and Gildenblat [18] which computes surface potential to the desired accuracy level of 10^{-12} V. Alternatively, efficient numerical algorithms of comparable accuracy and computational efficiency have been used in models such as HiSIM [10]. The strengths of a numerical approach lie in the very high degree of accuracy achieved. In numerical technique using the Newton–Raphson method, the number of iterations depend significantly on the initial guess [11].

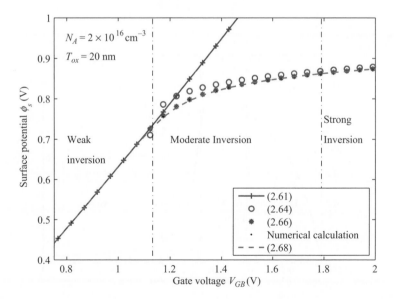

Figure 2.10 Plots of surface potential as a function of V_{GB} with various approximations

2.4.2 Slope Factor (n)

Equation (2.53) gives the relation between the gate voltage and the surface potential covering the depletion, weak, and strong inversion regions. The slope $dV_{GB}/d\phi_s$ is an important parameter, as it gives a physical insight about the way the surface potential varies with the change of gate voltage. The slope factor n is defined as:

$$n = \frac{dV_{GB}}{d\phi_s} \tag{2.69}$$

Differentiating Equation (2.53) w.r.t. ϕ_s, we get:

$$n = 1 + \frac{\gamma}{2} \frac{1}{\sqrt{\phi_s + \phi_t e^{\frac{\phi_s - 2\phi_f}{\phi_t}}}} \left(1 + e^{\frac{\phi_s - 2\phi_f}{\phi_t}}\right) \tag{2.70}$$

From Figure 2.11, which is a plot of $1/n$ as a function of V_{GB}, it is seen that in the weak inversion region it is a slowly varying function. It is, therefore, a normal practice to assume n to be constant in the weak inversion region. The value of n corresponding to $\phi_s = 1.5\phi_f$ is taken as this constant value in the weak inversion region. Physically, $1/n$ denotes the incremental voltage gain from the gate to the surface. At $\phi_s = 1.5\phi_f$, in weak inversion Equation (2.70) reduces to:

$$n = 1 + \frac{\gamma}{2\sqrt{1.5\phi_f}} \tag{2.71}$$

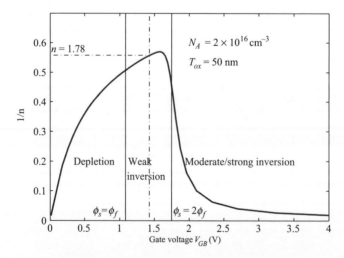

Figure 2.11 Plot of $1/n$ as a function of V_{GB}

2.5 Inversion Layer Thickness

We assumed that the inversion layer constituted by the mobile electrons forms a thin sheet of charge at the interface over the bulk fixed charge. We shall show below that the inversion layer thickness is negligible compared to the depletion width. Consequently, the potential drop across it can be ignored in the equation for potential balance across the MOS capacitor. This is a commonly used assumption in MOS models, and is the basis of *charge sheet approximation* [15]. Within the thin sheet of inversion layer, there is a spatial distribution of electron concentration.

The thickness of the inversion layer is represented by the centroid of the carrier distribution. The following approach is adopted for obtaining the distribution of inversion carriers and the expression for the inversion layer thickness W_i.

We start by considering the volume carrier distribution $n(x)$, which is known in terms of spatial dependence of potential $\phi(x)$ as given in Equation (2.4). For the determination of variation of potential within the inversion region we cannot use the depletion approximation as was done in Section 2.3. Since the carrier concentration has an exponential dependence on potential, an accurate determination of $\phi(x)$ in the presence of mobile carriers is essential. To calculate the potential distribution within the inversion layer, we use the expression for $F(x)$ given in Equation (2.12), retaining only the dominant terms corresponding to inversion and depletion charges:

$$F(x) = -\frac{d\phi}{dx} = \frac{\phi_t}{L_b}\left(\frac{\phi(x)}{\phi_t} + e^{-\frac{2\phi_f}{\phi_t}} e^{\frac{+\phi(x)}{\phi_t}}\right) \tag{2.72}$$

or:

$$-d\phi = \frac{\phi_t}{L_b}\left(\frac{\phi(x)}{\phi_t} + \frac{n_i^2}{N_A^2} e^{\frac{+\phi(x)}{\phi_t}}\right) dx \tag{2.73}$$

Equation (2.73) cannot be solved analytically. Solving numerically, however, we can obtain $\phi(x)$. Using the numerically calculated value of $\phi(x)$, we get the variation of electron concentration with distance, as shown in Figure 2.12. A simple expression for the inversion layer thickness can be obtained using a depletion approximation for the calculation of potential variation in the inversion layer. This assumption is valid up to the threshold voltage when the effect of the inversion charge on the potential variation is found to be negligible. The inversion charge distribution along the depth being nonuniform, the thickness of the inversion layer will be given by the centroid of the charge variation with distance, taking the interface as the origin. From the definition of the centroid for the charge distribution $n(x)\,dx$ [4], we get:

$$q\int_0^\infty (W_i - x)\,n(x)\,dx = 0 \tag{2.74}$$

$$qW_i\int_0^\infty n(x)\,dx = q\int_0^\infty xn(x)\,dx \tag{2.75}$$

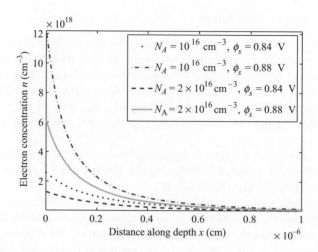

Figure 2.12 Variation of inversion charge volume density with x with doping and surface potential as parameters

where W_i is the centroid of the inversion layer thickness. For evaluation of the LHS, we use the equation $n(x) = n_{p0}e^{\phi(x)/\phi_t}$, where $\phi(x)$ is obtained from Equation (2.41). As we are considering a region $x \ll W_d$, performing a binomial expansion of $(1 - x/W_d)^2$, and neglecting the higher order term we obtain:

$$\phi(x) \approx \phi_s \left(1 - \frac{2x}{W_d} \right) \tag{2.76}$$

Putting the value of $n(x)$ and using $\phi(x)$ from Equation (2.76), we evaluate the LHS of Equation 2.75:

$$qW_i \int_0^\infty n(x)\, \mathrm{d}x = qW_i \int_0^\infty n_{p0} \exp\left[\frac{\phi_s}{\phi_t} \left(1 - \frac{2x}{W_d} \right) \right] \mathrm{d}x \tag{2.77}$$

or:

$$qW_i \int_0^\infty n(x)\, \mathrm{d}x = qW_i n_{po} \frac{\phi_t}{\phi_s} \frac{W_d}{2} \mathrm{e}^{\frac{\phi_s}{\phi_t}} \tag{2.78}$$

The RHS of Equation (2.75) equals:

$$q \int_0^\infty x n_{p0} \exp\left[\frac{\phi_s}{\phi_t} \left(1 - \frac{2x}{W_d} \right) \right] \mathrm{d}x = q n_{p0} \mathrm{e}^{\frac{\phi_s}{\phi_t}} \left(\frac{W_d \phi_t}{2\phi_s} \right)^2 \tag{2.79}$$

where, in the integration in Equation (2.79), we have used the relation:

$$\int_0^\infty z^n e^{-\alpha z}\, dz = \frac{n!}{\alpha^{n+1}} \tag{2.80}$$

Equating Equations (2.78) and (2.79), we get:

$$W_i = \frac{W_d \phi_t}{2\phi_s} \tag{2.81}$$

$$W_i = \frac{\phi_t}{F_{Si}(0)} \tag{2.82}$$

At the threshold, $F_{Si}(0) = 2\phi_f / W_{dm}$, and from Equation (2.81) we have $W_d/W_i = 2\phi_s/\phi_t \approx 4\phi_f/\phi_t$.

Substituting typical values of $\phi_f = 0.39$ V and $\phi_t = 0.025$ V, we get $W_{dm}/W_i = 60$. Thus, it is concluded that the inversion layer thickness is negligible compared to that of the space charge layer. The above result is the basis of the commonly used delta-depletion approximation, where the inversion layer is considered to be an infinitesimally thin layer of charge sheet at the Si–SiO$_2$ interface.

An increase in the inversion charge density enhances the surface field which, in strong inversion, is directly related to the inversion charge density. The enhanced surface field draws the carriers more towards the surface and thereby reduces the thickness.

2.6 Threshold Voltage

The threshold voltage V_{T0} is a fundamental parameter for MOS devices. It broadly signifies the transition from a nonconducting to a conducting mode of operation, corresponding respectively to an *off* and an *on* state of a MOS switch. In the literature, the threshold voltage has been conventionally defined in several equivalent ways.

- It is the gate voltage corresponding to which the surface potential is $2\phi_f$. This criterion will be considered to be *conventional*. From Figure 2.9 it is seen that $\phi_s = 2\phi_f$ indicates the transition when the device leaves the weak inversion region [7] and significant inversion charge starts building up.
- An alternative way is to express the above definition in terms of volume carrier concentration. It is defined as the gate voltage at which the volume minority carrier concentration at the surface equals the volume majority carrier concentration in the bulk as can be seen from the Figure 2.3. Thus the conductivity at the surface is *inverted* from a p-type to an n-type.
- It is also defined as the gate voltage at which the rate of increase of inversion layer charge density (cm^{-2}) equals the rate of increase of depletion surface charge density when the change is considered w.r.t. ϕ_s [21]. As the gate voltage is increased above the threshold voltage, the depletion charge remains relatively unchanged, as it has square root dependence on the surface potential. The inversion charge density, on the other hand, has an exponential dependence on surface potential and dominates the space charge density.

- The above definitions are based on theoretical models. In order to conform to agreement with experimental results over a wide range of doping concentrations, a more generalized definition is proposed modifying the conventional definition. The threshold voltage is defined as the gate voltage at which the surface potential is given by the following expression [29]:

$$\phi_s = 2\phi_f + \phi_t \ln\left(\frac{2\phi_s}{\xi\phi_t}\right) \tag{2.83}$$

where ξ is a parameter to be obtained by fitting with the experimentally extracted value of threshold voltage. The definition reduces to the conventional one for $\xi = (2\phi_f/\phi_t)$.

We can obtain an expression for threshold voltage using the potential balance condition across the MOS structure:

$$V_{GB} = \phi_s - \frac{Q'_{sc}}{C'_{ox}} \tag{2.84}$$

It may be noted that until the onset of inversion, the space charge in the semiconductor under the gate is constituted predominantly of bulk depletion charge. Hence, in the computation of threshold voltage it is valid to assume $Q'_{sc} \approx Q'_b$. With the above consideration, for the purpose of threshold voltage calculation, we can write at threshold $V_{GB} = V_{T0}$, and $Q'_b = Q'_{bm}$, leading to:

$$V_{T0} = 2\phi_f - \frac{Q'_{bm}}{C'_{ox}} \tag{2.85}$$

or:

$$V_{T0} = 2\phi_f + \gamma\sqrt{2\phi_f} \tag{2.86}$$

Theoretically, V_{T0} is identical with V_{MI} defined in Section 2.4, which represented the upper limit of the gate voltage for the weak inversion region of operation. It is important to note that while the threshold voltage is defined as the gate voltage at which a conducting inversion layer of mobile carriers is formed, the presence of the inversion layer is ignored in the quantitative estimation of the threshold voltage. Its inclusion will not significantly change the calculated value of the threshold voltage. Figure 2.13 shows the plot of threshold voltage as a function of substrate doping concentration with oxide thickness as a parameter. It may be noted that the term $2\phi_f = 2\phi_t \ln(N_A/n_i)$ has a relatively small variation with doping as it has logarithmic dependence. Therefore, the threshold voltage change is mainly affected by the body effect coefficient γ, which has square root dependence with doping and direct proportionality with oxide thickness.

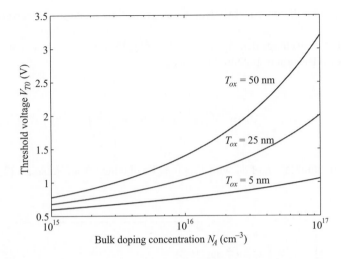

Figure 2.13   Variation of threshold voltage with substrate doping and gate oxide thickness

2.6.1 Inversion Charge Density Above and Below the Threshold

2.6.1.1 Strong Inversion

For $V_{GB} > V_{T0}$, we use the following equations for the charge and potential balance:

$$Q'_g = Q'_n + Q'_b \tag{2.87}$$

$$V_{GB} = \phi_s + V_{ox} \tag{2.88}$$

As

$$V_{ox} = -\frac{(Q'_n + Q'_{bm})}{C'_{ox}} \tag{2.89}$$

$$V_{GB} = 2\phi_f - \frac{(Q'_n + Q'_{bm})}{C'_{ox}} \tag{2.90}$$

Assuming ϕ_s to be pinned at $2\phi_f$ after threshold:

$$\frac{Q'_n}{C'_{ox}} = -\left[V_{GB} - \left(\frac{Q'_{bm}}{C'_{ox}} + 2\phi_f\right)\right] \tag{2.91}$$

$$Q'_n = -C'_{ox}(V_{GB} - V_{T0}) \tag{2.92}$$

Equation (2.92) is an approximation as it does not take into account the variation in the surface potential for $V_{GB} > V_{T0}$. This will lead to an error in the estimation of inversion charge density, and will consequently lead to an error in the modeling of current in the MOS transistor.

The actual inversion charge density, as a function of gate voltage, can be calculated indirectly from the the consideration of potential balance:

$$Q'_n = -C'_{ox}\left(V_{GB} - \phi_s + \frac{Q'_b}{C'_{ox}}\right) \qquad (2.93)$$

where the functional relation of ϕ_s with V_{GB} may be obtained from Equation (2.68).

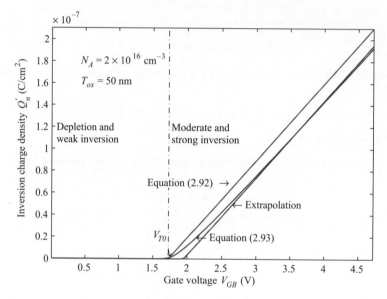

Figure 2.14   Plots of approximate, exact and extrapolated relations of Q'_n and V_{GB}

Figure 2.14 shows the plot of inversion charge density as a function of gate voltage, with Equation (2.92) giving the approximate and exact variation of Q'_n with V_{GB}. It may be concluded that:

- The approximate Q'_n versus V_{GB} relation given in Equation (2.92) overestimates the inversion charge density for a given gate voltage.
- *The inversion charge density is linearly related to the gate overdrive voltage ($V_{GB} - V_{T0}$) beyond the threshold voltage, with C'_{ox} as the proportionality constant.* Thus, for a higher inversion charge density for a given overdrive, the proportionality constant may be increased by reducing the oxide thickness.
- Better modeling of Q'_n versus V_{GB} beyond the threshold, with the linear relation may be done by changing the threshold voltage value as shown in Figure 2.14.

2.6.1.2 Weak Inversion

For $V_G < V_{T0}$, we recall Equation (2.32):

$$Q'_n = -\frac{Q'_x}{2}\left(\frac{\phi_t}{\phi_s}\right)^{\frac{1}{2}}\exp\left(\frac{\phi_s - 2\phi_f}{\phi_t}\right) \tag{2.94}$$

As the $\sqrt{\phi_s/\phi_t}$ term has a weak dependence on ϕ_s as compared to the exponential term, we may assume it to be a constant value taken for $\phi_s = 1.5\phi_f$:

$$Q'_n = -Q'_w\exp\left(\frac{\phi_s - 2\phi_f}{\phi_t}\right) \tag{2.95}$$

where $Q'_w = \frac{1}{2}Q'_x\sqrt{\phi_t/\phi_s}$. In weak inversion, referring to Figure 2.9, the slope factor n can be assumed to be constant:

$$n = \frac{dV_{GB}}{d\phi_s} = \frac{V_{GB} - V_{T0}}{\phi_s - 2\phi_f} \tag{2.96}$$

This gives the surface potential ϕ_s in weak inversion as:

$$\phi_s = \frac{V_{GB} - V_{T0}}{n} + 2\phi_f \tag{2.97}$$

Substituting for ϕ_s in Equation (2.95), we get:

$$Q'_n = -Q'_w\exp\left(\frac{V_{GB} - V_{T0}}{n\phi_t}\right) \tag{2.98}$$

The weak inversion charge density with gate voltage has the characteristic that the inversion charge density in weak inversion, with an exponential dependence on surface potential, has much more sensitivity to the gate voltage change as compared to the strong inversion situation. Therefore, both inversion charge and subthreshold parameters are very sensitive to device manufacturing processes.

Figure 2.15 shows the variation of inversion charge density with gate voltage, where it is observed that different regions of operation have characteristic inversion charge density–gate voltage functional dependence. From the nature of the dependence of inversion charge density on the gate voltage, the following broad classification can be made for the entire inversion region in terms of *gate voltage*. It may be mentioned that the criteria of labeling different regions are rather arbitrary.

- **Strong Inversion.** Here $V_{GB} \geq V_{T0} + 4\phi_t$ and $Q'_n = C'_{ox}(V_{GB} - V_{T0})$; the inversion charge density is directly proportional to the gate overdrive.
- **Moderate Inversion.** $V_{T0} - 4\phi_t < V_{GB} < V_{T0} + 4\phi_t$; Q'_n versus V_{GB} is in transition between exponential and linear gate voltage dependence.

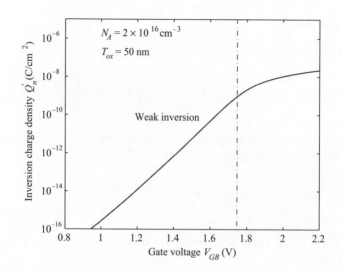

Figure 2.15   Variation of inversion charge density in weak inversion with gate voltage

- **Weak Inversion.** $V_{T0} - 4\phi_t \geq V_{GB} > V_{WI}$; $Q_n' \approx \exp(V_{GB})$; $Q_n' << Q_b'$; $\log |Q_n'|$ versus V_{GB} is linear. $\phi_s = k' V_{GB}$, where k' is the gate coupling coefficient representing the coupling between gate and surface potential.
- **Depletion.** $0 < V_{GB} < V_{WI} - 4\phi_t$; the inversion layer can be assumed to be absent.

2.6.2 Graphical Representation of Threshold Voltage

In the previous section we have outlined the various definitions of threshold voltage for a MOS capacitor. We also obtained the mathematical expression for estimating the threshold voltage in terms of doping and oxide thickness for an ideal MOS capacitor. It will be useful to provide a graphical representation of the threshold voltage following Taur [5].

Figure 2.16 shows the spatial dependence of electric field in the depletion region in silicon and in oxide in the F–x plane. The Si–SiO$_2$ interface is taken as the origin.

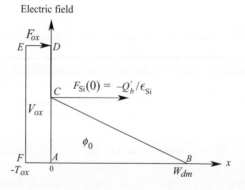

Figure 2.16   Graphical representation of threshold voltage

At the threshold voltage the depletion width is maximum, given by $W_{dm} = \sqrt{2\epsilon_{Si}\phi_0/qN_A}$, which is shown on the horizontal x-axis by the line AB. The electric field at the interface $(x = 0)$ in the silicon, $F_{Si}(0)$, is shown on the vertical axis, where $F_{Si}(0) = Q'_b/\epsilon_{Si}$ (following Gauss' law). The electric field varies linearly from $F_{Si}(0)$ at the interface to zero at the space charge edge where the neutral bulk region starts. Assuming the space charge to be depleted of any mobile carriers, the depletion charge density can be written as:

$$Q'_b = -qN_A W_{dm} = -\sqrt{2qN_A\epsilon_{Si}\phi_0} \tag{2.99}$$

If AC represents the field in silicon at the interface, $F_{Si}(0)$, we have:

$$AC = -\frac{Q'_b}{\epsilon_{Si}} = \frac{1}{\epsilon_{Si}}\sqrt{2qN_A\epsilon_{Si}\phi_0}$$

The area of the triangle CAB is given by $\frac{1}{2}AB \times AC$. Substituting for AB and AC we have:

$$\text{Area}\quad CAB = \frac{1}{2}\sqrt{\frac{2\epsilon_{Si}\phi_0}{qN_A}}\frac{1}{\epsilon_{Si}}\sqrt{2qN_A\epsilon_{Si}\phi_0} \tag{2.100}$$

$$= \phi_0 \tag{2.101}$$

Thus, the area shows the value of surface potential at the threshold, which is almost independent of doping because of logarithmic dependence. If the substrate doping is increased, the depletion width decreases for the area to remain constant. The electric field in silicon at the interface now needs to increase, which happens due to an increase in Q'_b.

The electric field at the interface in silicon is related to the electric field in the oxide by the continuity of displacement vector, given by:

$$F_{ox} = \frac{\epsilon_{Si}}{\epsilon_{ox}}F_{Si}(0) \tag{2.102}$$

Let the field in the oxide be represented by the height AD in the F–x plane. Thus:

$$AD \approx 3F_{Si}(0), \quad \left[\frac{\epsilon_{Si}}{\epsilon_{ox}} \approx 3\right] \tag{2.103}$$

and:

$$V_{T0} = V_{ox} + \phi_0 = 3F_{Si}(0)T_{ox} + \phi_0 \tag{2.104}$$

Expressing $F_{Si}(0)$ in terms of surface potential at the threshold given by $F_{Si}(0) = 2\phi_0/W_{dm}$, we have:

$$V_{T0} = \phi_0 + \left(\frac{6T_{ox}}{W_{dm}}\right)\phi_0 \tag{2.105}$$

Graphically, V_{T0} is represented by the area of the polygon $EFBCDE$. The rectangular area $ADEF$ represents the potential drop across the oxide.

2.7 Small Signal Capacitance

A MOS capacitor differs from a conventional capacitor in the sense that the gate electrode is made up of highly conducting material, but the other electrode is made of a semiconductor which has relatively high resistivity. When a voltage is applied across the MOS capacitor, a space charge is formed at the Si–SiO$_2$ interface. The space charge at the interface is associated with a potential drop across it (surface potential), which means that the entire gate voltage is not applied across the insulator. This phenomenon can be modeled by putting a capacitor in series with the oxide capacitor. The capacitance seen by a *small signal* variation between the gate and bulk electrode at a given bias depends on the series combination of bias independent linear oxide capacitance and bias dependent nonlinear space charge capacitance at the interface.

The equivalent small signal capacitance per unit area between the gate and reference bulk terminals is defined by:

$$C'_{eq} = \frac{\Delta Q'_g}{\Delta V_{GB}}$$

(2.106)

where $\Delta Q'_g$ is the change in the charge on the gate terminal in response to the change in gate voltage ΔV_{GB}. We shall make use of potential and charge balance conditions appropriate in a dynamic situation for the derivation of equivalent capacitance.

- Potential balance

$$\Delta V_{GB} = \Delta V_{ox} + \Delta \phi_s$$

(2.107)

- Charge balance

$$\Delta Q'_g = -\Delta Q'_{sc}$$

(2.108)

Taking the inverse of C'_{eq} and writing it in differential form we have:

$$\frac{1}{C'_{eq}} = \frac{dV_{GB}}{dQ'_g}$$

(2.109)

Using Equation (2.107), we have:

$$\frac{1}{C'_{eq}} = \frac{dV_{ox}}{dQ'_g} + \frac{d\phi_s}{dQ'_g}$$

(2.110)

We shall identify each term of the RHS in the above equation as components of a MOS capacitor structure called (i) the oxide capacitor C'_{ox}, and (ii) the semiconductor capacitor C'_{sc}

respectively, where the capacitances are expressed per unit area. The equivalent capacitance C'_{eq} can be written as:

$$\frac{1}{C'_{eq}} = \frac{1}{C'_{ox}} + \frac{1}{C'_{sc}}$$

(2.111)

Physically, the oxide capacitor and semiconductor capacitor are connected in series to give an equivalent capacitor between the gate and substrate terminals as shown in Figure 2.17.

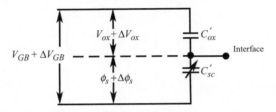

Figure 2.17 Equivalent circuit diagram of a MOS capacitor

The expression for C'_{sc} is obtained by differentiating Q'_{sc}, given in Equation (2.13), with respect to ϕ_s which gives:

$$C'_{sc} = \frac{\gamma C'_{ox}}{2} \frac{\left(1 - e^{-\frac{\phi_s}{\phi_t}}\right) + e^{-\frac{2\phi_f}{\phi_t}}\left(e^{\frac{\phi_s}{\phi_t}} - 1\right)}{\left[e^{-\frac{2\phi_f}{\phi_t}}\left(\phi_t e^{\frac{\phi_s}{\phi_t}} - \phi_t - \phi_s\right) + \phi_t e^{-\frac{\phi_s}{\phi_t}} + \phi_s - \phi_t\right]^{\frac{1}{2}}}$$

(2.112)

As the space charge consists of depleted ions and mobile carriers for $V_{GB} > 0$, we can write:

$$C'_{sc} = -\frac{\mathrm{d}(Q'_b + Q'_n)}{\mathrm{d}\phi_s}$$

(2.113)

where Q'_b and Q'_n are negative. It follows from Equation (2.113) that:

$$C'_{sc} = C'_b + C'_n$$

(2.114)

where C'_b is the depletion capacitance, and C'_n is the inversion capacitance. Combining Equations (2.111) and (2.114), we get:

$$\frac{1}{C'_{eq}} = \frac{1}{C'_{ox}} + \frac{1}{C'_b + C'_n}$$

(2.115)

The depletion and inversion capacitances can be obtained from Equations (2.22) and (2.23), by differentiating them w.r.t. ϕ:

$$C_b' = \frac{\gamma C_{ox}'}{2} \frac{1}{\left(\phi_s + \phi_t e^{\frac{\phi_s - 2\phi_f}{\phi_t}}\right)^{\frac{1}{2}}} \qquad (2.116)$$

and:

$$C_n' = \frac{\gamma C_{ox}'}{2} \frac{e^{\frac{\phi_s - 2\phi_f}{\phi_t}}}{\left(\phi_s + \phi_t e^{\frac{\phi_s - 2\phi_f}{\phi_t}}\right)^{\frac{1}{2}}} \qquad (2.117)$$

2.7.1 Regional Approximations

The value of C_{sc}' depends on the operating region (modes of operation) and frequency. Below, the space charge capacitances are calculated for (a) accumulation, (b) depletion, and (c) strong inversion regions. Figure 2.18 shows the small signal equivalent circuits for a capacitor for a two terminal MOS structure corresponding to the above regions of operation.

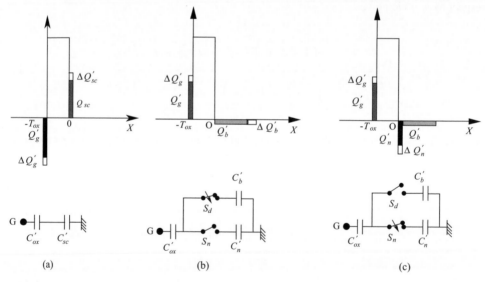

Figure 2.18   Charge balance and equivalent circuit diagrams of an MOS capacitor in (a) accumulation, (b) depletion, and (c) inversion conditions

2.7.1.1 Accumulation

In the accumulation region, the space charge capacitance can be calculated from the simplified expression given in Equation (2.18), which relates the space charge density with the surface potential. Differentiating Equation (2.18) for Q'_{sc} in the accumulation region with respect to ϕ_s, we have:

$$C'_{sc} = -\frac{d}{d\phi_s}\left(\frac{\epsilon_{Si}\phi_t}{L_b}e^{\frac{-\phi_s}{2\phi_t}}\right) \tag{2.118}$$

$$C'_{sc} = \frac{\epsilon_{Si}}{2L_b}e^{\frac{-\phi_s}{2\phi_t}} \tag{2.119}$$

In the deep accumulation region ϕ_s/ϕ_t is negative, and much greater than 1. So $C'_{sc} \gg C'_{ox}$. We can approximate $C'_{sc} \to \infty$, and Equation (2.111) reduces to:

$$C'_{eq} = C'_{ox} \tag{2.120}$$

Figure 2.18(a) shows the small signal capacitance equivalent circuit in accumulation.

2.7.1.2 Depletion

In the depletion mode, the space charge mainly consists of only the depletion charge due to ionized acceptor atoms and the space charge capacitance:

$$C'_{sc} = C'_b = -\frac{dQ'_{sc}}{d\phi_s} = -\frac{dQ'_b}{d\phi_s} \tag{2.121}$$

Substituting the value of Q'_b and W_d from Equations (2.30) and (2.42), we get:

$$C'_b = \frac{\sqrt{2q\epsilon_{Si}N_A}}{2\sqrt{\phi_s}} \tag{2.122}$$

$$C'_b = \frac{\epsilon_{Si}}{W_d} \tag{2.123}$$

With application of a positive gate voltage, the surface potential increases from the flat-band condition $\phi_s = 0$ (for $V_{GB} = 0$) to the maximum value of $2\phi_f$ for $V_{GB} = V_{T0}$. Hence, in this range of gate voltage, the terminal capacitance is given by:

$$C'_{eq} = \frac{1}{\frac{1}{C'_{ox}} + \frac{1}{C'_b}} \tag{2.124}$$

C_b' reaches the minimum value at $V_{GB} = V_{T0}$ as ϕ_s attains the maximum value at that bias voltage. Thus:

$$C_{eq_{min}}' = \frac{1}{\frac{1}{C_{ox}'} + \frac{1}{C_{b_{min}}'}} \tag{2.125}$$

$$C_{eq_{min}}' = \frac{1}{\frac{T_{ox}}{\epsilon_{Si}} + \frac{W_{dm}}{\epsilon_{Si}}} \tag{2.126}$$

The equivalent circuit for small signal capacitance in the depletion mode is shown in Figure 2.18(b). As there is no inversion region, the switch S_n, connected to C_n, is disabled. Since W_d depends on surface potential, which is controlled by gate voltage, the space charge capacitance is voltage dependent. Further, in the depletion mode of operation the change in depletion charge is accomplished in a very short time (dielectric relaxation time $\sim 10^{-12}$ s). Therefore, the depletion capacitance does not depend upon frequency in the practical frequency range of operation.

2.7.1.3 Strong Inversion

In the strong inversion mode of operation of the MOS capacitor, the space charge consists of Q_n' and Q_b', and the space charge capacitance is defined as:

$$C_{sc}' = -\frac{dQ_{sc}'}{d\phi_s} = -\frac{dQ_{n}'}{d\phi_s} - \frac{dQ_{b}'}{d\phi_s} \tag{2.127}$$

$$C_{sc}' = C_n' + C_b' \tag{2.128}$$

where C_n' is the inversion layer capacitance per unit area. The equivalent circuit of the MOS capacitor for the inversion region is represented in Figure 2.18(c).

The small signal capacitance in the strong inversion mode has bias dependence which is frequency sensitive. Depending on the frequency of operation, the switches S_n and S_d will be enabled. We shall obtain expressions for small signal capacitances for low and high frequencies as described below.

- **Low Frequency Characteristics**

  An incremental change in the gate charge due to the change in the gate voltage induces a corresponding change in the semiconductor space charge. At low frequency, as long as the time period of the applied signal is large enough compared to the generation time of the minority carriers in the depletion region, the incremental gate charge is balanced by the increase in the inversion charge. The inversion charge shields the incremental lines of force originating from the gate, and therefore, the depletion charge does not undergo any change. In the equivalent space charge capacitance in Figure 2.18(c), the depletion capacitance switch S_d is disconnected, as it does not participate in the process of balancing the charge neutrality. The term C_b' in Equation (2.115) becomes negligible compared to C_n'.

The charge balance condition requires:

$$\Delta Q'_g = -\Delta Q'_{sc} = -\Delta Q'_n \tag{2.129}$$

Figure 2.18(c) shows the low frequency charge balance diagram where the gate charge change is balanced by an equivalent inversion charge change. The space charge capacitance due to the inversion charge Q'_n is obtained from the relation of inversion charge density Q'_n given by Equation (2.33). Differentiating Equation (2.33), the space charge capacitance is given by:

$$C'_n = -\frac{dQ'_n}{d\phi_s} = \frac{\epsilon_{Si}}{2L_b} \exp\left(\frac{\phi_s - 2\phi_f}{2\phi_t}\right) = -\frac{Q'_n}{2\phi_t} \tag{2.130}$$

In the strong inversion region, $\phi_s > 2\phi_f$, which makes $C'_n \gg C'_{ox}$ and Equation (2.115) reduces to $C'_{eq} = C'_{ox}$.

- **High Frequency Characteristics**

If the frequency of the AC signal superimposed on the applied DC gate bias is very high compared to the generation time of the inversion carrier, there is not enough time available for the carriers in the inversion layer to follow the change in gate voltage, due to which $\Delta Q'_n \approx 0$. The change in the gate charge is accompanied by the change of depletion charge through the adjustment of majority carrier holes in the substrate. The majority carriers adjusting to satisfy the charge balance can respond at high frequency due to the small relaxation time of the substrate at normal doping. Under this condition we can write $\Delta Q'_g = -\Delta Q'_b$. Figure 2.18(c) with the switch S_d enabled and S_n disabled represents a high frequency capacitance equivalent circuit. The charge balance condition for the dynamic condition at high signal frequency will cause a change in the bulk charge equal to the gate charge change. The inversion charge does not respond to gate voltage changes at high frequency.

Since the surface potential gets pinned at $2\phi_f$, the depletion space charge capacitance value also gets fixed at the minimum value for $V_{GB} \geq V_{T0}$. Thus:

$$\frac{1}{C'_{eq_{HF}}} = \frac{1}{C'_{ox}} + \frac{1}{C'_{bmin}} \tag{2.131}$$

Since C'_{ox} can be obtained from the accumulation mode measurement, C'_{bmin} can be estimated from the high frequency capacitance measurement for $V_{GB} \geq V_{T0}$. As C'_{bmin} is related to doping, the high frequency measurement gives the estimate of substrate doping.

2.7.1.4 Flat-Band Capacitance

It is seen that at $V_{GB} = 0$, i.e. $\phi_s = 0$, which is the flat-band condition for an ideal capacitor, the capacitance value is less than the value in accumulation. This reduction is due to the contribution of the depletion capacitance at a flat-band, resulting from the depletion charge contained within the extrinsic Debye length given by $L_b = \sqrt{\epsilon_{Si}\phi_t/qN_A}$. Thus, the equivalent capacitance of a MOS structure at a flat-band condition is given by:

$$C'_{s_{fb}} = \frac{\epsilon_{Si}}{L_b}$$

(2.132)

Figure 2.19 illustrates the low and high frequency $C-V$ characteristics of an ideal two terminal MOS capacitor. It may be mentioned that the $C-V$ characteristics is a very important tool for characterizing important parameters, such as substrate doping concentration, threshold voltage, and flat-band voltage [6, 14].

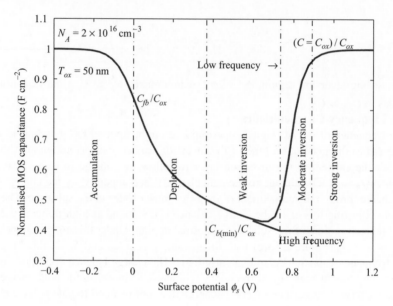

Figure 2.19   Normalized gate to bulk capacitance, C'_{gb}, as a function of surface potential

2.8 Three Terminal Ideal MOS Structures

The three terminal MOS structure is an extension of the two terminal MOS capacitor, where the third terminal is meant to act as an electrical probe to access the Si–SiO$_2$ interface when an inversion channel is formed by the gate voltage [1, 9, 20]. The conditions necessary for such an electrical probe are as follows.

- The channel and probe must be gate overlapped to provide the connectivity of the probe to the channel.
- The probe, which is made of doped silicon, must be of the same type of conductivity as that of the inversion channel.
- It should have high electrical conductivity to act as an ideal probe. To achieve this high conductivity, the probe is degenerately doped silicon. The diffused layer acting as the third terminal in the MOS structure is known in the literature as the *source* as it provides the

carriers to form the inversion channel. The three terminal MOS capacitor is also known as a *Sourced MOS Capacitor*. The source potential with respect to bulk is denoted by V_{SB}.

2.8.1 Cross-Section and Band Diagram

In a two terminal MOS capacitor the interface between the semiconductor and insulator is not accessible to the external world. The surface potential ϕ_s cannot be manipulated through any other external control voltage other than the gate. Figure 2.20(a) shows a three terminal MOS structure where, for the p-type substrate considered, the n^+ diffusion region overlapped by the gate acts as a third terminal *silicon probe* which can be used to modulate the surface potential. Figure 2.20(b) gives the energy band diagram for the inversion condition.

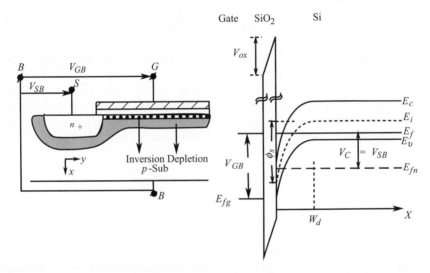

Figure 2.20 (a) A three terminal MOS capacitor in the inversion condition with an n^+ diffusion probe for controlling surface potential. (b) Energy band diagram

The physical mechanism governing the formation of the inversion channel under the gate can be understood from the following considerations.

- Assume that the gate voltage is adequate to drive the channel under the gate in deep inversion for $V_{SB} = 0$. In the strong inversion condition most of the potential drop is across the oxide, and the surface potential is a few ϕ_t higher than $2\phi_f$. The equilibrium carrier concentration in the inversion layer is described by the position of the single Fermi level corresponding to the equilibrium electron and hole concentrations.
- When we apply a positive bias V_{SB} at the source which is in contact with the channel, the n^+ diffused region and the channel potential are raised to V_{SB} as they are in electrical contact. The voltage V_{SB} drives the channel in a nonequilibrium condition, and the inversion carriers assume a new concentration which is lower than the equilibrium value. This reduction in the inversion charge can be physically explained in the following way. (i) The application of V_{SB} increases the surface potential, which in turn reduces the potential drop across the oxide, V_{ox}, to satisfy the potential balance condition for a given V_{GB}. It results in a corresponding

decrease in the gate charge Q_g, as can be seen from the relation $Q_g = C_{ox}V_{ox}$. (ii) From depletion approximation it is known that an increase in surface potential increases the negative bulk fixed charge. (iii) From charge conservation requirements, the reduction in positive gate charge and an increase in the negative bulk charge necessitates reduction in the negative inversion charge.

- The non-equilibrium inversion charge concentration now has to be defined by the quasi-Fermi level as discussed in Chapter 1. For a three terminal MOS structure, the *channel potential* is defined by the separation of the quasi-Fermi levels of the channel electrons from the equilibrium Fermi level of the majority carrier holes in the substrate. The Fermi potential level for electrons in the channel is increased by an amount V_{SB} from the Fermi potential level for holes. With the intrinsic level taken as a reference, we can write:

$$\phi_{fn} = \phi_f + V_{SB} \qquad (2.133)$$

where ϕ_{fn} is the quasi-Fermi potential for electrons in the channel [9]. From Figure 2.20 we can see that to attain an inversion condition in a three terminal MOS structure, the band bending has to be more, by an amount V_{SB} over the value of $2\phi_f$ required for a two terminal MOS capacitor. In other words, at threshold the surface potential in a three terminal MOS structure is $V_{SB} + 2\phi_f$, as against $2\phi_f$ for the two terminal MOS structure. The non-equilibrium electron concentration at the surface is now given by:

$$n_s = N_A \exp\left[\frac{\phi_s - (V_{SB} + 2\phi_f)}{\phi_t}\right] \qquad (2.134)$$

where ϕ_s, the surface potential, gives the extent of band bending at the surface.

- The surface potential is shown in the band diagram by the bending of the conduction/valence band or intrinsic level at the interface w.r.t. the substrate in the neutral region.
- The gate to bulk voltage is represented by the separation of the Fermi level of the gate from that of the equilibrium Fermi level in the substrate.
- In normal operation of MOS transistors $V_{SB} > 0$, i.e. the source to bulk junction is reverse biased.

2.8.2 Space Charge Density, Surface Potential and Threshold Voltage

The expression for space charge density Q'_{sc} and the inversion charge density at the Si–SiO$_2$ interface for a three terminal MOS structure may be obtained directly from Equation 2.13 by replacing $2\phi_f$ with $2\phi_f + V_{SB}$. The voltage $V_{SB} + 2\phi_f$ is the channel inversion potential in the presence of an externally applied voltage V_{SB} through the n$^+$ silicon probe, which reduces to $2\phi_f$ for a two terminal MOS capacitor. The equations for inversion charge density due to electrons, and depletion charge density due to acceptor ions, for ϕ_s greater than a few ϕ_t are obtained from Equations (2.20) (with suggested modification), and (2.30) respectively.

$$Q'_{sc} = -\sqrt{2q\epsilon_{Si}N_A\phi_t}\left[\frac{\phi_s}{\phi_t} + \exp\left(\frac{\phi_s - V_{SB} - 2\phi_f}{\phi_t}\right)\right]^{\frac{1}{2}} \qquad (2.135)$$

$$Q_b' = -\sqrt{2q\epsilon_{Si}N_A\phi_s} \tag{2.136}$$

From the relation $Q_{sc}' = Q_n' + Q_b'$ we have using Equations (2.135) and (2.136):

$$Q_n' = -\sqrt{2q\epsilon_{Si}N_A\phi_t}\left[\left(\frac{\phi_s}{\phi_t} + e^{\frac{\phi_s - V_{SB} - 2\phi_f}{\phi_t}}\right)^{\frac{1}{2}} - \left(\frac{\phi_s}{\phi_t}\right)^{\frac{1}{2}}\right] \tag{2.137}$$

The plot of Equation (2.137) is shown in Figure 2.21, where the onset of inversion is extended from $\phi_s = 2\phi_f$ to $\phi_s = V_{SB} + 2\phi_f$.

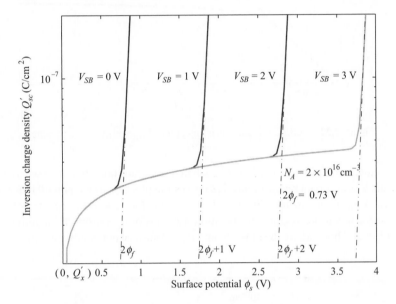

Figure 2.21 Variation of space charge density with channel potential and V_{SB} as a parameter

2.8.2.1 Surface Potential

The relation between gate, source, and surface potential can be obtained using the equation for potential balance, where Q_{sc}' is given by Equation (2.135):

$$V_{GB} = \phi_s - \frac{Q_{sc}'}{C_{ox}'} \tag{2.138}$$

Substituting for Q_{sc}' from Equation (2.135) gives:

$$V_{GB} = \phi_s + \gamma\left[\phi_s + \phi_t e^{\frac{\phi_s - (V_{SB} + 2\phi_f)}{\phi_t}}\right]^{\frac{1}{2}} \tag{2.139}$$

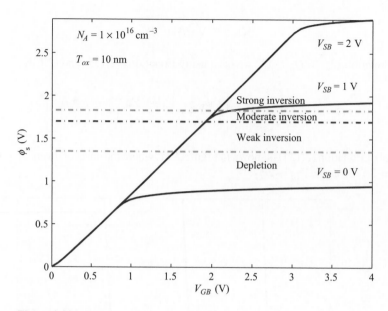

Figure 2.22   Surface potential as a function of V_{GB} with V_{SB} as a parameter

The above equation is plotted in Figure 2.22 where the surface potential is shown as a function of V_{GB} with V_{SB} as a parameter. Figure 2.23 gives the plot of surface potential as a function of V_{SB} with V_{GB} as parameter. These plots are useful for understanding the behavior of a MOS transistor in various regions of operation. Regions of operations are identified in these figures with dotted lines where the upper boundaries of surface potentials are $V_{SB} + \phi_f$, $V_{SB} +$

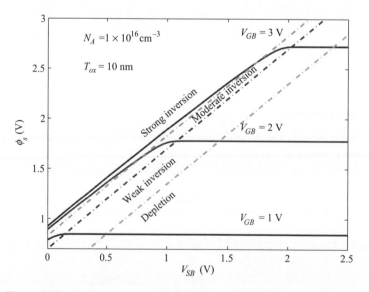

Figure 2.23   Surface potential as a function of V_{SB} with V_{GB} as a parameter

$2\phi_f$, and $V_{SB} + 2\phi_f + m\phi_t$ for depletion, weak inversion, and moderate inversion regions respectively. In Figure 2.22 the regions of operation are shown only for $V_{SB} = 1$V. From these figures we can make the following observations:

- The nature of the curve ϕ_s versus V_{GB} is similar to a two terminal MOS capacitor behavior, except that there is a shift in the gate voltage required to initiate onset of inversion. This shift in the gate voltage is equal to $V_{SB} + 2\phi_f + \gamma\sqrt{V_{SB} + 2\phi_f}$.
- The surface potential in strong inversion can be approximated to $V_{SB} + \phi_0$, as compared to ϕ_0 in a two terminal MOS capacitor where $\phi_0 = 2\phi_f + m\phi_t$.
- From Figure 2.23 it can be observed that for a given V_{GB} the surface potential follows V_{SB} with an approximate offset of ϕ_0 in strong inversion, and becomes almost independent of V_{SB} in weak inversion.

Having obtained the surface potential variation as a function of V_{GB} and V_{SB}, we are now in a position to find the inversion charge density as a function of applied terminal biases. Figure 2.24 shows such variations of magnitude in log as well as in linear scale. The following observations can be made.

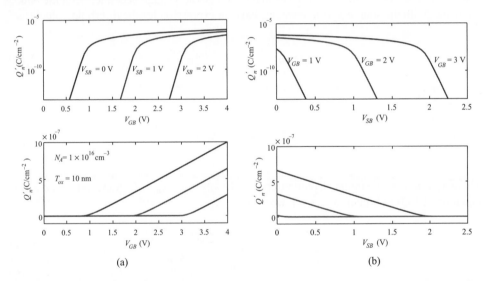

Figure 2.24   (a) Inversion charge density as a function of V_{GB} with V_{SB} as a parameter. (b) Inversion charge density as a function of V_{SB} with V_{GB} as a parameter

- It can be seen that the gate voltage required to create inversion increases with increase in V_{SB}.
- When the device enters strong inversion, the inversion charge has almost linear dependence on gate as well as source bias. It can also be observed that the gate and source voltages have an effect in opposite directions as regards change in inversion charge.
- In weak inversion the surface potential needs to be calculated very precisely, as even a small variation of surface potential with change in V_{SB} causes a significant relative change in inversion charge density.

2.8.2.2 Threshold Voltage

The threshold voltage for a two terminal MOS capacitor was defined by the gate voltage required to create a surface inversion potential of $2\phi_f$. In the case of a three terminal MOS capacitor, the surface inversion potential being $V_{SB} + 2\phi_f$, it requires a higher gate voltage to attain this value. This increases the threshold voltage. At threshold, from the equation of potential balance between gate and the substrate:

$$V_{GB} = (V_{SB} + 2\phi_f) - \frac{Q'_n + Q'_{bm}(V_{SB})}{C'_{ox}} \tag{2.140}$$

At threshold, $Q'_n \ll Q'_b$, which gives:

$$V_{TB}(V_{SB}) = V_{SB} + 2\phi_f - \frac{Q'_{bm}(V_{SB})}{C'_{ox}} \tag{2.141}$$

where the subscript 'B' in V_{TB} indicates that the threshold voltage is defined with the bulk as reference. With the source as reference, we have an expression for the threshold voltage as:

$$V_{TS}(V_{SB}) = 2\phi_f - \frac{Q'_{bm}(V_{SB})}{C'_{ox}} \tag{2.142}$$

For $V_{SB} = 0$:

$$V_{T0} = 2\phi_f - \frac{Q'_{b0}}{C'_{ox}} \tag{2.143}$$

where Q'_{b0} is the maximum depletion charge density for $V_{SB} = 0$. Substituting $2\phi_f$ from Equation (2.143) in Equation (2.142), we have:

$$V_{TB}(V_{SB}) = V_{T0} + V_{SB} + \frac{Q'_b(V_{SB}) - Q'_{b0}}{C'_{ox}} \tag{2.144}$$

Substituting for $Q'_{b0} = -\sqrt{2q\epsilon_{Si}N_A(2\phi_f)}$, and $Q'_b(V_{SB}) = -\sqrt{2q\epsilon_{Si}N_A(2\phi_f + V_{SB})}$ we have:

$$V_{TB}(V_{SB}) = V_{TB0} + V_{SB} + \gamma(\sqrt{V_{SB} + 2\phi_f} - \sqrt{2\phi_f}) \tag{2.145}$$

The change in threshold voltage due to applied V_{SB} is:

$$\Delta V_{TB} = V_{TB}(V_{SB}) - V_{TB0} = V_{SB} + \gamma(\sqrt{V_{SB} + 2\phi_f} - \sqrt{2\phi_f}) \tag{2.146}$$

Equation (2.146) shows that the change in threshold voltage in a substrate biased MOS capacitor has two components: (a) the change in the surface potential due to applied potential at the source terminal, and (b) the change in the voltage drop across the oxide due to additional depletion charge. Thus:

$$\Delta\phi_s = V_{SB} \tag{2.147}$$

$$\Delta V_{ox} = \gamma(\sqrt{V_{SB} + 2\phi_f} - \sqrt{2\phi_f}) \tag{2.148}$$

The threshold voltage has a non-linear dependence on V_{SB}. The slope of V_{TB} versus V_{SB} is obtained from Equation (2.145), and is given by:

$$\frac{dV_{TB}}{dV_{SB}} = 1 + \frac{\gamma}{2\sqrt{V_{SB} + \phi_0}} \tag{2.149}$$

Figure 2.25 shows the variation of threshold voltage V_{TB} with channel potential V_{CB} where the channel potential assumes the value of the source bias. From Equation (2.149), it is seen that from the plot of the slope of V_{TB} versus $1/\sqrt{V_{SB}}$, the body effect factor can be obtained experimentally.

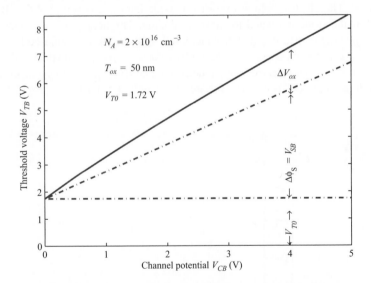

Figure 2.25 Threshold voltage dependence on channel potential

In Equation (2.146) we observe that γ is related to the change in the threshold voltage due to change in the source bias. A higher value of γ (by increasing N_A and/or T_{ox}) implies that a higher gate voltage is required for inversion at the same source bias. Thus, γ may be defined as the source bias coefficient. For historical reasons, when the source was conventionally taken as a reference terminal, V_{BS} implied the biasing of bulk w.r.t. the source. Hence, γ is conventionally known as the *bulk or substrate bias coefficient*.

Equation (2.145) gives the threshold voltage referred to the bulk. The threshold voltage referred to source follows directly from the said equation, and is given by:

$$V_{TS} = V_T(V_{SB}) - V_{SB} = V_{T0} + \gamma \left(\sqrt{-V_{BS} + 2\phi_f} - \sqrt{\phi_0} \right) \tag{2.150}$$

Equation (2.141) shows a nonlinear relationship between threshold voltage and substrate bias. In compact models, and for analytical purposes, it is useful to linearize the above relationship. The nonlinear relationship owes its origin to the square root dependence of the bulk depletion charge on surface potential. We shall use Taylor's expansion to linearize the bulk charge density expression $Q_b'(V_{SB})$ as given below:

$$Q_b'(V_{SB}) = Q_{b0}' + V_{SB} \times \left. \frac{\mathrm{d} Q_b'}{\mathrm{d} V_{SB}} \right|_{V_{SB}=0} \tag{2.151}$$

$$Q_b'(V_{SB}) = Q_{b0}' + C_b' V_{SB} \tag{2.152}$$

where $\left. \mathrm{d} Q_b'/\mathrm{d} V_{SB} \right|_{V_{SB}=0} = \left. C_b' \right|_{V_{SB}=0}$.

Figure 2.26 shows graphically the potential balance of a MOS capacitor with substrate bias V_{SB}. At threshold, the surface potential equals ($V_{SB} + 2\phi_f$) given by the triangle $AB'C'$ in the F–x plane. The electric field at the silicon–oxide interface is acting at C'. BC and B'C' are parallel as the slope is related to substrate doping. The height AC' is a measure of the field in *silicon* at the interface, which has increased compared to the unbiased field value AC. The electric field in the oxide gets scaled up by a factor $\epsilon_{Si}/\epsilon_{ox}$ compared to that in silicon at the interface $x = 0$. The electric field value AD' equals $((Q_{b0}' + C_b' V_{SB})/\epsilon_{Si})(\epsilon_{Si}/\epsilon_{ox})$. The threshold voltage, $V_T(V_{SB})$, is given by the area of the polygon $AB'C'D'E'FA$. The area of the

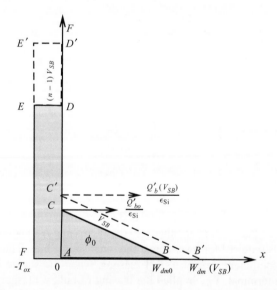

Figure 2.26   Graphical representation of potential balance at threshold with substrate bias

triangle $AB'C'$ is $V_{SB} + \phi_0$ which is the surface potential. The area of the rectangle $AD'E'F$ equals $AD' \times T_{ox}$ which accounts for the potential drop across oxide. Thus:

$$V_{TB}(V_{SB}) = \left[\left(\frac{Q'_{b0}}{C'_{ox}} + 2\phi_f \right) + V_{SB} \left(1 + \frac{C'_b}{C'_{ox}} \right) \right] \tag{2.153}$$

which can be put in the form:

$$V_{TB} = V_{T0} + n V_{SB} \tag{2.154}$$

where:

$$n = 1 + \frac{C'_b}{C'_{ox}}, \quad \text{with } C'_b = \left. \frac{dQ'_b}{dV_{SB}} \right|_{V_{SB}=0} \tag{2.155}$$

From Equation (2.154) it is seen that the bulk charge linearization with respect to channel potential leads to linear variation of threshold voltage with substrate bias, or equivalently with channel potential because at the threshold we can put $V_{SB} = V_{CB}$.

2.8.3 Pinch-off Voltage

For a given gate voltage in a three terminal MOS capacitor the level of inversion is determined by the source potential. For an inverted three terminal MOS capacitor, if the source potential is increased, holding the gate voltage constant, a situation will occur when the gate voltage is not able to sustain inversion.

Thus, for a given gate voltage V_{GB} there is a corresponding source or channel potential called the pinch-off voltage V_P, above which the surface is in the weak inversion region, and below which it is in strong inversion [16]. The pinch-off voltage V_P is thus a fundamental parameter uniquely defined by the gate voltage.

We shall now obtain an expression for the pinch-off voltage in terms of a given gate voltage V_{GB}. From the definition of pinch-off voltage, the surface potential at pinch-off will be $\phi_s = V_{SB} + 2\phi_f$. Substituting this value in Equation (2.139), we have:

$$V_{GB} = V_P + 2\phi_f + \gamma \left[V_P + 2\phi_f + \phi_t \right]^{\frac{1}{2}} \tag{2.156}$$

$$[V_{GB} - (V_P + 2\phi_f)]^2 = \gamma^2 [(V_P + 2\phi_f) + \phi_t] \tag{2.157}$$

Rearranging Equation (2.157):

$$(V_P + 2\phi_f)^2 - (2V_{GB} + \gamma^2)(V_P + 2\phi_f) + (V_{GB}^2 - \gamma^2 \phi_t) = 0 \tag{2.158}$$

The above equation is quadratic in $V_P + 2\phi_f$. Solving for V_P we get:

$$V_P = \left[\sqrt{V_{GB} + \left(\frac{\gamma}{2}\right)^2 + \phi_t} - \frac{\gamma}{2} \right]^2 - 2\phi_f - \phi_t \qquad (2.159)$$

Equation (2.159) is the expression for pinch-off voltage used in the Advanced Compact Model (ACM) for MOS transistors [24].

The following conclusions are drawn from the expression for pinch-off voltage.

- For $V_{GB} = 2\phi_f + \gamma\sqrt{2\phi_f + \phi_t}$, we have $V_P = 0$. The expression for V_{GB} for $V_P = 0$ is identical to the expression of threshold voltage neglecting ϕ_t compared to $2\phi_f$. Thus, for the gate voltage equalling the threshold voltage the corresponding pinch-off voltage is zero. This can be physically understood from the fact that to create inversion at $\phi_s = 2\phi_f$, the source to bulk potential should be zero.
- The pinch-off voltage depends upon process parameters such as substrate doping and oxide thickness.
- As per this definition of the pinch-off voltage *there is a finite amount of charge in the channel* corresponding to the surface potential $V_P + 2\phi_f$.

Below we will describe the definition adopted in EKV for defining the pinch-off voltage.

Making use of the potential balance equation in strong inversion with $\phi_s = V_{SB} + \phi_0$, the expression for inversion charge density is:

$$Q'_n = -C'_{ox}[V_{GB} - (V_{SB} + \phi_0) - \gamma\sqrt{V_{SB} + \phi_0}] \qquad (2.160)$$

The pinch-off voltage V_P is now defined as the source potential at which the inversion charge described by Equation (2.160) is reduced to zero. In other words, this approach extrapolates the strong inversion expression for Q'_n in the weak inversion region for the determination of V_P. It is observed that the strong inversion approximation overestimates the inversion charge.

Putting $Q'_n = 0$ at $V_{SB} = V_P$ in Equation (2.160), we have:

$$(V_P + \phi_0) + \gamma\sqrt{V_P + \phi_0} - V_{GB} = 0 \qquad (2.161)$$

The above equation is quadratic in $\sqrt{V_P + \phi_0}$. Solving for V_P we get:

$$\sqrt{V_P + \phi_0} = -\frac{\gamma}{2} \pm \frac{1}{2}\left(\gamma^2 + 4V_{GB}\right)^{\frac{1}{2}} \qquad (2.162)$$

Both V_P and ϕ_0 being positive, the admissible solution is:

$$\sqrt{V_P + \phi_0} = -\frac{\gamma}{2} + \frac{1}{2}\left(\gamma^2 + 4V_{GB}\right)^{\frac{1}{2}} \qquad (2.163)$$

or:

$$V_P = \left[-\frac{\gamma}{2} + \left(\frac{\gamma^2}{4} + V_{GB} \right)^{\frac{1}{2}} \right]^2 - \phi_0 \tag{2.164}$$

Equation (2.164) expresses the pinch-off voltage in terms of two parameters related to the process, such as γ and ϕ_0. With slight manipulation of the above expression, V_P can be expressed in terms of three parameters, i.e. γ, ϕ_0, and V_{T0} as shown below:

$$V_P = V_{GB} - V_{T0} - \gamma \left\{ \left[V_{GB} - V_{T0} + \left(\sqrt{\phi_0} + \frac{\gamma}{2} \right)^2 \right]^{\frac{1}{2}} - \left(\sqrt{\phi_0} + \frac{\gamma}{2} \right) \right\} \tag{2.165}$$

The above equation shows that for $V_{GB} = V_{T0}$, $V_P = 0$.

Since V_P is a function of the gate voltage, it is useful to observe the slope of V_P versus V_G dependence which is described by the slope factor n:

$$n = \frac{dV_{GB}}{dV_P} \tag{2.166}$$

From Equations (2.164) and (2.166) we get:

$$n = 1 + \frac{\gamma}{2\sqrt{V_P + \phi_0}} \tag{2.167}$$

The parameter n can also be expressed in terms of gate and threshold voltage from Equation (2.165), and is given by:

$$\frac{1}{n} = 1 - \frac{\gamma}{2\sqrt{V_{GB} - V_{T0} + \left(\frac{\gamma}{2} + \sqrt{\phi_0} \right)^2}} \tag{2.168}$$

Figure 2.27 gives the plot of pinch-off voltage and slope factor n as a function of gate voltage for a given process parameter. The value of n lies between 1.2 and 1.6 for normal doping of the MOS substrate. From the graph it is concluded that as n is a weak function of V_{GB}, i.e. the V_P–V_{GB} curve can be approximated to a straight line. A simple equation relating V_P and V_{GB} can be written as:

$$V_P = \frac{V_{GB} - V_{T0}}{n} \tag{2.169}$$

From Figure 2.22 we can see that the surface potential is almost independent of source potential once the weak inversion region of operation is entered. We had defined the pinch-off voltage such that at $V_{SB} \geq V_P$, the three terminal MOS structure enters weak inversion, and the surface potential can be approximated to $V_P + \phi_0$.

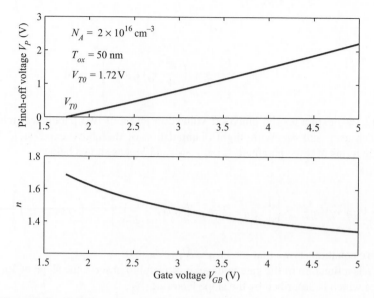

Figure 2.27 Variation of pinch-off voltage and slope factor against gate voltage

From Figure 2.23, which shows the variation of surface potential as a function of V_{SB} with V_{GB} as a parameter, we can see that in strong inversion the surface potential follows V_{SB} with an offset of ϕ_0. Thus, the surface potential for strong inversion is:

$$\phi_s \approx (V_{SB} + \phi_0) \qquad (V_{SB} \leq V_P) \tag{2.170}$$

In the weak inversion region, the surface potential follows the pinch-off voltage with an offset of ϕ_0:

$$\phi_s \approx (V_P + \phi_0) \qquad [V_{SB} \geq V_P] \tag{2.171}$$

2.8.4 Inversion Charge Density in Terms of V_P and V_{SB}

The charge density under the gate can be obtained by using the KVL as obtained before:

$$Q_n' + Q_b' = -C_{ox}'(V_{GB} - \phi_s) \tag{2.172}$$

or:

$$Q_n' = -C_{ox}'(V_{GB} - \phi_s - \gamma\sqrt{\phi_s}) \tag{2.173}$$

where we have substituted Q_b' by $-\gamma C_{ox}'\sqrt{\phi_s}$. The surface potential is now controlled by the applied gate and source bias as compared to only the gate bias in the case of the MOS capacitor. In Equation (2.173), we see that for a given gate to bulk potential V_{GB}, the inversion charge density is non-linearly dependent on surface potential through the term $\gamma\sqrt{\phi_s}$ related to the bulk depletion charge. The surface potential at which the inversion charge density is zero is defined as the *surface pinch-off potential* ϕ_{po}. Equating (2.173) to zero at $\phi_s = \phi_{po}$ gives:

$$\phi_{po} = V_{GB} - \gamma^2 \left(\sqrt{\frac{V_{GB}}{\gamma^2} + \frac{1}{4}} - \frac{1}{2} \right) \qquad (2.174)$$

Q'_b can be approximated as a linear function of surface potential by using a Taylor series expansion around ϕ_{po}. The inversion charge density can now be written as a linear function of the surface potential:

$$Q'_n = \left(\frac{dQ'_n}{d\phi_s} \right)_{\phi_s = \phi_{po}} (\phi_s - \phi_{po}) \qquad (2.175)$$

An alternative approach is proposed by Sallese [28] in which $dQ'_n/d\phi_s$ in Equation (2.175) is replaced by the slope of the secant joining $(Q'_n, 2\phi_f)$ to $(0, \phi_{po})$ which gives:

$$Q'_n = -C'_{ox} \left(1 + \frac{\gamma}{\sqrt{\phi_{po}}} \right) (\phi_{po} - \phi_s) \qquad (2.176)$$

The required surface potential, as a function of the applied voltages V_{GB} and V_{SB}, may be obtained either from numerical calculation or from the following analytical expression.

The surface potential is obtained explicitly in terms of V_{GB} and V_{SB} as shown in Section 2.4.1 by replacing $2\phi_f$ with $V_{SB} + 2\phi_f$. The final expression for ϕ_s is obtained as:

$$\phi_s = f(V_{GB})$$

$$+ \phi_t \ln \left\{ \left[\frac{V_{GB} - f(V_{GB}) - \frac{\phi_{s(wi)} - f(V_{GB})}{\sqrt{1 + [(\phi_{s(wi)} - f(V_{GB}))/(4\phi_t)]^2}}}{\gamma\sqrt{\phi_t}} \right]^2 - \frac{f(V_{GB})}{\phi_t} + 1 \right\} \qquad (2.177)$$

Equations (2.177) and (2.173) provide an explicit and continuous relation for the inversion charge density as a function of the gate voltage.

The channel charge density can be expressed directly in terms of the pinch-off voltage V_P and source potential V_{SB}, which is simpler and leads to compact analytical expressions. This approach can be divided into two categories, i.e (a) the piecewise charge based model for strong and weak inversion, and (b) the continuous charge model for strong, moderate, and weak inversion.

2.8.4.1 Strong Inversion

Using the approximate expression for surface potential in strong inversion given in Equation (2.170), and substituting in equation (2.173), we get:

$$Q'_n = -C'_{ox} \left(V_{GB} - V_{SB} - \phi_0 - \gamma \sqrt{\phi_0 + V_{SB}} \right) \qquad (2.178)$$

The above equation overestimates the inversion charge in the very strong inversion condition. This overestimation of charge results from the surface potential actually being more than $\phi_0 + V_{SB}$. The inversion charge density can be expressed in terms of V_P and V_{SB} by using the following relation:

$$\phi_0 = V_{GB} - V_P - \gamma \sqrt{V_P + \phi_0} \qquad (2.179)$$

The above equation is obtained by setting $Q'_n = 0$ in Equation (2.178), with $V_{SB} = V_P$. Substituting the value of ϕ_0 only in the linear term in Equation (2.178), we get:

$$Q'_n = -C'_{ox} \left(V_P - V_{SB} + \gamma \sqrt{V_P + \phi_0} - \gamma \sqrt{V_{SB} + \phi_0} \right) \qquad (2.180)$$

Equation (2.180) is found to overestimate the channel charge density apart from being a nonlinear function of the channel potential V_{SB}. The inversion charge density in the channel under the gate can, however, be approximated to a simpler linear equation relating the inversion charge density with V_P and V_{SB}. A Taylor's series expansion of Equation (2.180) around $V_{SB} = V_P$, and retaining the linear term, gives:

$$Q'_n = - \left(\frac{\partial Q'_n}{\partial V_{SB}} \right)_{V_{SB}=V_P} (V_P - V_{SB}) \qquad (2.181)$$

Using Equation (2.180), and taking the derivative of Q'_n w.r.t. V_{SB} at a given V_P:

$$Q'_n = -C'_{ox} \left(1 + \frac{\gamma}{2\sqrt{V_P + \phi_0}} \right) (V_P - V_{SB}) \qquad (2.182)$$

$$Q'_n = -nC'_{ox}(V_P - V_{SB}) \qquad (2.183)$$

It is noted that the slope factor n is defined here as $n = (C'_{ox})^{-1} \partial Q'_n / \partial V_{SB}$ [9], and the expression is identical to the expression given in Equation (2.167) obtained from the definition given in Equation (2.166).

Figure 2.28 shows the plot of the inversion charge density Q'_n as a function of the source potential V_{SB} (which equals the channel potential in the inversion condition) based on linear approximation. A linear relationship is obtained using the value of V_P obtained from extrapolation of strong inversion expression, and use of the slope factor derived in Equation (2.168).

An accurate estimation of charge is crucial to precise current modeling in a MOS transistor. It is also desirable to retain the simplicity of linear dependence of inversion charge on channel

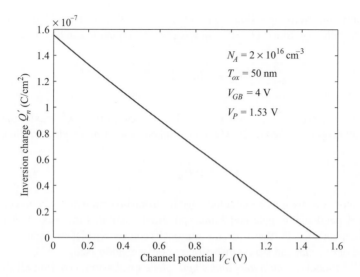

Figure 2.28 Inversion charge density as a function of channel potential

potential. These can be obtained by independently adjusting the values of V_{T0}, γ, and ϕ_0, even though they are correlated.

2.8.4.2 Weak Inversion

For calculating of the inversion charge in weak inversion we cannot use Equation (2.173), with an approximate surface potential given in Equation (2.171), as the resultant expression for inversion charge will be independent of V_{SB}. The above apparent nonphysical result is the outcome of the fact that in reality the surface potential is not independent of V_{SB} in the weak inversion, but has a small and finite variation in the weak inversion region. For calculating Q'_n in weak inversion using Equation (2.173) requires calculation of ϕ_s with a accuracy of less then a nanovolt [19]. We can, however still use the approximation of Equation (2.171) for surface potential in Equation (2.32) which is reproduced below for a three terminal capacitor:

$$Q'_n = -\frac{1}{2} Q'_x \left(\frac{\phi_t}{\phi_s} \right)^{\frac{1}{2}} e^{\frac{\phi_s - (V_{SB} + 2\phi_f)}{\phi_t}} \qquad (2.184)$$

In the weak inversion condition, we have $\phi_s = V_P + \phi_0$:

$$Q'_n = -\frac{1}{2} Q'_x \left(\frac{\phi_t}{V_P + \phi_0} \right)^{\frac{1}{2}} e^{\frac{\phi_0 - 2\phi_f}{\phi_t}} e^{\frac{V_P - V_{SB}}{\phi_t}} \qquad (2.185)$$

It is seen that in the weak inversion region ($V_{SB} > V_P$), the inversion charge density decays exponentially with V_{SB}. Since $Q_x' = \sqrt{2q\epsilon_{Si}N_A\phi_t}$, the equation can be written as:

$$Q_n' = -C_{ox}'\phi_t \left[(n-1)\,e^{\frac{\phi_0 - 2\phi_f}{\phi_t}} \right] e^{\frac{V_P - V_{SB}}{\phi_t}} \tag{2.186}$$

where $(n-1) = \frac{1}{2}\gamma/\sqrt{V_P + \phi_0}$. Since $\phi_0 = 2\phi_f + m\phi_t$, in the EKV model the term in the square brackets approximated to $2n$ where m is taken as a fitting parameter. Thus:

$$Q_n' = 2nC_{ox}'\phi_t e^{\frac{V_P - V_{SB}}{\phi_t}} \tag{2.187}$$

From the above discussions we conclude that in strong inversion the inversion charge density has a linear dependence to gate and source potentials, whereas in weak inversion it has an exponential dependence. Furthermore, the approach represents a piecewise and regional charge model where only weak and strong inversion regions are addressed.

In the ACM model a *continuous and single piece* expression, covering all the regions of operation, has been proposed for the inversion charge in Gouveia Filho *et al.* [33].

2.8.5 Small Signal Capacitor Model

We can represent the small signal capacitance model of a three terminal MOS structure through the charge control model [24, 25]. Figure 2.29 shows the proposed small signal capacitor model of a three terminal gated diode structure.

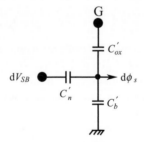

Figure 2.29   Small signal model of a three terminal MOS capacitor

The inversion charge, Q_n', is assigned to the source terminal as the supply of carriers to the channel for a change in the potential in the channel takes place from the source. Therefore, the inversion capacitance C_n' is connected between the source and the channel and is defined as given below:

$$dQ_n' = C_n'(dV_{SB} - d\phi_s) \tag{2.188}$$

$$\frac{d\phi_s}{dQ_n'} = C_n'\left(\frac{dV_{SB}}{d\phi_s} - 1\right)d\phi_s \tag{2.189}$$

The expression for C_n' has been obtained in Equation (2.130).

Similarly, Q'_b an Q'_g are charges assigned to the bulk and gate terminals as they become charged through the bulk and gate terminals respectively, due to change in the surface potential. C'_b and C'_{ox} are the corresponding capacitances shown in the equivalent circuit.

From the equivalent circuit, we see that:

$$\frac{d\phi_s}{dV_{SB}} = \frac{C'_n}{C'_n + C'_b + C'_{ox}} \tag{2.190}$$

Substituting for $d\phi_s/dV_{SB}$ from Equation (2.189) in the above equation gives:

$$dQ'_n = C'_n \left(\frac{C'_n + C'_b + C'_{ox}}{C'_n} - 1 \right) d\phi_s \tag{2.191}$$

Thus:

$$dQ'_n = (C'_{ox} + C'_b) \, d\phi_s \tag{2.192}$$

Since the inversion charge density can be approximately linearly related to the surface potential, Equation (2.192) can be put in the following form:

$$nC'_{ox} \, d\phi_s = (C'_{ox} + C'_b) \, d\phi_s \tag{2.193}$$

$$n = 1 + \frac{C'_b}{C'_{ox}} \tag{2.194}$$

The relationship of n, the slope factor with oxide and bulk capacitance, is thus obtained from the small signal model.

The three terminal equivalent capacitance model reduces to two terminal model as a special case if the inversion charge Q'_n is assigned to the bulk terminal.

References

[1] Y. Tsividis, *Operation and Modeling of MOS the Transistor*, McGraw Hill, New York, NY, 1988.
[2] S.M. Sze, *Physics of Semiconductor Devices*, 2nd Edition, John Wiley & Sons, Ltd, New York, NY, 1993.
[3] N.D. Arora, *MOSFET Models for VLSI Circuit and Simulation*, 2nd Edition, Springer-Verlag, Berlin, 1993.
[4] H.C. Casey, *Devices for Integrated Circuits: Silicon and III–V Compound Semiconductors*, John Wiley & Sons, Ltd, Chichester, UK, 1999.
[5] Y. Taur and T.H. Ning, *Fundamentals of Modern VLSI Devices, ABC*, Cambridge University Press, Cambridge, UK, 1998.
[6] D.K. Schroder, *Semiconductor Material and Device Characterization*, Wiley Interscience, New York, NY, 2006.
[7] J.J. Liou, *Advanced Semiconductor Device Physics and Modeling*, Artech House, Boston, MA, 1994.
[8] R.H. Kingston and S.F. Neustadter, Calculation of Space Charge, Electric Field and Free Carrier Concentration at the Surface of a Semiconductor, *J. App. Phys.*, **26**(6), 718–720, 1955.

[9] C.C. Enz and E.A. Vittoz, Charge-based MOS Transistor Modeling: The EKV Model for Low-Power and RF IC Design, John Wiley & Sons, Ltd, Chichester, UK, 2006.

[10] M. Miura-Mattausch, N. Sadachika, D. Navarro, G.Suzuki, Y. Takeda, M. Miyake, T. Warabino, Y. Mizukane, R. Imagaki, T. Ezaki, H. J. Mattausch, T. Ohguro, T. Iizuka, M. Taguchi, S. Kumashiro and S. Miyamoto, HiSIM2: Advanced MOSFET Model Valid for RF Circuit Simulation, *IEEE Trans. Electron Dev.*, **53**(9), 1994–2007, 2006.

[11] W.Z. Shangguan, M. Saeyes and X. Zhou, Surface-potential Solutions to the Pao–Sah Voltage Equation, *Solid-State Electron.*, **50**(7–8), 1320–1329, 2006.

[12] J. He, X. Zhang, G. Zhang and Y. Wang, Benchmark Tests on Surface Potential Based Charge-Sheet Models, *Solid-State Electron.*, **50**(2), 263–267, 2006.

[13] C.C. McAndrew and J.J. Victory, Accuracy of Approximations in MOSFET Charge Models, *IEEE Trans. Electron Dev.*, **49**(1), 72–81, 2002.

[14] K.H. Zaininger and F.P. Heinman, The *C–V* Technique as an Analytical Tool, *Solid-State Technol.*, **13**(6), 49–56, 1970.

[15] J.R. Brews, A Charge Sheet Model of the MOSFET, *Solid-State Electron.*, **21**, 345–355, 1978.

[16] C. Enz, F. Krummenacher and E.A. Vittoz, An Analytical MOS Transistor Model Valid in all Regions of Operation and Dedicated to Low-Voltage and Low-Current Applications, *Analog Integr. Circ. Signal Proc.*, **8**, 83–114, 1995.

[17] M. Bagheri and C. Turchelti, The Need for an Explicit Model Describing MOS Transistors in Moderate Inversion, *Electron. Lett.*, **21**(19), 873–874, 1985.

[18] T.L. Chen and G. Glidenblat, Analytical Approximation for the MOSFET Surface Potential, *Solid-State Electron.*, **45**(2), 335–339, 2001.

[19] T.L. Chen and G. Glidenblat, An Extended Analytical Approximation for the MOSFET Surface Potential, *Solid-State Electron.*, **49**(2), 267–270, 2005.

[20] A. Kapoor and R.P. Jindal, Modeling MOSFET Surface Capacitance Behavior under Non-Equilibrium, *Solid-State Electron.*, **49**, 976–980, 2005.

[21] M.C. Tobey and N. Gordon, Concerning the Onset of Heavy Inversion in MIS Devices, *IEEE Trans. Electron Dev.*, **21**(10), 649–650, 1974.

[22] N. Shigyo, An Explicit Expression for Surface Potential at High-End of Moderate Inversion, *IEEE Trans. Electron Dev.*, **49**(7), 1265–1273, 2002.

[23] D.J. Comer and D.T. Comer, Operation of Analog MOS Circuits in the Weak or Moderate Inversion Region, *IEEE Trans. Edu.*, **47**(4), 430–435, 2004.

[24] A.I.A. Cunha, M.C. Schneider and C.-G. Montoro, An Explicit Physical Model for the Long-Channel MOS Transistor Including Small-Signal Parameters, *Solid-State Electron.*, **38**(11), 1945–1952, 1995.

[25] A.I.A. Cunha, M.C. Schneider and C.-G. Montoro, Derivation of the Unified Charge Control Model and Parameter Extraction Procedure, *Solid-State Electron.*, **43**(3), 481–485, 1999.

[26] Y. Tsividis, Moderate Inversion in MOS Devices, *Solid-State Electron.*, **25**, 1099–1104, 1992.

[27] R.V. Langevelde and F.M. Klaassen, An Explicit Surface Potential Based MOSFET Model for Circuit Simulation, *Solid-State Electron.*, **44**, 409–418, 2000.

[28] J.M. Sallese, M. Bucher, F. Krummenacher and P. Fazan, Inversion Charge Linearization in MOSFET Modeling and Rigorous Derivation of the EKV Compact Model, *Solid-State Electron.*, **47**(4), 677–683, 2003.

[29] A. Ortiz-Conde, F.J. Garcia Sanchez and J.J. Liou, An Improved Definition for Modeling the Threshold Voltage for MOSFETs, *Solid-State Electron.*, **42**(9), 1743–1746, 1998.

[30] H. Wang, T.L. Chen and G. Gildenblat, Quasi-static and Nonquasi-static MOSFET Models Based on Symmetric Linearization of the Bulk and Inversion Charges, *IEEE Trans. Electron Dev.*, **50**(11), 2262–2272, 2003.

[31] M.-M. Mattausch, U. Feldmann, A. Rahm, M. Bollu and D. Savignac, Unified Complete MOSFET Model for Analysis of Analog and Digital Circuits, *IEEE Trans. Comput. Aided Des.*, **15**(1), 1–7, 1996.

[32] MOS11 (http://www.semiconductors.Philips.com/Philips.Models/mos models/model1101).

[33] O.C. Gouveia Filho, A.I.A. Cunha, M.C. Schnieder and C. Galup-Montoro, The ACM Model for Circuit Simulation and Equations for SMASH (Online) (http://www.dolphin.fr), September, 1997.

[34] H.R. Gummel and K. Singhal, Inversion Charge Modeling, *IEEE Trans. Electron Dev.*, **48**(8), 1585–1592, 2001.

Problems

1. Calculate the inversion surface charge density, Q'_n, as a function of surface potential for depletion and inversion modes of operation using Equations (2.27), (2.28), (2.29), and (2.31), assuming substrate doping concentration $N_A = 10^{16}$ cm^{-3}, $T_{ox} = 5.0$ nm. Vary the surface potential up to 1.0 V in steps of 0.1 V. Compare Q'_n obtained for each of the above equations, and interpret the results.

2. Find the gate charge of a two terminal MOS capacitor structure on a p-substrate in the accumulation region in the following form:

$$Q'_g = -2C'_{ox}\phi_t W \left[\frac{\gamma}{2\phi_t} \exp\left(-\frac{V_{GB}}{2\phi_t} \right) \right] \tag{2.195}$$

Hint: If $xe^x = a$, then the Lambert function $W(a) = x$.

Show the variation of the small signal capacitance as a function of gate voltage operating the capacitor in the accumulation region.

3. Calculate the change in the depletion and inversion charge densities in the space charge region of the substrate of a MOS capacitor for a change of surface potential $\phi_s = 2\phi_f$ to $\phi_s = 2\phi_f + m\phi_t$ for $m = 6$. Given $N_A = 10^{16}$ cm^{-3} and $T_{ox} = 5$ nm. For a two terminal MOS capacitor estimate the variation in V_{SI} for $m = 2$, 4, and 6 where V_{SI} represents the gate voltage corresponding to strong inversion level.

4. Calculate the electric field $F_{Si}(0)$ at the Si–SiO$_2$ interface of a MOS capacitor at the threshold voltage, given that the substrate doping concentration $N_A = 10^{17}$ cm^{-3}. What is the field in the oxide? Calculate the threshold voltage for an oxide thickness of 100 nm.

5. (a) Given that $N_A = 10^{15}$ cm^{-3} and $T_{ox} = 0.05\,\mu$m, obtain values for V_{SI}, V_{MI}, and V_{WI} assuming $m = 6$.

   (b) Calculate the maximum depletion width at an inversion surface potential of $2\phi_f + 5\phi_t$ in a MOS capacitor for the substrate doping $N_A = 10^{16}$ cm^{-3}. Compare it with the value for $m = 0$. Interpret the result.

   (c) For a two terminal MOS structure calculate the inversion charge density for $N_A = 10^{16}$ cm^{-3}, $T_{ox} = 0.1\,\mu$m and $V_{GB} = 4.0$ V. Compare the ratio of Q'_n/Q'_b.

   (d) Compare the values of C'_{sc} with C'_{ox} for $\phi_s = -4\phi_t$ in the accumulation region assuming $T_{ox} = 0.05\,\mu$m and $N_A = 5 \times 10^{16}$ cm^{-3}.

6. Assuming a strong inversion condition, using Equation (2.33), show that the inversion layer thickness is inversely related to the surface inversion charge density.

7. Derive expressions for $dQ'_n/d\phi_s$ and $dQ_b/d\phi_s$ and show that they are equal at the surface inversion potential $\phi_s = 2\phi_f$.

8. Calculate C'_{bmin}/C'_{ox} at the threshold voltage for $T_{ox} = 0.03\,\mu$m and $N_A = 6.5 \times 10^{16}$ cm^{-3}.

9. Show that the substrate doping concentration of a MOS capacitor can be obtained from measurements of (i) maximum capacitor value (C_{ox}) in accumulation, and (ii) high

frequency capacitance (C_{inv}) in strong inversion $(\phi_s = 2\phi_f)$. Obtain an equation in the following form:

$$N_A = C' \frac{C_{inv}^2}{\left(1 - \frac{C_{inv}}{C_{ox}}\right)^2}$$

where C' is a constant, assuming that the capacitance has an area A.

10. Estimate the magnitude of flat band capacitance for an ideal MOS structure for a substrate doping concentration $N_A = 10^{15}$ cm^{-3}, and $T_{ox} = 0.2\,\mu$m.

11. How will Figure 2.16 be modified to show the values of F_{ox} when there is an inversion layer at the interface beyond the threshold in the case of overdrive $(V_{GB} - V_t)$. (Note: In the definition of threshold voltage the inversion charge is neglected compared to the depletion charge.) Given: $N_A = 10^{17}$ cm^{-3}, a gate voltage overdrive $(V_{GB} - V_T)$ of 1.0 V and $T_{ox} = 0.1\,\mu$m. Calculate $F_{Si}(0)$ and F_{ox}.

12. Figure 2.20 shows an ideal three terminal MOS capacitor with an n^+ diffusion probe constituting a sourced MOS capacitor. Calculate the depletion layer width under the gate and the diffusion region for the following case: the diffusion layer is connected to the gate which is biased at 10 V. Calculate the electric field strength in the gate oxide assuming $T_{ox} = 0.2\,\mu$m, and $N_A = 10^{15}$ cm^{-3}. Is there an inversion layer created?

13. Plot threshold voltage as a function of doping concentration $N_A = 10^{15}$, 10^{16} and 10^{17} cm^{-3} assuming the conventional definition $\phi_s = 2\phi_f$ and an improved definition given by Equation (2.83). It is established that the fitting parameter $\xi = 10$ gives the closest match to the experimental value. Plot the percentage error in the prediction of threshold voltage as a function of doping concentration that will be incurred from the conventional definition.

14. Derive the expression for threshold voltage given in Equation (2.165) relating pinch-off voltage with gate voltage explicitly from the implicit relation in Equation (2.164). Construct the potential balance diagram for the sourced MOS capacitor presented in Figure 2.16, linearizing the bulk charge around the pinch-off voltage. Now calculate the pinch-off voltage in terms of gate voltage V_{GB}, threshold voltage V_{T0}, and slope factor n, given by $(V_{GB} - V_{T0})/n$.

15. A step pulse $V_{GB}(t)$ is applied between the gate and bulk terminals of a MOS capacitor driving the capacitor into the deep depletion condition. Obtain expressions for the variation of surface potential and depletion width with time during the relaxation process caused by the thermal generation of electron–hole pairs in the depletion region. Find the time for the capacitor to reach the steady state condition. Show the energy band diagram of the capacitor for $t = 0^+$ when the step pulse is applied and under steady state condition.

3

Non-ideal and Non-classical MOS Capacitors

3.1 Introduction

The metal–oxide–semiconductor structure discussed in the previous chapter was considered to be ideal due to several assumptions. It was assumed that the work functions of gate and substrate materials are identical, the oxide is free of charges, and the substrate is uniformly doped. In practical MOS structures, there is deviation from such idealized assumptions due to the reasons outlined below.

- *Gate–substrate work function difference*. The work function of the gate material will, in general, not be equal to that of the substrate, as the Fermi level of the semiconductor depends on its conductivity and doping. The type of polysilicon conductivity and its doping in polysilicon gate technology is governed by technological considerations. The work function difference, $\phi_{gs} = \phi_g - \phi_{Si}$, disturbs the flat-band condition assumed for the ideal case [1].
- *Fixed charges in oxide*. SiO_2 is not free from structural defects, particularly near the SiO_2 interface. In a MOS structure, there is a transition region at the oxide–silicon interface where a fixed positive charge is observed to be located. The location and magnitude of this charge make it dominant in disturbing the flat-band condition. In fact, any charge that enters within the insulator, or is located at the interface, affects the flat-band condition.
- *Inhomogeneity of the substrate doping*. The substrate is considered homogeneously doped, a condition hardly satisfied in MOS device technology. The adjustment for threshold voltage is done by ion-implantation, which brings about inhomogeneity in doping near the interface where the inversion channel is formed [3]. To optimize the performance of MOS transistors there is a need of substrate doping engineering, which also leads to inhomogeneity.
- *Non-degenerate polysilicon doping*. As the device sizes are scaled down, the level of doping of polysilicon constituting the gate electrode needs to be reduced due to technological constraints. The poorly doped polysilicon gate is not an ideal conductor which leads to the formation of depletion space charge at the polysilicon–SiO_2 interface [5].

Compact MOSFET Models for VLSI Design   A.B. Bhattacharyya
© 2009 John Wiley & Sons (Asia) Pte Ltd

- *Non-classical structures*. As the scaling process continues in MOS technology with innovations for improved performance, the *classical* polysilicon gate–SiO$_2$–Si structure is being
  subjected to *gate engineering, substrate or channel engineering, contact engineering*, and
  structural engineering, requiring new materials and device configurations. These cause
  deviations from the ideal device assumed earlier. The use of a SiGe–Si heterostructure at
  the gate dielectric and silicon interface [6], a high-*k* and stacked dielectric [8] as a gate
  insulator material, and undoped silicon for the substrate form non-classical MOS structures [7]. Such structures are projected to replace and build on the original MOS device.
  The non-idealities and non-classical structures, which are an integral part of present day
  MOS structures, require generalization or modification of the theory developed for an ideal
  MOS device. The following sections discuss these aspects.

3.2 Flat-Band Voltage

The gate–substrate work function difference and charges within the gate oxide introduce deviations from the ideal conditions assumed earlier. A finite amount of gate voltage is required
to be applied to the gate, to restore the MOS structure to the ideal flat-band condition. Thus
the gate voltage required to create inversion, in a real world MOS structure, has an additional
component compared to the ideal structure. This additional gate voltage required to achieve the
flat-band condition is called the *flat-band voltage*. The value of the flat-band voltage depends
upon the amount and sign of the work function *difference*, and charge in the gate oxide.

3.2.1 Gate Substrate Work Function Difference $\phi_g \neq \phi_{Si}$

We shall discuss, for the sake of illustration, the case of a non-ideal MOS structure as shown
in Figure 3.1(a), where the gate material (n^+ degenerate polysilicon with Fermi level assumed
coinciding with the conduction band edge), is shown to have a work function *less* then that of
a *p*-type substrate. After the formation of a MOS capacitor with an oxide layer between the
gate and the substrate, in the equilibrium condition, the Fermi levels of the gate and substrate
get aligned and there will be a band bending at the Si–SiO$_2$ interface as shown in Figure3.1(b).
This is due to flow of electrons from the gate to the silicon. Due to the fact that $\phi_g < \phi_{Si}$, there is
a positive sheet of charge on the gate, and a negative space charge at the Si–SiO$_2$ interface due
to the net transport of electrons from the gate to the semiconductor. The above phenomenon
results in an electric field and potential drop across the oxide. The flat-band condition, the
condition of no band bending in the substrate, can be restored by applying a voltage at the gate
with respect to the substrate of magnitude ϕ_{gs} [1]. Thus, V_{fb1}, the flat-band voltage component
for compensating the bend bending due to the work function difference between the gate and
the substrate is given by $V_{fb1} = \phi_{gs}$ as shown in Figure 3.1(c).

We have considered in our illustration a particular case. In CMOS technology, the substrate
has both *n*- and *p*-type conductivities for realizing *p*- and *n*-channel transistors respectively.
Further, for controlling the threshold voltage to the desired values, use of dual gate technology
is common, where use of both *n*- and *p*-type polysilicon gates is necessary. Therefore, the flat-
band voltage sign and magnitude are dependent on substrate doping and choice of polysilicon
gate conductivity.

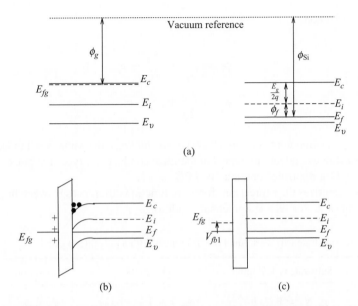

Figure 3.1 (a) Polysilicon (n^+) gate and p-silicon energy band diagrams before contact, (b) band diagram of MOS structure in equilibrium ($\phi_{gs} < 0$), and (c) MOS energy bend diagram with gate voltage $V_{GB} = V_{fb1}$

Normally, in MOS technology the heavily doped gate material is either n^+- or p^+-polysilicon. It is generally assumed that for the former case the Fermi level coincides with the conduction band edge, and for the latter it coincides with the valence band edge. It follows that:

- Gate: n^+-polysilicon

$$\phi_g = \frac{\chi_{Si}}{q} \tag{3.1}$$

  Substrate: p-silicon

$$\phi_{Si} = \frac{\chi_{Si}}{q} + \frac{E_g}{2q} + \phi_f \tag{3.2}$$

$$\phi_{gs} = (\phi_g - \phi_{Si}) = \phi_g - \left(\frac{E_g}{2q} + \phi_f + \phi_g\right) = -\left(\frac{E_g}{2q} + \phi_f\right) \tag{3.3}$$

- Gate: p^+-polysilicon

$$\phi_g = \frac{\chi_{Si} + E_g}{q} \tag{3.4}$$

Substrate: n-silicon

$$\phi_{Si} = \frac{\chi_{Si}}{q} + \frac{E_g}{2q} - \phi_f \tag{3.5}$$

$$\phi_{gs} = \frac{E_g}{2q} + \phi_f \tag{3.6}$$

In SPICE circuit simulators, for *polysilicon gate technology*, the parameter **TPG** takes care of the gate and substrate conductivity type. For identical conductivity types for gate and substrate, **TPG** = −1, and for opposite conductivity **TPG** = +1.

Table 3.1 summarizes the expressions for work function differences between the polysilicon gate and the substrate for selected practical conditions.

Table 3.1 Gate–substrate (Si) work function differences with a polysilicon gate

Gate	Substrate type/doping	Formula	Condition	ϕ_{gs}
n^+-poly-Si	p, $N_A = 10^{16}$ cm^{-3}	$\phi_{gs} = -\left(\frac{E_g}{2q} + \phi_f\right)$	$\phi_g < \phi_{Si}$	−0.9 V
n^+-poly-Si	n, $N_D = 10^{16}$ cm^{-3}	$\phi_{gs} = -\left(\frac{E_g}{2q} - \phi_f\right)$	$\phi_g < \phi_{Si}$	−0.2 V
p^+-poly-Si	n, $N_D = 10^{16}$ cm^{-3}	$\phi_{gs} = \left(\frac{E_g}{2q} + \phi_f\right)$	$\phi_g > \phi_{Si}$	+0.9 V

Figure 3.2 gives the plot of variation of gate–substrate work function difference for a p^+-polysilicon gate and n- and p-type substrates with doping concentrations varying from

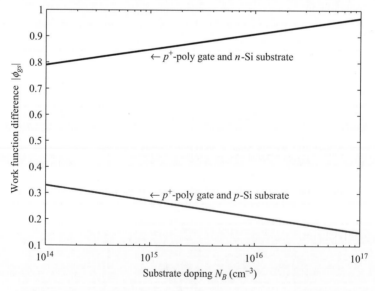

Figure 3.2 Variation of ϕ_{gs} with substrate doping N_B (n- and p-type) with degenerate p^+-polysilicon

10^{14} to $10^{17} cm^{-3}$. Similar plots can be created for an n^+-polysilicon gate as well [14]. As V_{fb1}, a component of threshold voltage, equals ϕ_{gs}, the discussion illustrates the option to select a proper gate material for meeting a specified threshold voltage.

The effect of the work function difference component on the flat-band voltage can be seen in the SCHRED [47] simulated $C-V$ plot given in Figure 3.3. In the given plot the substrate doping has been kept unchanged and the polysilicon gate has been changed from n^+- to p^+-type. As the position of the Fermi level moves from the edge of the conduction band edge for an n^+-polysilicon gate to the edge of the valence band edge for a p^+-polysilicon gate, the threshold voltage shifts by an amount equalling the silicon band gap of 1.1V. This example also illustrates that the proper selection of a gate material is an effective means of adjusting the threshold voltage in an MOS structure.

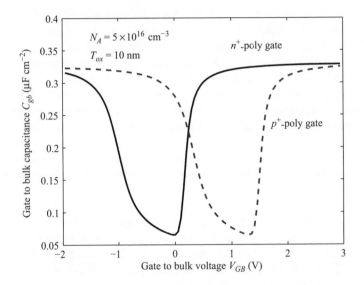

Figure 3.3   Shift in V_{fb} with n^+- and p^+-degenerate gate polysilicon gate

3.2.2 Oxide Charges

The insulating SiO_2 layer which is thermally grown on silicon at high temperature, being amorphous, is structurally not an ideal charge-free dielectric. It is found to have various types of charges [9] associated with it, as shown in Figure 3.4. The non-ideality due to charges has to be accounted for in the modeling of a MOS capacitor. The charges in the oxide will induce charges of opposite polarity in the substrate, and make the substrate deviate from the flat-band condition.

Comments

- **Fixed charge.** The number of fixed charge in the oxide, N_f, is minimum for the (100) orientation, and depends on oxidation temperature. It is very sensitive to processing conditions during the oxide growth. It is located within the oxide and very close to the interface. Therefore, it has a dominant effect on the threshold voltage. The process conditions are optimized to minimize its value.

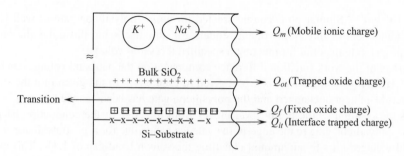

Figure 3.4   Location and nature of charges in SiO$_2$

- **Ionic charge.** It effects the device stability due to the migration of mobile ions in the oxide in the presence of a gate voltage induced electric field and operating temperature. Currently, the ionic charge effect is well controlled in MOS technology through mechanisms of growing oxide in ultra-clean ambiance, and adding ions such as chlorine or fluorine during oxidation to make nonionic complexes with sodium.
- **Interface charge.** Due to the trapping of carriers from the inversion layer it influences the value of transconductance and has significant effect on noise. Technologically, the interface charge is minimized significantly by low temperature annealing in the final stage of device processing.
- **Oxide trapped charges.** Electron–hole pairs generated by high energy radiation during processing of the device or otherwise may be partly trapped in the oxide giving rise to trapped charges.

Table 3.2 summarizes the basic features of thermally grown oxide charges.

Figure 3.5(a) shows the energy band diagram with fixed charge in the oxide located close to the oxide–silicon interface. The oxide charge causes a band bending at the interface, since a space charge is created due to mobile holes being pushed out exposing negative acceptor ions. In addition, minority carrier electrons also accumulate. The fixed acceptor negative ions and mobile electrons provide the negative charge on which the lines of force from the fixed positive charge terminate. We have assumed, for the sake of illustration, that all the lines of force from the fixed oxide charge terminate on the space charge near the interface because of its close proximity compared to the gate electrode. As a general case, however, part of the negative charge will be induced at the interface and part on the gate.

Figure 3.5(b) shows the energy band diagram at the flat-band condition with an externally applied gate voltage, V_{fb2}. In the flat-band condition, the voltage at the gate has to be of such a magnitude and polarity so that all the lines of force originating from the oxide charge terminate on the gate, and no space charge is created in the semiconductor. The balancing negative charge has to be concentrated only at the gate electrode, so that in the flat-band condition the semiconductor is electrically neutral throughout without any space charge. The external voltage applied to the gate is negative, given by:

$$V_{GB} = V_{fb2} = -\frac{Q'_f}{C'_{ox}} \tag{3.7}$$

Table 3.2 Charges in thermally grown oxide on silicon

Types of charges	Nature	Origin	Location
Fixed oxide charge, Q_f	• Positive • Does not communicate with silicon	• Structural defect at SiO_2 transition layer in silicon • Process dependent • $Q_f(111) > Q_f(100)$	Within 25 Å of interface in the oxide
Ionic charge, $Q_m = qN_m$	• Mobile and mostly positive • Does not communicate with silicon	• Process dependent • Ultra-clean processing and chlorine addition with oxygen counters the problem due to mobile ions	Distributed in the bulk
Interface charge, Q_{it}	• Transition from Si to SiO_2 lattice structure • Communicates with mobile carriers in silicon charge	• Dangling bonds at the Si–SiO_2 interface due to structural discontinuity	Si–SiO_2 interface
Oxide trapped charge, Q_{ot}	• Positive or negative depending on hole or electron trapped in the oxide	• Radiation, avalanche injection • Can be annealed out	Bulk oxide

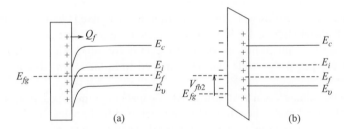

Figure 3.5 (a) Energy band diagram for a non-ideal capacitor with fixed oxide charge in the oxide near the interface, and (b) flat-band voltage component for a fixed charge at oxide–silicon interface

In the above case, while discussing the band bending in silicon due to oxide charge, it is assumed that $\phi_{gs} = 0$. Now, considering a general and practical case where both fixed oxide charge and gate–substrate work function difference are present, the voltage required to be applied to the gate for establishing a flat-band condition is given by:

$$V_{fb} = V_{fb1} + V_{fb2} = \phi_{gs} - \frac{Q'_f}{C'_{ox}} \tag{3.8}$$

In MOS technology, the interface charge and fixed oxide charges are now well controlled due to maturity developed in silicon-dioxide technology. The assumption that the charges can be considered to be concentrated at the interface is quite valid. But such an assumption may not hold good for modifications being introduced in the gate oxide for improved performance. For instance, introduction of a controlled amount of nitrogen improves performance by resisting boron penetration through the oxide from the p-polysilicon gate [4]. In such a case we shall assume a distribution for the fixed charge within the oxide. Incorporation of nitrogen is believed to introduce a distributed positive charge in the SiO_2 structure. We shall obtain below a general expression for flat-band voltage considering a generalized distribution of positive charge in the oxide.

Let us assume that the oxide charge density at any point x in the oxide is $q\rho(x)$, considering the gate terminal as the origin. Further, we assume that $\rho(x)\,dx$ constitutes the elemental charge density located at x. As we are considering the flat-band condition we shall assume that all the field lines due to oxide charge terminate on the gate for which we are to calculate the necessary amount of negative charge to be developed at the gate by the application of an external bias. The elemental charge requires an equal amount of negative charge on the gate terminal across the capacitor $C'_{ox}(x) = \epsilon_{ox}/x$. The gate voltage dV_{fb} provides the required charge $-C'_{ox}(x) \times dV_{fb}$.

The total voltage for the entire charge distribution can be obtained by integrating the contribution of the elemental charges, which gives:

$$V_{fb2} = -\frac{1}{C'_{ox}} \int_0^{T_{ox}} \frac{q\,x\,\rho(x)}{T_{ox}}\,dx \qquad (3.9)$$

It may be mentioned that V_{fb2} is always negative, whereas V_{fb1} will have a polarity depending on whether ϕ_g is greater or less than ϕ_{Si}.

The non-idealities, such as work function difference and fixed charge, result in the offset of the C–V characteristics of a MOS capacitor from the ideal characteristics as shown in Figure 3.3. Since V_{fb} can be measured experimentally from the shift of a real MOS capacitor, the fixed charge on the oxide can be estimated from the knowledge of work function difference.

The threshold voltage for a real world MOS capacitor is modified from the ideal case discussed in Chapter 2. Now, the effect of work function difference and oxide charge are to be included. Thus, we can rewrite the expression for threshold voltage as:

$$V_{T0} = V_{fb} + 2\phi_f + \gamma\sqrt{2\phi_f} \qquad (3.10)$$

It may be noted that in Chapter 2 we have not considered the flat-band voltage while writing the equation for potential balance. Whenever the flat-band voltage is not considered it will be assumed that if V_{GB} is replaced by $(V_{GB} - V_{fb})$, the potential balance equation will be valid when $V_{fb} \neq 0$.

3.3 Inhomogeneous Substrate

Localized manipulation of doping is a powerful tool available to device designers for manipulating the device property and characteristics. In MOS technology, an inhomogeneous substrate is now an integral part of the MOS device which has been introduced in pursuit of improving various aspects of device performance [3]. It is appropriate to take note of some of the implications of inhomogeneity that are related to device physics as compared to homogeneous substrate.

- The inhomogeneity is associated with a built-in field.
- The quasi-neutrality at any point, in the sense it is understood for the homogeneous substrate, can be maintained for a doping gradient less than a specific value which is functionally related to some power of doping concentration itself [10].
- For the definition of inversion or threshold voltage the average impurity concentration in the depletion region needs to be considered [2, 11, 16].

The inhomogeneity in the substrate, caused by the channel doping profile, is approximated to the following categories from a modeling viewpoint [12]:

  (i) Step profile: abrupt high–low doping.
 (ii) Step retrograde: abrupt low–high doping.
(iii) Graded retrograde: gradual low–high transition.

Ion implantation is an integral part of modern MOS technology used extensively for threshold voltage adjustment. The characteristic features of ion implantation technology are:

- A controlled amount of ions can be introduced into the substrate near the Si–SiO$_2$ interface. The amount, depth, and type of impurities (n or p) are determined by the target specification for the threshold voltage.
- The implantation can be carried out through the overlaying oxide film.
- The implantation depth can be adjusted by the energy of the implanted ions.
- The total number of implanted dopant D_I (cm^{-2}) is given by the product of the incident ion flux (cm^{-2} s^{-1}), and the duration of the implantation. D_I can be controlled with a high degree of precision.
- The implanted impurity profile is given by a Gaussian distribution which is often approximated to a box profile for the purpose of simplified modeling, as shown in Figure 3.6.

3.3.1 Types of Implant

We shall analyze two broad categories of substrate inhomogeneities resulting due to ion implantation.

3.3.1.1 Type I: Shallow Implant

The implant depth W_I is such that the depletion layer depth W_{dm} is greater than W_I for $V_{GB} = V_{TO}$ and $V_{SB} = 0$, as shown in Figure 3.7.

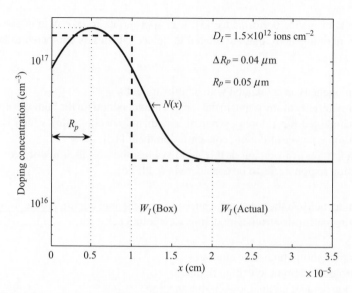

Figure 3.6   Approximated box distribution of an implanted Gaussian profile

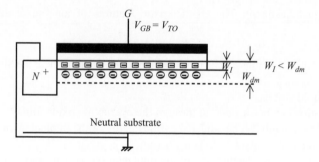

Figure 3.7   Shallow implant $W_I < W_{dm}$ ($V_{SB} = 0$)

3.3.1.2 Type II: Deep Implant

The implant depth W_I is greater than the maximum depletion width at $V_{GB} = V_{TO}$, but W_d crosses W_I for V_{SB} greater than V_{CR}, where V_{CR} is the value of V_{SB} at which the depletion depth coincides with the W_I under inversion condition. Figure 3.8 shows the stated conditions.

For obtaining the expression for the threshold voltage of a MOS capacitor with an inhomogeneous substrate of Type I, we need to solve Poisson's equation with appropriate boundary condition as outlined below. We shall consider high–low type inhomogeneity for the sake of illustration as shown in Figure 3.9, where the implanted region has higher concentration compared to the substrate.

Poisson's equation has to be solved considering the volume charge density of qN_S in Region I, and qN_A in Region II. The boundary conditions for solving Poisson's equation are outlined below.

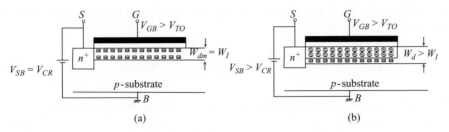

Figure 3.8 (a) The externally applied voltage to the channel for which $W_{dm} = W_I$ ($V_{SB} = V_{CR}$), and (b) the depletion depth for $V_{SB} > V_{CR}$

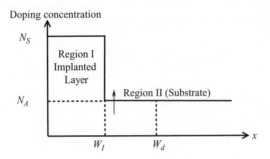

Figure 3.9 High–low doping profile of an inhomogeneous substrate

- Potential $\phi(x)$ and electric field $F(x)$ are zero at $x = W_d$, and $\phi(x) = \phi_s$ at $x = 0$.
- ϕ and F are continuous at $x = W_I$, and continuity of the displacement vector at the Si–SiO$_2$ interface gives $\epsilon_{ox} F_{ox} = \epsilon_{Si} F_{Si}|_{x=0}$.

For Region I, solving Poisson's equation with the above boundary conditions gives:

$$F(x) = \frac{qN_A}{\epsilon_{Si}}(W_d - W_I) + \frac{qN_S(W_I - x)}{\epsilon_{Si}} \tag{3.11}$$

$$\phi(x) = \phi_s - \frac{qx}{\epsilon_{Si}}\left[N_A(W_d - W_I) + N_S\left(W_I - \frac{x}{2}\right)\right] \tag{3.12}$$

In Region II:

$$F(x) = \frac{qN_A}{\epsilon_{Si}}(W_d - x) \tag{3.13}$$

$$\phi(x) = \frac{qN_A}{2\epsilon_{Si}}(W_d - x)^2 \tag{3.14}$$

The surface potential ϕ_s can be obtained from the continuity of the potential ϕ_i at $x = W_I$, which gives:

$$\phi_s = \frac{qN_S}{2\epsilon_{Si}}W_I^2 + \frac{qN_A}{2\epsilon_{Si}}(W_d^2 - W_I^2) \tag{3.15}$$

The depletion layer width can be directly obtained from the above equation as:

$$W_d = \sqrt{W_{du}^2 - \frac{N_S - N_A}{N_A}W_I^2} \tag{3.16}$$

where $W_{du} = \sqrt{2\epsilon_{Si}\phi_s/qN_A}$ is the depletion width for the non-implanted substrate. For $W_I \ll W_{du}$, Equation (3.16) can be expanded binomially to give:

$$W_d = W_{du}\left[1 - \frac{q(N_S - N_A)}{4\epsilon_{Si}\phi_s}W_I^2\right] \tag{3.17}$$

The depletion charge density can be written as:

$$Q_b' = -\left[q(N_S - N_A)W_I + qN_A W_d\right] \tag{3.18}$$

Substituting for W_d from Equation (3.16) in the expression for Q_b' in Equation (3.18), we have:

$$Q_b' = -\left\{q(N_S - N_A)W_I + qN_A\left[W_{du}^2 - \left(\frac{N_S - N_A}{N_A}\right)W_I^2\right]^{\frac{1}{2}}\right\} \tag{3.19}$$

which simplifies to:

$$Q_b' = -\left[qD_I + \sqrt{2q\epsilon_{Si}N_A}\left(\phi_s - \frac{qD_I W_I}{2\epsilon_{Si}}\right)^{\frac{1}{2}}\right] \tag{3.20}$$

where $D_I = (N_S - N_A)W_I$ is the implant dose (cm^{-2}). The expression for threshold voltage is obtained by substituting $\phi_s = 2\phi_f$ and $Q_b' = Q_{bm}'$ in Equation (2.86), which gives:

$$V_{TO} = V_{FB} + 2\phi_f + \frac{1}{C_{ox}'}\left[qD_I + \sqrt{2q\epsilon_{Si}N_A}\left(2\phi_f - \frac{qD_I W_I}{2\epsilon_{Si}}\right)^{\frac{1}{2}}\right] \tag{3.21}$$

The expressions obtained for a high–low step profile are valid for a low–high retrograde profile with the difference that for the later case $N_S < N_A$.

Figure 3.10 shows plots of electric field and potential with depth for a high–low inhomogeneous substrate with a box approximated impurity distribution. It is observed that in the highly doped implanted region (Region I), the slope of the electric field is higher compared to that in

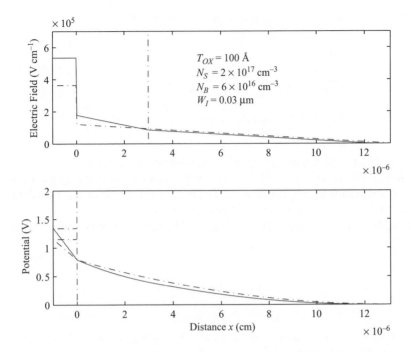

Figure 3.10 Field and potential variations for a high–low substrate: solid lines, implanted; dotted lines, unimplanted

the substrate. This results in an increase in the value of the electric field at $x = 0$, the Si–SiO$_2$ interface, compared to that in a uniformly doped substrate. The electric field in the oxide is $F_{ox} = (\epsilon_{Si}/\epsilon_{ox})F(0)$, where $F(0)$ is the electric field at $x = 0$ in silicon. The potential drop across the oxide, related to the field at the interface $F_{Si}(0)$, is given by:

$$V_{ox} = F_{ox}T_{ox} = \frac{\epsilon_{Si}}{\epsilon_{ox}}F_{Si}(0)T_{ox} \tag{3.22}$$

At the threshold we have $\phi_s = 2\phi_f$, which is marginally affected by doping, being logarithmically dependent. Hence, a high–low substrate results in a higher voltage drop across the oxide compared to the unimplanted substrate due to an increased field at the interface. Consequently, the threshold voltage, which contains the term due to voltage across the oxide, is also increased. The area under the solid line in the F–x plane in Figure 3.10 denotes V_T for the implanted case, and that with a discontinuous line shows V_T for the unimplanted case. A significant observation can be made that for a very shallow implant the depletion width is not affected significantly, which can be inferred from Equation (3.17) assuming $W_I \rightarrow 0$.

We shall now consider two typical commonly used cases with respect to implant depth.

Case I: *Impulse implant* ($W_I \rightarrow 0$).   The expressions for the depletion width W_d, bulk charge density Q_b, and threshold voltage V_{T0}, respectively are obtained from Equations (3.16), (3.20), and (3.21) corresponding to the impulse implant as given below:

$$W_d = W_{dui} \tag{3.23}$$

$$Q_b' = -(qD_I + \sqrt{2q\epsilon_{Si}N_A}\sqrt{\phi_s}) \tag{3.24}$$

$$V_{TO} = V_{fb} + 2\phi_f + \gamma\sqrt{2\phi_f} + \frac{qD_I}{C_{ox}'} = V_{TOu} + \frac{qD_I}{C_{ox}'} \tag{3.25}$$

where V_{TOu} denotes the threshold voltage for an unimplanted substrate.

Equation (3.25) shows that the threshold voltage with a delta doped implantation can be adjusted by controlling the implant dose D_I. Depending on the type of ion, D_I will be positive or negative. For a p-type dopant, D_I is taken as positive.

Case II: *Implant with finite depth* $(W_I \ll W_{dm})$.   The threshold voltage after implant can be written as:

$$V_{TO} = V_{TOu} - \frac{\Delta Q_b'}{C_{ox}'} \tag{3.26}$$

where $\Delta Q_b' = Q_b' - Q_{bu}'$, and $Q_{bu}' = -qN_A W_{du}$. Using Equations (3.17) and (3.18) for W_d and Q_b', with $W_{du} = \sqrt{2\epsilon_{Si}\phi_s/qN_A}$, we get:

$$V_{TO} = V_{TOu} + \frac{q(N_S - N_A)W_I}{C_{ox}'}\left(1 - \frac{W_I}{2W_{du}}\right) \tag{3.27}$$

Figure 3.11 show the variation of $\Delta V_{TO} = V_{TO} - V_{TOu}$ as per Equation (3.27), with implant depth W_I as a parameter. The curves indicate that the delta implant approximation becomes more accurate as $W_I \to 0$.

Figure 3.12 shows the effect of implant doping on the maximum depletion width in the substrate at the threshold voltage ($\phi_s = 2\phi_f$). Equation (3.27) is also valid with substrate bias, for which we are to substitute $2\phi_f + V_{SB}$ in place of $2\phi_f$.

Figure 3.13 shows the variation of threshold voltage with substrate bias for the following cases:

Case 1. Homogeneous substrate with doping concentration N_A (slope of V_T versus V_{SB} proportional to N_A).

Case 2. Homogeneous substrate with high doping concentration N_S.

Case 3. Delta implantation with substrate concentration N_A. Parallel shift of V_T from Case 1 by qD_I/C_{ox}'.

Case 4. High–low inhomogeneous substrate with implant depth W_I and corresponding critical voltage V_{CR}. Slope for $V_{SB} < V_{CR}$ is proportional to N_S, and that for $V_{SB} > V_{CR}$ is proportional to N_A.

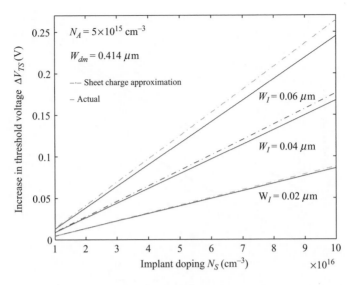

Figure 3.11   Increase in threshold voltage as a function of implant doping with depth as a parameter. The dotted line corresponds to the delta implant approximation

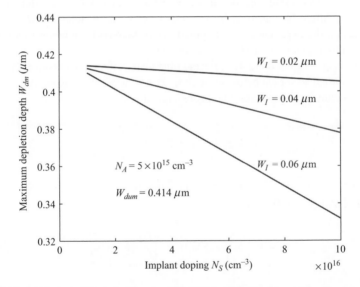

Figure 3.12   Variation of depletion layer width with implant doping and depth as a parameter

3.3.2 Pinch-off Voltage for an Inhomogeneous Substrate

We have obtained an expression for the pinch-off voltage in Chapter 2. The pinch-off voltage being an important parameter in EKV and ACM models, it will be useful to have its expression for an inhomogeneous substrate.

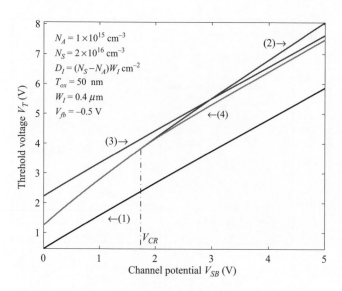

Figure 3.13   Variation of threshold voltage for cases 1, 2, 3 and 4

For the high–low substrate considered, we define a gate voltage V_{GBcr}, that corresponds to the pinch-off condition for substrate bias $V_{SB} = V_{CR}$. V_{GBcr} denotes the *upper limit of gate voltage* for which the expression for the pinch-off voltage corresponding to a homogeneous substrate given by Equation (2.163) holds good with $\gamma = \gamma_s$.

For $V_{GB} > V_{GBcr}$, Equation (2.163) for the pinch-off voltage obtained for uniform doping, needs to be modified as shown below to include the effect of inhomogeneity.

From the potential balance across the MOS capacitor:

$$V_{GB} = V_{SB} + V_{fb} + 2\phi_f - \frac{Q_b'}{C_{ox}'} - \frac{Q_n'}{C_{ox}'} \tag{3.28}$$

Now substitute for Q_b' from Equation (3.24) in Equation (3.28). Also at pinch-off $V_{SB} = V_P$, $\phi_s = \phi_0' + V_P$, and $Q_n' \to 0$. Thus:

$$V_{GB} = V_P + V_{fb} + 2\phi_f + \frac{qD_I}{C_{ox}'} + \gamma\sqrt{\phi_0' + V_P} \tag{3.29}$$

where $\phi_0' = 2\phi_0 - qD_IW_I/\epsilon_{Si}$. Solving for V_P, and retaining the physically acceptable root, we get for $V_{GB} > V_{GBcr}$:

$$V_P = V_{GB} - V_{T0}' - \gamma\left\{\left[V_{GB} + \phi_0' - V_{T0}' + \left(\frac{\gamma}{2}\right)^2\right]^{\frac{1}{2}} - \frac{\gamma}{2}\right\} \tag{3.30}$$

where $V_{T0}' = \phi_0 + qD_I/C_{ox}'$.

The slope factor n for an inhomogeneous substrate as a function of V_{GB} is obtained by differentiating Equation (3.30) w.r.t. V_{GB}, which gives:

$$\frac{1}{n} = 1 - \frac{\gamma}{2\left[V_{GB} - V'_{T0} + \phi'_0 + \left(\frac{\gamma}{2}\right)^2\right]^{\frac{1}{2}}} \tag{3.31}$$

Figure 3.14 shows the dependence of pinch-off voltage and Figure 3.15 shows the variation of slope factor n with gate voltages for homogeneous and inhomogeneous substrates.

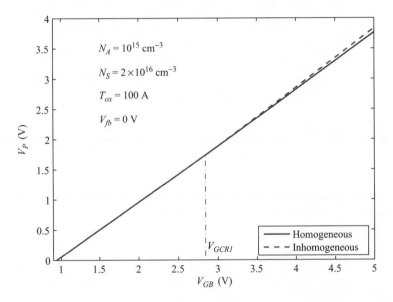

Figure 3.14 Pinch-off voltage for homogeneous and high–low inhomogeneous substrate

3.3.3 Vertical Substrate Inhomogeneity in Compact Models

Instead of using doping concentrations for implanted and substrate layers separately to obtain region-based solutions, an alternative approach has been in practice to obtain a transformed equivalent homogeneous doping concentration which gives identical results [13].

The transformation is based on the following considerations.

1. The total charge under the gate is conserved, which implies:

$$qN_E W_E = qN_S W_I + qN_A(W_d - W_I) \tag{3.32}$$

where W_E is the equivalent depletion layer width given by $W_E = \sqrt{2\epsilon_{Si}\phi_s/qN_E}$.

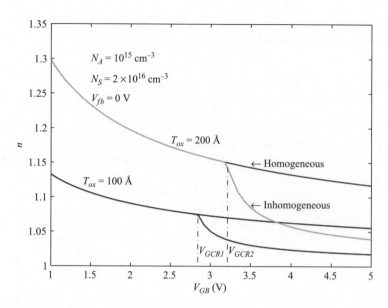

Figure 3.15 Variation of the slope factor n with gate voltage for homogeneous and nonhomogeneous substrates

2. Applying Poisson's equation and the boundary condition outlined in Section 3.3.1, the equivalent doping concentration can be shown to be:

$$N_E = N_S \frac{\phi_i}{\phi_s} \left\{ \left[1 - \frac{N_A}{N_S} + \frac{N_A}{N_S} \sqrt{1 + \frac{N_S}{N_A} \left(\frac{\phi_s}{\phi_i} - 1 \right)} \right] \right\}^2 \qquad (3.33)$$

where $\phi_i = q N_S W_I^2 / 2\epsilon_{Si}$, and stands for the surface potential at which the depletion layer depth coincides with the implant depth.

With the above transformation the threshold voltage for an inhomogeneous implanted substrate approximated to a box distribution can be given by:

$$V_{TO} = V_{fb} + 2\phi_f + \gamma_E \sqrt{\phi_0} \qquad (3.34)$$

where $\gamma_E = \sqrt{2q\epsilon_{Si}N_E}/C'_{ox}$.

There are several transformations, the broad features of which are outlined below:

- The bias dependent effective doping replacing an implanted Gaussian profile has been simple in approach but computationally involved [21].
- The transformation scheme where the vertical inhomogeneity is coupled with short channel effects is too complex relatively to be suitable for use in circuit simulators [19].
- EKV makes use of the relatively simple model proposed by Arora for threshold voltage and slope factor calculation, where the retrograde profile is approximated to a step

function [13, 20]. It has intrinsic simplicity built-in because the Poisson's equation is solved for respective regions, each of which is uniformly doped. The model uses an equivalent doping N_E, replacing the inhomogeneous substrate. It has been developed mostly for the prediction of substrate bias dependence on threshold voltage. Incorporated in circuit simulators, such a model encounters a discontinuity problem at the high–low transition point. The problem of continuity has been addressed in EKV literature [20] by modifying Arora's model to ensure a smooth transition near the region of discontinuity. The modified EKV doping is given as:

$$N_{new}(V_C) = \frac{N_s}{1 + \frac{1}{2}(G + \sqrt{G^2 + 4\epsilon^2})} \tag{3.35}$$

where:

$$G = \alpha \left(\frac{V_C}{V_{CR}} - 1 \right); \alpha \propto \sqrt{\frac{N_A}{N_S}} \tag{3.36}$$

and ϵ is a constant, the value of which is to be adjusted to ensure smooth transition.

- In BSIM3v3 the threshold voltage is a very important parameter to model the MOS transistor. The body effect factor γ is a constant quantity in an homogeneously doped substrate. In an inhomogeneous substrate, doping not being uniform in the vertical direction, γ becomes a function of substrate bias. BSIM3v3 approximates the implanted impurity distribution with a simplified box profile with a height representing an equivalent peak concentration N_{CH} and a depth X_T. It provides a simplified expression for the threshold voltage over a range of substrate bias with fitting parameters to be extracted [2]:

$$V_{TS} = V_{T0} + K_1 \left(\sqrt{\phi_s - V_{BS}} - \sqrt{\phi_s} \right) - K_2 V_{BS} \tag{3.37}$$

where K_1 and K_2 are model parameters. V_{T0} is the threshold voltage for $V_{SB} = 0$. In the absence of measured values for the model parameters, analytical expressions for the model parameters can be obtained assuming a box profile.

- PSP introduces the concept of gate voltage dependent doping $N(V_{GB})$, on the basis that an increase in gate voltage results in an increase in depletion width. In a retrograde profile, the increase in depletion depth also implies an increase in doping concentration at the edge of the depletion region. Thus, position dependent doping concentration, characterizing a retrograde profile, can be alternatively represented by gate voltage dependent substrate doping. An expression proposed is [22]:

$$N_{sub} = N_0 \left[1 + D_N f_\epsilon (V_{GB} - V_N) \right] \tag{3.38}$$

where:

$$f_\epsilon(w) = \frac{(w + \sqrt{w^2 + \epsilon})}{2} \tag{3.39}$$

and N_0, D_N, V_N, and ϵ are model parameters.

3.4 Polysilicon Depletion Effect

In an ideal MOS capacitor it was assumed that the gate electrode, being highly doped polysilicon, can be considered to be an ideal conductor like a metal. For a MOS structure with n-type polysilicon, the positive charge on the gate is provided by the depleted donor ions at the polysilicon–oxide interface. With the high doping of polysilicon the depletion width W_{pd} is negligible, and the voltage drop across it can be neglected. The charge in the depletion layer provides the gate charge which balances the charge in the substrate, if oxide charge is neglected.

In modern MOS technology, where the gate oxide is very thin, doping of polysilicon by ion implantation does not allow very high concentrations of doping to make it degenerate. The non-degenerately doped polysilicon gives rise to a depletion layer at the polysilicon–SiO_2 interface of sufficient width, across which the potential drop cannot be neglected. The significant potential drop across the depleted polysilicon reduces the effective gate voltage as compared to a MOS capacitor with an ideal metal gate [15]. The depletion layer in the polysilicon acts as a voltage dependent capacitor of value $C'_p = \epsilon_{Si}/W_{dps}$ in series with the ideal MOS equivalent capacitor discussed in Chapter 2. This series combination of C'_p with oxide capacitance decreases the total gate capacitance, which results in a decrease in induced inversion charge at the Si–SiO_2 interface for a given gate voltage. Figure 3.16 shows (a) the MOS capacitor structure with a non-degenerately doped polysilicon gate, and (b) the corresponding band diagram. The symbols used have the following meaning: ϕ_p, potential drop at the polysilicon–SiO_2 interface for applied voltage V_{GB}; W_{dps}, depletion charge width at the polysilicon–SiO_2 interface; N_P, doping concentration (donor) in the polysilicon gate. The discussion on the polydepletion effect follows the approach proposed in Sallese *et al.* [18].

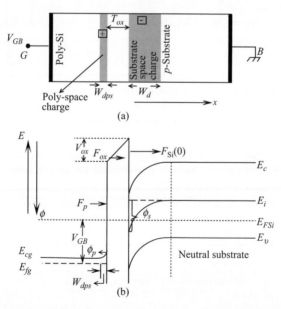

(a)

(b)

Figure 3.16 (a) Polysilicon–SiO_2–Si MOS structure, and (b) energy band diagram at inversion with band bending in polysilicon

The potential drop ϕ_p across the polysilicon can be obtained assuming a depletion condition in the polysilicon. This means that the gate to bulk potential is such that there is no possibility of inversion in polysilicon, which is generally satisfied in the normal operating condition. From the depletion approximation we can write the depletion charge density in polysilicon as:

$$Q'_p = C'_{ox} \gamma_p \sqrt{\phi_p} \tag{3.40}$$

where:

$$\gamma_p = \sqrt{2q\epsilon_{Si}N_P}/C'_{ox} \tag{3.41}$$

The polysilicon potential drop is simply written as:

$$\phi_p = V_{GB} - \phi_s - V_{ox} \tag{3.42}$$

The potential drop across oxide is $V_{ox} = Q'_p/C'_{ox}$. Substituting the expression for V_{ox} in Equation (3.42) using Equation (3.40), we get:

$$\gamma_p \sqrt{\phi_p} = V_{GB} - \phi_s - \phi_p \tag{3.43}$$

The above equation is quadratic in $\sqrt{\phi_p}$, whose solution retaining the physical root is:

$$\phi_p = \left[\sqrt{\frac{\gamma_p^2}{4} + (V_{GB} - \phi_s)} - \frac{\gamma_p}{2} \right]^2 \tag{3.44}$$

The increase in V_{GB} increases the space charge in the bulk, which requires a higher Q'_p in polysilicon, increasing the ϕ_p as given by the above relation.

3.4.1 Effect on Threshold Voltage

We shall obtain below the expression for the threshold voltage with the polydepletion effect using the conditions of potential balance, charge neutrality, and the assumption that at threshold voltage the inversion charge density is zero.

From the potential balance condition across the MOS structure, we have:

$$V_{GB} = \phi_p + V_{ox} + \phi_s \tag{3.45}$$

where we have neglected V_{fb}, the effect of which can be included in the expression for threshold voltage by substituting $V_{GB} - V_{fb}$ for V_{GB}. The condition for charge balance gives:

$$Q'_g = -(Q'_b + Q'_n) \tag{3.46}$$

When the gate is positively biased the positive charge on the gate is supported by the depletion charge due to the donor ions at the polysilicon–SiO$_2$ interface of an n^+ -polysilicon gate MOS structure. The potential drop ϕ_p across the depleted polysilicon is obtained from the relation:

$$Q'_g = \gamma_p C'_{ox} \sqrt{\phi_p} = V_{ox} C'_{ox} \tag{3.47}$$

where $\gamma_p = \sqrt{2q\epsilon_{Si} N_P}/C'_{ox}$ or $\phi_p = Q'^2_g/\gamma^2_p C'^2_{ox}$. Equivalently, ϕ_p can be expressed in terms of charges in the substrate using Equations (3.46) and (3.47), which gives:

$$\phi_p = \frac{(Q'_n + Q'_b)^2}{\gamma^2_p C'^2_{ox}} \tag{3.48}$$

Substituting the value of ϕ_p and V_{ox} in Equation (3.45):

$$V_{GB} = \frac{(Q'_n + Q'_b)^2}{\gamma^2_p C'^2_{ox}} - \frac{(Q'_n + Q'_b)}{C'_{ox}} + \phi_s \tag{3.49}$$

At the threshold voltage, $V_{GB} = V_{TP0}$, $\phi_s = \phi_0$, $Q'_b = Q'_{bm}$, and $Q'_n \rightarrow 0$, which gives:

$$V_{TP0} = \frac{Q'^2_{bm}}{\gamma^2_p C'^2_{ox}} - \frac{Q'_{bm}}{C'_{ox}} + \phi_0 \tag{3.50}$$

or:

$$V_{TP0} = \frac{\gamma^2}{\gamma^2_p}\phi_0 - \frac{Q'_{bm}}{C'_{ox}} + \phi_0 \tag{3.51}$$

It may be noted that due to the polysilicon depletion effect, the threshold voltage is increased by an amount $(\gamma^2/\gamma^2_p)\phi_0$ with respect to the value without a polydepletion effect, i.e. when the polysilicon gate is highly degenerate (when γ_p is large).

Figure 3.17 shows graphically the effect of the polydepletion effect on the threshold voltage. The diagram on the left shows the potential distribution in a MOS capacitor in the presence of the polydepletion effect. The triangle ABC represents the potential drop ϕ_p. The rectangular area signifies V_{ox}, while ϕ_0 represents the surface inversion potential at inversion. V_{TP0}, the gate voltage at threshold, is given by the sum of the areas mentioned. The diagram on the right shows the effect of inversion charge at the silicon–oxide interface. The inversion charge increases the interface field, $F_{Si}(0)$, from E to F', and balancing polysilicon field, F_{pi}, from C to C'. This results in an increase in the oxide field by the height PQ, and polysilicon depletion width by BB'. In effect, the inversion charge leads to a higher potential drop across the oxide and polydepletion layer.

Figure 3.17(a) shows further the electric field and potential drops in silicon and oxide at the threshold condition. The inversion charge increases the surface electric field, $F_{Si}(0)$, from the threshold value E to a higher value F' without adding to the area significantly as the line EF' does not add to the value ϕ_0 as shown in Figure 3.17(b). On the other hand, the width of the polydepletion layer AB' increases from the threshold value AB by an amount BB' increasing

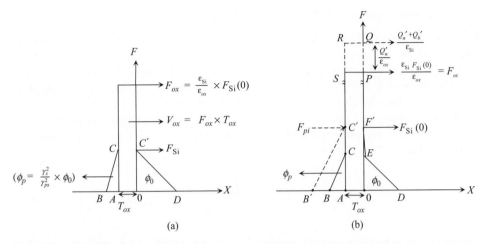

Figure 3.17 Graphical representation of the polydepletion effect (a) at the threshold and (b) in strong inversion

the area on polysilicon by an amount $BB'CC'$. The figure shows the changes that take place in the electric field in the oxide and silicon–oxide interface for the strong inversion condition. Figure 3.18 shows the effect of the polydepletion effect on the threshold voltage. It is noted that for the oxide thickness and substrate doping considered, the polydepletion effect becomes noticeable for polysilicon concentrations below 5×10^{19} cm^{-3}. When the polysilicon is highly degenerate, i.e. $\gamma_p \gg \gamma$, then $V_{TP0} = V_{T0}$. When T_{ox} increases γ/γ_p remains invariant. The polydepletion effect is less pronounced because an increase in T_{ox} results in a higher V_T and the relative change due to the polydepletion effect is small.

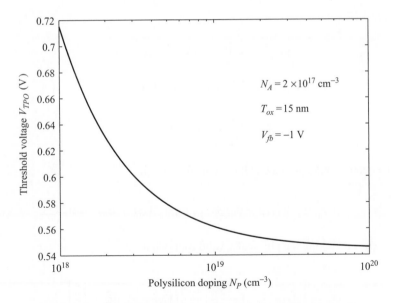

Figure 3.18 Threshold voltage dependence on polydepletion doping

3.4.2 Inversion Charge with Polydepletion Effect

When the gate voltage is greater than V_{TP0}, an inversion layer is created at the substrate–SiO$_2$ interface. We cannot, however, use the conventional formula relating inversion charge density and overdrive, i.e. we cannot assume that $Q'_n = -C'_{ox}(V_{GB} - V_{TB})$. It is because when the inversion charge is created, in order to satisfy the condition of charge balance, the gate charge has to increase. The gate charge is provided by the depletion charge in the polysilicon gate. The polydepletion charge under the inversion condition is more than that at the threshold condition.

The width of the depletion region in polysilicon in Figure 3.17 corresponding to a potential drop ϕ_p in this region is given by:

$$W_{dps} = \sqrt{\frac{2\epsilon_{Si}\phi_p}{qN_p}} \tag{3.52}$$

The field at the polysilicon–oxide interface, F_{pi}, can be obtained from the continuity of displacement vectors given by:

$$F_{pi} = \frac{\epsilon_{ox}F_{ox}}{\epsilon_{Si}} \tag{3.53}$$

Using Equation (3.53), and getting the field in the oxide from the potential balance condition across the MOS capacitor, we have:

$$F_{pi} = \left(\frac{V_{GB} - \phi_p - \phi_o}{T_{ox}}\right)\frac{\epsilon_{ox}}{\epsilon_{Si}} \tag{3.54}$$

From Figure 3.17:

$$\gamma_p\sqrt{\phi_p} = V_{ox} = \left(V_{GB} - \phi_o - \phi_p\right) \tag{3.55}$$

From the relation given in Equation (3.55), it follows that:

$$\phi_p = \frac{1}{\gamma_p^2}\left(V_{GB} - \phi_o - \phi_p\right)^2 \tag{3.56}$$

The above equation reduces to a quadratic equation for ϕ_p:

$$\phi_p^2 - \left[2(V_{GB} - \phi_o) + \gamma_p^2\right]\phi_p + (V_{GB} - \phi_o)^2 = 0 \tag{3.57}$$

Solving for ϕ_p, and taking the physically valid root, we have:

$$\phi_p = \left[(V_{GB} - \phi_o) + \frac{\gamma_p^2}{2}\right] - \left[(V_{GB} - \phi_o)\gamma_p^2 + \frac{\gamma_p^4}{4}\right]^{\frac{1}{2}} \tag{3.58}$$

For a three terminal MOS capacitor with substrate bias V_{SB}, replace ϕ_0 by $\phi_0 + V_{SB}$ in the above expression.

The inversion charge density Q'_n in the presence of the polydepletion effect is:

$$Q'_n = -C'_{ox}(V_{GB} - \phi_p - \phi_s - \gamma\sqrt{\phi_s}) \tag{3.59}$$

Substituting for ϕ_p, we have:

$$Q'_n = -C'_{ox}\left\{-\gamma\sqrt{\phi_s} + \frac{\gamma_p^2}{2}\left[\sqrt{1 + \frac{4(V_{GB} - \phi_s)}{\gamma_p^2}} - 1\right]\right\} \tag{3.60}$$

which converges to the equation for inversion charge without polydepletion effect, when the polysilicon doping is ideally infinite ($\gamma_p \to \infty$). This requires Binomial expansion of the square root term in Equation (3.60). Retain only the first two terms of this expansion, and substitute ϕ_0 for ϕ_s. The effect of the polysilicon depletion effect on the threshold voltage V_{TB} is to increase it by an amount $\gamma^2(V_{SB} + 2\phi_f)/\gamma_p^2$, which is the drop across the polysilicon at the threshold.

3.4.3 Effect on C–V Characteristics

The C–V of an ideal MOS capacitor has been discussed in Section 2.7. The MOS capacitor with a polydepletion effect shows C–V characteristics different from that of the conventional structure [16]. We shall qualitatively discuss the C–V characteristics of a MOS capacitor with polydepletion effect at low frequency [3].

The equivalent circuit of the MOS capacitor with polydepletion effect is shown in Figure 3.19. It is noted that the polydepletion effect adds an additional capacitance in series with the oxide and substrate space charge capacitances. This results in an effective reduction of the

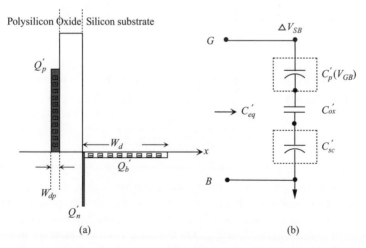

Figure 3.19 (a) Charge balance, and (b) equivalent circuit for a MOS capacitor with polydepletion effect in the inversion condition

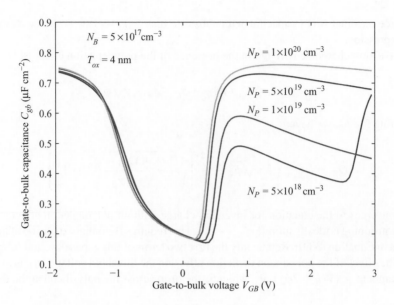

Figure 3.20 $C-V$ plot of a MOS capacitor with polysilicon doping

gate-to-bulk capacitance, and may be reflected as an increase in gate oxide thickness.

The typical low frequency $C-V$ characteristics with gate bias in the presence of polydepletion effect, as obtained from SCHRED, is shown in Figure 3.20. The first minima corresponds to the threshold voltage which is shifted to a higher value compared to the case without a polydepletion effect. In the inversion condition at the interface, the $C-V$ plot is characterized by a local maxima, after which the capacitance decreases with gate bias. This phenomenon can be qualitatively explained by considering the equivalent capacitance C'_{eq} which is given by:

$$\frac{1}{C'_{eq}} = \frac{1}{C'_p} + \frac{1}{C'_{ox}} + \frac{1}{C'_n} \tag{3.61}$$

In strong inversion $C'_n = Q'_n/2\phi_t$. Further, from the charge balance condition $Q'_p = -(Q'_n + Q'_b)$. In strong inversion $Q'_p \approx Q'_n$. It thus follows that $C'_n \propto Q'_p$. Further:

$$C'_p = \frac{qN_P\epsilon_{Si}}{Q'_p}, \quad C'_p \propto \frac{1}{Q'_p} \tag{3.62}$$

Thus, as the substrate enters into the strong inversion region, C'_n, the inversion layer capacitance increases, with inversion charge density being directly proportional to it. The increase of inversion charge results in an increase of the polydepletion width to satisfy the charge balance condition. Thus $C'_n \propto Q'_p$. On the other hand, an increase in polydepletion width means a decrease in the value of C'_p which is $\propto 1/Q'_p$. In the low frequency $C-V$ characteristics, therefore, there are two competing capacitances, the substrate space charge capacitance increasing with gate voltage and the polydepletion capacitance decreasing with gate voltage. A local maximum occurs when the two components are equal. After the point of maxima, an increase in gate voltage results in the decrease of the total capacitance because of dominance of the polydepletion capacitances which decrease with increasing gate bias. If the gate voltage is large enough to cause inversion in the polysilicon gate as well, the total capacitance is a series combination of two inversion capacitances in series with the oxide capacitance.The inversion capacitance being very large compared to oxide capacitance, the total capacitance approaches the oxide capacitance.

The parameters of interest for MOS modeling which get affected by the polysilicon depletion effect are threshold voltage, inversion charge density, pinch-off voltage, and surface potential.

3.4.4 Polydepletion Effect in Compact Models

The polydepletion effect manifests itself through the potential drop ϕ_p, which is addressed by various compact models as described below.

BSIM3v3. The model considers the polydepletion effect in strong inversion with the source as reference by considering an effective gate voltage. The gate voltage is expressed in terms of an effective value, V_{GBeff}, which is obtained by subtracting the drop of potential at the polysilicon–oxide interface, ϕ_p, given in Equation (3.44), from V_{GB}. The resulting expression is:

$$V_{GBeff} = V_{GB} - \phi_p = V_{GB} - \left[\sqrt{\frac{\gamma_p^2}{4} + (V_{GB} - \phi_s)} - \frac{\gamma_p}{2} \right]^2 \tag{3.63}$$

In the above equation, we have to substitute $V_{GBeff} = V_{GSeff} + V_{SB}$, $V_{GB} = V_{GS} + V_{SB}$ and $\phi_s = 2\phi_f + V_{SB}$. Also substituting γ_p from Equation (3.41), we obtain the equation for V_{GSeff} in the following form as used in BSIM3v3 [44]:

$$V_{GSeff} = \phi_s + \frac{q\epsilon_{Si}N_P T_{ox}^2}{\epsilon_{ox}^2} * \left[\sqrt{1 + \frac{2\epsilon_{ox}^2(V_{GS} - \phi_s)}{q\epsilon_{Si}N_P T_{ox}^2}} - 1 \right] \tag{3.64}$$

MM11. In this model, the space charge equation is solved at the oxide–polysilicon interface and the corresponding Kingston's relation is obtained [45]:

$$(V_{GB} - \phi_p - \phi_s)^2 = \gamma^2 \left[\phi_s + \phi_t \left(e^{\frac{-\phi_s}{\phi_t}} - 1 \right) + \phi_t e^{\frac{-(V_{SB} + 2\phi_f)}{\phi_t}} \left(e^{\frac{\phi_s}{\phi_t}} - 1 \right) \right] \tag{3.65}$$

The above equation can be solved numerically to obtain the surface potential in presence of the polydepletion effect. The obvious effect of polydepletion is to reduce the surface potential for a given V_{GB} especially in the strong inversion condition as shown in Figure 3.16.

HiSIM. As in the normal operating condition, the polysilicon–oxide interface does not enter into an inversion condition, a simplified expression for the electric field at the polysilicon–oxide interface due to the space charge in the above region can be expressed in the form [46]:

$$F_p = \sqrt{2qN_PL_{bp}}\sqrt{\left(\frac{\phi_p}{\phi_t} - 1\right)} \tag{3.66}$$

The above equation is obtained carrying out appropriate simplifications of Kingston's expression for the space charge in the polysilicon gate. Further:

$$V'_{GB} - \phi_s - \phi_p = -\frac{Q'_p}{C'_{ox}} = \frac{\epsilon_{Si}F_{Si}(0)}{C'_{ox}} \tag{3.67}$$

Equations (3.66) and (3.67) are solved iteratively to satisfy the condition $F_{Si}(0) = F_p$. The iterative procedure can be avoided by using fitting model parameters.

EKV. In this model, the pinch-off voltage V_p, and two parameters, the charge linearizing factor n_q, and the slope factor n_v, are affected by the polydepletion effect.

The expression for pinch-off voltage is obtained by putting $Q'_n = 0$ in Equation (3.60), where the substitution $\phi_s = V_{SB} + \phi_0$ is made. It follows that:

$$\gamma\sqrt{V_P + \phi_0} + \frac{\gamma_p^2}{2}\left\{1 - \sqrt{1 + \frac{4[V_{GB} - (V_P + \phi_0)]}{\gamma_p^2}}\right\} = 0 \tag{3.68}$$

Solving the above equation for V_P, we get:

$$V_P = \left\{\frac{\gamma_p}{2(\gamma_p^2 + \gamma^2)}\left[\gamma\gamma_p - \sqrt{4V_{GB}(\gamma_p^2 + \gamma^2) + \gamma^2\gamma_p^2}\right]\right\}^2 - \phi_0 \tag{3.69}$$

The pinch-off voltage can still be approximated as a linear function of V_{GB} as:

$$V_P = \frac{V_{GB} - V_{TB0}}{n_v} \tag{3.70}$$

where $n_v = dV_{GB}/dV_P$, which gives:

$$n_v = 1 + \frac{\gamma}{2\sqrt{2\phi_f + V_P}} + \frac{\gamma^2}{\gamma_p^2} \tag{3.71}$$

It may be noted that for $\gamma_p \to \infty$, $n_v = n$, where n is the slope factor for a MOS capacitor without polydepletion effect.

The plot of inversion charge density with the polydepletion effect shown in Figure 3.21 indicates that the inversion charge density though is a more nonlinear function of surface potential compared to the case without a polydepletion effect. It can, however, still be approximated as a linear function of surface potential ϕ_s,

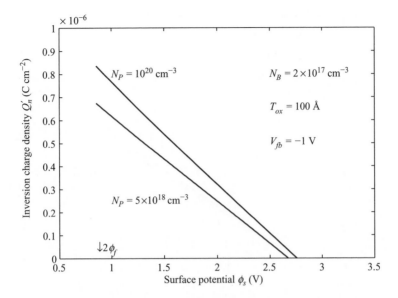

Figure 3.21   Inversion charge density with surface potential in the presence of the polydepletion effect

and it is found that the slope at $\phi_s = 2\phi_f + V_P/2$ gives the best fit [18]. The slope can be obtained using Equation (3.60) as given below:

$$n_q = \frac{1}{C'_{ox}} \left(\frac{\partial Q'_n}{\partial \phi_s} \right)_{\phi_s = 2\phi_f + V_P/2}$$

$$= \frac{\gamma}{2\sqrt{2\phi_f + \frac{V_P}{2}}} + \frac{\gamma_p^2}{\sqrt{(\gamma_p^2 + 2\gamma\sqrt{2\phi_f + V_P})^2 + 2\gamma_p^2 V_P}} \qquad (3.72)$$

It is observed that though n_q and n_v are used in different contexts, they assume the same expression in the absence of polydepletion effect, where it was labelled as n in Chapter 2. In the presence of polydepletion effect their values are different.

The dependence of pinch-off voltage on the gate voltage in the presence of polydepletion effect is shown in Figure 3.22. It is seen that it requires a higher gate voltage to annul the effect of substrate potential for the creation of inversion charge because of the potential drop across the polydepletion layer. In other words, the gate voltage has weaker control over the channel charge when the polydepletion effect is present.

3.5 Non-classical MOS Structures

We have so far considered the MOS capacitor which is used for conventional and mainstream CMOS structures. The MOS transistors are being constantly scaled to achieve higher performance in terms of power, speed, and cost. The conventional bulk CMOS structures are labeled

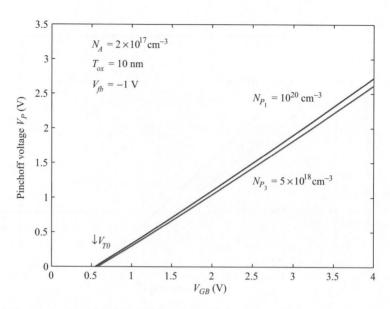

Figure 3.22   Pinch-off voltage dependence on gate voltage with polysilicon doping as a parameter

as classical in this text. In order to achieve the limits of scaling, extrapolating the road map given in Chapter 1, new or non-classical MOS structures are being proposed. Improvement through material engineering for enhancing the mobility has been a long term pursuit, and addition of germanium atoms in a silicon substrate has emerged as a practical technology to stretch the limits of frequency of operation of MOS transistors. An alternative option to enhance mobility is to use undoped silicon which eliminates mobility reduction of carriers due to Coulomb scattering. Another limiting factor in scaling has been the tunneling phenomenon through the ultra-thin gate oxide. The tunneling can be avoided using insulators having higher permittivity. In the following section the above issues will be briefly discussed.

3.5.1 Threshold Voltages for Si–SiGe–Si MOS Heterostructures

A strained Si–Si$_{1-x}$Ge$_x$–Si based substrate structure is emerging as a potential alternative in MOS technology because of the demonstrated stress induced improvements in the mobility values of electrons and holes [25–30]. For realizing such structures, the channel engineering involves introducing inhomogeneity at the interface due to changes in the composition of the silicon crystal. Such layers form a heterostructure at the interface between the gate and the substrate. The threshold voltage is an important parameter which has been studied for specific structures [31–33].

In this section we shall deal with an illustrative strained SiGe layered structure which is sandwiched between a silicon cap layer at the top and a silicon buffer layer at the bottom, of thicknesses W_{SiT} and W_{SiB} respectively. The strained SiGe layer has a thickness of W_{SiGe}. These stacked layers are grown on the n-type silicon substrate with doping density N_D. We are considering an n-substrate in view of its importance in the enhancement of hole mobility. The conditions for the strong inversion at the Si/SiGe and SiO$_2$/Si interfaces are of general interest as these inversion layers would contribute to the drain current of an MOS transistor. Then Poisson's equations are solved for the respective layers to arrive at the value of the threshold voltage, which is an important device parameter. A MOS strained layer heterostructure capacitor model provides a number of useful design and process related information. We shall describe below a model for the threshold voltage of an illustrative Si–SiGe–Si p-channel MOS capacitor [34].

3.5.2 Si–SiGe–Si MOS Heterostructure

As a negative gate voltage is applied, the substrate gets depleted of mobile charge carriers of depth W_{dep}, as shown in Figure 3.23. As the voltage is increased, strong inversion can occur at Si/SiGe (W_{SiGe}), or SiO$_2$(W_{SiT}), or at both of them. *It is, however, desirable that inversion occurs only at the Si/SiGe interface* as it enhances the hole mobility due to strong hole confinement, and reduced surface scattering. So in order to have the optimal device operation, the surface inversion layer at the SiO$_2$/Si interface needs to be avoided.

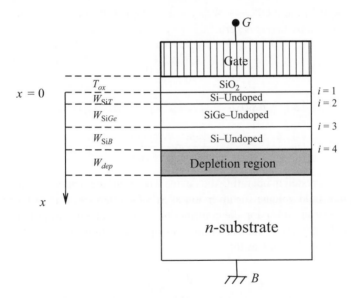

Figure 3.23 Representative MOS capacitor on a Si–SiGe–Si heterostructure substrate

The condition of strong inversion is defined as the gate voltage at which the concentration of minority carriers at the interface is equal to the doping concentration of the bulk substrate. Since there are two interfaces under consideration, the following two cases can arise:

Case 1. Si/SiGe interface:

$$\phi_{T2} = -2\phi_t \ln\left(\frac{N_D}{n_i}\right) + \frac{\Delta E_v}{q} - \phi_t \ln\left(\frac{N_{vSi}}{N_{vSiGe}}\right) \qquad (3.73)$$

Case 2. SiO$_2$/Si interface:

$$\phi_{T1} = -2\phi_t \ln\left(\frac{N_D}{n_i}\right) \tag{3.74}$$

where ϕ_{T1} and ϕ_{T2} are the threshold inversion potentials at the surface and top heterointerface respectively, ΔE_v is the valence band offset between SiGe and Si, and N_{vSi} and N_{vSiGe} are the effective density of states in the valence band for Si and SiGe respectively.

It is assumed that the device is so designed that $|V_{T2}| < |V_{T1}|$. In other words, the inversion occurs first at the Si/SiGe interface, and thus only Case 1 will be considered. Also, it is assumed that there are no interface charges trapped at the transition of the layers.

There are four layers that can be identified in this problem. The Poisson's equations are as follows:

1. For $(0 < x < W_{SiT})$ and $(W_{SiT} + W_{SiGe} < x < W_{SiT} + W_{SiGe} + W_{SiB})$, since the charge contained (ρ) in these layers is zero:

$$\frac{\mathrm{d}^2\phi}{\mathrm{d}x^2} = 0 \tag{3.75}$$

2. For $(W_{SiT} < x < W_{SiT} + W_{SiGe})$:

$$\frac{\mathrm{d}^2\phi}{\mathrm{d}x^2} = -\frac{qp(x)}{\epsilon_{SiGe}} \tag{3.76}$$

Since we are interested in operating the device where surface scattering is avoided we shall obtain the threshold voltage for inversion at Si/SiGe interface. It is further assumed that at the onset of strong inversion the contribution of the inversion carrier in determining the potential can be neglected. Hence, simplifying approximation can be made that $p(x) = 0$.

Thus Equation 3.76 reduces to:

$$\frac{\mathrm{d}^2\phi}{\mathrm{d}x^2} = 0 \tag{3.77}$$

3. For $(W_{SiT} + W_{SiGe} + W_{SiB} < x < W_{SiT} + W_{SiGe} + W_{SiB} + W_{dep})$:

$$\frac{\mathrm{d}^2\phi}{\mathrm{d}x^2} = -\frac{qN_D}{\epsilon_{Si}} \tag{3.78}$$

The Poisson's equations can be solved using the boundary conditions which are defined by:

1. Continuity of potential at the interfaces:

$$\phi\,|_{W_i^-} = \phi\,|_{W_i^+} \tag{3.79}$$

2. Gauss' Law:

$$\epsilon_1 F|_{W_i^-} = \epsilon_2 F|_{W_i^+} \tag{3.80}$$

where $i = 1, 2, 3$ is shown in Figure 3.23; x defines the distances of the various interfaces from the SiO$_2$–Si interfaces. In the case of $x_4 = W_{SiT} + W_{SiGe} + W_{SiB} + W_{dep}$, $\phi|_{x_4} = 0$ and $d\phi/dx|_{x_4} = 0$.

Now using Poisson's equations (Equations (3.75), (3.77), and (3.78)), and the boundary conditions given by Equations (3.79) and (3.80), the potential at the Si–SiGe interfaces ϕ_2, and the potential at the SiO$_2$–Si interface ϕ_1, are given as follows:

$$\phi_2 = \phi(x = W_{SiT}) = -\frac{qN_D W^2 dep}{2\epsilon_{Si}} - \frac{qN_D W_{dep} W_{SiB}}{\epsilon_{Si}} - \frac{qN_D W_{dep} W_{SiGe}}{\epsilon_{SiGe}} \tag{3.81}$$

and:

$$\phi_1 = \phi(x = 0) = \phi_2 - \frac{qN_D W_{dep} W_{SiT}}{\epsilon_{Si}} \tag{3.82}$$

From the condition of potential balance, $V_{GB} = V_{fb} + \phi_1 + V_{ox}$:

$$V_{GB} = V_{fb} + \phi_2 - qN_D W_{dep} \left[\frac{T_{ox}}{\epsilon_{ox}} + \frac{W_{SiT}}{\epsilon_{Si}} \right] \tag{3.83}$$

Now V_{T2} can be given as follows:

$$V_{T2} = V_{GB}(\phi_2 = \phi_{T2}) \tag{3.84}$$

$$V_{T2} = V_{fb} + \phi_{T2} - qN_D W_{depm} \left(\frac{T_{ox}}{\epsilon_{ox}} + \frac{W_{SiT}}{\epsilon_{Si}} \right) \tag{3.85}$$

where W_{depm} is the maximum depletion layer width, and is obtained from Equation (3.81) by substituting the following equation:

$$W_{depm} = W_{dep}(\phi_2 = \phi_{T2}) \tag{3.86}$$

$$W_{depm} = \sqrt{ \frac{2\epsilon_{Si}}{qN_D}(-\phi_{T2}) + \left(W_{SiB} + \frac{\epsilon_{Si}}{\epsilon_{SiGe}} W_{SiGe} \right)^2 } - W_{SiB} - \frac{\epsilon_{Si}}{\epsilon_{SiGe}} W_{SiGe} \tag{3.87}$$

Figure 3.24 shows the variation of threshold voltage using Equation (3.84) with the percentage concentration of germanium in Si$_{1-x}$Ge$_x$. It is observed that with the increase of germanium content, the magnitude of the threshold voltage decreases or it becomes less negative. The effect of doping on the threshold voltage in an SiGe heterostructure substrate is also shown in the figure. The increase of doping in the n-type substrate requires a higher negative

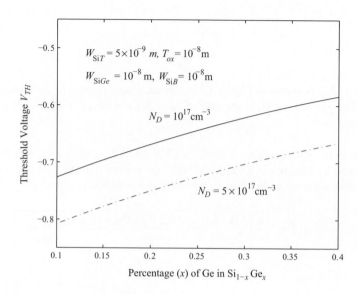

Figure 3.24   Variation of threshold voltage with percentage of Ge in $Si_{1-x}Ge_x$

voltage for the inversion condition. We can obtain the expression for V_{T1} adopting the procedure outlined for V_{T2}, with the difference that it becomes necessary to consider the mobile charge at the Si/SiGe interface. It is worth mentioning that in the MOSFET structure containing an Si/SiGe/Si heterostructure, the thickness of the various layers are designed such that the difference between V_{T1} and V_{T2}, which is the threshold voltage window, is maximized.

3.5.3 Undoped Silicon

Though conventional or classical MOS devices are mostly based on doped silicon (*n*- or *p*-type), undoped or lowly doped silicon substrates are drawing a lot of attention due to the possibility of (i) eliminating threshold voltage fluctuation due to statistical variation in dopant concentration which gets aggravated at high doping [3, 23], and (ii) increasing mobility due to the absence of coulomb scattering of the dopant ions. Figure 3.25(a) shows a MOS capacitor on an intrinsic silicon substrate. Figure 3.25(b) shows the energy band diagram for a positive voltage applied to the gate. For a positive voltage applied at the gate, electrons will be attracted towards the interface and holes will be repelled, due to which $n > n_i$ and $p < p_i$. Thus a space charge region is created near the interface due to deviation of concentrations of electrons and holes from the values in the absence of bias. The space charge, in turn, creates an electric field and potential drop in that region. Evaluation of $\rho(x)$, $F(x)$, and $\phi(x)$ in the space charge region for an intrinsic silicon substrate is of direct relevance for device modeling. We shall obtain below an explicit analytical expression for the threshold voltage of a MOS capacitor with intrinsic silicon as the substrate [24].

For intrinsic silicon, the volume space charge density due to the applied gate voltage, constituted by non-equilibrium holes $p(x)$ and electron concentrations $n(x)$, is expressed by:

$$\rho(x) = q[p(x) - n(x)] \tag{3.88}$$

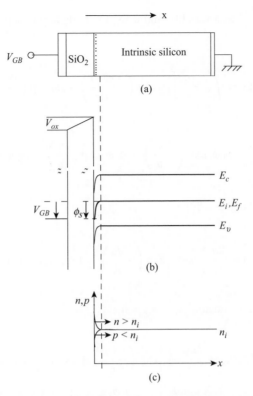

Figure 3.25 (a) MOS capacitor on intrinsic silicon, (b) energy band diagram for positive voltage applied to the gate, and (c) electron and hole concentrations

From Boltzmann's statistics, the carrier concentrations $p(x)$ and $n(x)$ are related to the equilibrium concentrations p_0 and n_0 respectively by: $n(x) = n_0 \exp(\phi(x)/\phi_t)$ and $p(x) = p_0 \exp(-\phi(x)/\phi_t)$, where $\phi(x)$ is the potential in the space charge region with a neutral bulk taken as reference, considering the interface as the origin:

$$\frac{d^2\phi}{dx^2} = -\frac{q}{\epsilon_s}\left[p_0 \exp\left(-\frac{\phi(x)}{\phi_t}\right) - n_0 \exp\left(\frac{\phi(x)}{\phi_t}\right)\right] \tag{3.89}$$

By application of the 'Chain Rule' we get:

$$\frac{d}{d\phi}\frac{d\phi}{dx}\frac{d\phi}{dx} = -\frac{q}{\epsilon_s}\left[p_0 \exp\left(\frac{-\phi(x)}{\phi_t}\right) - n_0 \exp\left(\frac{\phi(x)}{\phi_t}\right)\right] \tag{3.90}$$

Now, treating $\alpha = \frac{d\phi}{dx}$ as a variable, and using the boundary condition $\alpha = 0$ when $\phi = 0$, we get:

$$\int_0^\alpha \alpha \, d\alpha = \int_0^\phi -\frac{q}{\epsilon_s} \left[p_0 \exp\left(-\frac{\phi(x)}{\phi_t} \right) - n_0 \exp\left(\frac{\phi(x)}{\phi_t} \right) \right] d\phi \qquad (3.91)$$

$$\frac{\alpha^2}{2} = \frac{q}{\epsilon_s} \left[p_0 \phi_t \exp\left(-\frac{\phi}{\phi_t} \right) + n_0 \phi_t \exp\left(\frac{\phi}{\phi_t} \right) - p_0 \phi_t - n_0 \phi_t \right] \qquad (3.92)$$

Using the definition, $F(x) = -\frac{d\phi}{dx}$:

$$F^2(x)\Big|_{x=0} = F^2(\phi_s) = \frac{2q}{\epsilon_s} \left\{ p_0 \phi_t \left[\exp\left(-\frac{\phi_s}{\phi_t} \right) - 1 \right] + n_0 \phi_t \left[\exp\left(\frac{\phi_s}{\phi_t} \right) - 1 \right] \right\} \qquad (3.93)$$

For an intrinsic semiconductor, $p_0 = n_0 = n_i$:

$$F^2(\phi_s) = \frac{2q\phi_t n_i}{\epsilon_s} \left[\exp\left(-\frac{\phi_s}{\phi_t} \right) + \exp\left(\frac{\phi_s}{\phi_t} \right) - 2 \right] \qquad (3.94)$$

$$F^2(\phi_s) = \frac{4q n_i \phi_t}{\epsilon_s} \left[\cosh\left(\frac{\phi_s}{\phi_t} \right) - 1 \right] \qquad (3.95)$$

From the potential balance across the MOS capacitor, when the gate voltage corrected for the flat-band voltage equals the potential drop across the oxide and the space charge layer:

$$(V_{GB} - V_{fb} - \phi_s)^2 = \frac{\epsilon_s^2}{C_{ox}'^2} F(\phi_s)^2 \qquad (3.96)$$

Combining Equations (3.95) and (3.96):

$$(V_{GB} - V_{fb} - \phi_s)^2 = \frac{4q\epsilon_s n_i \phi_t}{C_{ox}'^2} \left[\cosh\left(\frac{\phi_s}{\phi_t} \right) - 1 \right] \qquad (3.97)$$

$$(V_{GB} - V_{fb} - \phi_s) = \text{sgn}\,(\phi_s)\gamma \sqrt{2\phi_t} \left[\cosh\left(\frac{\phi_s}{\phi_t} \right) - 1 \right]^{\frac{1}{2}} \qquad (3.98)$$

where $\gamma = \sqrt{2q\epsilon_s n_i}/C_{ox}'$ and the sign for the RHS is decided by the sign of the surface potential; $\text{sgn}(\phi_s)$ indicates that a minus sign is used when bands bend up, and a plus sign when bands bend down. With the hyperbolic function sinh, expressions for potential containing both positive and negative signs can be handled. Hence, the solution is put in the following forms:

$$VGB - V_{fb} = \phi_s + 2\gamma\sqrt{\phi_t}\sinh\left(\frac{\phi_s}{2\phi_t}\right) \tag{3.99}$$

$$VGB - V_{fb} = \phi_s + \gamma\sqrt{\phi_t}\left[\exp\left(\frac{\phi_s}{2\phi_t}\right) - \exp\left(-\frac{\phi_s}{2\phi_t}\right)\right] \tag{3.100}$$

For a positive gate voltage, electrons accumulate and we shall designate the interface as n-type enhancement. Only the positive exponential term is dominant when $\phi_s \gg \phi_t$. Equation (3.100) reduces to:

$$VGB - V_{fb} = \phi_s + \gamma\sqrt{\phi_t}\exp\left(\frac{\phi_s}{2\phi_t}\right) \tag{3.101}$$

For the strong inversion region $\phi_s/\phi_t \gg 1$. Therefore, the exponential term will be very large, and we can write:

$$VGB - V_{fb} = \gamma\sqrt{\phi_t}\exp\left(\frac{\phi_s}{2\phi_t}\right) \tag{3.102}$$

For the subthreshold (weak inversion) region ϕ_s will be more dominant than $\gamma\sqrt{\phi_t}\exp(\frac{\phi_s}{2\phi_t})$:

$$VGB - V_{fb} = \phi_s \tag{3.103}$$

The onset of strong inversion is called the threshold. Combining Equations (3.102) and (3.103) and replacing $VGB = V_T$ we get:

$$V_T - V_{fb} = \gamma\sqrt{\phi_t}\exp\frac{V_T - V_{fb}}{2\phi_t} \tag{3.104}$$

Dividing both sides by $-2\phi_t$ we get:

$$\frac{-(V_T - V_{fb})}{2\phi_t}\exp\frac{-(V_T - V_{fb})}{2\phi_t} = -\frac{\gamma}{2\sqrt{\phi_t}} \tag{3.105}$$

Taking the W_{-1} negative branch of the Lambert W-function, we have:

$$W_{-1}\left(-\frac{\gamma}{2\sqrt{\Phi_t}}\right) = \frac{-(V_T - V_{fb})}{2\phi_t} \tag{3.106}$$

Rearranging the above equation:

$$V_T = V_{fb} - 2\phi_t W_{-1}\left(-\frac{\gamma}{2\sqrt{\phi_t}}\right) \tag{3.107}$$

The negative branch of the Lambert W-function can be approximated in terms of the elementary logarithm function for very small negative values of the argument:

$$W_{-1}(x) \approx \ln \left[\frac{x}{\ln(-x)} \right] \qquad (3.108)$$

Therefore, the threshold voltage for an undoped body MOSFET can be given by the expression:

$$V_T = V_{fb} - 2\phi_t \ln \left[\frac{\frac{-\gamma}{2\sqrt{\phi_t}}}{\ln\left(\frac{\gamma}{2\sqrt{\phi_t}}\right)} \right] \qquad (3.109)$$

Figure 3.26 illustrates the surface potential ϕ_s versus positive gate voltage V_{GB} for a specific gate oxide thickness. The projection of the intersection of asymptotes in the subthreshold and superthreshold regions, to the horizontal axis graphically defines the 'threshold voltage'.

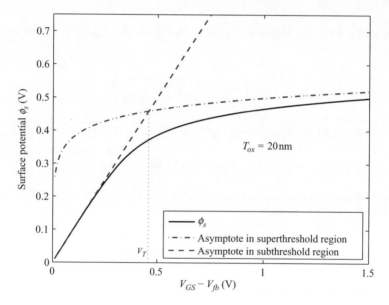

Figure 3.26   Extraction of threshold voltage for undoped silicon

3.6 MOS Capacitor With Stacked Gate

As MOSFET devices approach deep-submicron feature size and less, the gate terminal which is conventionally considered to be non-conducting, shows finite conductivity and current flows through the gate terminal under a biased condition. The leakage through the gate is a matter of serious concern as it contributes to heat dissipation in the chip in the off-state with attendant reliability problems. The current flows through the gate insulator is attributed to tunneling phenomenon. The technology roadmap projects that with SiO_2 as the gate dielectric that there has to be a maximum permissible limit of the tunneling current density [35]. The limit is quantitatively prescribed by:

$$J < \frac{10 I_{subth}}{L} \tag{3.110}$$

where I_{subth} is the subthreshold leakage current/width expressed in $A\,cm^{-1}$. For a 30 nm technology node, the SiO_2 thickness is estimated to be around 0.1 nm only, which is barely one atomic layer of the insulating gate dielectric layer. Tunneling through the gate oxide degrades the performance of basic logic and storage building blocks of CMOS digital circuits [37]. Therefore, there is a lower limit of thickness for the SiO_2, if tunneling is to be avoided. In this section we shall only outline the basic issues related to tunneling through the gate SiO_2 dielectric so that technological solutions being explored to overcome its effect can be discussed.

3.6.1 Thin Gate Oxide and Tunneling

Tunneling through the gate oxide can broadly be divided into two categories: (i) Fowler–Nordheim and (ii) direct [40]. Figure 3.27 shows the energy band diagram of a MOS capacitor of n^+ -poly–SiO_2–n-Si substrate type with a positive voltage applied to the gate terminal with respect to the substrate. The capacitor is in the accumulation region and has been considered as an illustration. The electrons in the conduction band of the substrate see a barrier height of ϕ_b at the Si–SiO_2 interface. For simplification it is assumed that both the substrate and the n^+ -poly gate are degenerately doped, and the Fermi levels coincide with the conduction band edge. We also ignore the polydepletion effect as the gate is taken to be very highly doped. The substrate surface potential is ϕ_s. Electrons can tunnel through the thin oxide layer separating the gate and the substrate. Tunneling through a gate oxide will be discussed in detail in Chapter 7.

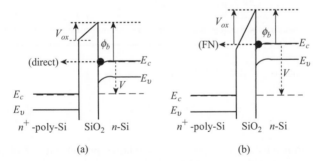

(a)   (b)

Figure 3.27   Energy band diagrams of a polysilicon gate capacitor in the accumulation mode of operation: (a) direct; (b) FN tunneling

3.6.2 High-k Dielectric and Stacked Gate

The tunneling through a thin SiO_2 gate dielectric is now proving to be an issue of serious concern and a limitation to future scaling in spite of its excellent compatibility of SiO_2 with silicon and established performance record. The thickness of silicon-dioxide has reached only a few atomic layers and tunneling is unavoidable. Therefore, there is an intense search for alternative dielectrics with high-k (permittivity) which has properties as close to SiO_2 as possible, but offers the opportunity to use a higher thickness for the gate insulator. In other words, the requirement is to find a substitute dielectric which gives the same capacitance with higher thickness [39]. A large number of materials have been investigated which, to replace SiO_2, should have the following attributes [36, 38]:

- The material must have high permittivity.
- The high-k oxide bandgap must be large with good band offset with respect to silicon. Most practical high-k oxides considered to be a possible substitute for SiO_2 have a band offset of ≈ 5.0 eV.
- It should have good bonding capability with Si, SiO_2, and metals to be compatible with MOS processing.
- It should be thermodynamically stable to be compatible with high temperature processing associated with MOS technology.
- It should have low electrically active defects in order to avoid instability in threshold voltage due to charge trapping.

Table 3.3 below summarizes the properties of some of the dielectrics which have shown promise as a substitute for SiO_2 [41], though none of the materials have gone through the level of benchmarking experienced by SiO_2.

Table 3.3   Physical parameters for representative high-k dielectric materials

Parameter	Material				
	ZrO_2	HfO_2	Si_3N_4	$HfSiO_4$	SiO_2
Permittivity ($F\,cm^{-2}$)	$(20\text{--}25)\epsilon_0$	$20\epsilon_0$	$7\epsilon_0$	$12\epsilon_0$	$3.9\epsilon_0$
CB offset (eV)	1.4–1.5	1.5	1.9	1.5	3.1
Bandgap (eV)	5.7–5.8	4.5–6.0	5.3	≈ 6	9.0

The gate capacitance of a MOSFET device is given by:

$$C = \frac{\epsilon_0 K_d A}{T_d} \tag{3.111}$$

where $\epsilon_0 =$ the permittivity of free space, $K_d =$ relative permittivity of dielectric material, $A =$ area, and T_d is the gate dielectric thickness.

The tunneling problem associated with scaling of an SiO_2 gate oxide can be overcome by replacing it with a material of higher dielectric constant, which enables use of a thicker gate dielectric. The thickness of the high-k dielectric insulator is obtained from the relation:

$$T_{ox} = \mathbf{E}ffective\ \mathbf{O}xide\ \mathbf{T}hickness\,(EOT) = \frac{K_{Si}}{K_d} T_d \tag{3.112}$$

Figure 3.28 shows the energy band diagram of a stacked dielectric under conditions when tunneling may yet occur.

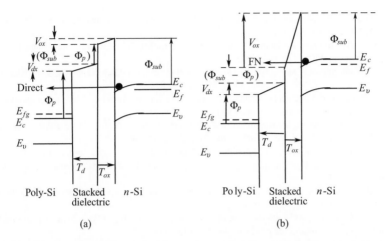

Figure 3.28   Energy band diagram of a high-k stacked gate: (a) direct; (b) FN tunneling

The new materials and structures discussed above required to stretch the capability of MOS structures have yet many unresolved issues to sort out. Si–Ge material technology is required to be compatible with matured silicon technology in terms of material perfection, thermal budget, etc. For the high-k dielectric the compatibility of a polysilicon gate is considered as an area of concern, and search of metal gate is in progress. Poor interface integrity with silicon affects the surface scattering and reduces mobility. Apart from the above mentioned variations around the material, the structural changes in non-classical MOSFET structures is envisaged to be around Ultra Thin Body (UTB) Silicon-On-Insulator (SOI) structures to eliminate bulk substrate related limitations [42], and double or multigate structures to enhance the control of the gate on the MOSFET performance [43].

References

[1] D. Foty, *MOSFET MODELING with SPICE, Principles and Practice*, Prentice Hall, Upper Saddle River, NJ, USA, 1997.

[2] Y. Cheng and C. Hu, *MOSFET Modeling and BSIM3 User's Guide*, Kluwer Academic Publishers, Boston, MA, 1999.

[3] Y. Taur and T.H. Ning, *Fundamentals of Modern VLSI Devices*, Cambridge University Press, Cambridge, UK, 1998.

[4] D.A. Buchanan, Scaling the Gate Dielectric: Materials, Integration, and Reliability, *IBM J. Res. Develop.*, **43**(3), 245–264, 1999.

[5] N.D. Arora, R. Rios and C.-L. Huang, Modeling the Polysilicon Depletion Effect and its Impact on Submicrometer CMOS Circuit Performance, *IEEE Trans. Electron. Dev.*, **42**(5), 935–943, 1995.

[6] P.M. Zeitzoff, J.A. Hutchby, G. Bersuker and H.R. Huff, Integrated Circuit Technologies: From Conventional CMOS to The Nanoscale Era in *Nano and Giga Challenges in Microelectronics,* J. Greer, A. Korkin and J. Labanowski (Eds), pp. 1–25, Elsevier, 2003.

[7] S. Adee, Transistors go Vertical, *IEEE Spectrum*, 13–14, 2007.

[8] M. Bohr, R.S. Chau, T. Ghani and K. Mistry, The High-k Solution, *IEEE Spectrum*, **44**, 23–29, 2007.

[9] B.E. Deal, Standardized Terminology for Oxide Charges Associated with Thermally Oxidized Silicon, *IEEE Trans. Electron Dev.*, **27**(3), 606–608, 1980.

[10] I.P. Stepanenko, *Osnovi Teorii Transistorov i Transistornikh Skhem*, 4th Edition, Energiya Publishing House, Moscow, USSR (in Russian) 1977.

[11] H. Feltl, Onset of Heavy Inversion in MOS Devices Doped Nonuniformly Near Surface, *IEEE Trans. Electron Dev.*, **24**(3), 288–289, 1977.

[12] J.B. Jacobs and D. Antoniadis, Channel Profile Engineering for MOSFETs with 100 nm Channel Lengths, *IEEE Trans. Electron Dev.*, **42**(5), 870–875, 1995.

[13] N.D. Arora, Semi-empirical Model for the Threshold Voltage of a Double Implanted MOSFET and its Temperature Dependence, *Solid-State Electron.*, **30**(5), 559–569, 1987.

[14] W.M. Werner, The Work Function Difference of the MOS-System with Aluminum Field Plates and Polycrystalline Silicon Field Plates, *Solid-State Electron.*, **17**(8), 769–775, 1974.

[15] C.L. Huang and N.D. Arora, Measurements and Modeling of MOSFET *I–V* Characteristics with Polysilicon Depletion Effect, *IEEE Trans. Electron Dev.*, **40**(12), 2330–2337, 1993.

[16] C.Y. Lu, J.M. Sung, H.C. Kirsch, S.J. Hillenius, T.E. Smith and L. Manchanda, Anomalous *C–V* Characteristics of Implanted Poly MOS Structure in n^+/p^+ Polysilicon in a Dual–Gate CMOS Process, *IEEE Electron Dev. Lett.*, **10**(5), 192–194, 1989.

[17] R. Rios, N.D. Arora and C.-L. Huang, An Analytical Polysilicon Depletion Effect Model for MOSFETs, *IEEE Electron Dev. Lett.*, **15**(4), 129–131, 1994.

[18] J.-M. Sallese, M. Bucher and C. Lallement, Improved Analytical Modeling of Polysilicon Depletion in MOSFETs for Circuit Simulation, *Solid-State Electron.*, **44**, 905–912, 2000.

[19] P. Ratnam, C. Andre and T. Salama, A New Approach to the Modeling of Non-uniformly Doped Short Channel MOSFETs, *IEEE Trans. Electron Dev.*, **31**(9), 1289–1298, 1984.

[20] C. Lallement, M. Bucher and C. Enz, Modeling and Characterization of Non-uniform Substrate Doping, *Solid-State Electron.*, **41**(12), 1857–1861, 1997.

[21] P.K. Chatterjee, J.E. Leiss and G.W. Taylor, A Dynamic Average Model for the Body Effect in Ion Implanted Short Channel ($L = 1\mu$m) MOSFETs, *IEEE Trans. Electron Dev.*, **28**(5), 606–607, 1981.

[22] G. Gildenblat, X. Li, W. Wu, H. Wang, A. Jha, R. van Langevelde, G.D.J. Smit, A.J. Scholten, and D.B.M. Klaassen, PSP: An Advanced Surface-Potential-Based MOSFET Model for Circuit Simulation, *IEEE Trans. Electron Dev.*, **53**(9), 1979–1993, 2006.

[23] H.S. Wong and Y. Taur, Three-dimensional Atomistic Simulation of Discrete Random Dopant Distribution Effects in Sub-0.1μm MOSFETs, *IEDM Tech. Digest*, 705–708, 1993.

[24] A. Ortiz-Conde, F.J. Garcia and M. Guzman, Exact Analytical Solution of Channel Surface Potential as an Explicit function of Gate Voltage in Undoped-body MOSFET Using the Lambert W Function and a Threshold Voltage Definition Therefrom, *Solid-State Electron.*, **47**, 2067–2079, 2003.

[25] K. Rim, S. Koester, M. Hargrove, J. Chu, P.M. Mooney, J. Ott, T. Kanarsky, P. Ronsheim, M. Ieong, A. Grill and H.S.P. Wong, Strained Si NMOSFETs for High Performance CMOS Technology, *Symposium on VLSI Technology Digest of Technical Papers*, 59–60, Kyoto, Japan, 2001.

[26] K. Rim, J. Chi, H. Chen. K.A. Jenkins, K. Kanarsky, K. Lee, A. Mocuta, H. Zhu, R. Roy, J. Newbury, J. Ott, K. Petrarca, P.M. Mooney, D. Lacey, K. Koester, K. Chan, D. Boyd, M. Ieomg and H.S. Wong, Characteristics and Device Design of sub-100 nm Strained Si N- and P-MOSFETs, *Symposium on VLSI Technology Digest of Technical Papers*, 98–99, 2002.

[27] D.A. Antoniadis, I. Aberg, C.Ni Chléirigh, O.M. Nayfeh, A. Khakifirooz and J.L. Hoyt, Continuous MOSFET Performance Increase with Device Scaling: The Role of Strain and Channel Material Innovation, *IBM J. Res. Dev.* **50**(4/5), 363–376, 2006.

[28] S. Dhar, H. Kosina, S.E. Ungersboeck, T. Grasser and S. Selberherr, High Field Electron Mobility Model for Strained-Silicon Devices, *IEEE Trans. Electron Dev.*, **53**(12), 3054–3062, 2006.

[29] S.H. Olsen, K.S.K. Kwa, L.S. Driscolli, S. Chattopadhyay, and Q.G. Oneil, Design, Fabrication and Characterization of Strained Si/SiGe MOS Transistors, *IEEE Proc. Circ. Dev. Syst.*, **151**(5), 431–437, 2004.

[30] S.G. Badock, A.G.O' Neill, and E.G. Chester, Device and Circuit Performance of SiGe/Si MOSFETs, *Solid-State Electron.*, **46**, 1925–1932, 2002.

[31] H.M. Nayfeh, J.L. Foyt and D.A. Antoniadis, A Physically Based Analytical Model for the Threshold Voltage of Strained-Si *n*-MOSFETs, *IEEE Trans. Electron Dev.*, **51**(12), 2069–2072, 2004.

[32] K. Chandrasekaran, X. Zhou, S.B. Chiah, W.Z. Shangguan, G.H. See, L.K. Bera, N. Balasubramanian and S.C. Rustagi, Effect of Substrate Doping on the Capacitance–Voltage Characteristics of Strained-Silicon MOSFETS, *IEEE Electron Dev. Lett.*, **27**(1), 62–64, 2006.

[33] Y.L. Tsang, S. Chattopadhyay, S. Uppal, E.E.-Cousin, H.K. Ramakrishnan, S.H. Olsen and A.G. O'Neill, Modeling of the Threshold Voltage in Strained $Si/Si_{1-x}Si_{1-y}Ge_y$ ($x \geq y$) CMOS Architectures, IEEE Trans. Electron Dev., **54**(11), 3040–3048, 2007.

[34] K. Iniewski, S. Voinigescu, J. Atcha and C.A.T. Salama, Analytical Modeling of Threshold Voltages in p-Channel Si/SiGe/Si MOS Structures, *Solid-State Electron.*, **36**(5), 775–783, 1993.

[35] International Technology Roadmap for Semiconductors (ITRS) (online). (http://public.itrs.net) 2003.

[36] G.D. Wilk, R.M. Wallace and J.M. Anthony, High-k Gate Dielectrics: Current Status and Materials Properties Considerations, *J. Appl. Phys.* **89**(10), 5243–5275, 2001.

[37] P.J. Wright and K.C. Saraswat, Thickness Limitations of SiO_2 Gate Dielectrics for MOS LSI, *IEEE Trans. Electron Dev.*, **37**(8), 1884–1892, 1990.

[38] J. Robertson, High Dielectric Constant Oxides, *Eur. Phys. J. Appl. Phys.*, **28**, 265–291, 2004.

[39] M.D. Groner and S.M. George, High-k Dielectrics Grown by Atomic Layer Deposition: Capacitor and Gate Applications, In *Interlayer Dielectrics for Semiconductor Technologies*, S.P. Murarka, M. Eizenberg and A.K. Sinha (Eds), pp. 327–348, Elsevier Academy Press, Amsterdam, 2003.

[40] M. Depas, B. Vermeire, P.W. Mertens, R.L. Van Meihaeghe and M.M. Heyns, Determination of Tunnelling Parameters in Ultra-Thin Oxide Layer Poly-Si/SiO2/Si Structures, *Solid-State Electron.*, **38**(8), 1465–1471, 1995.

[41] C.M. Osburn, I. Kim, S.K. Han, I. De, K.F. De, K.F. Yee, S. Gannavaram, S.J. Lee, C.H. Lee, Z.J. Luo, W. Zhu, J.R. Hauser, D.L. Kwong, G. Lucovsky, T.P. Ma and M.C. Oztürk, Vertically Scaled MOSFET Gate Stacks and Junctions: How Far are We Likely to Go?, *IBM Res. Dev.*, **46**(2/3), 299–315, 2002.

[42] C. Yang-Kyu, K. Asano, N. Lindert, V. Subramanian, K. Tsu-Jae, J. Bokor and H. Chenming, Ultrathin-body SOI MOSFET for Deep-Sub-Tenth Micron Era, *IEEE Electron Dev. Lett.*, **21**(5), 254–255, 2000.

[43] F. Balestra, S. Cristoloveanu, M. Benachir, J. Brini and T. Elewa, Double-Gate Silicon-on-Insulator Transistor with Volume Inversion: A New Device with Greatly Enhanced Performance, *Electron Dev. Lett.*, **8**(9), 410–412, 1987.

[44] W. Liu, *MOSFET Models for SPICE Simulation, including BSIM3v3 and BSIM4*, John Wiley & Sons, Inc., New York, NY, 2001.

[45] R. van Langevelde, A.J. Scholten and D.B.M. Klaassen, *Physical Background of MOS model 11, Level* 1101, (http://www.semiconductors.philips.com) Koninklijke Philips Electronics, NV, The Netherlands, 2003.

[46] K. Machida, B. D. Hu, W. Liu, F. Fujimoto, A. Kobayashi, S. Ito, K. Morikawa and H. Masuda, *HiSIM 1.1.1*, (http://home.hiroshima-u.ac.jp) Hiroshima, Japan, 2002.

[47] (http://www.nanohub.org/tools/schred).

Problems

1. (a) Calculate the flat-band voltage for a MOS capacitor for the following process parameters: $N_f = 5.5 \times 10^{10}$ cm^{-2} and $T_{ox} = 30$ nm.
   (b) Obtain the threshold voltage of a MOS capacitor given that $N_A = 10^{16}$ cm^{-3}, $T_{ox} = 30$ nm and $Q_f = 1.1 \times 10^{-8}$ C cm^{-2}.

2. (a) Show from the balance of drift and diffusion current in an inhomogeneous semiconductor that the expression for the internal built-in field in an n-type inhomogeneous semiconductor is given by:

$$F(x) = -\phi_t \frac{1}{N(x)} \frac{dN(x)}{dx} \qquad (3.113)$$

where the doping concentration has a spatial gradient in the x-direction. Draw the energy band diagram for the inhomogeneous silicon assuming that the doping concentration N is decreasing with x.
   (b) Show that for quasi-neutrality to hold good in an inhomogeneous semiconductor, the

following condition needs to be satisfied:

$$\left|\frac{dn}{dx}\right| \ll \sqrt{\frac{q}{\epsilon_{Si}\phi_t}} n^{\frac{3}{2}} \tag{3.114}$$

where n is the electron concentration at a given point x.

Hint: substitute the value of the built-in field $F(x)$ derived in part (a) in Poisson's equation $dF/dx = \rho/\epsilon_{Si}$.

3. Calculate the threshold voltage V_{T0} for an MOS capacitor on a p-type silicon substrate which has been subjected to boron impulse doping for a threshold voltage adjustment of $D_I = 5 \times 10^{11}$ ions cm^{-2}. Assume that $T_{ox} = 30$ nm , $\phi_{gs} = -0.9$ V, $N_A = 10^{15}$ cm^{-3} and $Q_f = 1.6 \times 10^{-9}$ C cm^{-2}.

4. Plot the variation of a polysilicon gate to the silicon work function difference, ϕ_{gs} with a substrate doping (n- and p-type) with a degenerate n^+ -polysilicon gate.

5. Assume an implanted Gaussian profile for threshold voltage adjustment as given below (see Figure 3.6):

$$N(x) = \frac{Q_I}{(2\pi)^{\frac{1}{2}}\Delta R_p} \exp\left[\frac{-1}{2}\left(\frac{x - R_p}{\Delta R_p}\right)^2\right]$$

where R_p is the range defined as the mean depth in the semiconductor measured from the oxide–Si interface, ΔR_p is the standard deviation from the mean, and Q_I is the implanted dose. Assume that the depletion depth W_d is less than the implant depth, W_I. Obtain the expression for potential variation with depth and threshold voltage.

6. Assuming a box distribution for an implanted profile as shown in Figure 3.9, give the derivation for the equivalent doping shown in Equation (3.33).

7. Assume that a MOS capacitor is made on a n-type substrate with a boron implant. The substrate profile is as shown in Figure 3.29. With the box distribution for the implanted profile shown, obtain the expression for the electric field and potential as a function of distance, x, assuming that the depletion depth penetrates the entire implanted region.

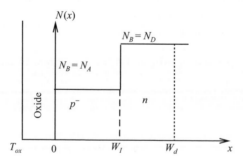

Figure 3.29 Approximated box profile for boron implant modeling of a p^-–n substrate

8. Assume that the boron dose in Problem 7 is small such that at the oxide–silicon interface a lightly doped n^- -region is formed over the n-substrate as shown in Figure 3.30. Derive the expressions for the electric field and potential in the n^-–n substrate.

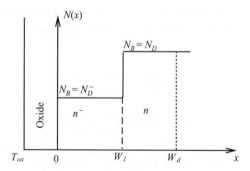

Figure 3.30   Approximated box profile for light boron implant modeling of an n^-–n substrate

9. Calculate the difference in threshold voltage considering the polydepletion effect for a MOS capacitor given that $N_A = 2 \times 10^{17}$ cm^{-3}, $T_{ox} = 7.0$ nm, $N_p = 5 \times 10^{18}$ cm^{-3} and $V_{fb} = -1.0$ V.

10. Draw the energy band diagram of an MOS capacitor for the following cases of gate and substrate work functions:

(a) $\phi_g > \phi_{Si}$: substrate n-type, $V_{GB} = 0$.
(b) $\phi_g < \phi_{Si}$: substrate p-type, $V_{GB} = 0$.
(c) $\phi_g = \phi_{Si}$: substrate n-type, (i) $V_{GB} < 0$, (ii) $V_{GB} = 0$, (iii) $V_{GB} > 0$, (iv) $V_{GB} \gg 0$.

Assume that the oxide is ideal with no trapped charge.

11. For the MOS capacitor on an Si/SiGe/Si layered substrate shown in Figure 3.23 obtain the expression for threshold voltage, V_{T1} at the SiO$_2$/Si interface in the following form:

$$V_{T1} = V_{GB}(\phi_1 = \phi_{T1}) = V_{fb} + \phi_{T1} - \frac{qN_D W_{DEPL}}{C'_{ox}}(1 + M(\phi_2)) \qquad (3.115)$$

where:

$$M(\phi) = \frac{2\epsilon_{SiGe}\phi_t}{qN_D W_{dep}^2}\left[\exp\left(\frac{\phi_{T2} - \phi}{\phi_t}\right) - \exp\left(\frac{\phi_{T2} - \phi_{x1}}{\phi_t}\right)\right], x1 = W_{SiT} + W_{SiGe} \qquad (3.116)$$

The symbols have the meanings as explained in Section 3.5.2. (Hint: use the condition $\phi_{T1} = -2\phi_t \ln(N_D/n_i)$ at the threshold condition $V_{GB} = V_{T1}$.)

12. Assume that the threshold voltage of a three terminal MOS capacitor is given by:

$$V_T(V_{SB}) = V_{T0} + nV_{SB} \qquad (3.117)$$

where n is defined by the equation $n = 1 + C'_d/C'_{ox}$ (Equation (2.156)). Show graphically the potential balance for a MOS capacitor assuming that V_{fb} is the flat-band voltage for a substrate bias varying from 0 to 10.0 V. Assume $T_{ox} = 0.05$ mm and $N_A = 10^{16}$ cm^{-3}, and $V_{fb} = -0.8$ V.

13. For a MOS capacitor with undoped silicon as the substrate obtain the potential variation as a function of distance (depth) in the space charge region near the oxide–silicon interface in the following form:

$$\phi(x) = 2\phi_t \ln \frac{N_r}{D_r} \tag{3.118}$$

$$N_r = 1 + \tanh\left(\frac{\phi_s}{4\phi_t}\right) \exp\frac{-x}{L_{bi}} \tag{3.119}$$

$$D_r = 1 - \tanh\left(\frac{\phi_s}{4\phi_t}\right) \exp\frac{-x}{L_{bi}} \tag{3.120}$$

Simplify the above expressions for a small value of surface potential ϕ_s, to obtain the variation of potential in the space charge region with depth x. Hence, calculate the value of the Debye length for an intrinsic silicon.

14. Obtain an expression for surface carrier charge density induced in the channel for an undoped silicon substrate in terms of gate voltage and oxide/capacitance area.
    Hint: The surface charge density, $Q'_s \equiv V_{ox}C'_{ox}$, $V_{ox} = C'_{ox}(V_{GB} - \phi_s - V_{fb})$. Obtain ϕ_s using Lambert's function from Equation (3.99).
    Plot the variation of Q'_s with gate voltage.

4

Long Channel MOS Transistor

4.1 Introduction

The MOS transistor may be viewed as an extension of the three terminal MOS structure where two diffused regions known as *source* and *drain*, acting as electrical probes, establish contacts with the inversion layer formed under the gate. MOS transistors can be broadly classified into four categories:

1. *n*-channel enhancement.
2. *p*-channel enhancement.
3. *n*-channel depletion.
4. *p*-channel depletion.

For an *n*-channel transistor the substrate is of *p*-type with source and drain made of n^+ diffused regions. The inversion carriers, induced at the Si–SiO$_2$ interface by the positive gate voltage, form a conducting channel connecting highly conducting source and drain diffused layers. The carriers in the inversion layer forming the channel under the gate are supplied by the source, and collected by the drain which is biased at a higher potential than that at the source. Therefore, for the *n*-channel MOS transistor, the terminal at higher potential is called the drain and the one which is at lower potential is called the source. Conventionally, the current is considered positive, when the current enters the terminal. Structurally, the MOS device is symmetrical. For a *p*-channel MOS transistor, which is the complement of the *n*-channel device and has *n*-Si as the substrate, the conducting channel formed under the gate due to the negative voltage applied on the gate, has holes as inversion carriers in the channel. The drain is biased at a lower potential than the source. The channel electric field is directed from source to drain which, in turn, causes holes to flow from source to drain. The drain current flows out of the drain terminal, and following the convention mentioned before, the drain current is considered to be negative. The drain current is designated as I_{DS}. The enhancement type MOS transistor is normally 'OFF' for zero gate voltage. In contrast, the depletion type MOS transistor is normally 'ON' under zero gate bias. In the depletion type a conducting channel is technologically created between source and drain by means of implantation.

Compact MOSFET Models for VLSI Design   A.B. Bhattacharyya
© 2009 John Wiley & Sons (Asia) Pte Ltd

Figure 4.1 shows a schematic representation of the four types of MOS transistors, and their symbols.

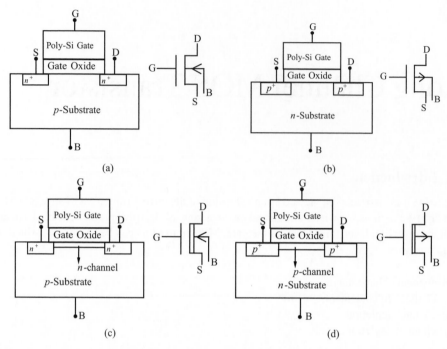

Figure 4.1   Four types of MOS transistors: (a) n-channel enhancement, (b) p-channel enhancement, (c) n-channel depletion, and (d) p-channel depletion

Historically, the source terminal has been the conventional voltage reference mostly due to dominance of the digital gate circuit driving the MOS technology where the source is the reference voltage terminal. From physical consideration, however, the bulk terminal seems to be a more natural reference from the structural symmetry of the device with respect to source and drain. Models for MOS transistors have been developed with both source or bulk as reference. Throughout this chapter our discussion will revolve around n-channel enhancement transistors.[1] In general, the conclusions arrived at for the NMOS applies to the p-channel transistor as well. The development of MOS technology has been due to the some unique features of the CMOS gate, which is configured by the integration of enhancement mode NMOS and PMOS transistors. Figure 4.2(a) represents a vertical cross-section of the CMOS inverter (NOT gate) structure, while Figure 4.2(b) shows the corresponding schematic diagram. In the CMOS, the gates are connected, which form the input to the inverter. The source of the PMOS is connected to V_{DD} and that of the NMOS is connected to ground. The output is taken from the drains of the NMOS and PMOS transistor, which are connected. Such a combination ideally consumes no power in the 'ON' or 'OFF' state, has full dynamic range of the supply voltage, and has good noise margin.

---

[1] We will refer to n-channel enhancement transistors as NMOS and p-channel enhancement transistors as PMOS.

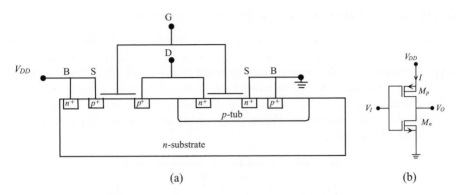

(a)   (b)

Figure 4.2   Cross-section of a CMOS inverter structure

Figure 4.3(a) and Figure 4.3(b) represent the output and transfer characteristics of NMOS transistors respectively. Similarly, Figure 4.3(c) and Figure 4.3(d) give the output and transfer characteristics of PMOS transistors respectively. It may be noted that the polarity of the bias is reversed in the PMOS compared to the polarity used for the NMOS.

Since the usage of depletion mode transistors nowadays is limited, they are not in the purview of discussion in this text.

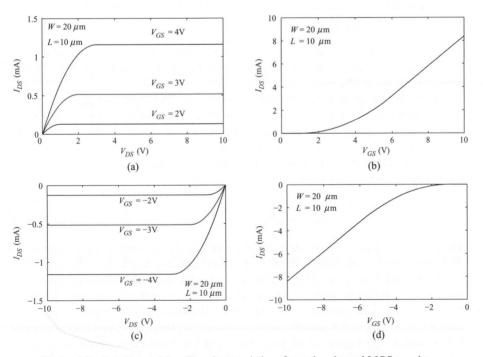

Figure 4.3 I_{DS}–V_{DS} and I_{DS}–V_{GS} characteristics of n-and p-channel MOS transistors

4.2 Layout and Cross-Section of Physical Structure

To understand the physical mechanism governing the operation of MOS transistors and develop compact models, it is essential to know the structure of the device. The circuit performances are strongly dependent on device dimensions. Figure 4.4 shows the physical structure of a MOS transistor in three dimensions. We mention below some of the salient points related to the physical structure of a MOS transistor. The dimensions quoted are only indicative for a *long channel* transistor. The dimensions shrink progressively as the technology node progresses towards smaller feature sizes.

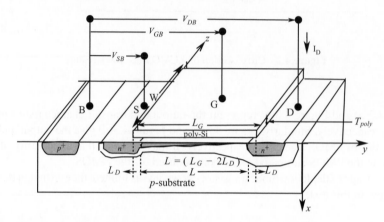

Figure 4.4   Three dimensional view of a MOS transistor with the bulk as reference

- The gate insulator above the channel region is realized by thermally grown thin silicon-dioxide called the gate oxide (GOX) of thickness T_{ox}, typically in the range of 30–500 Å depending on the device feature size.
- The mask gate length of designers is designated as L_M. The printed length on polysilicon is labelled as L_G, which determines the source–drain edges, determined by the self aligned silicon gate technology. The high temperature processing results in a lateral diffusion of length L_D at both source and drain edges under the gate oxide. Therefore, the channel length of the MOS transistor which is defined by the maximum length of the inversion layer under the gate is given by:

$$L = L_G - 2L_D \tag{4.1}$$

The lateral penetration of the diffusion L_D towards the channel is responsible for the deviation of the actual channel length from the printed gate length L_G. The gate overlap over source and drain ensures the electrical contact of the terminals with the channel formed under the gate. It should, however, be as small as possible to reduce the overlap capacitance.
- The conducting gate material on the gate oxide is usually heavily doped polysilicon of thickness $T_{poly} \sim 1\,\mu m$. Normally, the doping is n-type, but in advanced CMOS processes both n- and p-type doping for the polysilicon gate are used to have proper control of the threshold voltage of MOS transistors of complementary conductivity.

- The source–drain diffusion regions are heavily doped. The junction depth X_j is typically in the range of $\sim$0.3–1 μm.
- The source, drain, and substrate contacts are located in the so-called active area which is surrounded by the thick layer of SiO_2 ($\sim$1 μm), called the Field Oxide (FOX), to provide isolation (no inversion layer is formed in the underlying substrate with the supply voltage running over the FOX due to its large thickness).
- The doping of the substrate under the field or isolation oxide by implantation dose is meant to increase the threshold voltage and prevent inversion under it.
- The inversion layer under the gate forms an inversion diode (n^+–p) with a depletion region which electrically isolates the inversion layer from the surrounding p-substrate. The n^+–p inversion diode is always reverse biased. Therefore, a MOS transistor is electrically self isolated.
- The gate voltage V_{GB} determines the conductivity of the inversion channel by varying the population of mobile carriers. The drain-to-source voltage V_{DS} causes the inversion electrons under the gate oxide to flow from source-to-drain. This flow of electrons constitutes the drain current I_{DS}. The drain current may be represented in a functional form by:

$$I_{DS} = k(W, L, T_{ox}, \mu_{neff}) \left[f(V_{GB}, V_{SB}) - f(V_{GB}, V_{DB}) \right] \qquad (4.2)$$

where μ_{neff} is the effective carrier mobility. The function $f(V_{GB}, V_{SB}(V_{DB}))$ depends on the operating mode of the transistor which, in turn, determines the transport mechanism of the carrier along the channel. The transport mechanism can be either drift or diffusion, or both.

- The dimensions given are only *indicative* as they are constantly undergoing changes due to scaling of devices with continuous advancements in technology.
- Figure 4.5(a) describes the device layout in the y–z plane. The horizontal and vertical cross-sections are also shown in Figure 4.5(b) and Figure 4.5(c) respectively The substrate thickness is in the x-direction, length of the channel is along the y-direction, and width in the z-direction. Contact windows provide access to source, drain, substrate, and polysilicon gate for metallurgical contact with aluminum—the common interconnect material.

4.3 Static Drain Current Model

The principle of operation of a long channel MOS transistor regarding the flow of drain current as a function of terminal voltages is qualitatively explained below. We consider a p-substrate for the sake of illustration. The applied gate voltage is assumed large enough to induce an inversion layer under the gate. The modulation of the channel conductivity due to the change in the gate voltage results in a change in the drain current when a bias is applied across drain and source terminals, resulting in the *field effect transistor* action. The potential difference across the source and drain causes carriers to flow, constituting the drain current I_{DS}, which is dependent on the terminal voltages V_{GB}, V_{SB}, V_{DB}. The drain-to-source potential difference is distributed along the channel. Therefore, any point y along the channel can be viewed as a three terminal MOS structure, with the local source potential of value $V_{CB}(y)$. As V_{CB} increases with y, the local threshold voltage $V_{TB}(y)$ increases along the channel resulting in a corresponding decrease in the inversion charge density per unit area. For a given gate voltage

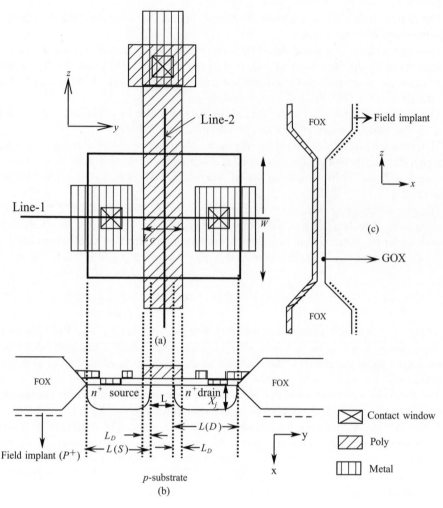

Figure 4.5 (a) Layout of a MOS transistor. (b) Vertical cross-section along Line-1. (c) Vertical cross-section along Line-2

there is a specified channel potential V_P at which inversion is not sustained. If $V_{CB}(L)$, that is the drain potential V_{DB}, is less than V_P the entire channel is in inversion, and the transistor is operating in the linear or ohmic region. If at any point near the drain $V_{CB} \geq V_P$, the channel is pinched-off. The drain voltage at which the pinch-off condition occurs is designated as the drain saturation voltage V_{DBsat}. If the drain saturation voltage is measured w.r.t. source it is designated as the drain-to-source saturation voltage V_{DSsat}.

The potential drop along the conducting channel has two effects in relation to current transport mechanism:

1. The potential gradient results in a longitudinal electric field along the channel, causing a drift component of the channel current.

2. The variation of inversion charge density along the channel results in a concentration gradient causing the diffusion component of the channel current.

It may be mentioned here qualitatively that if the channel is strongly inverted ($V_{GB} > V_{TB}$), the drift component dominates. On the other hand, in the weak inversion condition of the channel ($V_{GB} < V_{TB}$), the potential variation along the channel is negligible, resulting in the absence of drift current and domination of the diffusion component.

It is of significant importance to point out that the MOS transistor is electrically *self-isolated*, because the inversion layer and the diffusion probes, constituting the source and drain terminals of the MOS transistor structure, are isolated from the substrate by a non-conducting depletion layer surrounding them.

Equation (4.2) shows only the qualitative functional dependence of the drain current on various terminal potentials. For circuit analysis and design it is necessary to have expressions for the drain current in terms of the physical parameters associated with device physics, layout defining the device geometry, and terminal voltages determining the operating conditions.

In general, exact solutions for charges, currents, and potentials determining the device operation can only be obtained through numerical methods [2]. For an analytical solution various simplifying assumptions are made which will be discussed in the following sections.

4.3.1 Gradual Channel Approximation

The general nature of the electric field at the Si–SiO$_2$ interface in the channel of a MOS transistor under the gate due to the terminal voltages applied, is obtained from the solution of a two-dimensional Poisson's equation given by:

$$\frac{\partial F(x, y)}{\partial x} + \frac{\partial F(x, y)}{\partial y} = \frac{\rho(x, y)}{\epsilon_{Si}} \tag{4.3}$$

where ρ is the volume space charge density consisting of bulk depletion charge and mobile charges (electrons and holes) under the gate.

For obtaining an analytical solution for the drain current, which is related to the inversion charge, the *Gradual Channel Approximation* (GCA) is commonly used. The GCA assumes that in the channel the rate of change of the vertical field component $|\partial F_x/\partial x|$ set up by the gate-to-substrate voltage is much greater than the rate of the change of the lateral field component $|\partial F_y/\partial y|$ set up by the drain-to-source voltage, where F_x and F_y denote respectively, the vertical and horizontal components of the channel electric field. With this simplification, intrinsically, the two-dimensional MOSFET Poisson's equation reduces to a one-dimensional form which governs the space charge density. Applying the GCA assumption, Poisson's equation governing the channel space charge is described by:

$$\frac{\partial F(x, y)}{\partial x} = \frac{\rho(x)}{\epsilon_{Si}} \tag{4.4}$$

From Equation (4.4) we may conclude that:

• The effect of vertical and horizontal fields set by the gate and drain–source voltage are thus separable. It may be considered that the vertical field set by the gate inverts the substrate at the interface and the lateral field along the channel F_y determines the drift component of the current. The Gradual Channel Approximation holds good in the inversion region under the gate up to the pinch-off point in the case of saturation. Beyond pinch-off, the variation of lateral electric field becomes comparable to that of the vertical field.
• Equation (4.4) also leads to the commonly adopted view that the inversion layer is confined near the Si–SiO$_2$ interface until the pinch-off point. Beyond the pinch-off point, the gate loses control over the inversion charge and the current path is dictated more by the drain voltage.

In the presence of a two-dimensional electric field the above GCA concept becomes questionable, especially at source and drain boundaries. But the error in the calculation of charges can be neglected in a sufficiently long channel MOS transistor. Therefore, the validity of the GCA is dependent on channel length and applied drain bias. The minimum device length and maximum applied V_{DS} for which the GCA is applicable is decided by the oxide thickness, substrate doping, etc. The purpose of changing the process parameters of a device is to make the gate to have more control on channel charge than drain, which can be done, for example, by decreasing the gate oxide thickness.

One can examine the validity of the GCA assumption for a long channel MOSFET to a first order approximation by obtaining simplified expressions for $\partial F_x/\partial x$ and $\partial F_y/\partial y$ in terms of process parameters and terminal voltages.

4.3.2 Continuity Equation

For the formulation of a continuity equation related to the current flow along the channel we consider an elementary volume element of length dy, width W, and channel thickness W_i.

We refer to Figure 4.6 for applying the continuity equation for the channel current in a MOSFET. Neglecting gate[2] and recombination–generation current components in the elemental volume, the one-dimensional continuity equation (Chapter 1) reduces to:

$$\frac{d}{dt}(\text{total charge in volume element}) = I_c(y + dy) - I_c(y) \qquad (4.5)$$

where $I_c(y + dy)$ and $I_c(y)$, denote the channel current entering and leaving the elemental volume at $y + dy$ and y respectively which are in the same direction as the drain current.

$$-\frac{d}{dt}W dy \int_0^{W_i} qn(x)\,dx = I_c(y + dy) - I_c(y) \qquad (4.6)$$

---

[2] Gate oxide thickness is assumed such that there is no tunneling mechanism through the oxide for a given bias.

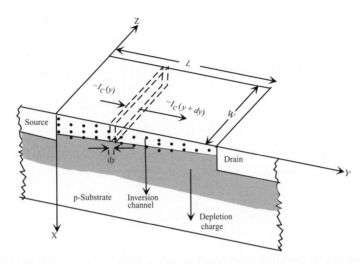

Figure 4.6   Volume element in the channel under the gate oxide for an *n*-channel MOSFET

The integral $-\int_0^{W_i} qn(x)\,\mathrm{d}x$ is nothing but the surface electron charge density per unit area (Q'_n):

$$W\mathrm{d}y\frac{\mathrm{d}Q'_n}{\mathrm{d}t} = \left[I_c(y) + \frac{\mathrm{d}I_c(y)}{\mathrm{d}y}\mathrm{d}y\right] - I_c(y) \qquad (4.7)$$

$$W\frac{\mathrm{d}Q'_n}{\mathrm{d}t} = \frac{\mathrm{d}I_c(y)}{\mathrm{d}y} \qquad (4.8)$$

For a static current $\frac{\mathrm{d}Q'_n}{\mathrm{d}t} = 0$, which reduces Equation (4.8) to:

$$\frac{\mathrm{d}I_c(y)}{\mathrm{d}y} = 0 \qquad (4.9)$$

The above equation signifies that the static current in the channel has zero divergence and, therefore, $I_c(y) = I_{DS}$ at any point in the channel.

4.3.3 Transport Equation

In Chapter 1 we have shown that the transport mechanism responsible for the carrier flow in a semiconductor is constituted of drift and diffusion, due to potential and carrier concentration gradients respectively. For calculating the channel current in a MOSFET we consider a thin strip of the inversion layer at *y* of thickness d*x* and width *W* through which the elemental current dI_c flows along the channel. This elemental current integrated over the channel thickness W_i gives the channel current at position *y*.

The expression for the drain current, considering both drift and diffusion mechanisms, can be expressed through a quasi-Fermi potential gradient as given in Equation (1.103) *(current flow considered in the positive y-direction).* We replace ϕ_n by V_{CB} which is the channel potential w.r.t. bulk representing the quasi-Fermi potential for the channel electrons. *Considering that the drain current is taken as positive as it flows from drain to source in the negative y-direction along the channel,* the sign of Equation (1.103) has to be reversed which gives:

$$dI_c(x, y) = qW\mu_n(x, y)\,n(x, y)\,dx\frac{dV_{CB}(y)}{dy} \tag{4.10}$$

Integrating over the channel thickness W_i, which gives the channel current $I_c(y)$, and recognizing that $I_c(y) = I_{DS}$, we get:

$$I_{DS} = qW\mu_{neff}\int_o^{W_i} n(x, y)\,dx\left[\frac{dV_{CB}(y)}{dy}\right] \tag{4.11}$$

In Equation (4.11) the mobility term has been taken outside the integral assuming it to be independent of x and y, though strictly speaking it is field dependent. For the sake of simplicity an average of the value at the source and drain can be assumed designating it as an *effective mobility* μ_{neff}.

Now, using the continuity condition $I_c(y) = I_{DS}$, and that $-q\int_0^{W_i} n(x, y)\,dx = Q_n'(y)$ is the inversion charge density at y, Equation (4.11) reduces to:

$$I_{DS} = -W\mu_{neff}Q_n'\frac{dV_{CB}}{dy} \tag{4.12}$$

which can also be written in integral form as:

$$I_{DS} = -\mu_{neff}\frac{W}{L}\int_{V_{SB}}^{V_{DB}} Q_n'\,dV_{CB} \tag{4.13}$$

4.3.4 Pao–Sah Exact Model

In the previous section we obtained an expression for the drain current in an integral form. For the evaluation of this integral we need to know the inversion charge density at a position y where the channel potential is $V_{CB}(y)$. In Chapter 2 we have shown that this can found precisely by numerical techniques. The Pao–Sah model [6] follows the above procedure, which is recognized as the benchmark model for the drain current of a MOS transistor. From Equation (2.4) we can write:

$$Q_n'(V_{CB}) = q\int_{x_c}^0 n(x, V_{CB})\,dx = \int_{\phi_f}^{\phi_s} \frac{n_i^2}{N_A}\,e^{\frac{\phi(x)-V_{CB}}{\phi_t}}\frac{dx}{d\phi}\,d\phi \tag{4.14}$$

In the above equation, it has been assumed that the inversion carriers are negligible beyond the position $x = x_c$, where the band bending equals ϕ_f.

Substituting Equation (4.14) into Equation (4.13) we have:

$$I_{DS} = -\mu_{neff}\frac{W}{L}\int_{V_{SB}}^{V_{DB}}\left(\int_{\phi_f}^{\phi_s}\frac{n_i^2}{N_A}e^{\frac{\phi(x)-V_{CB}}{\phi_t}}\frac{1}{F(x, V_{CB})}\,d\phi(x)\right)dV_{CB} \qquad (4.15)$$

which is known as *Pao–Sah double integral formula* for MOSFET drain current. The following broad observations apply to the Pao–Sah model:

- It considers both drift and diffusion current transport mechanism in the channel, and therefore is general, and not restrictive in terms of an assumed transport mechanism.
- It is based on the estimation of inversion charge density Q_n' numerically by integrating the inversion carrier concentration along the depth x without restoring to any approximation, except the Boltzmann's distribution of carriers as given in Equation (4.14).
- It is the most comprehensive model, and therefore, is considered a benchmark MOSFET drain current model.
- As the expression for current is computationally involved, it has not been a popular device model for use in circuit simulators.

4.3.5 MOSFET Modeling Approaches for Circuit Simulation

An accurate analytical and compact device model is a basic prerequisite in computer aided design activity, as it faithfully represents device behavior and characteristics through a set of *model parameters* acceptable by circuit simulators. As there are diverse groups of designers specializing in analog, digital, mixed signal spread over a wide range of frequency and power, modeling has specific demands depending on accuracy, computational effort, and applications. Therefore, there have been several approaches to MOSFET modeling [3–5] adopted by various modeling groups. Currently available models can be classified into the following broad categories.

1. Surface potential based models [7, 9–14].
2. Charge based models [16, 27, 28].
3. Threshold voltage based models [17–19].

4.3.5.1 Surface Potential (SP) Based Models

The surface potential approach involves the computation of surface potential at the source and drain ends of the channel from the applied terminal voltages. Subsequently, channel charge is related to surface potential at a given point of the channel from which expressions of current are obtained.

In surface potential based methods, Equation (4.13) is recast in the following form, where the integrating variable is changed from channel potential V_C to surface potential ϕ_s as follows:

$$I_{DS} = -\mu_{neff}\frac{W}{L}\int_{\phi_{sS}}^{\phi_{sD}}Q_n'(\phi_s)\frac{dV_C}{d\phi_s}\,d\phi_s \qquad (4.16)$$

SP based models have the following broad features:

- The majority of surface potential based models consider, in general, the substrate/bulk as the reference terminal, which ensures symmetric device operation.
- The surface potentials at the source and drain ends are computed either analytically or numerically for given terminal voltages. Though the task of calculating surface potential, using either complex analytic expression or numerical techniques, is computationally involved, improved algorithms make the computational time acceptable for circuit simulation.
- The approach has the potential to characterize various phenomena resulting from the technology scaling. It addresses the moderate inversion region *naturally*, due to the inclusion of both drift and diffusion components of the current. It claims to be powerful in RF applications.
- The notable advantage of this modeling approach is its inherent capability to accurately describe the drain current and its higher derivatives without discontinuity in the entire range of device operation.
- Examples: Philips MM11 [13], HiSIM [7], SP [14], PSP [15], etc.

4.3.5.2 Charge Based (CB) Models

The emphasis in the charge based modeling approach is on the simplified but accurate modeling of inversion charge in the channel, and relating it to channel potential for finally obtaining compact analytical current–voltage relationships. In the charge based modeling approach, the general drain current expression given in Equation (4.13) is expressed in terms of inversion charge density as the variable in the integral, as given below:

$$I_{DS} = -\mu_{neff}\frac{W}{L}\int_{Q'_{nS}}^{Q'_{nD}} Q'_n\frac{\mathrm{d}V_C}{\mathrm{d}Q'_n}\,\mathrm{d}Q'_n \qquad (4.17)$$

The features of this class of models are:

- Most of the charge based models are developed with the bulk as the reference making it consistent with the physical symmetry of the device.
- The continuity of current and its derivatives in transition from different modes of operation is generally assured using mathematical interpolation functions.
- With suitable linearization of the inversion charge density, a compact relation for small signal parameters in terms of bias voltages, in any operating region can be obtained analytically, which is particularly suitable for analog design.
- In some charge based models numerical algorithms are used for accurate modeling of channel charge.
- Examples: EKV [27], ACM [28], BSIM5 [30], etc.

4.3.5.3 Threshold Voltage (V_T) Based Models

Threshold voltage based models are built on the regional representation of the MOS transistor operation in strong and weak inversion. The approach happens to be one of the earliest to be implemented in the SPICE circuit simulator. In the VT based approach we basically solve Equation (4.13), namely:

$$I_{DS} = -\mu_{neff} \frac{W}{L} \int_0^{V_{DS}} Q_n' \, dV_{CS} \qquad (4.18)$$

where Q_n' is expressed in terms of channel potential and threshold voltage V_T by using regional expressions. The general features of V_T based modeling are:

- The device operation is modeled with source as reference.
- Smoothing (interpolating) functions have been incorporated later to provide continuity in expressions for current and its derivatives at the transition regions of operation.
- These models are available for wide range of technological variation related to MOSFETs.
- It has the history of sustained model upgradation which enables it to evolve continuously keeping pace with the refinements in MOSFET technology.
- Examples: Berkeley MOSFET model family LEVEL 1–3, BSIM1–4 [19], and Philips MM9 [46], etc.

In the modeling approaches mentioned above, the integrals in the respective drain current equations have been evaluated by assuming a charge sheet approximation, which has been a major simplifying factor, as discussed in the next section.

Figure 4.7 summarizes various approaches implemented for modeling of MOS transistors.

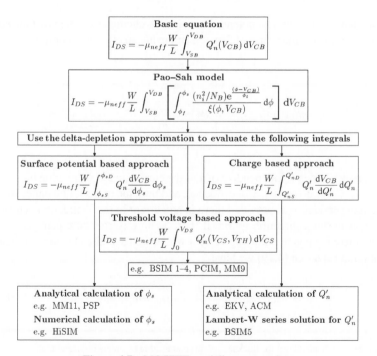

Figure 4.7   MOSFET modeling approaches

4.3.6 Charge Sheet Approximation

For the evaluation of drain current using the Pao–Sah model it is required to solve a double integral numerically, which is not suitable for circuit simulation. To simplify the numerical computational burden, Brews introduced a *Charge Sheet Model* which makes the following assumptions [23]:

- The channel inversion charge layer located at the Si–SiO$_2$ interface is infinitesimally thin, which can be viewed as a sheet of charge without any potential drop across it.[3]
- For the estimation of the bulk charge underlying the inversion layer, the depletion approximation is applicable.
- The drain current is considered to be constituted by both drift and diffusion mechanisms for which the carrier concentration in the channel is expressed in terms of a quasi-Fermi potential. In order to maintain the current continuity, $I_c = constant$ along the channel: the quasi-Fermi potential, and correspondingly the carrier concentration, adjusts itself along the channel. Such an adjustment leads to a gradient of quasi-Fermi potential along the channel, using which, a *single transport equation* is able to provide an expression for the drain current which includes both drift and diffusion mechanisms.

It may be recalled that the charge sheet approximation is implied in the delta-depletion approximation introduced in Chapter 2 for evaluating depletion charge density in a MOS structure. Because of its elegance and simplicity, the charge sheet approximation has been a sort of common platform for all of the current compact MOSFET models briefly described below.

Expanding the electrochemical potential term into its chemical and electrostatic components in the expression for the drain current density, and using an effective mobility μ_{neff} value, we have:

$$J = -qn(x,y)\mu_{neff}\frac{\partial}{\partial y}\left[\underbrace{\phi_t \ln\frac{n(x,y)}{n_i}}_{\text{chemical potential}} - \underbrace{\phi(x,y)}_{\text{electrostatic potential}}\right] \tag{4.19}$$

In Equation (4.12) V_{CB} is independent of x as the current in the x-direction is zero. However, in Equation (4.19) we do have x, y dependence, and both the terms will balance together such that V_{CB} is independent of x, provided there is no current in the x-direction. The assumption of zero current in the x-direction is true if there is no electron–hole pair generation in the channel, and there is no gate tunneling current which we have already assumed. Carrying out differentiation in Equation (4.19) we have:

$$J = -kT\mu_{neff}\frac{\partial n(x,y)}{\partial y} + q\mu_{neff}n(x,y)\frac{\partial \phi(x,y)}{\partial y} \tag{4.20}$$

---

[3] It has been shown in Chapter 2 that the thickness of inversion is thin by orders of magnitude when compared to the depletion width.

Equation (4.20) can be written for current by multiplying both sides by dx and dz, where dz is along the width of the channel:

$$J \, dx \, dz = -kT\mu_{neff} \frac{\partial n(x, y)}{\partial y} \, dx \, dz + q\mu_{neff} n(x, y) \frac{\partial \phi(x, y)}{\partial y} \, dx \, dz \tag{4.21}$$

Performing double integration over x and z we get:

$$I_{DS} = -kT\mu_{neff} W \int \frac{\partial n(x, y)}{\partial y} \, dx + q\mu_{neff} W \int n(x, y) \frac{\partial \phi(x, y)}{\partial y} \, dx \tag{4.22}$$

Introducing the charge sheet approximation, by which the potential drop across the sheet of inversion charge can be ignored, we can write:

$$\phi(x, y) \cong \phi_s(y) \tag{4.23}$$

Therefore, Equation (4.22) takes the following form:

$$I_{DS} = -kT\mu_{neff} W \int \frac{\partial n(x, y)}{\partial y} \, dx + q\mu_{neff} W \int n(x, y) \frac{\partial \phi_s(y)}{\partial y} \, dx \tag{4.24}$$

Reordering the integration and differentiation, we have:

$$I_{DS} = -kT\mu_{neff} W \frac{\partial}{\partial y} \int n(x, y) \, dx + q\mu_{neff} W \frac{\partial \phi_s(y)}{\partial y} \int n(x, y) \, dx \tag{4.25}$$

$$I_{DS} = \underbrace{\phi_t \mu_{neff} W \frac{\partial Q'_n(y)}{\partial y}}_{\text{diffusion component}} - \underbrace{\mu_{neff} W Q'_n(y) \frac{\partial \phi_s(y)}{\partial y}}_{\text{drift component}} \tag{4.26}$$

The first term on the RHS of Equation (4.26) represents the diffusion component of the drain current, as it is dependent on the spatial variation of charge sheet density. The second term indicates the drift component of the current, as it is proportional to the gradient of the surface potential, i.e. channel electric field.

Assuming individual components are present such that $I_{DS} = I_{drift} + I_{diff}$, we can write:

$$I_{drift} = -\mu_{neff} W Q'_n(y) \frac{\partial \phi_s(y)}{\partial y} \tag{4.27}$$

$$I_{diff} = \phi_t \mu_{neff} W \frac{\partial Q'_n(y)}{\partial y} \tag{4.28}$$

For evaluating the drift and diffusion components of the current we need the expression for the charge density at a point y in the channel which is given as:

$$Q'_n(y) = -C'_{ox}\left(V_{GB} - \phi_s(y) - \gamma\sqrt{\phi_s(y)}\right)$$
(4.29)

where we have assumed depletion approximation for the estimation of Q'_b. Substituting Equation (4.29) into Equation (4.27) and integrating, we have:

$$I_{drift} = \mu_{neff}C'_{ox}\frac{W}{L}\left(V_{GB}\phi_s - \frac{\phi_s^2}{2} - \frac{2}{3}\gamma\phi_s^{\frac{3}{2}}\right)\Bigg|_{\phi_{sS}}^{\phi_{sD}}$$
(4.30)

and:

$$I_{diff} = \phi_t\mu_{neff}C'_{ox}\frac{W}{L}\left(\phi_s + \gamma\sqrt{\phi_s}\right)\Bigg|_{\phi_{sS}}^{\phi_{sD}}$$
(4.31)

Equations (4.30) and (4.31) together form the basis for the *Surface Potential Based MOSFET* model in which the surface potentials at the source and drain boundaries are calculated by approximate explicit expressions (e.g., SP MOSFET Model), or by efficient numerical algorithms (e.g., HiSIM MOSFET Model).

Equations obtained for drain current using the charge sheet approximation have been compared with that of the Pao–Sah model and accuracy within 1 % is achieved [24]. When at least one end of the channel is strongly inverted the transistor is said to be operating in strong inversion mode.[4] With the source end in strong inversion, if the potential at the drain terminal is raised such that the drain end of the channel enters into the weak inversion region at $V_{DB} = V_{DBsat}$, the transistor is said to be operating in the saturation mode. As we know from Chapter 2, in weak inversion the surface potential gets almost locked, which means that the upper limit of Equation (4.30) is constant. The current gets saturated for $V_{DB} > V_{DBsat}$, and that is the reason for its label as the *saturation region*. The region before saturation, i.e. for $V_{DB} < V_{DBsat}$, the drain current is found to be an *almost* linear function of V_{DB}, due to which the region is called *linear*.

The magnitude of the drift and diffusion components of the drain current will depend on the region of operation of the MOS transistor. In Figure 4.8 the gate voltage is swept such that the operating mode of the transistor travels from the weak to strong inversion mode. It is clearly seen that the drift component is much less compared to the diffusion component in weak inversion. This fact is the result of negligible *field assisted* carrier transport from source to drain, as the surface potential at the source and drain ends is almost equal resulting in a small electric field. The drift component dominates over the diffusion component in strong inversion because of the lateral field along the channel induced by the drain-to-source potential difference.

---

[4] In the strong inversion operation the source end of the channel is more strongly inverted than the drain since the channel potential is higher at the drain than at the source in the case of an NMOS transistor.

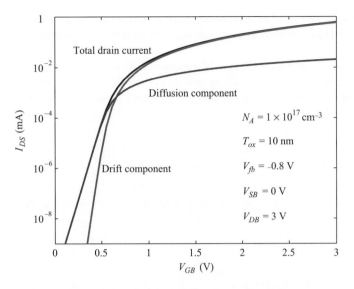

Figure 4.8 Drift and diffusion components of the drain current

We can also have an expression for drift and diffusion currents in terms of inversion charge density at the source and drain ends of the channel. Linearizing the inversion charge density $Q'_n(y)$ w.r.t. surface potential $\phi_s(y)$ as shown in Chapter 2, we have the relation between incremental charge and surface potential, given by:

$$dQ'_n(y) = nC'_{ox}\, d\phi_s(y) \tag{4.32}$$

where n is the linearizing slope factor for the channel charge density. Substituting Equation (4.32) into Equation (4.27), we can have an alternate expression for the drift component in terms of channel charge as the variable:

$$I_{drift} = -\mu_{neff} W Q'_n(y)\frac{1}{nC'_{ox}}\frac{\partial Q'_n(y)}{\partial y} \tag{4.33}$$

Integrating we have:

$$I_{drift} = \mu_{neff}\frac{W}{L}\frac{1}{2nC'_{ox}}\left(Q'^2_{nS} - Q'^2_{nD}\right) \tag{4.34}$$

Further, for the diffusion component, integrating Equation (4.28) we have:

$$I_{diff} = \phi_t\mu_{neff}\frac{W}{L}\left(Q'_{nD} - Q'_{nS}\right) \tag{4.35}$$

Equations (4.34) and (4.35) form the basis for the *Charge Based* MOSFET model in which the drain current is related to the channel inversion charge densities at the source and drain ends.

These charges are expressed in terms of applied bias with explicit expressions (eg. ACM, EKV MOSFET model) as given in Chapter 2, or efficient numerical algorithms (BSIM5 MOSFET model).

The ACM model gives a continuous expression for the inversion charge in terms of applied bias for all regions of operation above depletion, and therefore a continuous single piece analytical expression for the drain current [28]. Adding Equations (4.34) and (4.35), we have:

$$I_{DS} = \frac{\mu_{neff}}{C'_{ox}} \frac{W}{L} \frac{1}{2n} \left(Q'^2_F - Q'^2_R \right) \tag{4.36}$$

where $Q'_{F(R)} = Q'_{nS(D)} - nC'_{ox}\phi_t$. Equation (4.36) is usually reported in the literature in the following form:

$$I_{DS} = \mu_{neff} C'_{ox} \frac{W}{L} \frac{\phi_t^2}{2} \left[\frac{1}{n} \left(\frac{Q'_F}{C'_{ox}\phi_t} \right)^2 - \frac{1}{n} \left(\frac{Q'_R}{C'_{ox}\phi_t} \right)^2 \right] \tag{4.37}$$

An alternative approach for the charge based model is based on regional approximations for the inversion charge density as adopted in the EKV model, and then having an interpolation function to connect weak and strong inversion asymptotically to maintain continuity during transition from weak to strong inversion.

In strong inversion, where the drift current dominates, we have the expression for inversion charge density as derived in Chapter 2:

$$Q'_{n(S,D)} = -nC'_{ox}(V_P - V_{(S,D)B}) \tag{4.38}$$

Substituting Equation (4.38) into the expression for the drift current in Equation (4.34), we have:

$$I_{drift} = \mu_{neff} \frac{W}{L} \frac{1}{2nC'_{ox}} \left\{ [nC'_{ox}(V_P - V_{SB})]^2 - [nC'_{ox}(V_P - V_{DB})]^2 \right\} \tag{4.39}$$

$$I_{drift} = \underbrace{\frac{n\beta}{2} [(V_P - V_{SB})]^2}_{I_F} - \underbrace{\frac{n\beta}{2} [(V_P - V_{DB})]^2}_{I_R} \tag{4.40}$$

In weak inversion, where the diffusion current dominates, we have:

$$Q'_{n(S,D)} = -2nC'_{ox}\phi_t e^{\frac{V_P - V_{(S,D)B}}{\phi_t}} \tag{4.41}$$

Substituting Equation (4.41) into the expression for the diffusion current in Equation (4.35), we have:

$$I_{diff} = \phi_t \mu_{neff} \frac{W}{L} \left(2nC'_{ox}\phi_t e^{\frac{V_P - V_S}{\phi_t}} - 2nC'_{ox}\phi_t e^{\frac{V_P - V_D}{\phi_t}} \right) \tag{4.42}$$

$$I_{diff} = \underbrace{2n\beta\phi_t^2 e^{\frac{V_P - V_S}{\phi_t}}}_{I_F} - \underbrace{2n\beta\phi_t^2 e^{\frac{V_P - V_D}{\phi_t}}}_{I_R} \tag{4.43}$$

Equations (4.40) and (4.43) represent the regional current models for the strong and weak inversion regions respectively. A smart mathematical interpolation function [26], though *non-physical*, connects the two regions and provides a *single-piece* analytical expression for the drain current. The resulting expression includes the transitional moderate inversion region sandwiched between the strong and weak regions as discussed later in Section 4.9.

In the next section we will discuss threshold voltage based models which have been *de facto* industry standards since inception.

4.4 Threshold Voltage (V_T) Based Model

Threshold voltage models are inherently regional in approach, considering the threshold voltage as the transition gate voltage between weak and strong inversion modes of operation. The threshold voltage based models have gone through extensive evolution. Versions offering capabilities tailored for different technology nodes are available. These models have been popular as they support (i) second order phenomena due to scaling, and (ii) adaptability to a variety of MOS technologies.

The following section considers the drain current expressions of several versions of threshold voltage based models for long channel MOSFETs.

4.4.1 Strong Inversion

We start with the expression for the drift current (Equation (4.30)), as in strong inversion it is the drift mechanism that dominates the carrier transport:

$$I_{drift} = \mu_{neff} C'_{ox} \frac{W}{L} \left(V_{GB}\phi_s - \frac{\phi_s^2}{2} - \frac{2}{3}\gamma\phi_s^{\frac{3}{2}} \right) \Big|_{\phi_{sS}}^{\phi_{sD}} \tag{4.44}$$

As discussed in Chapter 2, in strong inversion the surface potential at the source and drain channel ends can be approximated by $\phi_{sS} = V_{SB} + \phi_0$ and $\phi_{sD} = V_{DB} + \phi_0$ respectively. Due to the assumption that the surface potential gets pinned in strong inversion, V_T based models are also known as *pinned surface potential based models*. Therefore, we have:

$$I_{drift} = \mu_{neff} C'_{ox} \frac{W}{L} \left(V_{GB}\phi_s - \frac{\phi_s^2}{2} - \frac{2}{3}\gamma\phi_s^{\frac{3}{2}} \right) \Big|_{V_{SB}+\phi_0}^{V_{DB}+\phi_0} \tag{4.45}$$

Historically, V_T based models are source referred. Therefore, we can write $V_{SB} = -V_{BS}$, $V_{GB} = V_{GS} - V_{BS}$, $V_{DB} = V_{DS} - V_{BS}$. Using these transformations in Equation (4.45), we have:

$$I_{drift} = \mu_{neff} C'_{ox} \frac{W}{L} \left[(V_{GS} - V_{BS})\phi_s - \frac{\phi_s^2}{2} - \frac{2}{3}\gamma\phi_s^{\frac{3}{2}} \right] \Bigg|_{-V_{BS}+\phi_0}^{V_{DS}-V_{BS}+\phi_0} \tag{4.46}$$

After putting limits of integration, we get:

$$I_{drift} = \mu_{neff} C'_{ox} \frac{W}{L} \left\{ \left(V_{GS} - \phi_0 - \frac{V_{DS}}{2} \right) V_{DS} \right.$$

$$\left. - \frac{2}{3}\gamma \left[(V_{DS} - V_{BS} + \phi_0)^{\frac{3}{2}} - (-V_{BS} + \phi_0)^{\frac{3}{2}} \right] \right\} \tag{4.47}$$

Equation (4.47) can be derived alternatively, conforming directly to the framework outlined in Equation (4.18), where the inversion charge density is required to be expressed as a function of channel potential, $Q'_n = f(V_{CS}, V_{GS})$. The expression for inversion charge density in strong inversion, Q'_n, as a function of the source referred channel potential V_{CS}, follows directly from the equation $Q'_n = -C'_{ox}(V_{GB} - V_{TB})$, with voltages referred to source:

$$Q'_n = -C'_{ox} \left(V_{GS} - \phi_0 - V_{CS} - \gamma\sqrt{V_{CS} + V_{SB} + \phi_0} \right) \tag{4.48}$$

Substitution of Equation (4.48) into Equation (4.18), and subsequent integration, results in the same expression as given in Equation (4.47).

Equation (4.47) represents the SPICE LEVEL 2 MOSFET model equation for drain current in the linear region with $\phi_0 = 2\phi_f$. The current in the saturation region is obtained by putting V_{DSsat} for V_{DS} in the above equation, at which the channel gets pinched-off. The condition for pinch-off voltage[5] was derived in Section 2.8.3. The pinch-off voltage was defined with respect to bulk as the reference. For the source as reference, we can write:

$$V_{DSsat} = V_P + V_{BS} \tag{4.49}$$

which gives:

$$V_{DSsat} = V_{BS} - 2\phi_f + \frac{\gamma^2}{4} \left\{ 1 - \left[1 + \frac{4}{\gamma^2}(V_{GS} - V_{BS}) \right]^{\frac{1}{2}} \right\}^2 \tag{4.50}$$

It has been found that Equation (4.47) is computationally inefficient because of the $\frac{3}{2}$ power involved in the expression. The origin of the term with the power of $\frac{3}{2}$ can be traced to the square

---

[5] $V_P = \left(-\frac{\gamma}{2} + \sqrt{\frac{\gamma^2}{4} + V_{GB}} \right)^2 - \phi_0.$

root dependence of the bulk charge with the surface potential, as can be seen in Equations (4.29) and (4.30).

A computationally efficient simpler expression for the drain current can be obtained by linearizing the bulk charge as a function of surface potential using Taylor's expansion around $\phi_s = \phi_{sS}$:

$$Q_b' = Q_{b0}' + (\phi_s - \phi_{sS}) \left. \frac{\partial Q_b'}{\partial \phi_s} \right|_{\phi_s=\phi_{sS}} \tag{4.51}$$

$$Q_b' = -\gamma C_{ox}' \left[\sqrt{\phi_{sS}} + \delta(\phi_s - \phi_{sS}) \right] \tag{4.52}$$

where:

$$Q_{b0}' = -\gamma C_{ox}' \sqrt{\phi_{sS}} \tag{4.53}$$

and:

$$\delta = \frac{1}{2\sqrt{-V_{BS} + \phi_0}} \tag{4.54}$$

It may be noted that the symbol ϕ_{sS} represents *the surface potential at the source end of the channel measured w.r.t. bulk.*

Substituting Equation (4.52) for Q_b', Q_n' can be evaluated from the following expression:

$$-\frac{Q_n'}{C_{ox}'} = V_{GB} - \phi_s + \frac{Q_b'}{C_{ox}'} \tag{4.55}$$

Substituting Q_n' obtained from the above Equation (4.52) and (4.55) into Equation (4.27) and integrating, the expression for the drift current is given by:

$$I_{drift} = \mu_{neff} C_{ox}' \frac{W}{L} \left(V_{GB}\phi_s - \frac{\phi_s^2}{2} - \gamma\sqrt{\phi_{sS}}\phi_s - \delta\gamma\frac{\phi_s^2}{2} + \delta\gamma\phi_{sS}\phi_s \right) \Bigg|_{V_{SB}+\phi_0}^{V_{DB}+\phi_0} \tag{4.56}$$

After rearrangement, we get:

$$I_{drift} = \mu_{neff} C_{ox}' \frac{W}{L} \left\{ \left[V_{GB} - \gamma(\sqrt{\phi_{sS}} - \delta\phi_{sS}) - \frac{(1 + \gamma\delta)}{2}\phi_s \right] \phi_s \right\} \Bigg|_{V_{SB}+\phi_0}^{V_{DB}+\phi_0} \tag{4.57}$$

Transforming the above equation from bulk referred to source referred considering the potential difference between source and bulk and assuming $I_{DS} = I_{drift}$, we have:

$$I_{DS} = \mu_{neff} C_{ox}' \frac{W}{L} \left[\left(V_{GS} - \phi_0 - \gamma \sqrt{-V_{BS} + \phi_0} \right) V_{DS} - \frac{(1 + \delta\gamma)}{2} V_{DS}^2 \right] \qquad (4.58)$$

This can be alternatively written as:

$$I_{drift} = I_{DS} = \mu_{neff} C_{ox}' \frac{W}{L} \left[(V_{GS} - V_{TS}) V_{DS} - \frac{n V_{DS}^2}{2} \right] \qquad (4.59)$$

where:

$$n = 1 + \gamma\delta = 1 + \frac{\gamma}{\sqrt{2\phi_f - V_{BS}}} \qquad (4.60)$$

Equation (4.58) is an expression for the drain current in the LEVEL 3 long channel MOSFET model in the linear region, and transforms to a saturation current equation with $V_{DS} = V_{DSsat}$.

To obtain an expression for V_{DSsat} for LEVEL 3 we again use the pinch-off condition at the drain considering the bulk charge linearization. Under the pinch-off condition $Q_n' = 0$, we have:

$$Q_b' = -C_{ox}' [V_{GB} - (V_{DBsat} + \phi_0)] \qquad (4.61)$$

Using Equations (4.52) and (4.61), we get:

$$\gamma \sqrt{\phi_{sS}} + \gamma\delta(\phi_{sD} - \phi_{sS}) = V_{GB} - V_{DBsat} - \phi_0 \qquad (4.62)$$

Substituting the following transformations into Equation (4.62):

$$V_{GB} = V_{GS} + V_{SB}, \qquad\qquad V_{DBsat} = V_{DSsat} + V_{SB}$$

$$\phi_{sD} = V_{DBsat} + \phi_0, \qquad\qquad \phi_{sS} = V_{SB} + \phi_0$$

the expression for V_{DSsat} is:

$$V_{DSsat} = \frac{(V_{GS} - V_{TS})}{1 + \gamma\delta} \qquad (4.63)$$

where:

$$V_{TS} = \phi_0 + \gamma\sqrt{V_{SB} + \phi_0} \qquad (4.64)$$

V_{TS} denotes the threshold voltage with source as the reference.

Linearization of bulk charge thus simplifies the drain current and drain saturation voltage expressions, and it is a common practice to linearize the bulk charge in developing MOSFET compact models. The accuracy of such a model depends on the location of channel potential at which linearization is carried out, and also the slope δ. Later sections will discuss more about such linearization.

A relatively simplified drain current model, useful for hand calculation, follows from Equation (4.58) by assuming that there is no variation of bulk charge along the channel which corresponds to $\delta = 0$, namely:

$$I_{drift} = \mu_{neff} C'_{ox} \frac{W}{L} \left[\left(V_{GS} - \phi_0 - \gamma \sqrt{-V_{BS} + \phi_0} \right) V_{DS} - \frac{V_{DS}^2}{2} \right] \qquad (4.65)$$

$$I_{drift} = I_{DS} = \mu_{neff} C'_{ox} \frac{W}{L} \left[(V_{GS} - V_{TS}) V_{DS} - \frac{V_{DS}^2}{2} \right] \qquad (4.66)$$

Equation (4.66) is the SPICE LEVEL 1 MOSFET model for the drain current in the linear region, and was the first model to be implemented in a SPICE circuit simulator. In the saturation region the drain current equation may be given as:

$$I_{drift} = I_{DS} = \frac{1}{2} \mu_{neff} C'_{ox} \frac{W}{L} (V_{GS} - V_{TS})^2 \qquad (4.67)$$

where we have used $V_{DSsat} = V_{GS} - V_{TS}$ as the drain saturation voltage, which can be derived from Equation (4.63) with $\delta = 0$: For $\delta \neq 0$:

$$I_{drift} = I_{DS} = \mu_{neff} C'_{ox} \frac{W}{L} \frac{(V_{GS} - V_{TS})^2}{2n} \qquad (4.68)$$

Due to the quadratic dependence of the current on gate overdrive $V_{GS} - V_{TS}$, the LEVEL 1 model is also referred to as the *square law model*.

The accuracy of the current predicted by all of the three MOSFET models is compared with that given by the Brews' model (Equation 4.30) [23, 43]. The comparison is shown in Figure 4.9[6] where output characteristics are depicted. From this figure it is seen that the LEVEL 1 model overestimates the current. This can be physically understood from the assumption involved that the threshold voltage variation along the channel is constant which, in turn, overestimates the channel inversion charge.[7] In spite of this inaccuracy the LEVEL 1 model is widely used for tutorial purposes and hand calculation for circuit design, due to its simplicity and compact form. It may be mentioned that it is common practice to modify the parameters involved in the respective model equations to get better fits with the experimental results. These analytical and compact model equations are incorporated in the SPICE simulation engine. The *model parameters* are to be provided in the SPICE netlist code.

Table 4.1 outlines the LEVEL 1 model parameters required for the calculation of static drain current. The number of model parameters varies from model to model in an attempt to accurately describe the device behavior [18].

---

[6] Throughout this chapter the Brews' model will be used as a benchmark and will be represented by circles (o) in plots.

[7] The threshold voltage along the channel increases from source to drain as the surface potential, and consequently the depletion charge density along the channel increases from source to drain.

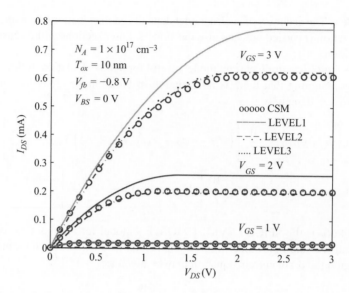

Figure 4.9   Comparison of the LEVELS 1, 2, and 3 drain current models with Equation (4.30).

The model parameters supplied by the foundry and device dimensions (W, L) selected by the designer together decide the $I-V$ characteristics of the device. *Thus, model parameters provide the key interface between the foundry and design houses.* For the optimal use of a device for a given technology it is of utmost importance that the physical significance of model parameters are understood clearly. To faithfully reproduce the measured characteristics of the device, an accurate determination of model parameters is as important as the development of the model itself.

The DC equivalent circuit of a MOSFET is shown in Figure 4.10 with source as reference [20]. The parasitic resistances r_s and r_d are the ohmic resistances between the source/drain contacts and the channel edge under the gate. A non-linear current source I_{DS} which is dependent on terminal voltages replaces the conducting channel under the gate. As the source/drain

Table 4.1   SPICE LEVEL 1 static model parameters

Model Parameter	Symbol	Description
VTO	V_{TS}	Zero-bias threshold voltage
KP	$\mu_{neff}C'_{ox}$	Process transconductance
GAMMA	γ	Body effect coefficient
PHI	$2\phi_f$	Inversion surface potential
LAMBDA	λ	Channel length modulation factor (Section 4.6)
LD	L_D	Lateral diffusion under gate
RS	r_s	Source ohmic resistance
RD	r_d	Drain ohmic resistance
IS	I_S	Junction reverse saturation current

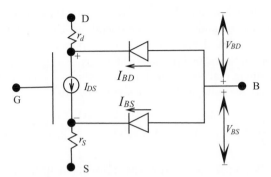

Figure 4.10   DC equivalent circuit for an MOS transistor with gate voltage controlled current sources

to bulk junction is always reverse biased under normal operating conditions, there will be a reverse saturation junction leakage current. These are designated by I_{BS} and I_{BD}.

The conditions of operation for the three regions are:

- $V_{GS} \leq V_{TH}$, $I_{DS} = 0$: cut off region.
- $V_{GS} \geq V_{TH}$, $V_{DS} \leq V_{GS} - V_{TH}$: linear region.
- $V_{GS} \geq V_{TH}$, $V_{DS} \geq V_{GS} - V_{TH}$: saturation region.

It is worth mentioning here that in digital circuits where the MOS transistor acts as a switch, during transition from one state to other, the operating region passes through both linear and saturation regions of operation. For estimation of the transient performance of the switch, such as delay, etc., a *resistor model* is commonly used and considered to be adequate. The MOSFET being a nonlinear resistor, it is common practice to represent the MOS transistor by a linear equivalent resistor, R_E, of a suitably *averaged* value that gives the best fit to the simulation result for the switch using the SPICE MOSFET model [21]. The resistor R_E, approximating the MOSFET switch in the conducting state, is obtained by averaging two values of the channel resistance R taken at two points of its operation [22].

$$R_E = \frac{1}{2} \left(R|_{V_{DS}=V_{DD}} + R|_{V_{DS}=\frac{V_{DD}}{2}} \right) \quad \text{for} \quad V_{GS} = V_{DD} \tag{4.69}$$

4.4.2 Smoothing Function for Piecewise Current Models in Strong Inversion

The output drain current characteristics for the MOS transistor operated in the strong inversion region are described by Equations (4.66) and (4.67), which correspond to the linear (ohmic) and saturation regions of operation respectively. The current in the ohmic region, being described by an equation which is parabolic, reaches a maximum value at drain voltage given by Equation (4.63). After that, the current is supposed to fall which is contrary to experimental results. The current attains a saturation value after the drain saturation voltage. Equations (4.59) and (4.68), though continuous for the current and its first derivative, are not continuous for higher derivatives. For compact models, this is considered to be undesirable. The problem is overcome by introducing a smoothing function, the use of which provides a single-piece equation which covers both the linear and saturation regions of operation. The function defines an equation relating V_{DSeff} with V_{DS}. It has the property that (i) for low values of the drain voltage V_{Dseff} asymptotically assumes the value of V_{DS}, and (ii) for $V_{DS} > V_{Dssat}$, V_{DSeff} assumes the value of V_{DSsat}. In the smoothing function the real, physical drain–source voltage V_{DS}, is replaced by an effective drain–source voltage V_{DSeff}, described by [19]:

$$V_{DSeff} = V_{DSsat} - \frac{1}{2}\left(V_{DSsat} - V_{DS} - \Delta + \sqrt{(V_{DSsat} - V_{DS} - \Delta)^2 + 4\Delta V_{DSsat}}\right) \quad (4.70)$$

The parameter Δ is to be properly selected to have a close fit in the transition region. The above smoothing function has been used in the BSIM compact model [19]. An alternative function has been proposed by Arora for ensuring continuity of the derivative at the transition from the linear to saturation region [47]. PSP uses a compact expression for the same purpose [14].

4.4.3 Subthreshold or Weak Inversion Region

The drain current in weak inversion ($V_{GS} < V_{TS}$), which is due to a diffusion mechanism, can be expressed in terms of inversion charge density at source and drain:

$$I_{diff} = -\phi_t \mu_{neff} \frac{W}{L}\left(Q'_{nS} - Q'_{nD}\right) \quad (4.71)$$

Using Equation (2.98):

$$Q'_{nS} = -Q'_w e^{\frac{V_{GB}-V_{TB}}{n\phi_t}} \quad (4.72)$$

where $Q'_w = \frac{1}{2}Q'_x\sqrt{\phi_t/1.5\phi_f}$. Similarly, we have the following expression for Q'_{nD} as:

$$Q'_{nD} = -Q'_w e^{\frac{V_{GB}-V_{TB}-V_{DS}}{n\phi_t}} \quad (4.73)$$

Replacing $V_{GB} - V_{TB}$ with $V_{GS} - V_{TS}$ in the above two equations, and substituting in Equation (4.71), we get the following expression for I_{DS}:

$$I_{DS} = \phi_t \mu_{neff} Q'_w \frac{W}{L} e^{\frac{V_{GS}-V_{TS}}{n\phi_t}}\left(1 - e^{-\frac{V_{DS}}{\phi_t}}\right) \quad (4.74)$$

The above equation is plotted in Figure 4.11 in which the output characteristics are displayed. The distinct feature of this region of operation is the drain-to-source voltage required to keep the transistor in saturation is independent of V_{GS}. In contrast, in strong inversion it is equal to $V_{GS} - V_{TS}$ as per the LEVEL 1 model. We can assume that $V_{DSsat} \approx 5\phi_t$, beyond which the drain current given in Equation (4.74) saturates to:

$$I_{DSsat} = \phi_t \mu_{neff} Q'_w \frac{W}{L} e^{\frac{V_{GS}-V_{TS}}{n\phi_t}} \quad (4.75)$$

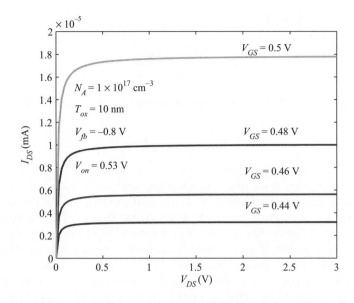

Figure 4.11 Output characteristics of an MOS transistor in weak inversion

In order to maintain the continuity of I_{DS} in transition from weak to strong inversion, the above equation for the drain current is modified by replacing V_{TS} with $V_{on} = V_{TS} + n\phi_t$ and $\phi_t \mu_{neff} Q'_w \frac{W}{L}$ with I_{on}, where I_{on} is that value of current at which $V_{GS} = V_{on}$ [31]:

$$I_{diff} = I_{on} e^{\frac{V_{GS} - V_{on}}{n\phi_t}} \tag{4.76}$$

In the LEVEL 1 model, the current in weak inversion is assumed to be zero.

For providing continuity between the subthreshold and linear regions in the piecewise model, a smoothing function is used with an effective gate-to-source voltage, V_{GSeff}, given below [19]:

$$V_{GSeff} = \frac{2n\phi_t \ln\left[1 + \exp\left(\frac{V_{GS} - V_{TS}}{2n\phi_t}\right)\right]}{1 + 2nC'_{ox}\sqrt{\frac{2\phi_s}{q\epsilon_{Si}N_A}}\exp\left(-\frac{V_{GS} - V_{TS} - 2V_{off}}{2n\phi_t}\right)} \tag{4.77}$$

4.4.4 Subthreshold Slope

Equation (4.76) indicates that the drain current is an exponential function of V_{GS} in weak inversion (subthreshold region). For a MOS transistor to be used as an ideal switch, the current should go abruptly to zero for $V_{GS} < V_{TS}$. From Equation (4.76) we see that the term $n\phi_t$ is the determining factor in characterizing the on–off switching behavior of the MOS transistor. Conventionally the above fact is reflected through a term S, designated as *subthreshold slope* [1, 36] which is defined as:

$$S = \left(\frac{\mathrm{d} \log_{10} I_{DS}}{\mathrm{d} V_{GS}} \right)^{-1} \tag{4.78}$$

Differentiating Equation (4.76) after taking the log, we have an expression for S as:

$$S = 2.3\phi_t n = 2.3\phi_t \left(1 + \frac{C_b'}{C_{ox}'} \right) \tag{4.79}$$

Typical values of S range from 70 to 100 mV/decade at room temperature, and are directly proportional to temperature. They also depend on process parameters and substrate bias through its influence on n. For example, lowering the gate oxide thickness T_{ox} increases the oxide capacitance C_{ox}', and higher substrate bias V_{BS} decreases the channel depletion capacitance which, in turn, result in sharper subthreshold characteristics. On the other hand, an increase in substrate doping increases depletion capacitance, thus increasing the value of the subthreshold slope S. The lower the value of the subthreshold slope, the better is the MOSFET as a switch. Substrate heating degrades the value of S due to increase in temperature. Low temperature operation of a MOS transistor for sharper subthreshold characteristics for switching off the MOS transistor is an attractive option.

4.5 Memelink–Wallinga Graphical Model

We shall discuss below the Memelink–Wallinga model [25], which gives a graphical visualization of the drain current in a MOS transistor. It has a useful pedagogical value at the introductory level. It is basically a strong inversion model having simplified expression through linearization of the bulk depletion charge w.r.t. the channel potential V_{CB}. Starting with the expression for drain current in the integral form (Equation (4.13)) with channel potential as the variable, we have:

$$I_{DS} = \mu_{neff} C_{ox}' \frac{W}{L} \int_{V_{SB}}^{V_{DB}} \frac{-Q_n'(V_{CB})}{C_{ox}'} \, \mathrm{d} V_{CB} \tag{4.80}$$

which can be written as:

$$I_{DS} = \mu_{neff} C_{ox}' \frac{W}{L} \left[\int_{0}^{V_{DB}} \frac{-Q_n'(V_{CB})}{C_{ox}'} \, \mathrm{d} V_{CB} - \int_{0}^{V_{SB}} \frac{-Q_n'(V_{CB})}{C_{ox}'} \, \mathrm{d} V_{CB} \right] \tag{4.81}$$

$$I_{DS} = \mu_{neff} C_{ox}' \frac{W}{L} \left\{ \int_{0}^{V_{DB}} [V_{GB} - V_{TB}(V_{CB})] \, \mathrm{d} V_{CB} \right.$$
$$\left. - \int_{0}^{V_{SB}} [V_{GB} - V_{TB}(V_{CB})] \, \mathrm{d} V_{CB} \right\} \tag{4.82}$$

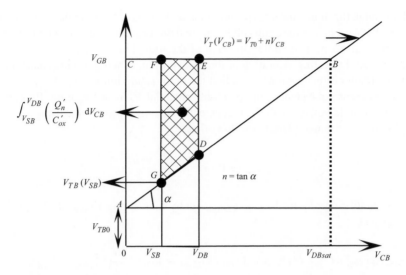

Figure 4.12 Memelink–Wallinga graphical representation of drain current

Equation (4.82) is graphically represented in Figure 4.12, where the shaded area multiplied by a factor $\mu_{neff}C'_{ox}\frac{W}{L}$ represents the drain current. From this figure the functional dependence of the area related to the drain current is also visualized:

$$V_{TB}(V_{CB}) = V_{TB0} + nV_{CB} \tag{4.83}$$

Substituting Equation (4.83) into Equation (4.82), and integrating we have:

$$I_{DS} = \mu_{neff}C'_{ox}\frac{W}{L}\frac{1}{2n}\left[(V_{GB} - V_{TB0} - nV_{SB})^2 - (V_{GB} - V_{TB0} - nV_{DB})^2\right] \tag{4.84}$$

Equation (4.84) can be put in an alternative form as:

$$I_{DS} = \mu_{neff}C'_{ox}\frac{W}{L}\left[(V_{GB} - V_{TB0})(V_{DB} - V_{SB}) - \frac{n}{2}(V_{DB}^2 - V_{SB}^2)\right] \tag{4.85}$$

For $V_{SB} = 0$, we have $V_{GB} = V_{GS}$, $V_{DB} = V_{DS}$, and $V_{TB0} = V_{T0}$, which reduces Equation (4.85) to the source referred SPICE LEVEL 3 drain current model as shown below:

$$I_{DS} = \mu_{neff}C'_{ox}\frac{W}{L}\left[(V_{GS} - V_{T0})V_{DS} - \frac{n}{2}V_{DS}^2\right] \tag{4.86}$$

4.5.1 Drain Current Saturation

With bulk charge linearization, it may be seen from Figure 4.12 that the channel potential increases linearly towards the drain due to the applied drain source voltage. Consequently, the inversion charge decreases along the channel. For a given gate voltage, there is a channel potential corresponding to which the inversion charge is zero. At a drain voltage equal to this channel potential, the drain current is proportional to the area of the triangle ABC for substrate bias $V_{SB} = 0$. If we increase the drain voltage beyond this channel potential, where the inversion charge is pinched-off near the drain, the area of the triangle enclosing the inversion charge remains constant, indicating saturation of the drain current. Therefore, this channel potential at which the drain current saturates is called the drain saturation voltage V_{DBsat}. This drain saturation voltage is equivalent to the pinch-off potential V_P near the drain. From the linear variation of threshold voltage with channel potential having a slope n, and using a geometrical relationship for the triangle (ABC), it follows that:

$$V_{DBsat} = V_P = \frac{V_{GB} - V_{TB0}}{n} \tag{4.87}$$

The drain saturation current for $V_{DB} \geq V_{DBsat}$ got by substituting V_{DBsat} for V_{DB} in Equation (4.85), is given by:

$$I_{DSsat} = \mu_{neff} C'_{ox} \frac{W}{L} \left[(V_{GB} - V_{TB0})(V_{DBsat} - V_{SB}) - \frac{n}{2}(V^2_{DBsat} - V^2_{SB}) \right] \tag{4.88}$$

4.5.2 Geometrical Interpretation

Equation (4.84) highlights the intrinsic symmetry of the Memelink–Wallinga MOSFET drain current model with the substrate used as a reference terminal. The drain current has two components:

- Forward saturation current $I_F = f(V_{GB} - V_{SB})$, source inverted, drain pinched-off.
- Reverse saturation current $I_R = f(V_{GB} - V_{DB})$, drain inverted, source pinched-off.

On substituting the value for V_{DBsat} from Equation (4.87) in Equation (4.88), the expression for the individual components are:

$$I_F = \mu_{neff} C'_{ox} \frac{W}{L} \frac{1}{2n} (V_{GB} - V_{TB0} - nV_{SB})^2 \tag{4.89}$$

and similarly:

$$I_R = \mu_{neff} C'_{ox} \frac{W}{L} \frac{1}{2n} (V_{GB} - V_{TB0} - nV_{DB})^2 \tag{4.90}$$

In the linear or ohmic region:

$$I_{DS} = I_F - I_R = f(V_{GB}, V_{SB}, V_{DB}), \text{ where both source and drain ends are inverted} \tag{4.91}$$

The conduction mode of operation is illustrated in Figure 4.13, where it is seen that in the ohmic or linear region of operation both source and drain regions are inverted. In this operating conduction, in the conduction mode, the inversion charge profile is described by the polygon $AA'B'B$. It can be visualized that the polygon $AA'B'B$ is, in fact, the superposition of two

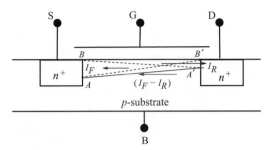

Figure 4.13 MOSFET operation in the conduction mode as a superposition of forward and reverse saturation modes

triangles $AB'B$ and $A'BB'$. Now, $AB'B$ corresponds to the inversion charge profile when the source is inverted and the drain is pinched-off and a forward saturation current, I_F, flows from drain to source. Similarly, $A'BB'$ corresponds to the condition for flow of the reverse saturation current I_R flowing from source to drain. Therefore, the conduction mode charge profile $AA'B'B$ represents the flow of net conduction current, $(I_F - I_R)$, which is the superposition of forward and reverse saturation currents.

The term $(V_{GB} - V_{TB0} - nV_{SB})$ is proportional to inversion charge density at the source. We may, therefore, conclude that the forward saturation drain current is proportional to the square of the charge density at the source. Similarly, for the reverse drain saturation current the source is pinched-off, and the current is proportional to the square of the charge density induced at the drain $(V_{GB} - V_{TB0} - nV_{SB})$. Such interpretation holds good for the symmetrical models such as ACM and EKV.

When the drain current is independent of drain voltage, the device enters the saturation region. In other words $I_{DS} = f(V_{GB}, V_{SB})$ physically signifies a forward saturation drain current; similarly $I_{DS} = f(V_{GB}, V_{DB})$ indicates a reverse saturation drain current component. Figure 4.14 shows qualitatively the drain current characteristics of a bulk referred symmetric model. From Figure 4.14(a) it is observed that for $V_{DB} < V_{SB}$ the net drain current is negative, i.e. the current flows from source to drain. In other words $I_R > I_F$. In Figure 4.14(b) the effect of substrate bias on drain current is reflected. It is seen that an increase in source voltage reduces the drain current for a given gate and drain voltage. We observe that as we increase the source voltage it also requires a larger gate voltage to initiate the drain current. Both the above observations can be explained by the phenomenon of an increase in the value of threshold voltage with source to bulk bias. An alternate way of looking at this effect is to observe the variation of drain current with source voltage using the gate voltage as a parameter, as shown in Figure 4.14(c). The transition of a MOS transistor from cut-off to saturation to the linear region of operation is shown in Figure 4.14(d). In the cut-off region for $V_{GB} < V_{TB}$, the drain current is zero.[8] As the gate voltage is increased there is inversion only at the source end, and the transistor is in saturation with the drain current independent of V_{DB}. With further increase in V_{GB} the drain end also gets inverted, and the transistor enters into a linear region with the drain current dependent on V_{GB} and V_{DB}. For V_{DB} much smaller than V_{GB}, the curve become

---

[8] Note that in reality there is finite drain current for V_{GB} below the threshold which is neglected in the present explanation.

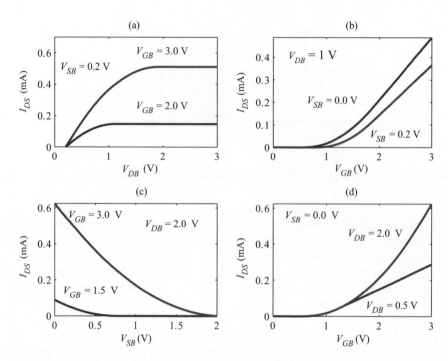

Figure 4.14 Plots of drain currents from the Memelink–Wallinga model: (a) I_{DS} versus V_{DB} with V_{GB} as the parameter; (b) I_{DS} versus V_{GB} with V_{SB} as the parameter; (c) I_{DS} versus V_{SB} with V_{GB} as the parameter; (d) I_{DS} versus V_{GB} with V_{DB} as the parameter

almost linear as can be seen from Equation (4.86), and the range of V_{GB} for which the transistor stays in saturation becomes smaller.

4.6 Channel Length Modulation

As explained earlier, the MOS transistor enters the saturation region when the channel is pinched-off near the drain. The inversion channel between the source to pinch-off point supports the saturation voltage. When the drain-to-source voltage V_{DS} is in excess of the drain saturation voltage V_{DSsat}, the excess voltage $V_{DS} - V_{DSsat}$ acts across the edge of the drain diffusion region and the pinch-off point, as shown in Figure 4.15. With further increase in V_{DS} beyond V_{DSsat}, the depletion width extends into the channel region, and the pinch-off point moves towards the source. This results in channel length shortening equal to the depletion width ΔL. Thus, the the effective channel length L_e is $L - \Delta L$. The reduction in channel length by ΔL is known as the *Channel Length Modulation* (CLM). The expression for the drain current beyond saturation with $V_{DB} > V_{DBsat}$, considering the channel length modulation, gets modified to:

$$I_{DS} = \frac{1}{2n}\mu_{neff}C'_{ox}\frac{W}{L - \Delta L}(V_{GB} - V_{TB0} - nV_{SB})^2 \qquad (4.92)$$

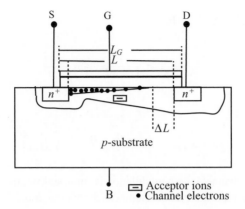

Figure 4.15   A MOS transistor in saturation with channel length modulation

which can be written as:

$$I_{DS} = \frac{1}{2n} \mu_{neff} C'_{ox} \frac{W}{L - \Delta L} (V_{GB} - V_{TB})^2 \tag{4.93}$$

With source as a reference we have:

$$I_{DS} = \frac{1}{2n} \mu_{neff} C'_{ox} \frac{W}{L - \Delta L} (V_{GS} - V_{TS})^2 \tag{4.94}$$

From the above expression it is seen that the drain current is not saturated to a constant value for $V_{DB} > V_{DBsat}$, but increases because of a decrease in the channel length by ΔL (which is a function of V_{DB}).

An accurate estimation of ΔL involves solving the complex 2-D field problem near the drain. In this region the gate is not effective in confining the inversion charge at the interface. In other words, the condition of gradual channel approximation, which was assumed to be valid in the inversion region, does not hold good in this region. To a first-order approximation, considering the lateral field to be dominant, we can have a one-dimensional approximation of the problem and obtain a simple expression for ΔL. In the 1-D approach we may use a standard expression for depletion layer width with an n^+-p abrupt junction approximation. The effect of mobile charges due to the flow of current is assumed to be negligible. The potential drop across the depletion region is that between the pinch-off point in the channel and drain terminal:

$$\Delta L = \sqrt{\frac{2\epsilon_{Si}}{qN_A} [V_{DS} + \phi_{bi} - (V_{DSsat} + \phi_0)]} \tag{4.95}$$

where ϕ_{bi} is the built-in junction potential. Assuming $\phi_0 \approx \phi_{bi}$:

$$\Delta L = \alpha \sqrt{V_{DS} - V_{DSsat}} \tag{4.96}$$

where $\alpha = \sqrt{\frac{2\epsilon_{Si}}{qN_A}}$. For $\frac{\Delta L}{L} \ll 1$ we have:

$$\frac{L}{L - \Delta L} = \frac{1}{1 - \frac{\Delta L}{L}} \approx 1 + \frac{\Delta L}{L} \qquad (4.97)$$

which gives:

$$I_{DS} = I_{DSsat}\left(1 + \frac{\alpha}{L}\sqrt{V_{DS} - V_{DSsat}}\right) \qquad (4.98)$$

In SPICE LEVEL 1 the channel length modulation is empirically expressed by a dimensionless parameter λ (channel length modulation factor) such that the drain current beyond saturation can be represented by:

$$I_{DS} = \frac{1}{2}\mu_{neff}C'_{ox}\frac{W}{L}(V_{GS} - V_{TS})^2(1 + \lambda V_{DS}) \qquad (4.99)$$

Comparing Equations (4.98) and (4.99) we have:

$$\lambda V_{DS} = \frac{\alpha}{L}\sqrt{V_{DS} - V_{DSsat}} \qquad (4.100)$$

$$\lambda = \frac{\alpha}{V_{DS}L}\sqrt{V_{DS} - V_{DSsat}} \qquad (4.101)$$

The above equation shows that $\lambda = f(L, N_A, V_{DS}, V_{GS})$ [32]. The parameter λ is inversely related to channel length. For a very long channel MOSFET the channel length modulation factor is small. and its effect can be neglected. It is also seen that a MOSFET is less affected by channel length modulation for higher substrate doping, because of less penetration of depletion width in the channel for a given drain voltage. In spite of its dependence on channel length, λ is assumed to be constant in the LEVEL 1 SPICE model.

The effect of CLM on the output characteristics of an MOSFET is shown in Figure 4.16 with the help of a simple LEVEL 1 SPICE model. Introduction of a ΔL correction beyond saturation results in unwanted discontinuity at $V_{DS} = V_{DSsat}$. Such discontinuity should be avoided for numerical convergence in a SPICE simulation. By multiplying the CLM term $(1 + \lambda V_{DS})$ with the current for a linear region, the discontinuity can be removed. However, the current can be overestimated in the linear region, but this overestimation will not be appreciable as in the linear region V_{DS} is small.

4.6.1 Early Voltage

'Early voltage' is an alternative way to model the effect of channel length modulation on the drain current beyond the saturation voltage. The phenomenon is similar to the base width modulation effect on the collector current of a bipolar transistor as proposed by Early. If a tangent is drawn at the saturation drain voltage of the $I_{DS}-V_{DS}$ characteristics beyond

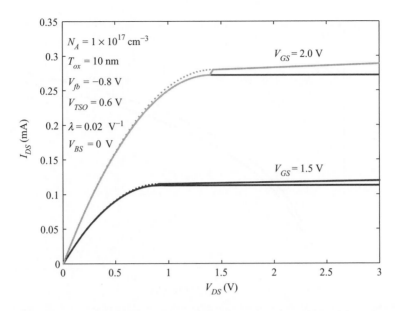

Figure 4.16   Finite slopes in output characteristics beyond saturation due to CLM

saturation, the slope intersects the voltage axis at a point which gives the 'Early voltage'. It is seen that a large 'Early voltage' is indicative of the extent of channel length modulation leading to change in the drain current with drain voltage beyond saturation. The length and doping dependence of 'Early voltage' can be empirically represented by the following relation [32]:

$$V_E \approx \frac{1}{\lambda} = \left(5\frac{V}{\mu\,m}\right) L \sqrt{\left(\frac{N_A}{10^{15}cm^{-3}}\right)} \tag{4.102}$$

Thus, the channel length modulation factor is inversely related to channel length, and has a square root dependence on substrate doping concentration. More realistic results for CLM can be had from 2-D analysis of the pinch-off region [43].

Figure 4.17 shows the variation of 'Early voltage' with channel length for a given substrate doping.

4.7 Channel Potential and Field Distribution Along Channel

The potential distribution along the channel is an important parameter of the MOS device as it determines the electric field in the channel, mobile charge distribution, transit time, etc. In this section we shall use the equation for the drain current in strong inversion to find potential and field distributions along the channel. Making use of the expression for current in Equation (4.12), which is related to the channel potential V_{CB}, we get:

$$\frac{I_D}{W\mu_{neff}C'_{ox}} = (V_{GB} - V_{TB0} - nV_{CB})\frac{dV_{CB}}{dy} \tag{4.103}$$

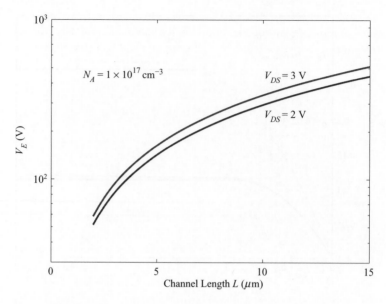

Figure 4.17 Dependence of 'Early voltage' on channel length

where the term in parentheses is $-Q'_n/C'_{ox}$. Rearranging and integrating with the boundary condition that at $y = 0$, $V_{CB} = V_{SB}$, we have:

$$n \int_{V_{SB}}^{V_{CB}} V_{CB}\, dV_{CB} - \int_{V_{SB}}^{V_{CB}} (V_{GB} - V_{TB0})\, dV_{CB} + \frac{I_D}{W\mu_{neff}C'_{ox}} \int_0^y dy = 0 \qquad (4.104)$$

$$n V_{CB}^2 - 2(V_{GB} - V_{TB0})V_{CB}$$
$$+ 2\left[\frac{I_D}{W\mu_{neff}C'_{ox}} y - \frac{n}{2}V_{SB}^2 + (V_{GB} - V_{TB0})V_{SB} \right] = 0 \qquad (4.105)$$

The above equation is quadratic in V_{CB}, in which the negative root satisfies the boundary condition. Thus:

$$V_{CB}(y) = \frac{(V_{GB} - V_{TB0})}{n}$$
$$- \frac{\sqrt{(V_{GB} - V_{TB0})^2 - 2n\left[\frac{I_D}{W\mu_{neff}C'_{ox}} y - \frac{n}{2}V_{SB}^2 + (V_{GB} - V_{TB0})V_{SB} \right]}}{n} \qquad (4.106)$$

Equation (4.106) is a general expression for the channel potential distribution along the channel in strong inversion for arbitrary voltages at the source and drain terminals. We shall study the special case for the sake of simplicity, assuming source and bulk are connected ($V_{SB} = 0$), then Equation (4.106) reduces to:

$$V_{CB}(y) = \frac{(V_{GB} - V_{TB0}) - \sqrt{(V_{GB} - V_{TB0})^2 - 2n\left(\dfrac{I_D}{W\mu_{neff}C'_{ox}}\right)y}}{n} \tag{4.107}$$

$$V_{CB}(y) = V_{DBsat} - \sqrt{V^2_{DBsat} - \frac{2}{n}\frac{I_D}{W\mu_{neff}C'_{ox}}y} \tag{4.108}$$

The above equation involves drain current and, therefore, the positional dependence of V_{CB} can be obtained in the linear region of operation of the MOS transistor, by substituting corresponding expressions for I_{DS}. In the case of saturation we can use the above equation to find the field in the channel till the pinch-off point with $V_{DB} = V_{DBsat}$. Now, substituting for I_{DSsat} and V_{DBsat} in saturation:

$$V_{CB}(y) = \left(\frac{V_{GB} - V_{TB0}}{n}\right) - \sqrt{\left[\left(\frac{V_{GB} - V_{TB0}}{n}\right)^2 - \frac{(V_{GB} - V_{TB0})^2}{n^2 L}y\right]} \tag{4.109}$$

$$V_{CB}(y) = V_{DBsat}\left[1 - \left(1 - \frac{y}{L}\right)^{\frac{1}{2}}\right] \tag{4.110}$$

The channel potential increases non-linearly along the channel in the saturation region of the strong inversion mode of operation of an MOS transistor.

Differentiating Equation (4.110) w.r.t. y we obtain the following expression for the electric field along the channel in saturation condition:

$$F(y) = -\frac{dV_{CB}}{dy} = -\frac{V_{DBsat}}{2L}\frac{1}{\left(1 - \frac{y}{L}\right)^{\frac{1}{2}}} \tag{4.111}$$

Figure 4.18 shows the variation of channel potential, electric field, and inversion charge density along the length of the channel in saturation.

From Equation (4.111) is it seen that at the point of saturation the electric field goes to infinity, which is physically not possible, as shown in Figure 4.18. The reason for this anomaly is due to the assumption that the inversion charge is zero at the pinch-off point. In fact there should be a finite inversion charge at and beyond the pinch-off point for a finite current to flow. Assuming it to be zero is only valid if the electric field at the pinch-off point is infinitely large such that it sweeps the electron instantaneously with infinite velocity, whereas in reality there is a limit set by the electron saturation velocity v_{sat}. The saturation velocity for an electron is around 10^5 cm s^{-1}.

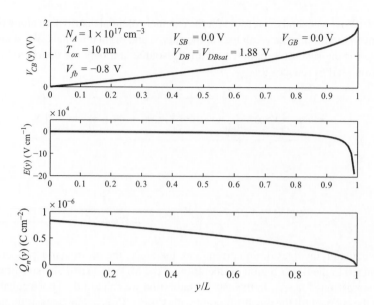

Figure 4.18 Channel potential, field, and inversion charge density variation along the channel with $V_{DB} = V_{DBsat}$

4.8 Carrier Transit Time

The transit time is defined as the average time required by the channel carrier to traverse the channel length. In the strong inversion mode of operation, the carrier transport mechanism being dominated by the drift mechanism, the transit time depends on the nature of the electric field in the channel, which in turn, is dependent on the mode of operation of the transistor. The transit time can be defined in two different forms as follows [33]:

$$\tau = \int_0^L \frac{dy}{v_{dr}(y)} \tag{4.112}$$

$$\tau = \frac{Q_n}{I_{DS}} \tag{4.113}$$

For using the first definition (Equation (4.112)), we need to use the spatial variation of electric field along the channel as $v_{dr}(y) = \mu_{neff} E(y)$. The second definition (Equation (4.113)) needs an expression for the total inversion charge, which will be used below as an illustration. We shall estimate the transit time for the saturation mode of operation by using Equation (4.113).

The total channel inversion charge, Q_n, under the gate is given by:

$$-Q_n = C'_{ox} W \int_0^L \left[V_{GB} - V_{TB0} - n V_{CB}(y) \right] dy \tag{4.114}$$

Substituting for $V_{CB}(y)$ with $V_{SB} = 0$ from Equation (4.110) for the saturation condition we get:

$$-Q_n = C'_{ox} W \int_0^L \left\{ (V_{GB} - V_{TB0}) - nV_P \left[1 - \left(1 - \frac{y}{L} \right)^{\frac{1}{2}} \right] \right\} dy \qquad (4.115)$$

Integrating, we get:

$$-Q_n = \frac{2}{3} C_{ox}(V_{GB} - V_{TB0}) \qquad (4.116)$$

where $C_{ox} = WLC'_{ox}$ is the total gate oxide capacitance.

Substituting the values of Q_n and I_{DSsat} from Equations (4.116) and (4.88) (with $V_{SB} = 0$) respectively, we get:

$$\tau_{sat} = \frac{\frac{2}{3} WLC'_{ox}(V_{GS} - V_{T0})}{\frac{1}{2} \frac{\mu_{neff}C'_{ox}}{n} \left(\frac{W}{L} \right) (V_{GS} - V_{T0})^2} \qquad (4.117)$$

$$\tau_{sat} = \frac{4}{3} \frac{nL^2}{\mu_{neff}(V_{GS} - V_{T0})} \qquad (4.118)$$

Thus, the transit time in saturation for a MOS transistor operating in strong inversion has a quadratic dependence on L. The quadratic dependence of transit time on channel length can be physically explained by the fact that a decrease in length say, by half, will reduce the distance to be traveled by the carrier by half with a drift velocity that is doubled because of doubling of the electric field. The inverse relationship of transit time with $V_{GS} - V_{T0}$ can be understood from the fact that $(V_{GS} - V_{T0})/n$ is the potential drop across the inversion channel in saturation. A higher $(V_{GS} - V_{T0})/n$ signifies a higher electric field which shortens the transit time of electrons across the channel.

4.9 EKV Drain Current Model

In threshold voltage based models, expressions have been obtained for drain current for operation of a MOS transistor in the weak and strong inversion regions assuming the dominant transport mechanism being either drift or diffusion in the respective regions. In other words, it was assumed there was an abrupt transition of the transport mechanism for $V_{GS} > V_{on}$. Physically however a region exists sandwiched between the weak and strong inversion regions where both drift and diffusion mechanisms effectively contribute to the drain current. This region is known as the moderate inversion region, and occurs around $V_{GS} = V_{TS}$. For low voltage, low-power, and analog design this region becomes important. As physical modeling of this region is complex, the problem is circumvented in the EKV model by mathematical interpolation so that a smooth transition exists for drain current and its derivative between the two regions [27]. The continuity of the derivative is a specific requirement of analog design. The EKV MOSFET model is a charge based compact interpolation model dedicated primarily for low-power analog design [40]. The distinguishing features of the EKV model are given below.

- The model considers the bulk as the reference terminal for all of the terminal potentials such as drain, source, and gate.
- The above approach correspondingly ensures the structural symmetry of the source and drain terminals with respect to bulk.
- The number of parameters necessary to accurately describe the current voltage character-istics of MOSFET devices are relatively small.
- The model describes the device operation in all the modes of operation of interest to de-signers such as weak, strong, and moderate inversion regions.
- With the use of interpolation functions the static drain current and its first derivative are made continuous in the entire range of operation, reducing simulation time.

4.9.1 Drain Current Interpolation

In Section 4.3.6, the drain current for the EKV model was expressed as:

$$I_{DS} = I_F - I_R = I_S \left(i_f - i_r \right) \tag{4.119}$$

where $I_S = 2n\beta\phi_t^2$ is called the *specific current*, and i_f and i_r are normalized forward and reverse drain currents with the normalization factor as I_S. The normalized currents i_f and i_r have a different dependence on the applied bias in the strong and weak inversion regions, which are given below:

$$i_f(r) = \left(\frac{V_P - V_{SB(DB)}}{2\phi_t} \right)^2 \qquad \text{In strong inversion} \tag{4.120}$$

$$i_f(r) = e^{\frac{V_P - V_{SB(DB)}}{\phi_t}} \qquad \text{In weak inversion} \tag{4.121}$$

Let $(V_P - V_{SB})/\phi_t = \xi_f$. Therefore, the equations for i_f become:

$$i_f = \frac{\xi_f^2}{4} \qquad \text{In strong inversion} \tag{4.122}$$

$$i_f = e^{\xi_f} \qquad \text{In weak inversion} \tag{4.123}$$

In moderate inversion, there is a transition of i_f from a quadratic dependence in strong inver-sion to an exponential dependence in weak inversion on ξ_f. The property of the interpolation function should therefore be such that in the asymptotic limits the function should converge from ξ_f^2 to e^{ξ_f}. The function proposed to meet the above requirement is given below [26]:

$$i_f = \left[\ln \left(1 + e^{\frac{\xi_f}{2}} \right) \right]^2 \tag{4.124}$$

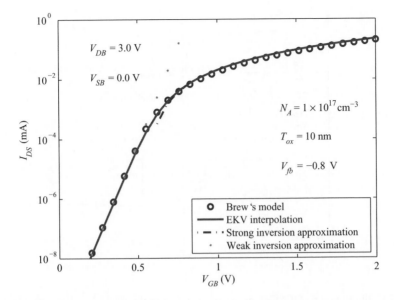

Figure 4.19 EKV interpolated drain current plot and comparison with the Brews' charge sheet model

The interpolation function plot is shown in Figure 4.19 where it is compared with the physical Brews' charge sheet model. The interpolation function is a key and powerful component of the EKV model framework.

Similarly, we can write the expression for the reverse current component. In spite of the fact that the moderate inversion region is described by a non-physical mathematical function, it has been found that it does match with the experimental observation in the moderate inversion region with acceptable accuracy.

The level of inversion can now be represented through a factor called the *inversion coefficient* IC, which is defined by the ratio of the forward saturated drain current to specific current I_S:

$$IC = \frac{I_F}{I_S} \qquad (4.125)$$

The different regions of operation can be expressed in terms of inversion coefficient as shown in Table 4.2.

Table 4.2 Inversion coefficients in various regions of operation

Regions of operation	IC
Weak inversion	≤ 0.1
Moderate inversion	0.1–10
Strong inversion	≥ 10

Since I_S is related to the aspect ratio $\frac{W}{L}$ of the MOS transistor, it is seen that *for a given saturated drain current I_F, the device can be operated at a specified level of inversion, by adjusting the dimensions of the device.*

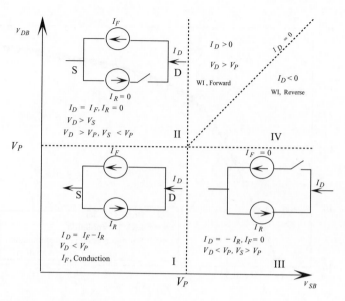

Figure 4.20 Illustration of MOSFET modes of operation in the $V_{SB}-V_{DB}$ plane (after Figure 4.3 of Enz and Vittoz [40])

Once the pinch-off voltage is known for a given gate voltage, the various regions of operation can be identified and are represented graphically in Figure 4.20. This figure shows the four quadrants around the pinch-off voltage, V_P, in the V_{SB} and V_{DB} plane. In quadrant (I), $V_{SB}, V_{DB} < V_P$. Source and drain regions are in the inversion; the device operates in strong inversion forward conduction mode. In quadrant (II), $V_{SB} \leq V_P$, $V_{DB} \geq V_P$. The drain is pinched-off, the device operating in the strong inversion forward saturation mode and $I_{DS} = I_F$, $I_R = 0$. In quadrant (III), $V_{SB} \geq V_P$, $V_{DB} \leq V_P$, $I_F = 0$, and $I_{DS} = -I_R$. The device operates in the strong inversion reverse saturation mode. In quadrant (IV), $V_{SB}, V_{DB} > V_P$. The device operates in the weak inversion mode; it is in forward saturation if $V_{DB} > V_{SB} + 3\phi_t$, and in reverse saturation if $V_{SB} > V_{DB} + 3\phi_t$. Figure 4.21 illustrates the transfer characteristics of a MOS transistor with V_{SB} as a parameter, as obtained from the EKV model.

4.9.2 Drain to Source Saturation Voltage

The drain to source saturation voltage is an important design parameter for analog circuits. It is an indicator of the voltage required at drain with respect to source, beyond which the transistor acts as a current source for the given level of inversion defined by the inversion coefficient. This drain to source saturation voltage is found to be $V_P - V_{SB}$, or $2\phi_t\sqrt{IC}$ in strong inversion, since V_P is the drain to bulk saturation voltage.

To derive a general expression for V_{DSsat}, inverting Equation (4.124), we have:

$$V_P - V_{SB} = 2\phi_t \ln\left(e^{\sqrt{i_f}} - 1\right) \tag{4.126}$$

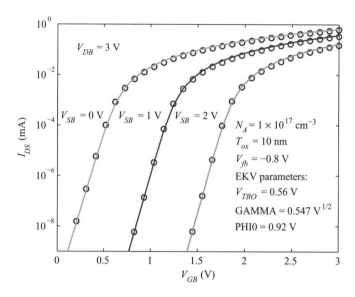

Figure 4.21 Simulated EKV transfer characteristics and comparison with the Brews' Charge sheet model

Similarly:

$$V_P - V_{DB} = 2\phi_t \ln\left(e^{\sqrt{i_r}} - 1\right) \tag{4.127}$$

From the above two equations we have:

$$V_{DS} = 2\phi_t \left[\ln\left(e^{\sqrt{i_f}} - 1\right) - \ln\left(e^{\sqrt{i_r}} - 1\right)\right] \tag{4.128}$$

In saturation, we assume that the condition $i_r = \epsilon i_f$ is satisfied, where $\epsilon \ll 1$. For a more accurate condition, refer to Cunha *et al.* [29]. With $V_{DS} = V_{DSsat}$, the above equation gives:

$$V_{DSsat} = 2\phi_t \left[\ln\left(e^{\sqrt{i_f}} - 1\right) - \ln\left(e^{\sqrt{\epsilon i_f}} - 1\right)\right] \tag{4.129}$$

The above equation is plotted in Figure 4.22, where the weak and strong inversion asymptote for V_{DSsat} is satisfied by Equation (4.127). It is seen that for $IC > 10$, V_{DSsat} has a strong dependence on IC which is related to the applied gate voltage, and indicates operation in the strong inversion region. For $IC < 0.1$, $V_{DSsat} \approx 5\phi_t$, which is a constant, and the operation relates to the weak inversion region.

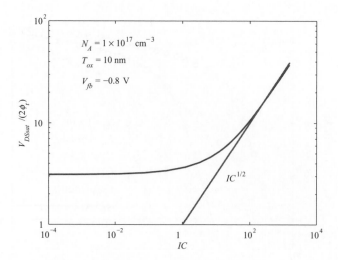

Figure 4.22 Drain to source saturation voltage as a function of the inversion coefficient IC

4.10 ACM and BSIM5 Models

The ACM model expression for the drain current has already been derived in Equation (4.37), and is reproduced below:

$$I_{DS} = I_{SA} \left[\underbrace{\left[\left(\frac{Q'_{nS}}{nC'_{ox}\phi_t} \right)^2 - \frac{2Q'_{nS}}{nC'_{ox}\phi_t} \right]}_{i_{fa}} - \underbrace{\left[\left(\frac{Q'_{nD}}{nC'_{ox}\phi_t} \right)^2 - \frac{2Q'_{nD}}{nC'_{ox}\phi_t} \right]}_{i_{ra}} \right] \qquad (4.130)$$

$I_{SA} = \frac{n}{2}\beta\phi_t^2$ is a normalization current, and is four times smaller then I_S as defined in the EKV model. i_{fa} and i_{ra} are normalized forward and reverse saturation currents with I_{SA} as a normalization factor, and are four times larger then their counterparts in the EKV model. The accuracy of ACM in predicting the drain current of a long channel NMOS is shown in Figure 4.23 with the Brews' charge sheet model taken as a benchmark. Using the simple interpolated expression for Q'_{nS} from Chapter 2 in terms of terminal voltage, we obtain an expression for the forward saturation current as:

$$i_{fa} = \left[\ln \left(1 + e^{\frac{V_{Pa}-V_{SB}}{\phi_t}} \right) \right]^2 + 2\ln \left(1 + e^{\frac{V_P-V_{SB}}{\phi_t}} \right) \qquad (4.131)$$

which can be recast in the following form:

$$i_{fa} = \left[1 + \ln \left(1 + e^{\frac{V_P-V_{SB}}{\phi_t}} \right) \right]^2 - 1 \qquad (4.132)$$

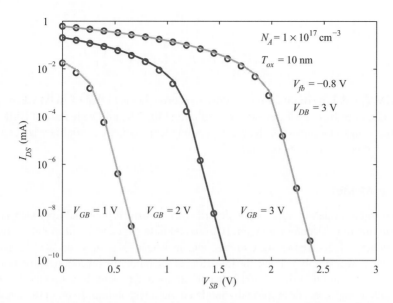

Figure 4.23 Simulated MOSFET output characteristics using the ACM and Brews' charge sheet models

Similarly i_{ra} is given as:

$$i_{ra} = \left[1 + \ln \left(1 + e^{\frac{V_P - V_{DB}}{\phi_t}} \right) \right]^2 - 1 \tag{4.133}$$

From the above equation we can write:

$$\ln \left(1 + e^{\frac{V_P - V_{S(D)B}}{\phi_t}} \right) = \sqrt{1 + i_{f(r)a}} - 1 \tag{4.134}$$

The LHS of Equation (4.134) can be related to charge as shown below [29]:

$$-\left(\frac{Q'_{nS(D)}}{nC'_{ox}\phi_t} \right) = \sqrt{1 + i_{f(r)a}} - 1 \tag{4.135}$$

Substituting the above equation into the implicit unified charge control expression, we can write:

$$\frac{V_P - V_{S(D)B}}{\phi_t} = \sqrt{1 + i_{f(r)a}} - \sqrt{1 + i_{pa}} + \ln \left(\frac{\sqrt{1 + i_{f(r)a}} - 1}{\sqrt{1 + i_{pa}} - 1} \right) \tag{4.136}$$

where i_{pa} is the value of i_{fa} at $V_{SB} = V_P$. The above equation can be used to derive the expression for V_{DSsat} for a given inversion coefficient IC. V_{DSsat} is defined alternatively as the drain to source voltage at which Q'_{nD} is a small fraction of Q'_{nS}. Let the fraction be ϵ_a, then we have:

$$V_{DSsat} = \phi_t \left[\sqrt{1 + i_f} - 1 + \ln\left(\frac{1}{\epsilon_a}\right) \right] \tag{4.137}$$

In the BSIM5 model the drain current expression corresponds to that of ACM where Q_{nS} and Q_{nD} are calculated from the Lambert series solution [30]. The unique feature of BSIM5 is that unlike other compact models the channel inversion charge density is not linearized either with respect to channel or surface potential.

4.11 PSP Model

Out of the three modeling approaches described, the surface potential approach is gaining momentum due to its accuracy and ability to model field dependence of mobility, symmetry, high frequency effects, tunneling currents, etc. in MOSFETs on a physical basis. Due to scaling of the power supply voltage, the modeling of the moderate inversion region is assuming importance as operation in this region is becoming more frequent. In a V_T based model, this region is either neglected, or empirically modeled via interpolation. Even in the EKV model, which claims to accurately characterize the moderate inversion region, it is handled through interpolation, which is a non-physical approach. In the ACM model, however, the attempt has been made to model this region physically, but assumptions were made which overestimate the current calculation in this region.

In Section 4.3.6 we derived the expression for the drain current in terms of surface potential at the source and drain ends. The drain current equation requires the calculation of ϕ_s, its square root $\sqrt{\phi_s}$, and $\phi_s^{3/2}$. Linearizing the bulk charge is found to be effective in simplifying the expression for the drain current, as in the V_T based or charge based model. In the surface-potential based PSP model discussed below, the technique of symmetric bulk-charge linearization [34, 35] with respect to surface potential is adopted in order to meet the important criterion of *symmetry* required by compact models.

4.11.1 Symmetric Bulk and Inversion Charge Linearization

The bulk charge density is expressed using the delta-depletion approximation as:

$$Q_b' = -\gamma C_{ox}' \sqrt{\phi_s - \phi_t} \tag{4.138}$$

and the inversion charge density as:

$$Q_n' = -C_{ox}'(V_{GB} - \phi_s) - Q_b' \tag{4.139}$$

From the above equations we see that linearizing the bulk depletion charge linearizes the inversion charge density as well with respect to surface potential. This is similar to the approach followed in the V_T based model. Such linearization w.r.t. surface potential is shown in Chapter 2 where series expansion of Equation (4.139) is made around $\phi_s = \phi_p$. In the PSP model, on the other hand, the symmetric bulk depletion and inversion charge linearization is carried out with the Taylor series expansion at ϕ_{sm}, the mean of the surface potential between source and drain ends [34].

$$Q_b'(\phi_s) = Q_b'(\phi_{sm}) + (\phi_s - \phi_{sm}) \left(\frac{dQ_b'}{d\phi_s} \right)_{\phi_s = \phi_{sm}} \tag{4.140}$$

$$Q_b'(\phi_s) = Q_b'(\phi_{sm}) + (\phi_s - \phi_{sm}) \left(\frac{-C_{ox}'\gamma}{2\sqrt{\phi_{sm} - \phi_t}} \right) \tag{4.141}$$

With the above linearization the expression for the inversion charge density is given as:

$$Q_n'(\phi_s) = Q_n'(\phi_{sm}) + \alpha_m(\phi_s - \phi_{sm})C_{ox}' \tag{4.142}$$

where $\alpha_m = 1 + \gamma/2\sqrt{\phi_{sm} - \phi_t}$. Now we use the drain current expression (Equation (4.26)), which is reproduced below after integration:

$$I_{DS} = -\mu_{neff} \frac{W}{L} \left[\phi_t(Q_{nS}' - Q_{nD}') + \int_{\phi_{sS}}^{\phi_{sD}} Q_n'(\phi_s)\,d\phi_s \right] \tag{4.143}$$

Substituting for Q_n' from Equation (4.142), we get:

$$I_{DS} = -\mu_{neff} \frac{W}{L} (Q_{nm}' - \alpha_m\phi_t C_{ox}')(\phi_{sD} - \phi_{sS}) \tag{4.144}$$

The above equation is the expression for the drain current after symmetric linearization of the inversion charge w.r.t. surface potential. Now what remains to be done is to calculate the inversion charge density at $\phi_s = \phi_{sm}$, i.e. Q_{nm}'. It seems that one can use Equation (4.139) for calculation of Q_{nm}' by substituting for $\phi_s = \phi_{sm}$. It is, however, found that the term $(V_{GB} - \phi_{sm})$ in Equation (4.139), if calculated from Equation (2.53), leads to an improvement in precision in the calculation of Q_{nm}', especially in the weak inversion region.

$$V_{GB} - \phi_{sm} = \gamma \left(\phi_{sm} - \phi_t + \phi_t e^{\frac{\phi_s - 2\phi_f - V_{CBm}}{\phi_t}} \right)^{\frac{1}{2}} \tag{4.145}$$

The use of the above equation requires the calculation of $e^{(\phi_s - 2\phi_f - V_{CBm})/\phi_t}$, which can be obtained from the boundary conditions of the channel potentials V_{CB} at the source and drain ends.

$$V_{GB} - \phi_{sS} = \gamma \left(\phi_{sS} - \phi_t + \phi_t e^{\frac{\phi_s - 2\phi_f - V_{SB}}{\phi_t}} \right)^{\frac{1}{2}} \tag{4.146}$$

and:

$$V_{GB} - \phi_{sD} = \gamma \left(\phi_{sD} - \phi_t + \phi_t e^{\frac{\phi_s - 2\phi_f - V_{DB}}{\phi_t}} \right)^{\frac{1}{2}} \tag{4.147}$$

Solving the above three equations for $e^{(\phi_s - 2\phi_f - V_{CBm})/\phi_t}$, we have:

$$e^{\frac{\phi_s - 2\phi_f - V_{CBm}}{\phi_t}} = \frac{1}{2}\left(e^{\frac{\phi_s - 2\phi_f - V_{SB}}{\phi_t}} + e^{\frac{\phi_s - 2\phi_f - V_{DB}}{\phi_t}}\right) - \frac{(\phi_{sD} - \phi_{sS})^2}{4\gamma^2\phi_t} \qquad (4.148)$$

The Equations (4.144), (4.145), and (4.148) together with analytical algorithms to calculate surface potentials at the source and drain ends, form a complete PSP model of drain current for a long channel MOS transistor. Figure 4.24 presents the output characteristics as computed using Equations (4.30), (4.31), and (4.144).

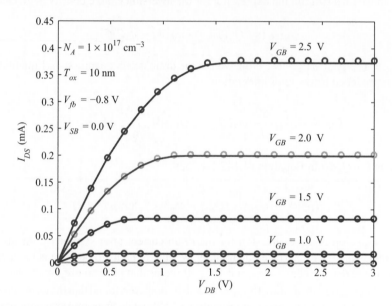

Figure 4.24 Simulated output characteristics using the PSP model with symmetric charge linearization and comparison with CSM

4.12 HiSIM (Hiroshima University STARC IGFET Model) Model

This is another compact MOSFET model aimed at satisfying the core criteria of accuracy and computational efficiency required for circuit simulation, with adaptability to accommodate technological diversities, and various scaling related phenomena [7]. The salient features are:

- Computation of surface potential ϕ_{sS} and ϕ_{sD} is carried out by solving Poisson's equation iteratively in terms of process, device geometry, and terminal potentials while ensuring the prescribed degree of precision. In saturation, ϕ_{sD} implies the potential at the end of the inversion channel L_e, where, L_e is the effective channel length and indicates the length of the GCA region. In other words, $\phi_{sD} = \phi_{sL_e}$. For a given drain to source voltage V_{DS}, the excess value over ϕ_{sL_e}, $V_{DS} - \phi_{sL_e}$, gives the total potential drop across the high field or pinch-off region and that in the gate–drain overlap region. The potential at the drain terminal is related to the potential ϕ_{sL_e} and $V_{DS} + \phi_{sS}$ empirically through a fitting parameter CLM1 characterizing the junction hardness with respect to abruptness.

- With the strategy of using GCA and charge sheet approximations, coupled with the capability of computing surface potentials at the source and drain ends with a high degree of accuracy numerically, enables obtaining physical, and yet pragmatic, analytical expressions for MOSFET related equations in terms of surface potentials ϕ_{sS} and ϕ_{sL_e} [8].
- The drain current is computed considering both drift and diffusion which is characteristic of the surface potential approach and, therefore, a single expression is valid for all regions of operation. This considerably reduces the number of model parameters.

The expression for drain current, with source as a reference, is reproduced below:

$$
\begin{aligned}
I_{ds} = \left(\phi_t \mu \frac{W}{L} \right) &\left\{ C'_{ox} \left(1 + \frac{V_{GS}}{\phi_t} \right) (\phi_{sD} - \phi_{sS}) - \frac{C'_{ox}}{2\phi_t} (\phi_{sD}^2 - \phi_{sS}^2) \right. \\
&- qN_A L_b \frac{2\sqrt{2}}{3} \left[\left(\frac{\phi_{sD}}{\phi_t} - 1 \right)^{\frac{3}{2}} - \left(\frac{\phi_{sS}}{\phi_t} - 1 \right)^{\frac{3}{2}} \right] \\
&\left. + qN_A L_b \sqrt{2} \left[\left(\frac{\phi_{sD}}{\phi_t} - 1 \right)^{\frac{1}{2}} - \left(\frac{\phi_{sS}}{\phi_t} - 1 \right)^{\frac{1}{2}} \right] \right\}
\end{aligned}
\tag{4.149}
$$

where ϕ_{sS} and ϕ_{sD} are the surface potentials at the source and drain ends of the channel respectively.

It is seen from the drain current expression that its value is obtained in terms of surface potentials at the source end ϕ_{sS}, and that at the drain end ϕ_{sD}. These potentials can be related to technological parameters such as doping, gate oxide, thickness, flat band voltage, and applied terminal voltages. Two iteration procedures have been adopted, one for calculating ϕ_{sS}, and the other for ϕ_{sD}.

The equation for ϕ_{sS}, for $V_{BS} = 0$, is given below:

$$
\begin{aligned}
&C'_{ox}(V'_{GS} - \phi_{sS}) = \\
&\left(\frac{\sqrt{2\epsilon_{si}\phi_t}}{L_b} \right) \sqrt{ \exp\left(-\frac{\phi_{sS}}{\phi_t} \right) + \left(\frac{\phi_{sS}}{\phi_t} \right) - 1 + \frac{n_{p0}}{N_A} \left[\exp\left(\frac{\phi_{sS}}{\phi_t} \right) - \left(\frac{\phi_{sS}}{\phi_t} \right) - 1 \right] }
\end{aligned}
\tag{4.150}
$$

For $V_{BS} \neq 0$, ϕ_{sS} is replaced by $\phi_{sS} - V_{BS}$. It is claimed that the above implicit equation for ϕ_{sS} for a given V_{GS} has been solved numerically with a high degree of precision with only a few iterations. This approach is claimed to be superior to the alternative analytical approaches available.

The quantity ϕ_{sD} can be expressed in terms of the already computed ϕ_{sS} using the concept that V_{DS} represents the quasi-potential difference across the channel ϕ_{f_n}, the gradient of which along the channel represents the total channel current comprising both drift and diffusion transport mechanisms:

$$\frac{\phi_{sD}}{\phi_t} = \frac{\phi_{sS}}{\phi_t} + \frac{V_{DS}}{\phi_t} + \ln \frac{Q'_n(D)}{Q'_n(S)} \qquad (4.151)$$

4.13 Benchmark Tests for Compact DC Models

This chapter has given a brief introduction to the framework that defines the features of representative MOSFET models belonging to charge, threshold voltage, and surface potential categories. While different models have their characteristic strength, they are not without drawbacks. Each modeling structure defines its potential, limitations, and application specific requirements [37]. Digital, analog, memory, RF, high-speed, low-voltage, and mixed-signal circuits have their specific requirements. With the growing diversity of MOSFET models to meet a wide ranging field of applications, benchmarking and defining standards has become essential for using them in SPICE-like simulators [41]. The IEEE benchmark criteria [42] briefly outlined below relate to the evaluation procedure for model performance for features such as symmetry, discontinuity in the regions of transition of mode of operation, parameter extraction complexity, etc.

Some typical and indicative test and characterization procedures recommended are:

- TS (Triode Saturation, I_{DS} vs. V_{DS}) characteristics.
- TH (Threshold, I_{DS} vs. V_{GS}) characteristics.
- ST (Subthreshold, $\log(I_{DS})$ vs. V_{GS}) characteristics.
- g_m/I_{DS} (Transconductance-to-drain current ratio) characteristics.
- SR (Gummel Slope Ratio) characteristics $(SR = (I_2 + I_1) * (V_2 - V_1)/(I_2 - I_1)*(V_2 + V_1))$.

Figure 4.25 shows a simple circuit which enables operating a MOSFET symmetrically in the forward and reverse directions by sweeping one independent voltage source V_X [38, 39]. From the circuit diagram it may be seen that the drain–bulk bias $V_{DB} = V_{B0} + V_X$, and the source–bulk bias is $V_{SB} = V_{B0} - V_X$. The MOSFET under the symmetry test can be driven

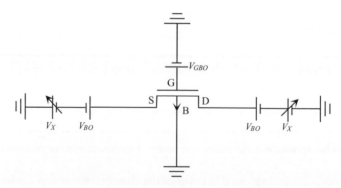

Figure 4.25   Circuit diagram for the test of drain current symmetry of a MOSFET model

symmetrically by sweeping the value of V_X from a given negative value to a positive value of the same magnitude. For the device to be symmetric, the current voltage characteristics should satisfy the following condition:

$$I_D(V_X) = -I_D(-V_X) \qquad (4.152)$$

For the model to be continuous it is essential that the second derivative of the I_D-V_X plot be zero around $V_X = 0$. Mathematically, a truly symmetrical model should as well satisfy the following condition:

$$\frac{\delta^2 I_D}{\delta V_X^2} = 0 \qquad (4.153)$$

It may be mentioned that apart from physical structural symmetry, the way parameters such as mobility, etc. are modeled for the description of current influences the symmetry test [44].

From these set of benchmark tests modeling accuracy for $I-V$ characteristics, short channel effects, mobility degradation effect, substrate effect, moderate inversion region, continuity of derivatives of current over various operating conditions, and output conductance g_d, etc., can be evaluated.

References

[1] S.M. Sze, *Physics of Semiconductor Devices*, 2nd Edition, John Wiley & Sons Inc., New York, NY, p. 447, 1981.

[2] S. Selberherr, A. Schütz and H.W. Pötzl, MINIMOS-A Two Dimensional Analyzer, *IEEE Trans. Electron Dev.*, **27**(8), 1540–1550, 1980.

[3] J. Watts, C. McAndrew, C. Enz, C. Galup-Montoro and G. Gildenblat, Advance Compact Models for MOSFETs, in *Proceedings of the NSTI Nanotechnology Conference and Trade Show*, pp. 3–11, 2005.

[4] M. Chan, X. Xi, J. He, K.M. Cao, M.V. Dunga, A.M. Niknejad, P.K. Ko and C. Hu, Practical Compact Modeling Approaches and Options for Sub-0.1 μm CMOS Technologies, *Microelectron. Reliabil.*, **43**(3), 399–404, 2003.

[5] C.T. Sah, A History of MOS Transistor Compact Modeling, in *Technical Proceedings of the 2005 Workshop on Compact Modeling*, Ch. 5, pp. 347–390, 1992.

[6] H.C. Pao and C.T. Sah, Effects of Diffusion Current on Characteristics of Metal Oxide (Insulator) Semiconductor Transistors, *Solid-State Electron.*, **9**(10), 927–937, 1966.

[7] M. Miura-Mattausch, Unified Complete MOSFET Model for Analysis of Analog and Digital Circuits, *IEEE Trans. Comput.-Aided Des.*, **15**(1), 1–7, 1996.

[8] M. Miura-Mattausch, H.J. Mattausch, N.D. Arora and C.Y. Yang, MOSFET Modeling Gets Physical, *IEEE Circ. Dev.*, **17**(6), 29–36, 2001.

[9] R. van Langevelde and F. M. Klaassen, Explicit Surface Potential Based MOSFET Model for Circuit Simulation, *Solid-State Electron.*, **44**(3), 409–418, 2000.

[10] K. Joardar, K.K. Gullapalli, C.C. McAndrew, M. E. Bumham and A. Wild, An Improved MOSFET Model for Circuit Simulation, *IEEE Trans. Electron Dev.*, **45**(1), 134–148, 1998.

[11] A.R. Boothroyd, S.W. Tarasewicz, and C. Slaby, MISNAN—A Physically Based Continuous MOSFET Model for CAD Applications, *IEEE Trans. Comput-Aided Des.*, **10**, (12), 1512–1529, 1991.

[12] S.D. Cho, H.T. Kim, S.J. Song, S.S. Chi, M.S. Kim, W.S. Chang, H.J. Kang, H.T. Shin, T.E. Kim, D.J. Kim and D.M. Kim, A Physics-Based Continuous Charge-Sheet MOSFET Model Using a Balanced Bulk-Charge-Sharing Method, *J. Kor. Phys. Soc.*, **42**(2), 214–223, 2003.

[13] R. van Langevelde, A.J. Scholten and D.B.M. Klaassen, Recent Enhancements of MOS Model 11, in *Proceedings of the NSTI Nanotechnology Conference and Trade Show*, Vol. 2, pp. 60–65, 2004.

[14] G. Gildenblat, H. Wang, T.-L. Chen, X. Gu and X. Cai, SP: An Advanced Surface-Potential-Based Compact MOSFET Model, *IEEE J. Solid-State Circ.*, **39**(9), 1394–1406, 2004.

[15] G. Gildenblat, X. Li, H. Wang, W. Yu, R. van Langevelde, A.J. Scholten, G.D.J. Smit and D.B.M. Klaassen, Introduction to PSP MOSFET Model, in *Proceedings of the Technical Workshop on Compact Modeling*, Annaheim, CA, pp. 19–24, 2005.

[16] Y.H. Byun. K. Lee and M. Shur, Unified Charge Control Model and Subthreshold Current in Heterostructure Field Effect Transistors, *IEEE Electron Dev. Lett.*, **11**(1), 50–53, 1990.

[17] B.J. Sheu, D.L. Scharfetter, P.K. Ko and M.C. Jeng, BSIM: Berkeley Short Channel IGFET Model for MOS Transistors, *IEEE J. Solid-State Circ.*, **22**(4), 558–566, 1987.

[18] G. Massobrio and P. Antognetti, *Semiconductor Device Modeling with SPICE*, 2nd Edition, McGraw Hill, New York, NY, 1993

[19] W.Liu, *MOSFET Models for SPICE simulation, Including BSIM3v3 and BSIM4*, John Wiley & Sons, New York, NY, 2001.

[20] S. Shichman and D.A. Hodges, Modeling and Simulation of Insulated-Gate Field-Effect Transistor Switching Circuits, *IEEE J. Solid-State Circ.*, **3**(3), 285–289, 1968.

[21] W.Wolf, Modern *VLSI Design-System on Silicon*, 3rd Edition, Prentice Hall PTR, Upper Saddle River, NJ, USA, 2002.

[22] L.Wei, Z. Chen, K. Roy, M.C. Johnson and V.K. De, Design and Optimization of Dual-Threshold Circuits for Low-Voltage Low-Power Application, *IEEE Trans. Very Large Scale Integration (VLSI) Syst*, **7**(1), 16–24, 1999.

[23] J.R. Brews, A Charge Sheet Model of the MOSFET, *Solid-State Electron.*, **21**, 345–355, 1978.

[24] A. Nussbaum, R. Sinha and D. Docus, Theory of the Long Channel MOSFET, *Solid-State Electron.*, **27**(1), 97–106, 1984.

[25] H. Wallinga and K. Bult, Design and Analysis of CMOS Analog Circuits by Means of a Graphical MOST Model, *IEEE J. Solid-State Circ.*, **24**(3), 672–680, 1989.

[26] H.J. Oguey and S. Cserveny, Modele du Transistor MOS Valable dans un Grand Domaine de Courants, *BULL. SEV/VSE*, No. 2, 1982.

[27] C.C. Enz, F. Krummenacher and E.A. Vittoz, An Analytical MOS Transistor Model Valid in All Regions of Operation and Dedicated to Low-Voltage and Low-Current Application, Special Issue on Analog Integrated Circuits and Systems Processing, J. Low Voltage Low Power Circ., **8**(1), 83–114, 1995.

[28] A.I.A. Cunha, M.C. Schneider and C. Galup-Montoro, An Explicit Physical Model for The Long Channel MOS Transistor Including Small Signal Parameters, *Solid-State Electron.*, **38**, (11), 1945–1952, 1995.

[29] A.I.A. Cunha, M.C. Schneider and C. Galup-Montoro, An MOS Transistor Model for Analog Circuit Design, *IEEE J. Solid-State Circ.*, **33**(10), 1510–11519, 1998.

[30] J. He, X. Xi, H. Wan, M. Dunga, M. Chan and A.M. Niknejad, BSIM5: An Advanced Charge Based MOSFET Model for Nanoscale VLSI Circuit Simulation, *Solid-State Electron*, **51**(3), 433–444, 2007.

[31] R. Swanson and J. Meindl, Ion Implanted Complementary MOS Transistors in Low Voltage Circuits, *IEEE J. Solid-State Circ.*, **7**(2), 146–153, 1972.

[32] F.S. Shoucair, Small Signal Drain Conductance of MOSFET in Saturation Region–A Simple Model, *Electron. Lett.*, **22**(5), 239–241, 1986.

[33] J.R.F. McMacken and S.G. Chamberlain, Analytic and Iterative Transit-Time Models for VLSI MOSFETs in Strong Inversion, *IEEE J. Solid-State Circ.*, **25**(5), 1257–1367, 1990.

[34] T.L. Chen and G. Gildenblat, Symmetric Bulk Charge Linearization in Charge-Sheet MOSFET Model, *Electron. Lett.*, **37**(12), 791–793, 2001.

[35] H. Wang, T. Chen and G. Gildenblat, Quasi-static and Nonquasi-static Compact MOSFET Models Based on Symmetric Linearization of the Bulk and Inversion Charges, *IEEE Trans. Electron Dev.*, **50**(11), 2262–2272, 2003.

[36] J.R. Brews, Subthreshold Behaviour of Uniformly and Nonuniformly Doped Long-Channel MOSFET, *IEEE Trans. Electron Dev*, **26**, 1282–1291, 1979.

[37] Y.P. Tsividis, MOSFET Modeling for Analog Circuit CAD: Problems and Prospects, *IEEE J. Solid-State Circ.*, **29**, 210–216, 1994.

[38] K. Joardar, K.K. Gullapalli, C.C. McAndrew, M.E. Burnham and A. Wild, An Improved MOSFET Model for Circuit Simulation, *IEEE Trans. Electron Dev.*, **45**(1), 134–148, 1998.

[39] C.C. McAndrew, Validation of MOSFET Model Source–Drain Symmetry, *IEEE Trans. Electron Dev.*, **53**(9), 2202–2206, 2006.

[40] C.C. Enz, E.A. Vittoz, *Charge-Based MOS Transistor Modeling: The EKV Model for Low-Power and RFIC Design*, John Wiley & Sons, Chichester, UK, 2006.

[41] (< http://www.eigroup.org/cmc >)

[42] *Test Procedures for Micro-electronic MOSFET Circuit Simulator for Model Validation*, Group IEEE standard working Draft, http://ray.eeel.nist.gov/modval/database/contents/reports/micromosfet/standard.html, 2001.

[43] Y. Tsividis, *Operation and Modeling of The MOS Transistor*, 2nd Edition, *McGraw-Hill*, New York, NY, 1999.

[44] G,H. See, X. Zhou, K. Chandrasekaran, S.B. Chiah, Z. Zhu, C. Wei, S. Lin, G. Zhu and G.H. Lim, A Compact Model Satisfying Gummel Symmetry in Higher Order Derivatives and Applicable to Asymmetric MOSFETs, *IEEE Trans. Electron Dev.*, **55**(2), 624–631, 2008.

[45] J.A. Power and W.A. Lane, Enhanced SPICE MOSFET model for analog application including parameter extraction schemes, in *Proceeding of the IEEE 1990 International Conference on Microelectronics Test Structures*, pp. 129–134, 1990.

[46] D. Foty, *MOSFET Modeling with SPICE: Principles and Practice*, Prentice-Hall, New York, NY, 1997.

[47] N. Arora, *MOSFET Models for VLSI Circuit Simulation: Theory and Practice*, Springer-Verlag, New York, NY, 1993.

Problems

1. In a MOS transistor for Gradual Channel Approximation, it is assumed that the normal field set up by the gate voltage V_{GS}, is much greater than the lateral electric field set up by the drain–source voltage V_{DS}, i.e. $dE_x/dx \gg dE_y/dy$. Assuming that the average lateral field along the channel is V_{DS}/L, where L is the channel length, show that for the gradual channel approximation to hold good, the following relation needs to be satisfied:

$$(V_{GS} - V_T)^2 >> \phi_t V_{DS} \left(\frac{3T_{ox}}{L} \right)^2 \qquad (4.154)$$

The symbols have their usual meanings.

2. Compare the static drain characteristics of an enhancement MOS transistor obtained using LEVEL 1 and LEVEL 2 SPICE model parameters assuming device dimensions as L= 10 μm and $W = 50\,\mu$m. Ignore the parameter related to dependence of mobility on the gate voltage of LEVEL 2. The given model parameters for LEVEL 1 and LEVEL 2 are as follows: NSUB = E16, UO = 650, TOX = 650E − 9, TPG = +1. Sweep the drain voltage from 0 to 10 V, and change the gate voltage V_{GS} from 0 to 10 V in steps of 2.5 V. Explain the difference in drain saturation voltages V_{DSsat} and drain saturation currents I_{DSsat} obtained for the LEVEL 1 and LEVEL 2 models.

3. Obtain the value of R_E of the MOS transistor to be used for an MOSFET model in a digital design. Show the dependence of R_E, representing the transistor, on threshold voltage, with supply voltage as a parameter. Plot R_E as the threshold voltage varies from 0.3 V to 0.9 V for two values of the supply voltage $V_{DD} = 5.0$ V and 3.0 V respectively, for a minimum size transistor, $W = L$. Assume $K'_n = \mu C'_{ox} = 100\,\mu$A V^{-2}.

4. Consider an NMOS inverter configured with enhancement mode MOS transistors as shown in Figure 4.26 where the load transistor drain is tied to the drain. Using the Memelink–Wallinga diagram show the output voltage, and obtain the expression for the output voltage of the inverter in terms of driver and load device transconductances β_D, β_L, V_{DD}, V_{IN}, and V_{T0}.

5. For the circuit diagram shown in Figure 4.27 a current source $I_0 = 100\,\mu$A is used to set the bias current. The NMOS transistor has the following parameters: $N_{SUB} = 10^{16}$ cm^{-3},

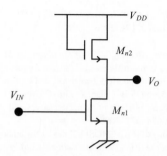

Figure 4.26   Inverter circuit diagram with enhancement mode MOS transistors

$T_{ox} = 0.5\,\mu$m, $V_T = 1$ V, $\mu_n = 600$ cm^2 V^{-1}s^{-1}. The output voltage is monitored at the source terminal. Using LEVEL 1 MOS parameters, obtain the DC transfer characteristics $V_0 = f(V_i)$. Observe the difference in the transfer characteristics, if the current source value is doubled to 200 μA. Interpret the result with the aid of a Memelink–Wallinga diagram.

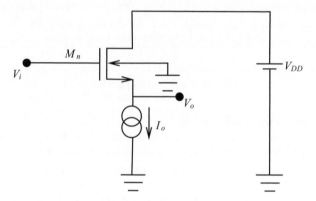

Figure 4.27   Current biased source follower

6. Show that the expression for the channel length modulation factor, λ, is given by:

$$\lambda = \frac{L_e}{(L_e - \Delta L)^2} \sqrt{\frac{\epsilon_{Si}}{2qN_A}} \frac{1}{\sqrt{V_{DS} + \phi_{bi}}} \tag{4.155}$$

Hint: Use a one-sided approximation for the drain–substrate junction.

(a) Plot the dependence of 'Early voltage' on channel length L_e, varying it from 3 μm to 15 μm.

(b) Show the variation of λ on V_{DS} in the range of 2 to 10 V for channel lengths of 3 μm and 10 μm. Given: $N_A = 10^{16}$ cm^{-3}, $\phi_{bi} = 0.75$ V.

7. Sketch the layout of the smallest NMOS transistor using the MOSIS scalable design rule in terms of lambda.

Assuming $\lambda = 1.0\,\mu$m, calculate:

(a) Drawn channel length.
(b) Gate width.
(c) Source and drain areas.
(d) Source and drain perimeter length (excluding channel edge).
(e) Total transistor area.

8. For the transistor obtained in Problem 7, the following process related parameters are provided: $N_A = 10^{15}\,\text{cm}^{-3}$, $T_{ox} = 20\,\text{nm}$, $\mu_n = 486\,\text{cm}^2\,\text{V}^{-1}\text{s}^{-1}$, source–drain lateral diffusion length, $L_D = 0.1\,\mu$m, implant depth, $W_i = 0.1 \times 10^{-4}$ cm, implant concentration, $N_i = 6 \times 10^{16}\,\text{cm}^{-3}$, $Q'_f = 10^{-8}\,\text{C cm}^{-2}$.

   Obtain ϕ_{gs} for a silicon gate device with an n^+-poly-gate.

   Obtain the operating gate voltage from hand calculation without considering the channel length modulation for $I_{DS} = 20\,\mu$A. Assume $V_{SB} = 0$, and that the transistor operates in saturation. Compare the result with a SPICE simulation using LEVEL 1 model parameters by using the given data.

9. An NMOS transistor has the following parameters: $W = 10\,\mu$m, $L = 5\,\mu$m, $V_{TO} = 0.6$ V, $\gamma = 1.0$, $\phi = 0.7$ V, $\mu_n C'_{ox} = K_n = 50\,\mu\text{AV}^{-2}$. Plot the transfer characteristics ($I_D$vs.V_{GB}) of the transistor using the EKV model showing the conduction and saturation branch of the drain current.

10. Plot the smoothing function given by Equation (4.70), and study the effect of Δ on the smoothness of the function near the transition between the linear and saturation regions. Plot the $I-V$ characteristics of a MOS transistor using the LEVEL 3 expressions for current given by Equations (4.59) and (4.68), using V_{DSeff} for the smoothing function.

    Study the characteristics of an alternative smoothing function given below [45, 47]:

$$V_{DSeff} = V_{DSsat} \left\{ 1 - \frac{1}{B} \ln \left[1 + \exp A \left(1 - \frac{V_{DS}}{V_{DSsat}} \right) \right] \right\} \qquad (4.156)$$

    where $B = \ln(1 + \exp A)$. Study the effect of the parameter A on the smoothness of the function in the transition region.

11. Obtain the expressions for conduction and saturation (forward and reverse) currents of the EKV model in terms of pinch-off voltage, the MOS transistor operating in weak inversion, for the parameters $V_{TO} = 0.75$ V, $\gamma = 1\sqrt{V}$ and $\phi_0 = 0.7$ V.

(a) Plot the transfer characteristics $\ln(I_{DS}/I_{D0})$ against V_{GB}/ϕ_t with V_{SB} as a parameter.
(b) Plot the transfer characteristics $\ln(I_{DS}/I_{D0})$ against V_{SB}/ϕ_t with V_{GB} as a parameter.
(c) Plot the forward current normalized w.r.t. forward saturation current against V_{DS} normalized w.r.t. ϕ_t for a constant V_{SB}.

12. Show the steps involved in arriving at the expression for drain saturation voltage V_{DSsat} in terms of forward normalized current, as given in Equation (4.137).

    Hint: Obtain the expression for $(V_P - V_C)$ using (1) the linearized relation between channel charge dQ'_n and surface potential $d\phi_s$, (2) the definition of pinch-off that at the pinch-off voltage V_p the inversion charge is Q'_{np}, and (3) Equation (4.136).

13. Give the steps that lead to the final expression given in Equation (4.148).

14. Write a program for calculating and plotting the output I_D-V_{DS} characteristics of a MOS transistor with V_{GS} as a parameter, using the HiSIM compact MOSFET model. Compare the above characteristics with that obtained from the Brews' charge sheet model (i) without considering the diffusion component, and (ii) including the diffusion component of drain current. Given: $N_A = 5.10^{16}\,\text{cm}^{-3}$, $T_{ox} = 15\,\text{nm}$, $Q'_f = q \times 1.6 \times 10^{10}\,\text{C cm}^{-2}$, $\phi_{gs} = -0.6\,\text{V}$.

15. Obtain the I_D vs. V_x characteristics for a long channel MOS transistor for the Gummel symmetry test using the circuit diagram shown in Figure 4.25. Assume: $L = 5\,\mu\text{m}$, $W = 20\,\mu\text{m}$, $V_{T0} = 1\,\text{V}$, $\mu_n = 650\,\text{cm}^2\,\text{V}^{-1}\,\text{s}^{-1}$, $T_{ox} = 50\,\text{nm}$. Use SPICE LEVEL 3, and EKV parameters for simulation in benchmarking the model symmetry.

5

The Scaled MOS Transistor

5.1 Introduction

The MOS transistor feature size has been subjected to scaling down for the last several decades [1–4, 6]. The scaling phenomenon has more or less followed the prediction of Moore's Law, according to which the complexity of MOS device integration is approximately doubled every eighteen months [4]. The extraordinary success in scaling or miniaturization of the feature size of MOS transistors is attributed to the following factors [4]

- High level of component integration (large scale) with the prospect of realizing a complete functional electronics subsystem/system on silicon, which improves system reliability.
- Cost reduction due to increased device and subsystem count in a given area of silicon estate, as chip area is directly related to minimum feature size.
- Short transit time of carrier in the channel and faster speed of charging and discharging of capacitive loads. The maximum clock frequency in a digital system is governed by the current drive and capacitive load, which improves with scaling.

Figure 5.1 shows the cross-section of a reference long channel MOS transistor and its scaled version. The long or scaled (short) MOSFET classification is broadly based on the relation of the channel length to the depletion widths under the gate for $V_{DS} \approx 0$. When the channel length is considerably larger than the sum of the source and drain depletion widths, the transistor is considered to be long. The characteristics of a scaled MOS transistor differs from that of a long channel structure due to the following reasons:

2-D Field. The electric field in the channel under the gate is two-dimensional in a scaled MOSFET where the components in the vertical and horizontal directions are comparable. This is in contrast with the assumption of a dominant vertical field gradient assumed in the case of a long channel device under GCA approximation, where the lateral field gradient

Compact MOSFET Models for VLSI Design   A.B. Bhattacharyya
© 2009 John Wiley & Sons (Asia) Pte Ltd

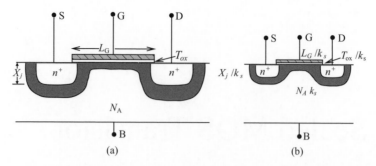

Figure 5.1   (a) Reference long channel MOS transistor, and (b) scaled MOS transistor

is considered to be negligible. The vertical field owes its origin to the gate voltage and, is therefore, considered to be *gate induced*. The solution of the 1-D Poisson's equation in the vertical direction gives the *input voltage equation* relating gate voltage to the surface potential at the Si–SiO$_2$ interface. The drain to source voltage, on the other hand, is related to the lateral field in the channel and provides the *output current equation* which is related to the gate voltage through the charge induced by the vertical field [5]. The lateral field, which is superimposed on the gate induced field, is termed as the *junction induced field* and is responsible for short channel effects in scaled devices.

High Electric Field. As the supply voltage does not scale down as aggressively as device dimensions due to practical reasons, the electric field in the channel is high leading to a number of phenomena related to a high electric field. The reduction of supply voltage is an obvious step to counter a high field effect which, however, adversely affects the propagation delay unless the threshold voltage is also reduced. A reduction in V_T, however, directly increases static power dissipation due to off-state leakage current. Though scaling is the driving motivation for improved CMOS circuits, performance, it is also accompanied by a large number of undesirable higher order effects. Hence, a certain degree of trade-off is essential to get the optimal benefit of scaling advantages.

Technology Related Phenomena. Miniaturization of feature size of a MOS transistor is associated with channel and gate engineering. This, however, introduces unwanted parasitics and new physical phenomena such as hot electron effects, tunneling, quantization of energy levels, leakage, nonuniform substrate, etc. [54, 55]. When scaling approaches physical limits, the solution may lie in new materials and non-conventional structures.

5.2 Classical Scaling Laws

The process of scaling brings about changes in the field pattern within the device which affects the device physics. This gives rise to a number of second order effects which are undesirable from the viewpoint of overall performance improvement and reliability. Therefore, classically the scaling has been guided by a few approaches with specific application objectives in view. Based on the assumption of (i) dimensional miniaturization of the device through the improvement of lithography, and (ii) invariance of the basic structure of the MOS transistor, the scaling laws may be classified into the following broad categories.

5.2.1 Constant Field (CF) Scaling [1]

Geometrical dimensions (horizontal and vertical) and supply voltage are scaled down by a factor k_s, and substrate doping concentration is scaled up by the same factor, where k_s is a dimensionless scaling factor greater than unity ($k_s > 1$). With this dimensional and doping transformation, the internal field patterns in the device remains invariant and the long channel characteristics of the MOS device are preserved. The increase in doping concentration by scaling factor is meant to avoid merging of source/drain depletion layers. The increased doping of the substrate reduces the depletion layer width of the source/drain regions. The reduction of supply voltage maintains the field pattern unchanged in spite of scaling.

5.2.2 Constant Voltage (CV) Scaling [10]

Constant voltage scaling retains the supply voltage as constant so that the device remains compatible to prevailing electronic systems regarding supply voltage standards. The geometrical dimensions and doping, however, are scaled as in the case of CF scaling, to take advantage of miniaturization. Constant voltage is associated with the penalty of high electric field and related reliability problems. However, there may be improvement in speed if the threshold voltage is scaled down, due to an increase in V_{DD}/V_T ratio. The relatively high field issues involved are addressed through specific source drain engineering such as lowly doped drain structures [60].

5.2.3 Quasi-Constant Voltage (QCV) Scaling

In quasi-constant voltage scaling, the voltage is reduced less aggressively compared to dimensions, thereby attempting a trade-off and a practical compromise. The dimensions are scaled down linearly by a factor k_s, whereas the voltage is scaled with square root dependence $\sqrt{k_s}$.

5.2.4 Generalized Scaling (GS) [11]

Generalized scaling proposes that the scaling factor need not be the same for the geometrical feature sizes and the potentials. Thus assigning k_s as the scaling factor for the dimensions and ϵ the scaling factor for the potential, it is possible to optimize the trade-off between miniaturization and optimal circuit performance.

5.2.5 Flexible Empirical Scaling [8]

The subthreshold behavior which was shown to be independent of drain bias in a long channel MOS transistor, however, has significant dependence in a short channel MOSFET as will be discussed later. The empirical relation for scaling is obtained from extensive 2-D numerical simulation and processing data, which give the minimum channel length for which long channel subthreshold behavior can be preserved. Control of leakage current in the subthreshold operating condition is an important consideration in high performance digital circuits. The defining criteria for setting $L = L_{min}$ is such that the subthreshold current changes by 10%

for drain voltage V_{DS} varying from 0.5 to 1.0 V. The *empirical relation* proposed by Brews *et al.* for minimum channel length in microns, is given by:

$$L_{min} = A \left[X_j T_{ox}(W_s + W_d)^2 \right]^{\frac{1}{3}} \tag{5.1}$$

where T_{ox} is the oxide thickness in angstroms, X_j the junction depth in microns, and W_s and W_d the widths of source and drain depletion regions in microns respectively. A is a proportionality constant, $A = 0.41 \, \text{Å}^{-1/3}$. W_s and W_d are given by:

$$W_{s(d)} = \sqrt{\frac{2\epsilon_{Si}}{qN_A}(V_{S(D)B} + V_{bi})} \tag{5.2}$$

The above guideline is flexible in view of the fact that device parameters can be tailored independently to ensure insensitivity of subthreshold leakage current to drain bias. A long channel index M is defined by:

$$M = A \frac{\left[X_j T_{ox}(W_s + W_d)^2 \right]^{\frac{1}{2}}}{L} \tag{5.3}$$

which is an indicator of the boundary value between short and long channel behavior. $M > 1$ denotes predominance of short channel effects, and $M \leq 1$ ensures long channel subthreshold characteristics. Figure 5.2 shows the combination of T_{ox} and N_A in the design of a scaled MOSFET following Brews' guideline for a given (fixed) X_j and V_{dd}, with channel length as parameter. The straight line indicates the boundary between long and short channel behavior for a given technological parameter, such that any point with (N_A, T_{ox}) coordinates lying below the line for corresponding L_{min}, ensures long channel behavior.

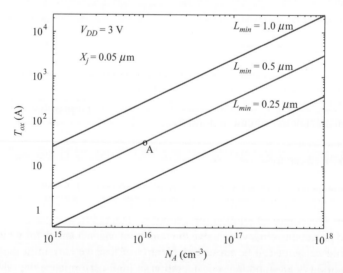

Figure 5.2   Scaled MOSFET design following Brews' guideline for a fixed X_j and V_{DD}

Table 5.1 Effect of scaling laws on MOSFET device/circuit performance

Device/circuit parameter	Scaling law			
	CF	CV	QCV	GS
Dimensions (L, W, T_{ox}, X_j)	$1/k_s$	$1/k_s$	$1/k_s$	$1/k_s$
Supply voltage (V_{DD})	$1/k_s$	1	$1/\sqrt{k_s}$	$1/\epsilon$
Gate oxide field (F_{ox})	1	k_s	$k_s^{1/2}$	k_s/ϵ
Substrate doping (N_A)	k_s	k_s	k_s	k_s^2/ϵ
Current (I_{DS})	$1/s$	k_s	1	k_s^2/ϵ^2
Transconductance (g_m)	1	k_s	$k_s^{1/2}$	k_s^2/ϵ
C_d/C_{ox}	1	$k_s^{-1/2}$	$k_s^{-1/4}$	1
Gate delay	$1/k_s$	$1/k_s^2$	$k_s^{3/2}$	ϵ/k_s^2
Power dissipation	$1/k_s^2$	k_s	$1/k_s^{1/2}$	k_s/ϵ^3
Power density	1	k_s^3	$k_s^{3/2}$	k_s^3/ϵ^3
Breakdown voltage (V_{BR})	$k_s^{-3/4}$	$k_s^{-3/4}$	$k_s^{3/2}$	$\epsilon^{3/4}/k_s^{3/2}$
M (Brews' factor)	$k_s^{-1/3}$	1	$k_s^{-1/6}$	$k_s^{-1/3}$

From a practical viewpoint, the attractive feature of the above relationship is that the parameters involved need not be scaled down by the same factor, which offers a good deal of flexibility for technologists.

While the above relationship served as a scaling guideline at the submicron level, the extension of the equation to and beyond deep submicron levels encountered the following limitations:

- As T_{ox} or X_j tend to zero during scaling, Equation (5.1) indicates that long channel behavior can be maintained with $L_{min} \rightarrow 0$ which is physically inconsistent.
- The scaling consideration is application specific, in the sense of acceptability of the limit of increase in subthreshold current for specific types of circuits, such as logic or memory.

Through an extensive use of a device simulator, an improved scaling guideline has been proposed [9], as given below:

$$L_{min} = Bf_1 \left(\frac{\partial V_{TS}}{\partial V_{DS}} \right) [f_2(T_{ox} + C)][f_3(W_{sd}) + D][f_4(X_j) + E] \qquad (5.4)$$

where $\partial V_{TS}/\partial V_{DS}$ is the threshold voltage shift in the device operating in the subthreshold region for the same current. The constants B, C, D, and E ensure that L_{min} does not converge to zero as T_{ox}, $W_{S,D}$, and X_j approach zero in the limit. Table 5.1 outlines the effect of various scaling procedures on important MOSFET device and circuit parameters. Detailed assessment of MOSFET scaling on performance of analog circuits shows improvements in bandwidth, gain, and transient response [12–14]. The above performance evaluation has been made assuming that long channel theory for current–voltage relations remain unchanged. This is an oversimplification and the assessment is only indicative of the trend. The scaling down of the device necessarily results in a number of higher order effects which significantly affect the device operations and must be considered in a compact model. The important effects in scaled MOS transistors reported are:

1. Short Channel Effect (SCE) and Narrow Width Effect (NWE). Effect due to threshold voltage dependence on channel length and width respectively.
2. Drain Induced Barrier Lowering (DIBL).
3. Reverse Short Channel Effect (RSCE).
4. Mobility degradation due to gate field.
5. Carrier velocity saturation.
6. Velocity overshoot effect.
7. Parasitic source and drain resistance effect.
8. Polysilicon depletion layer effect.
9. Substrate current due to hot electron effect.
10. Punchthrough effect.
11. Gate tunneling current.
12. Gate Induced Drain Leakage (GIDL).
13. Bias dependent parasitic capacitances.
14. Carrier energy quantization.
15. Ballistic transport.

The International Technology Roadmap for Semiconductors (ITRS) given in Table. 1.1 is an indicator of the performance target for future nodes of MOSFET technology in terms of cost, speed, power dissipation, and functional density [75]. It helps the VLSI community to identify areas of development and research related to process integration, system architecture, materials, and device structure.

There are numerous issues related to development of products, equipments, EDA tools, process integration, materials, test procedures, etc. in the pursuit of miniaturization of MOSFET devices. For using such devices in VLSI design, an accurate and compact device modeling, including scaling related phenomena, is considered to be a key component in providing the design–technology interface. The number of MOSFET model parameters has been growing in complexity with scaling. Various approaches to MOSFET modeling have already been discussed in Chapter 4. The short channel effects need to be described within the framework of the selected modeling framework. In the following sections various short channel phenomena will be discussed along with their adaptations to different modeling approaches. Though numerical solution of the 2-D Poisson's equation gives detailed insight into the physics of short channel effects, such elaborate numerical models are not suitable for applications to circuit simulators. Physics based analytical models have been proposed with various degree of approximations, and engineering models have been obtained with fitting parameters.

Below we shall qualitatively describe how the two-dimensional field in a short channel MOS transistor is likely to affect the results of one-dimensional field perception which characterize the long channel device. Further, analytical models of various short channel effects will be presented to emphasize the physical mechanisms involved in short channel MOS devices.

5.3 Lateral Field Gradient

In a short channel MOS transistor, the field in the channel induced by the gate voltage, is perturbed due to the proximity of the source and drain induced field. Figure 5.3 shows the cross section of long and short channel MOS transistors along the length showing the effect of scaling on the distribution of the equipotential line in the channel obtained from the 2-D simulator MINIMOS-NT.

It is assumed that only the length of the channel is shortened. L_R is the gate length of the reference long channel transistor, and $L_S = L_R/k_s$, is the corresponding gate length of the scaled transistor. It is also assumed that the device has a long width W, which eliminates any fringing effect in the z-direction. The nature of the field pattern in the channel region under the gate of the short channel MOS transistor can be best obtained from the solution of the 2-D Poisson's equation with appropriate boundary conditions in the x- and y-directions (normal and longitudinal directions) of the channel. We shall confine ourselves to a qualitative effect of the 2-D field in the channel in the subthreshold condition. As the discussion relates primarily to the effect of channel length shortening on threshold voltage, we shall make the standard assumption that at threshold voltage, the inversion charge density Q'_n is negligible, i.e. the space charge, defining the field pattern in the channel under the gate is defined solely by the depletion charge.

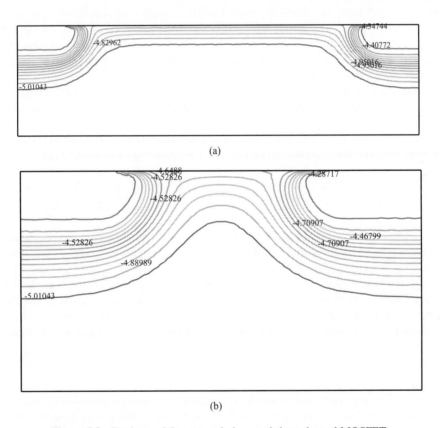

(a)

(b)

Figure 5.3 Equipotential contours in long and short channel MOSFETs

Under the above assumption, we have the following Poisson's equation in the channel region under the gate [20, 21]:

$$\frac{\partial F_y}{\partial y} + \frac{\partial F_x}{\partial x} = -\frac{q}{\epsilon_{Si}} N_A \tag{5.5}$$

$$\frac{\partial F_x}{\partial x} = -\left(\frac{q}{\epsilon_{Si}} N_A + \frac{\partial F_y}{\partial y}\right) \tag{5.6}$$

$$\frac{\partial F_x}{\partial x} = -\frac{q}{\epsilon_{Si}} (N_A - N_J) = -\frac{q}{\epsilon_{Si}} N_{eff} \tag{5.7}$$

where $N_J(x, y) = -(\epsilon_{Si}/q)\partial F_y/\partial y$ is a positive quantity, as the term $\partial F_y/\partial y$ is negative, and $N_{eff}(x, y) = N_A - N_J(x, y)$ is an *effective reduced bulk doping*. Thus, by replacing the substrate doping by an effective reduced value, the two-dimensional field description in a short channel MOSFET can be visualized through a one-dimensional gate field as given in Equation (5.7). The quantity qN_J may be interpreted as the charge density consumed by the electric field originating laterally from the junctions.

The two-dimensional effect reduces the normal field supporting the depletion layer in the channel. Physically, one may interpret that the gate induced normal field 'sees' less amount of effective depletion charge. An effective reduction of depletion charge seen by the gate induced field is responsible for a reduction in the threshold voltage, as the threshold voltage is directly related to channel depletion charge density.

In a long channel device the term $\frac{\partial F(y)}{\partial y}$ is zero as the equipotential lines are parallel to the channel. It may also be noted that a reduction in the effective doping implies a larger depletion width for a given channel potential. It can be shown by solving the 2-D Poisson's equation with appropriate boundary condition that the lateral component of the electric field, $F(y)$, originating from the source/drain can be given by the following expression [21]:

$$F(y) = \sin\left[\frac{\pi(x + 3T_{ox})}{W_C + 3T_{ox}}\right] \left\{ F_S \cosh\left[\frac{\pi(L - y)}{W_C + 3T_{ox}}\right] - F_D \cosh\left(\frac{\pi y}{W_C + 3T_{ox}}\right) \right\} \tag{5.8}$$

where F_S and F_D are, respectively, the magnitudes of source and drain junction fields, and are functions of source/drain voltages, channel length, oxide thickness, etc. W_C is the minimum depletion layer thickness for the short channel device.

Though numerical solution of the 2-D Poisson's equation gives a detailed insight into the physics of short channel effects, it is not suitable for circuit simulators. Analytical solutions have been attempted with various geometrical approximations in *charge sharing models* of the depletion charge between the gate and source/drain junctions with varying degrees of success.

5.3.1 Effect on Threshold Voltage

In the previous section it was shown that the presence of a lateral field gradient results in an apparent decrease in the bulk doping concentration. The decrease in effective bulk doping seen by the gate field results in a decrease of the threshold voltage. This implies that the above effect can be taken into account by an appropriate modification of the threshold voltage model. In the model framework where threshold voltage is not a parameter, as in surface potential based MOSFET models, the above effect can be taken into account through an apparent increase in the gate voltage by an amount equal to the decrease in the threshold voltage.

The 2-D nature of the electric field under the gate of a scaled MOSFET, is a reality which makes it difficult for the model developers to arrive at a physical, accurate, and yet compact model for the threshold voltage. The two phenomena adopted for demonstrating the effect of the 2-D field under the gate in the channel are (i) charge sharing, and (ii) barrier lowering, the latter being more physical. Barrier lowering due to drain voltage is relatively involved which is discussed in the next section. The charge sharing concept is dealt with in Section 5.3.3.

5.3.2 Barrier Lowering Model

In the discussion related to the operation of a long channel MOS transistor, it was implied that there exists a potential barrier between the source and the channel at the Si–SiO_2 interface, as shown in Figure 5.4, which is controlled by gate voltage. When the gate voltage is increased the surface potential increases, thereby decreasing the energy barrier difference between the source and the channel. This results in an increase of carrier injection from the source to the channel over the lowered energy barrier. From the definition of threshold voltage, the potential barrier difference at threshold is $\phi_{ssT} = \phi_{bi} - 2\phi_f$, where ϕ_{ssT} is the surface potential w.r.t. source at threshold [56].

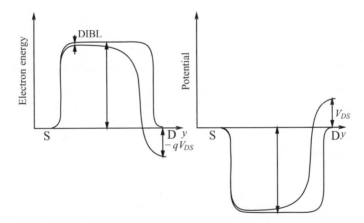

Figure 5.4   Energy and potential variation along y at the interface of an n-channel MOSFET

When the channel length is shortened in a scaled MOS transistor the drain junction field penetrates towards the source, contributing to lowering of the potential barrier that exists between the source and the channel. The effect of increasing the drain potential on the barrier height between the source and the channel is shown in Figure 5.4. It may be mentioned that such a barrier lowering may even happen in the absence of drain to source voltage in aggressively scaled MOSFETs. In such MOSFETs, drain and source induced channel fields interact due to overlap of the depletion layers associated with the junctions. The drain functions as a sort of additional gate terminal controlling the barrier height between the source and the channel. Such a phenomenon is known as *Drain Induced Barrier Lowering* (DIBL) and is characteristic of short channel MOSFETs. In the presence of the DIBL effect, the gate voltage required to attain a barrier difference height of ϕ_{ssT} will be less as compared to the long channel case [57, 58]. In other words, there is a lowering of threshold voltage, which is now also a function of the drain bias V_{DS}, attributed to the DIBL effect.

There have been attempts to explain the DIBL effect as a reduction in the threshold voltage due to the charge sharing mechanism arising from the contribution from the drain voltage. Such models, however, though comparatively simple, do not adequately explain the experimental results of threshold voltage dependence under a wide range of operating conditions. Therefore, solution of the 2-D Poisson's equation under the gate, with depletion approximation in the subthreshold condition, is considered to be a more appropriate approach to understand DIBL phenomena. Such 2-D equations are either solved numerically or analytically with suitable simplifying assumptions. Below, the approach of Liu *et al.* [23] for modeling the DIBL effect is outlined, which holds good up to the range of deep submicron MOS devices.

Figure 5.5 shows the cross-section of a MOS transistor at threshold condition. The depletion region is under the influence of vertical field due to the gate bias and horizontal field due to drain-to-source voltage. The rectangular Gaussian box is bounded between (i) source, $y = 0$, and drain, $y = L$, in the y-direction along the channel, and (ii) the Si–SiO$_2$ interface, $x = 0$, and the edge of the neutral substrate region, $x = W_{dm}$, where W_{dm} stands for the depletion width at the point of minimum surface potential.

The depletion width is given by the expression:

$$W_{dm} = \sqrt{\frac{2\epsilon_{Si}}{qN_A} (V_{SB} + 2\phi_f)} \tag{5.9}$$

Such a point of minimum potential in the channel is designated as the *virtual cathode*, as this is the location in the channel which determines the carriers injected from the source to the channel. Thus DIBL analysis, in effect, means the determination of the lowering of the barrier height at the virtual cathode due to the lateral field. It is assumed that the 2-D Poisson's equation in the channel is described by:

$$\frac{\partial F_x}{\partial x} + \frac{\partial F_y}{\partial y} = -\frac{qN_A}{\epsilon_{Si}} \tag{5.10}$$

The boundary conditions for the potential at the ends of the Gaussian box, with source as reference and $V_{SB} = 0$, gives:

$$\phi_{ss}(0) = \phi_{bi} \tag{5.11}$$

$$\phi_{ss}(L) = V_{DS} + \phi_{bi} \tag{5.12}$$

where ϕ_{bi} is the junction built-in potential.

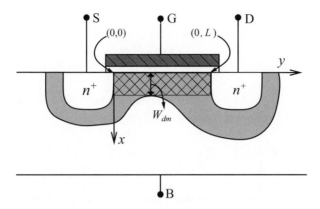

Figure 5.5 Gaussian box of depth W_d for 2-D analysis

The gradient of the x-component of the electric field at a point along the y-axis of the elemental Gaussian box in silicon is computed from the average field in the depletion region. The average electric field is obtained from the difference of the field in the gate oxide at $(0, y)$ and (W_{dm}, y), divided by the depth of the Gaussian box, W_{dm}. From the potential balance equation between gate and substrate, the electric field in the oxide at $(0, y)$ is:

$$F_{ox}(0, y) = \frac{V_{GS} - V_{fb} - \phi_{ss}(y)}{T_{ox}} \tag{5.13}$$

From the continuity of the displacement vector $(\epsilon_{ox} F_{ox}(0, y) = \epsilon_{Si} F_x(0, y))$, where $F_x(0, y)$ represents the field in silicon at the interface, we have:

$$F_x(0, y) = \frac{\epsilon_{ox}}{\epsilon_{Si}} \left(\frac{V_{GS} - V_{fb} - \phi_{ss}(y)}{T_{ox}} \right) \tag{5.14}$$

Using the boundary condition that the normal electric field component leaving the Gaussian element at $x = W_{dm}$, i.e. $F_x(W_{dm}, y)$ is zero, the gradient of the x-component of electric field is obtained as:

$$\frac{\partial F_x}{\partial x} = \frac{[F_x(0, y) - F(W_{dm}, y)]}{W_{dm}} = \frac{\epsilon_{ox}}{\epsilon_{Si}} \frac{(V_{GS} - V_{fb} - \phi_{ss}(y))}{T_{ox} W_{dm}} \tag{5.15}$$

For the Gaussian box under consideration, the field gradient $\partial F_y/\partial y$ is a function of x along depth. However, the term $\partial F_y/\partial y$ can be treated as uniform along depth, with an average value given as $(1/\eta) dF_{sy}/dy$ without loss of generality. F_{sy} denotes the value of F_y at the surface $(x = 0)$, and η is a fitting parameter [60].

Substituting the values of divergence of electric fields obtained above, and with rearrangement of terms we get:

$$\frac{dF_{sy}}{dy} + \left(\frac{V_{GB} - V_{fb} - \phi_s(y)}{l^2} \right) = -\frac{\eta q N_A}{\epsilon_{Si}} \tag{5.16}$$

where:

$$l = \sqrt{\frac{\epsilon_{Si} T_{ox} W_{dm}}{\eta \epsilon_{ox}}} \qquad (5.17)$$

is defined as the *characteristic length* of the lateral electric field in the channel. Equation (5.16) can be recast in terms of surface potential through the relation $F_{sy} = -d\phi_{ss}/dy$:

$$\frac{d^2\phi_{ss}(y)}{dy^2} - \frac{\phi_{ss}(y)}{l^2} = \frac{qN_A\eta}{\epsilon_{Si}} - \frac{(V_{GB} - V_{fb})}{l^2} \qquad (5.18)$$

The above equation has the form of a standard second order linear differential equation where the RHS is constant for a given V_{GB}. $\phi_{ssL} = V_{GS} - V_{T0} + \phi_s$ is the long channel surface potential. The solution of Equation (5.18) is:

$$\phi_{ss}(y) = \phi_{ssL} + (\phi_{bi} + V_{DS} - \phi_{ssL})\frac{\sinh(y/l)}{\sinh(L/l)} + (\phi_{bi} - \phi_{ssL})\frac{\sinh[(L-y)/l]}{\sinh(L/l)} \qquad (5.19)$$

Figure 5.6(a) shows the plot of the variation of the surface potential along the channel for a given set of process and device parameters. It is seen that there is a minima in the potential, which is identified as the virtual cathode as mentioned earlier.

It is observed that the barrier height between the source and the channel decreases for a shorter channel as was qualitatively discussed earlier. Further, the barrier height decreases with the increase in drain voltage for a given channel length. The lateral component of electric field at the Si–SiO$_2$ interface can be obtained directly from Equation (5.19) by differentiating it w.r.t. y, and is plotted in Figure 5.6(b). It is seen that the lateral component of the electric field reduces to zero due to cancellation of the lateral field from source and drain, as they are acting in the opposite direction along the channel. For a long channel MOSFET the location of the virtual cathode lies at the center of the channel. It moves towards the source as the channel

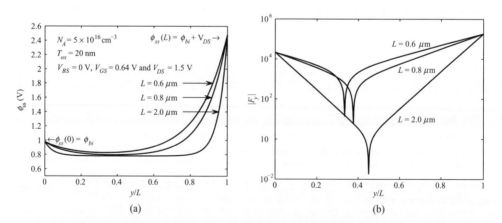

Figure 5.6 Simulated variations of (a) surface potential, and (b) junction induced surface lateral field along the channel with channel length as the parameter for fixed drain bias

length reduces for a given drain voltage V_{DS}. It can also be seen that the magnitude and the gradient of the lateral electric field are higher for short channel MOSFETs. The lateral electric field is found to decrease exponentially along the channel from source and drain.

The location y_0 in the channel where the surface potential has a minimum value, ϕ_{ssmin}, can be obtained using the condition:

$$\left.\frac{d\phi_{ss}}{dy}\right|_{y=y_0} = 0 \tag{5.20}$$

At threshold voltage, $\phi_{ssmin} = 2\phi_f$. ϕ_{ssmin} can be obtained explicitly from Equation (5.19) for $V_{DS} \ll (\phi_{bi} - \phi_{ssL})$, assuming $y_0 = L/2$.

From the above considerations it can be shown that the DIBL induced reduction in threshold voltage is given by:

$$V_T(L) = V_{T0} - \Delta V_T \tag{5.21}$$

where:

$$\Delta V_T = \frac{2(\phi_{bi} - \phi_{smin}) + V_{DS}}{2\cosh(L/2l) - 2} \tag{5.22}$$

For $L \gg l$, Equation (5.22) reduces to:

$$\Delta V_T(L) = [2(\phi_{bi} - \phi_{smin}) + V_{DS}]\left(e^{\frac{-L}{2l}} + 2e^{\frac{-L}{l}}\right) \tag{5.23}$$

For $L > 5l$, the first exponential term dominates, and ΔV_T has an exponential dependence on channel length L. For shorter channel length, the reduction in threshold voltage due to drain voltage is determined by both the exponential terms shown in Equation (5.23).

The reduction in the threshold voltage as obtained from pseudo-2-D analysis taking the contribution of source drain junction fields is shown in Figure 5.7. In Figure 5.7(a) it is seen

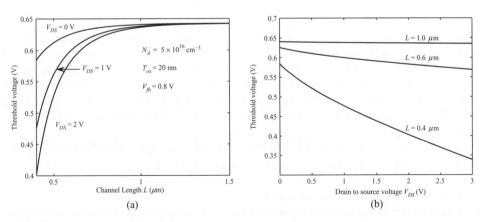

Figure 5.7 Simulated variations of threshold voltage (a) with channel length and V_{DS} as a parameter, and (b) with V_{DS} and channel length as a parameter

that there is roll-off in threshold voltage with the reduction in channel length. The simulation reveals that the roll-off is higher for greater drain voltage. Figure 5.7(b) demonstrates an approximate linear dependence of threshold voltage with drain bias for a given channel length.

From the derivative of Equation (5.19), denoted by F_{sy}, it may be concluded that the characteristic length may be considered to be an indicator of the rate of decay of lateral electric field along the channel. Hence, the magnitude of l is a measure of the severity of short channel effects on the threshold voltage. From the expression of the characteristic length given by Equation (5.17), it is noted that the characteristic length, l, has a square root dependence on gate oxide thickness (T_{ox}) and depletion thickness (W_{dm}). While W_{dm} is assumed constant in the analysis, it depends on channel length, drain voltage, and junction depth X_j. The junction depth determines the effective channel length through its contribution to lateral diffusion in mask defined polysilicon gate length [61, 62]. Because of processing related uncertainties, l is preferred to be considered as a fitting parameter.

The measure of the DIBL effect is quantitatively expressed through a parameter $R = \partial V_{TS}/\partial V_{DS}$, i.e. the rate of change of threshold voltage with applied drain bias [64, 73, 74]. This parameter needs to be minimized during optimization of device design.

From Brews' criteria on subthreshold current, demarcating the boundary between long and short channel MOSFET [8], it is found that minimum acceptable channel length is approximately five times the characteristic length [23]. An empirical relation has been proposed keeping in view the above approach, which is given below:

$$l = 0.1 \left(X_j T_{ox} W_{dm}^2 \right)^{\frac{1}{3}} \tag{5.24}$$

5.3.3 Charge Sharing Model

The conventional expression for the threshold voltage of a long channel MOS transistor is given by:

$$V_T = V_{fb} + 2\phi_f - \frac{Q_b'}{C_{ox}'} \tag{5.25}$$

where Q_b' is the depletion charge per unit area under the gate. The above expression is independent of channel length or width. The expression in Equation (5.25) has been obtained under the implicit assumption that the entire depletion charge in the channel under the gate is available to the gate field for termination. While the model is valid for long and wide MOS transistors, it is inadequate for MOS transistors with small dimensions. In order to concentrate on the effect of scaling the channel length alone, we shall consider a MOS transistor with large width so that any fringe effect along the width can be ignored.

Consider now a short channel MOS transistor as shown in Figure 5.8. The following simplifying assumptions are made along the lines in reference [18]:

- The substrate is homogeneously doped.
- Source and drain are at the same potential.
- The sidewall part of source and drain are circular with a radius equal to the junction depth X_j.

- The depletion layer depth, W_{dm}, under the gate which supports the inversion surface potential $2\phi_f$, can be considered to be equal to the one-sided depletion layer width $W_{dx(x=s,d)}$ of source–drain junctions corresponding to the junction barrier potential ϕ_{bi}. The approximation is valid as the difference between $2\phi_f$ and ϕ_{bi} is only a few mV.

From Figure 5.8 it is seen that the depletion charge of negative acceptor ions under the gate due to the surface potential at inversion has a width W_{dm}, and a vertical cross-section defined by the area $CC'D'D$. The depletion charge is formed of negative acceptor ions in a p-type substrate. The field lines originating from positive charges which terminate on the negative acceptor ions confined in the area $CC'D'D$ has two origins:

1. The positive charge induced on the gate electrode due to positive bias applied on it.
2. The donor ions ($+$) within the depletion layer of the n^+-region forming the source and drain junctions with the substrate.

Therefore, the negative depletion charge under the gate is *shared* between the gate and the source/drain induced fields. The lines of force of these fields originate from the positive charges associated with the gate and the donor ions of the source/drain depletion regions. Intuitively, it can be presumed that the negative depletion charge in the channel in proximity to the source/drain will be used by the lateral field lines emanating from them. The remaining depletion charge would be available to the normal field lines originating from the gate. Due to the complex nature of the two dimensional field profile in a short channel device, an analytical evaluation of the extent of charge sharing is not feasible. One has to make an intelligent approximation and look for best fit with the experimental results. In Figure 5.8, a triangular area ACD, containing negative acceptor ions, has been curved out for the charge balance for

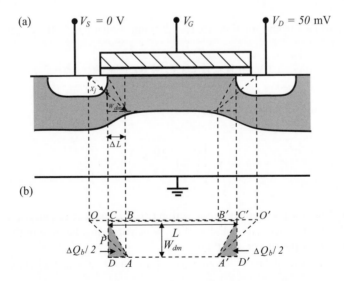

Figure 5.8 (a) The cross-sectional view of a MOS transistor showing the sharing of channel depletion charge between gate and source/drain. (b) Trapezoidal approximation of the depletion charge $AA'C'C$ assigned to the gate field

the positive donor ions associated with the source. Again, a general geometrical shape does not lead to a tractable analytical approximation of charge sharing. The simplest approximation proposed by Poon *et al.* [17] and Yau [18] are the triangular regions ACD for the source, and $A'D'C'$ for the drain for charge sharing from the rectangular area $C'CDD'$ under the gate. This leaves the charge within the trapezoidal area $C'CAA'$ for the gate field termination. The approach, though speculative, has been the basis of subsequent modified models, and is able to provide an analytical expression and physical insight for the dependence of threshold voltage on channel length.

For obtaining the trapezoid area, the triangular areas have to be computed from the known parameters. The height DC ($D'C'$) is known in terms of surface potential and substrate doping concentration N_A:

$$W_{dm} = \frac{\sqrt{2q\epsilon_{Si}(2\phi_f)}}{qN_A} \tag{5.26}$$

The base $DA(A'D') = \Delta L$ remains to be determined in terms of known parameters from geometrical considerations. In $\triangle OAB$:

$$OA = OP + PA = X_j + W_{DX} = X_j + W_{dm} \tag{5.27}$$

$$OB = OC + CB = X_j + \Delta L \tag{5.28}$$

$$AB = W_{dm} \tag{5.29}$$

$$OA^2 = (OC + CB)^2 + BA^2 \tag{5.30}$$

Substituting the values from Equations (5.27–5.29) into Equation (5.30) we get:

$$(X_j + W_{dm})^2 = (X_j + \Delta L)^2 + W_{dm}^2 \tag{5.31}$$

Solving the above equation for ΔL we get:

$$\Delta L = X_j \left[\left(1 + \frac{2W_{dm}}{X_j} \right)^{\frac{1}{2}} - 1 \right] \tag{5.32}$$

The loss of area from the rectangular cross-section $C'CDD'$ under the gate is given by the sum of the area of the two triangles ACD and $A'D'C'$, which are equal because of the equal biasing condition assumed for the source and drain.

Now, the depletion charge under the gate Q_{b0} at threshold voltage, without charge sharing, is given by:

$$Q_{b0} = -qN_A WLW_{dm} \tag{5.33}$$

The effective charge Q_{be}, available to the gate after charge sharing, under the trapezoidal area is:

$$Q_{be} = -qW\frac{L + (L - 2\Delta L)}{2} W_{dm} N_A \qquad (5.34)$$

The loss of charge to the gate field equals $\Delta Q_b = Q_{b0} - Q_{be}$, from Equations (5.33) and (5.34):

$$\Delta Q_b = -qW_{dm}W\Delta L N_A \qquad (5.35)$$

Assuming that the loss of bulk depletion charge is uniformly distributed over the entire gate length, the effective or average depletion charge loss to the gate per unit area Q'_{be} is given by:

$$\Delta Q'_{be} = -q\frac{W_{dm}\Delta L N_A}{L} \qquad (5.36)$$

Substituting the value of L from Equation (5.32), we get:

$$\Delta Q'_{be} = -qN_A W_{dm}\frac{X_j}{L}\left[\left(1 + \frac{2W_{dm}}{X_j}\right)^{\frac{1}{2}} - 1\right] \qquad (5.37)$$

Therefore, the change in threshold voltage ΔV_{TS} is given by:

$$\Delta V_{TS} = \frac{\Delta Q'_{be}}{C'_{ox}} = -\frac{qN_A W_{dm}}{C'_{ox}}\frac{X_j}{L}\left[\left(1 + \frac{2W_{dm}}{X_j}\right)^{\frac{1}{2}} - 1\right] \qquad (5.38)$$

The negative sign in the expression indicates that the charge sharing of the depletion charge under the gate by source/drain is responsible for a reduction in the threshold voltage.

The expression for the threshold voltage including the short channel effect due to charge sharing is given by:

$$V_{TS} = V_{fb} + 2\phi_f - \left[1 - \left(\sqrt{1 + \frac{2W_{dm}}{X_j}} - 1\right)\frac{X_j}{L}\right]\frac{Q'_b}{C'_{ox}} \qquad (5.39)$$

The above expression for V_{TS} can be put in the following form:

$$V_{TS} = V_{fb} + 2\phi_f - f_s\frac{Q'_b}{C'_{ox}} \qquad (5.40)$$

where $f_s = \left[1 - \left(\sqrt{1 + 2W_{dm}/X_j} - 1\right)X_j/L\right]$ is known as the *short channel form factor*. The nature of the form factor for a given set of device data has been plotted as a function of the channel length in Figure 5.9, which shows that the form factor ($f_s < 1$) decreases with scaling down of channel length. The short channel correction for the threshold voltage obtained for

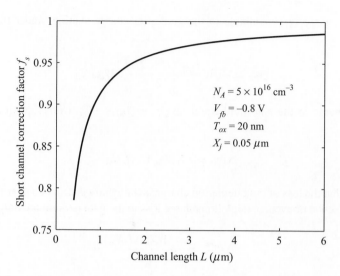

Figure 5.9 Dependence of short channel form factor on channel length

the particular case of $V_{SB} = V_{DB} = 0$, can be generalized for bias applied to the source and drain voltages as given below.

We shall now obtain the expression for the short channel correction factor involving source and drain bias V_{SB}, V_{DB} with $V_{DS} = 0$, for which the threshold voltage reduction can be expressed as $\Delta V_{TS} = f_1(V_{SB}) + f_2(V_{DB})$. For $V_{DB} > V_{SB}$, the depletion width W_{dd} at the drain end will be larger than W_{ds} at the source. The reduction in V_{TS} due to charge sharing by source and drain is given by:

$$\Delta V_T = \frac{1}{C'_{ox}} \left(\frac{\Delta Q_S + \Delta Q_D}{L} \right) \tag{5.41}$$

where ΔQ_S and ΔQ_D are the reduction in depletion charge available to the gate field:

$$\Delta V_T = -\frac{q N_A X_j}{2 C'_{ox} L} \left[\left(\sqrt{1 + \frac{2 W_{ds}}{X_j}} - 1 \right) W_{ds} + \left(\sqrt{1 + \frac{2 W_{dd}}{X_j}} - 1 \right) W_{dd} \right] \tag{5.42}$$

It is of practical interest to study the effect of substrate bias on threshold voltage, which can be obtained from Equation (5.41). The effect is demonstrated through the simulation curve plotted in Figure 5.10 for various channel lengths. It may be concluded that the sensitivity of threshold voltage w.r.t. substrate bias is reduced in short channel MOSFETs. Physically, the observation can be interpreted to be due to an effective reduction of substrate doping due to charge sharing.

The threshold voltage expression with substrate bias can now be written as:

$$V_{TS} = V_{fb} + 2\phi_f - \gamma f_s \sqrt{2\phi_f - V_{BS}} \tag{5.43}$$

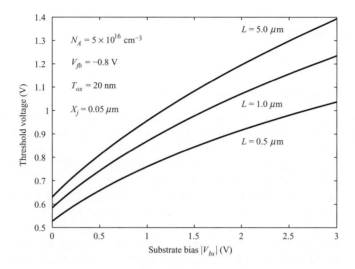

Figure 5.10 Variation of threshold voltage with substrate bias with channel length as a parameter

where $f_s = 1 - (X_j/2L)\left[\left(\sqrt{1 + 2W_{ds}/X_j} - 1\right) + \left(\sqrt{1 + 2W_{dd}/X_j} - 1\right)\right]$. The above equation is used in the Berkeley LEVEL 2 MOSFET model [16]. In the model described above, the length of the channel was not considered short enough to include the possibility of overlap of source/drain depletion regions. For more aggressively scaled channel lengths, when further correction is called for, the LEVEL 3 model adopts the approach of Dang [19] with modification. The new short channel correction factor is given as:

$$f_s = 1 - \frac{1}{L}\left[(W_C + LD)\sqrt{1 - \left(\frac{W_P}{X_j + W_P}\right)^2} - LD\right] \qquad (5.44)$$

where $W_C = 0.063\,135\,3X_j + 0.801\,329\,2W_P - 0.011\,107\,77W_P^2/X_j$ and $W_P = \sqrt{2\epsilon_{Si}V_x/qN_A}$. W_P is the depletion width corresponding to the planar part of the junction. The value of W_C, the source–substrate depletion width for the curved part of the junction, which is assumed to be cylindrical, has been obtained numerically making a one-sided step junction approximation, and solving Poisson's equation for the cylindrical junction. V_x denotes the potential barrier across the source/drain junction to be calculated corresponding to the biasing condition of the junctions, and $LD = 0.7X_j$. In a scaled MOS transistor, apart from estimating the value of threshold voltage, the gradient of the threshold voltage w.r.t. to channel length, i.e. dV_{TS}/dL, happens to be an important criterion demonstrating the effect of charge sharing on the threshold voltage [64].

Both SCE and DIBL demonstrate the manifestation of a two-dimensional field in the channel of scaled MOS transistors. SCE is the phenomenon of V_T decrease at relatively low drain voltage, whereas DIBL is estimated by measurements of threshold voltage both at low and at an assigned value of V_{DS} of nominal higher magnitude. The resistance of a scaled MOSFET to the influence of the 2-D field may be reflected through a quantity known as *Electrical Integrity* (EI) [7], which can be expressed in terms of process and design parameters of the device. A low value of EI indicates the dominance of the gate field and 1-D field pattern corresponding to a long channel MOSFET.

5.3.4 V_T Implementation in Compact Models

In the previous two sections we have discussed two different conceptual approaches for modeling the effect of lateral field gradient on threshold voltage. The charge sharing models by Yau [18] and Dang [19] look arbitrary. Their acceptance, however, arise from their simplicity and successful implementation in models such as LEVEL 2, LEVEL 3, and BSIM1 which were in wide use for several earlier nodes of MOS technology.

In the advanced compact MOSFET models such as BSIM3v3, Philips MM11, and HiSIM, the lateral field gradient effect is generally taken into account by approximate solution of the 2-D Poisson equation under various simplifying assumptions. The approximate solution has been dealt with in Section 5.3.2, where it is shown that ΔV_T has an exponential dependence on channel length L. Though the solution is categorized as a DIBL effect, it is general practice to identify only the drain bias dependent term in the ΔV_T expression as DIBL components, and the bias independent term with a short channel component. We shall discuss below the implementation of threshold voltage reduction, ΔV_T, in various compact MOSFET models.

5.3.4.1 BSIM3v3

From Liu's model, referring to Equation (5.23), we have:

$$\Delta V_T = \left(e^{\frac{-L}{2l}} + 2e^{\frac{-L}{l}} \right) [2(V_{bi} - \phi_s)] + \left(e^{\frac{-L}{2l}} + 2e^{\frac{-L}{l}} \right) V_{DS} \qquad (5.45)$$

where the first term is identified as the short channel component, and the second term as the DIBL component. The above expression which underlines the basic physical approach, however, has to be further modified with empirical fitting parameters to be extracted experimentally for precise matching with experimental data. The change in threshold voltage due to short channel and DIBL effects is modeled as:

$$\Delta V_T = \mathbf{DVT0} \left(e^{\frac{-\mathbf{DVT1} \cdot L}{2l_{SC}}} + 2e^{\frac{-\mathbf{DVT1} \cdot L}{l_{SC}}} \right) [2(\phi_{bi} - \phi_s)]$$

$$+ \left(e^{\frac{-\mathbf{DSUB} \cdot L}{2l_{DIBL}}} + 2e^{\frac{-\mathbf{DSUB} \cdot L}{l_{DIBL}}} \right) V_{DS} \qquad (5.46)$$

where $l_{SC} = \sqrt{(\epsilon_{Si} T_{ox}/\epsilon_{ox}) X_d} \, (1 + \mathbf{DVT2} \cdot V_{BS})$, $l_{DIBL} = \sqrt{(\epsilon_{Si} T_{ox}/\epsilon_{ox}) X_d}$, $\mathbf{DVT0}$, $\mathbf{DVT1}$, $\mathbf{DSUB}$, and $\mathbf{DVT2}$ are empirical model parameters.

5.3.4.2 HiSIM

The 2-D Poisson's equation gives:

$$\frac{F_{xs}(y)}{W_{dm}} + \frac{dF_y(y)}{dy} = -\frac{qN_A}{\epsilon_{Si}} \tag{5.47}$$

where F_{xs} is the vertical component of the electric field at the Si–SiO$_2$ interface. For a given V_{GS} in the subthreshold region, we have for normal surface electric field in long channel transistor:

$$F_{xs} = \frac{C'_{ox}}{\epsilon_{Si}} \left(V_{GS} - V_{fb} - \phi_{ss}\right) \tag{5.48}$$

where ϕ_{ss} is the surface potential w.r.t. source. As the presence of the lateral field gradient affects the threshold voltage in the case of short channel MOSFETs, we have for the normal surface electric field:

$$F_{xs} = \frac{C'_{ox}}{\epsilon_{Si}} \left(V_{GS} - V_{fb} - \Delta V_{TS} - \phi_{ss0}\right) \tag{5.49}$$

where ΔV_{TS} is the change in threshold voltage due to channel shortening, and ϕ_{ss0} is the surface potential at threshold w.r.t. source for long channel transistor.

Substituting Equation (5.49) into Equation (5.47), we have:

$$\frac{C'_{ox}}{\epsilon_{Si} W_{dm}} \left(V_{GS} - V_{fb} - \Delta V_{TS} - \phi_{ss0}\right) + \frac{dF_y}{dy} = \frac{-qN_A}{\epsilon_{Si}} \tag{5.50}$$

which can be rearranged as:

$$\left(V_{GS} - V_{fb} - \Delta V_{TS} - \phi_{ss0}\right) + \frac{\epsilon_{Si} W_{dm}}{C'_{ox}} \frac{dF_y}{dy} = -qN_A \frac{W_{dm}}{C'_{ox}} \tag{5.51}$$

The RHS of the above equation can be identified as the voltage drop across the oxide in the case of a long channel transistor, i.e. $V_{ox} = V_{GS} - V_{fb} - \phi_{ss0}$, and therefore Equation (5.51) yields:

$$\Delta V_{TS} = \frac{\epsilon_{Si} W_{dm}}{C'_{ox}} \frac{dF_y}{dy} \tag{5.52}$$

where:

$$W_{dm} = \sqrt{\frac{2\epsilon_{Si}(2\phi_f - V_{BS})}{qN_A}} \tag{5.53}$$

From Figure 5.6(a) we get the potential distribution along the channel, and for $V_{DS} = 0$ a parabolic distribution may be assumed. Assuming the value of the potential at the ends of the channel to be $V_{bi} - 2\phi_f$ above that at the center, where the value is minimum $(2\phi_f)$, it follows the position independent lateral field gradient $\partial F_y/\partial y$ may be given by [43, 44]:

$$\frac{\mathrm{d}F_y}{\mathrm{d}y} = \frac{2(V_{bi} - 2\phi_f)}{L^2} \tag{5.54}$$

Therefore, the change in the threshold voltage due to the lateral field gradient is given as:

$$\Delta V_{TS} = \frac{\epsilon_{Si}}{C'_{ox}} W_{dep} \frac{2(V_{bi} - 2\phi_f)}{L^2} \tag{5.55}$$

In deriving the above equation it was assumed that $V_{DS} = 0$. The effect of V_{DS} can be introduced in Equation (5.55) by noting from Figure 5.6(b) that the decrease in the threshold voltage is a linear function of V_{DS}. Therefore, the expression for ΔV_{TS}, with functional dependence on V_{DS} and V_{BS}, may be described by the following expression [45]:

$$\Delta V_{TS} = \frac{\epsilon_{Si}}{C'_{ox}} W_{dep} \frac{2(V_{bi} - 2\phi_f)}{(L - \mathbf{PARL2})^2} (\mathbf{SC1} + \mathbf{SC2} \cdot V_{DS}) \tag{5.56}$$

where **PARL2** represents the depletion width of the source/drain junction along the length of the channel. It is implied that through the term $(L - \mathbf{PARL2})$, the depletion penetration in the channel from source and drain, and thereby the 2-D effect due to the lateral field, has been taken care of. The model parameter **SC1** models the effect of small V_{DS} and V_{BS} on the threshold shift, whereas **SC2** models the gradient of ΔV_{TS} with V_{DS} [45].

5.3.4.3 Philips MM11

In Philips MM11 also, V_T is inversely related to the square of the channel length [36]. The model is based on the so-called voltage–doping transformation [59] through which the effective doping due to presence of lateral field gradient is shown to be:

$$N_{Aeff} = N_A(1 - f) = N_A \left(1 - \frac{\epsilon_{Si}}{qN_A} \frac{2V_{DS}^*}{L^2} \right) \tag{5.57}$$

The approximated expression for V_{DS}^* is given by:

$$V_{DS}^* = 2V_{DS} + 4(\phi_{bi} - 2\phi_f) \tag{5.58}$$

It follows that:

$$\Delta V_T = V_T(\text{long}) - V_T(\text{short}) \tag{5.59}$$

$$\Delta V_T = \left(V_{fb} + 2\phi_f + V_{SB} + \gamma\sqrt{V_{SB} + 2\phi_f} \right)$$
$$- \left(V_{fb} - 2\phi_f + V_{SB} + \gamma_{eff}\sqrt{V_{SB} + 2\phi_f} \right) \tag{5.60}$$

$$\Delta V_T = (\gamma - \gamma_{eff})\sqrt{V_{SB} - 2\phi_f} \tag{5.61}$$

$$\Delta V_T = \gamma\sqrt{V_{SB} - 2\phi_f}\left(1 - \sqrt{\frac{N_{Aeff}}{N_A}} \right) \tag{5.62}$$

Substituting for N_{Aeff} from Equation (5.57) and expanding the square root term, it turns out that:

$$\Delta V_T = \gamma\sqrt{V_{SB} + 2\phi_f}\left(\frac{\epsilon_{Si} V_{DS}^*}{2 N_A L^2} \right) \tag{5.63}$$

Substituting for V_{DS}^* from Equation (5.58):

$$\Delta V_T = \gamma\sqrt{V_{SB} + 2\phi_f}\frac{\epsilon_{Si}}{q N_A L^2}\left[2V_{DS} + 4(\phi_{bi} - 2\phi_f) \right] \tag{5.64}$$

$$\Delta V_T = \gamma\sqrt{V_{SB} + 2\phi_f}\frac{\epsilon_{Si}}{q N_A L^2}\left[4(\phi_{bi} - 2\phi_f) \right] + \sigma_{DIBL}\sqrt{V_{SB} + 2\phi_f}\, V_{DS} \tag{5.65}$$

Like the BSIM model, MM11 also contains a term which is independent of drain voltage characterizing the short channel phenomenon, and the second term is a function of drain voltage through which DIBL effect is manifested.

In the PSP model the lateral field gradient is modeled as:

$$f = 1 - \frac{2\epsilon_{Si}}{q N_A L^2}\xi\left(2 + \frac{V_{DS}}{\xi} + \sqrt{1 + \frac{V_{DS}}{\xi}} \right) \tag{5.66}$$

where $\xi = (V_{SB} + \phi_{bi} - \phi_{fc})$, and ϕ_{fc} is the surface potential at a certain point in the channel.

The effective doping using the lateral field gradient factor is subsequently used in the computation of surface potential for calculating short channel effects in surface potential based compact MOSFET models.

5.4 Narrow and Inverse Width Effects

The dimensional scaling of MOS transistors involves not only reduction of length but also of width. The effect of the former has been discussed in the previous section. The present section will discuss the effect of reducing or narrowing the width. The narrow width or narrow channel effect, however, depends on the nature of field oxide isolation technology. In the case of conventional LOCOS[1] technology, as the width of the MOS transistor is decreased, there is a spread out of depletion charge in the direction of width beyond the channel defined by the thin gate oxide due to a fringing effect. Figure 5.11(a) shows the cross-section of the MOS transistor in the direction of width.

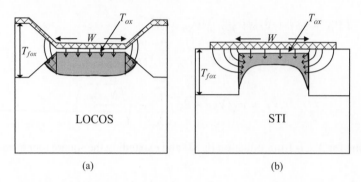

LOCOS   STI

(a)   (b)

Figure 5.11   Cross-sections of a MOS transistor in the direction of width showing the field lines for (a) LOCOS and (b) STI technologies

The cross-section shows the fringing fields in LOCOS technology from the gate to the substrate where there is a tapered transition from the thin gate oxide (GOX) to the thick field oxide (FOX). The depletion in the hashed shaded area in Figure 5.11(a) represents the additional fringing charge seen to the gate. As the device width is scaled down, the depletion charge under the gate is reduced but the fringing charge under the tapered oxide region remains relatively unchanged. Hence, the fringe charge assumes a significant proportion as the channel width is shortened. The region under the tapered oxide is approximated by a circular geometry. There are other approximations for the fringing depletion region [77]. The model for the narrow width effect on threshold voltage, considering a circular geometry, is discussed below.

- The gate induced depletion charge per unit area under the gate (thin) oxide is Q'_b, and that for the fringing part per unit area is Q'_{be}.
- The depletion charge spreading beyond the gate oxide in the direction of width in the tapered region can be modeled to be contained in two quarter cylinders with a radius equal to the depth of the depletion region under the gate, W_{dm}. The depletion width W_{dm} is given by:

$$W_{dm} = \sqrt{\frac{2\epsilon_{Si}}{qN_A}(V_{SB} + 2\phi_f)}$$ (5.67)

- The additional fringe charge/length in each quarter cylinder is given by:

$$\left(\frac{\pi}{4}W_{dm}^2\right)qN_A$$ (5.68)

---

[1] LOCal Oxidation of Silicon.

- Considering the two quarter cylinders and averaging the charge per unit length over the width W, we get the fringing charge per unit area as:

$$\Delta Q'_{bf} = \frac{2 \left(\frac{\pi}{4} W^2_{dm}\right) q N_A}{W} \tag{5.69}$$

Substituting for W_{dm} from Equation (5.67), we get:

$$\Delta Q'_{bf} = \frac{\pi \epsilon_{Si}}{W}(V_{SB} + 2\phi_f) \tag{5.70}$$

The additional fringe charges increase the threshold voltage. We can write:

$$\Delta V_T = \frac{\Delta Q'_{bf}}{C'_{ox}} = \frac{\pi \epsilon_{Si}}{W}\left(\frac{T_{ox}}{\epsilon_{ox}}\right)(V_{SB} + 2\phi_f) \tag{5.71}$$

$$\Delta V_T = \frac{\pi}{2} \frac{q N_A W^2_{dep}}{C'_{ox} W} \tag{5.72}$$

The approximation for the fringe charge is also dictated by the profile of the recessed oxide.

Figure 5.12 shows the increase in threshold voltage with decrease in channel width as per the quarter circle model. Thus, the threshold voltage of a minimum-size device can be given by:

$$V_T(\text{Minimum Size Device}) = V_T(\text{Long Channel and Wide Width Device})$$

$$- V_T(\text{Short Channel Effect}) + V_T(\text{Narrow Width Effect}) \tag{5.73}$$

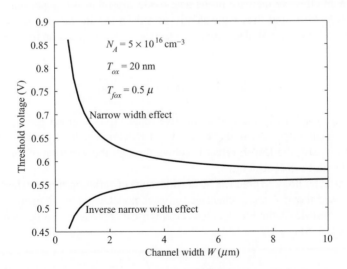

Figure 5.12   Narrow and inverse narrow width effect on threshold voltage

The drive towards high packing density in CMOS technology resulted in the search for an alternative structure to LOCOS technology, where the 'bird's beak' form, due to lateral oxidation is considered area inefficient. Shallow Trench Isolation (STI) with a vertical field oxide, shown in Figure 5.11(b), improves the area efficiency in device isolation. The structure is obtained through a recessed oxide. Figure 5.11(b) shows the fringing field in the gate in recessed STI technology. It is observed that there is considerable crowding of the fringing field terminating on the depletion charge under the gate, which can be interpreted as an effective increase in gate oxide field. Alternatively, the phenomena can be modeled as a effective reduction in gate oxide as perceived by the gate terminal.

Fully recessed isolation oxide in STI technology shows threshold voltage dependence on channel width, which is different from that of LOCOS technology. The V_T dependence on W for STI type technology is termed as the Inverse Narrow Width Effect (INWE) or Inverse Narrow Channel Effect (INCE). It is so named because the threshold voltage decreases with decrease in channel width, which is the inverse of the variation pattern corresponding to LOCOS technology where V_T increases with decrease of channel width.

For an STI structure, with depth of recessed oxide T_{fox}, width W, and gate oxide thickness T_{ox}, the total gate capacitance for Figure 5.11(b) is:

$$C_g = WLC'_{ox} + 2C_f \tag{5.74}$$

where C_f is the fringing capacitance on each side. Using conformal mapping the fringe oxide capacitance can be shown [78, 79] to be equal to $(2\epsilon_{ox}L/\pi)\ln(2T_{fox}/T_{ox})$. Thus:

$$C_g = \left(1 + \frac{F^*}{W}\right)(WL)C'_{ox} \tag{5.75}$$

where $F^* = (4T_{ox}/\pi)\ln(2T_{fox}/T_{ox})$ is the fringing field factor. Mathematically the fringing field results in an effective increase in the gate oxide capacitance C'_{ox} per unit area to a new value $C'_{ox}(1 + F^*/W)$, and thus equivalently to a decrease in the gate oxide thickness to $T'_{ox} = T_{ox}/(1 + F^*/W)$. As W decreases, T'_{ox} decreases. The expression for threshold voltage gets modified to:

$$V_T = V_{fb} + 2\phi_f + \gamma_{eff}\sqrt{2\phi_f} \tag{5.76}$$

where $\gamma_{eff} = \sqrt{2q\epsilon_{Si}N_A}\,T'_{ox}/\epsilon_{ox}$ is the effective reduced body effect parameter due to the inverse narrow width effect. As W decreases, V_T also decreases due to an effective reduction in T'_{ox}. In Figure 5.12, the INWE effect is shown through the variation of threshold voltage with channel width for STI structures.

The isolation technology, in practice, has a wide range of variations around the representative structure discussed above. A comprehensive compact model, valid for all operating conditions, is not available. BSIM3v3, therefore, adopts an empirical approach to model NWE and INWE as given below [15, 51]:

$$\Delta V_T = (\mathbf{K3} + \mathbf{K3B}\,V_{BS})\left(\frac{T_{ox}}{W + \mathbf{W0}}\right)2\phi_f \tag{5.77}$$

where **K3**, **K3B**, and **W0** are empirical fitting parameters. Here, the channel width is the effective width W_{eff}. **W0** is an offset introduced to make room for adjustments for fitting of the model that NWE is related inversely to transistor width ($W_{eff} + $ **W0**) [15]. **K3** is taken as positive for NWE and negative for INWE.

In the HiSIM model, INWE is handled through an effective gate capacitance, including fringe effect, through one fitting parameter **WFC**, as described below [45]. Concentrating on the fringe capacitance term:

$$C_f = \left[\frac{4\epsilon_{ox}}{\pi} \ln \left(\frac{2T_{fox}}{T_{ox}} \right) \right] \frac{L}{2} \tag{5.78}$$

The term in square brackets is treated as a fitting parameter **WFC** for a given process technology. Therefore, we can write [63]:

$$C_f = \mathbf{WFC} \, \frac{L}{2} \tag{5.79}$$

5.4.1 Scaling Effect on Pinch-off Voltage

In the EKV and ACM models, the concept of pinch-off voltage V_P is used, which is a function of gate and threshold voltage. In these models the short channel and narrow width effects are considered by introducing a correction factor in the expression for V_P [22].

A correction factor α_L, is multiplied to the depletion charge density value without any charge sharing effect, Q'_b. The factor α_L is the effective depletion charge density Q'_{bSC} for a short channel MOSFET normalized to Q'_b, which is represented by:

$$\alpha_L = \frac{Q'_{bSC}}{Q'_b} = 1 - \frac{X_j}{2L} \left(\sqrt{1 + \frac{2W_s}{X_j}} + \sqrt{1 + \frac{2W_d}{X_j}} - 2 \right) \tag{5.80}$$

This model adopts Yau's charge sharing model [18]. The above equation is simplified in EKV's model [80] by assuming $W_{s(d)} \ll X_j$, which after expanding the square root term results in:

$$\alpha_L = 1 - \frac{W_s + W_d}{2L} \tag{5.81}$$

Substituting for the depletion width at source and drain ends of the channel by the expressions, $W_{s(d)} = \sqrt{2\epsilon_{Si}(V_{S(D)B} + \mathbf{PHI})/qN_A}$, in Equation (5.81), we get:

$$\alpha_L = 1 - \left(\frac{1}{2} \sqrt{\frac{2\epsilon_{Si}}{qN_A}} \right) \frac{1}{L} \left(\sqrt{(V_{SB} + \mathbf{PHI})} + \sqrt{(V_{DB} + \mathbf{PHI})} \right) \tag{5.82}$$

where **PHI** is the bulk Fermi potential. The effective reduced depletion charge density is given as:

$$Q'_{bSC} = -C'_{ox}\alpha_L\gamma\sqrt{\phi_s} = -C'_{ox}\gamma^o\sqrt{\phi_s} \tag{5.83}$$

where the $\gamma^o = \alpha_L \gamma$ is the reduced body effect parameter due to charge sharing effect, and can be written from Equation (5.82) as:

$$\gamma^o = \gamma - \left(\frac{\epsilon_{Si}}{C'_{ox}}\right) \frac{1}{L} \left(\sqrt{(V_{SB} + \mathbf{PHI})} + \sqrt{(V_{DB} + \mathbf{PHI})}\right) \qquad (5.84)$$

A fitting parameter **LETA** is introduced, which is used to get agreement with experimental results:

$$\gamma^o = \gamma - \left(\frac{\epsilon_{Si}}{C'_{ox}}\right) \frac{\mathbf{LETA}}{L} \left(\sqrt{(V_{SB} + \mathbf{PHI})} + \sqrt{(V_{DB} + \mathbf{PHI})}\right) \qquad (5.85)$$

From the above equation it can be seen that γ^o can take an imaginary value for some negative bias condition, which is non-physical. This is avoided by writing $V'_{S(D)B}$ for $V_{S(D)B}$, as shown below:

$$V'_{S(D)B} = \frac{1}{2} \left[V_{S(D)B} + \mathbf{PHI} + \sqrt{(V_{S(D)B} + \mathbf{PHI})^2 + (4\phi_t)^2} \right] \qquad (5.86)$$

Equation (5.85) can also result in negative values of γ which is taken care of by using the following equation:

$$\gamma' = \frac{1}{2} \left(\gamma^o + \sqrt{\gamma^{o2} + 0.1\phi_t} \right) \qquad (5.87)$$

From the treatment of the narrow width effect given in Section 5.4, the narrow width correction factor can be put in the form:

$$\alpha_W = 1 + \frac{\Delta Q'_b}{Q'_b} = 1 + \left(\frac{\pi \epsilon_{Si}}{C'^2_{ox} \sqrt{2q\epsilon_{Si} N_A}}\right) \frac{1}{W} \sqrt{V_{SB} + \mathbf{PHI}} \qquad (5.88)$$

The term in the brackets can be taken as three times the fitting parameter **WETA**. The resultant γ^o, taking short and narrow channel effects into consideration, is:

$$\gamma^o = \gamma - \left(\frac{\epsilon_{Si}}{C'_{ox}}\right) \frac{\mathbf{LETA}}{L} \left(\sqrt{(V_{SB} + \mathbf{PHI})} + (\sqrt{(V_{DB} + \mathbf{PHI})}\right.$$

$$\left. - \frac{3\,\mathbf{WETA}}{W} \sqrt{V_P + \mathbf{PHI}}\right) \qquad (5.89)$$

In the ACM model a similar approach is used for $V_{SB} = V_{DB} = 0$, and the effect of source and drain bias is added as an additional term as shown below:

$$V_P = \left(\sqrt{V'_{GB} + \left(\frac{\gamma'}{2}\right)^2} - \frac{\gamma'}{2}\right)^2 - \mathbf{PHI} + \frac{\sigma}{n}(V_{SB} + V_{DB}) \qquad (5.90)$$

where $\sigma = \mathbf{SIGMA}/L^2$, with $\mathbf{SIGMA}$ being the DIBL coefficient, and n is the slope factor. γ' is given by:

$$\gamma' = \alpha\gamma = \gamma - \frac{\epsilon_{Si}}{C'_{ox}} \left[\frac{2 \cdot \mathbf{LETA}}{L} - \frac{3 \cdot \mathbf{NP} \cdot \mathbf{WETA}}{W} \right] \sqrt{\mathbf{PHI}} \qquad (5.91)$$

In the above equation the value of α is for $V_{SB} = V_{DB} = 0$.

5.5 Reverse Short Channel Effect

In the normal short channel effect there is a threshold voltage roll-off with decrease of channel length as discussed in Section 5.3.3. In certain type of ultra-short channel MOS technology, it has been observed that there is first roll-up or enhancement of threshold voltage with decrease of channel length which is known as the *reverse short channel effect* (RSCE). This is subsequently followed by the conventional V_T roll-off with decrease in channel length [24–26, 52]. The reason attributed to the threshold voltage increase is boron redistribution/pile-up in the vicinity of source/drain junctions under the gate, induced by Oxidation Enhanced Diffusion (OED) and Transient Enhanced Diffusion (TED) [24] in a *p*-type substrate. OED/TED results in pile-up of doping concentration near the source/drain edges under the gate. The net effect of this phenomenon is to increase the *average doping concentration* with the decrease in channel length as seen by the gate, leading to V_T roll-up. The V_T roll-up, incidentally, is considered a desirable feature in sub-micron and lower dimension MOS devices as it initiates normal V_T roll off at a relatively low short channel, and also resists source drain punch-through. In order to exploit the above advantages in modern scaled MOS transistors, a deliberate lateral channel engineering is implemented by the so-called *pocket* or *halo implant* under the gate adjacent to source and drain junctions [27, 67].

This unintentional or intentional enhanced doping concentration in the vicinity of the source/drain edges results in doping inhomogeneity along the length of the channel, as shown in Figure 5.13. The lateral inhomogeneity of doping concentration introduces two-dimensional inhomogeneity in the substrate under the gate which affects the threshold voltage [47, 53]. Irrespective of the mechanism of lateral channel inhomogeneity, the general approach is to model the inhomogeneity near the source and drain edges with some characteristic length defining the range, and simplifying approximations for the actual impurity profile. The various approaches adopted in different MOSFET models are outlined below. In such analysis, at the first instance, the vertical substrate inhomogeneity is generally ignored.

5.5.1 Arora's Model

A simple quantitative analytical model for RSCE was first reported by Arora [94], assuming an exponential decaying ionic charge buildup at the source/drain edges along the channel. The OED related channel ionic charge pile-up at the source/drain edge is modeled by:

$$Q'_f(y) = Q'_0 e^{\frac{-y}{K_0}} \qquad (5.92)$$

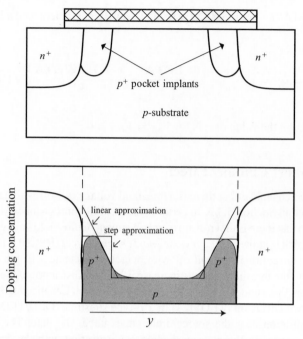

Figure 5.13   Doping concentration variation in the presence of a pocket implant

where Q_0' is the peak charge density at the source drain edges, and K_0 is the characteristic length signifying the extent of inhomogeneity along the channel.

The total enhanced charge in the channel ΔQ_{fs}, is obtained from the following expression:

$$\Delta Q_{fs} = 2W \int_0^L Q_0' e^{\frac{-y}{K_0}} \, dy \tag{5.93}$$

$$\Delta Q_{fs} = 2W Q_0' K_0 \left(1 - e^{\frac{-L}{K_0}} \right) \tag{5.94}$$

which lead to a threshold voltage increase by ΔV_T:

$$\Delta V_T = \frac{\Delta Q_{fs}}{WLC_{ox}'} \tag{5.95}$$

The above equation shows rolling-up of the threshold voltage with decrease in channel length, contrary to conventional roll-off characteristics explained by charge sharing or lateral field gradient model. The conventional roll-off of threshold voltage overtakes the reverse short channel effect at deeper channel length scaling.

Figure 5.14 shows plots of reverse and normal short channel effects as a function of channel length. It is noted that the normal V_T roll-off is initiated earlier as compared to the case when there is impurity pile-up which initiates RSCE.

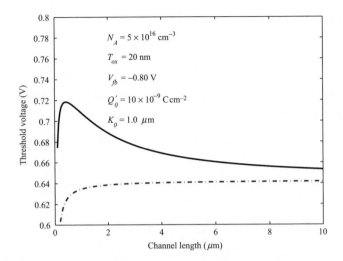

Figure 5.14 Threshold voltage variation dependence on channel length with (solid line) and without (dashed line) a reverse short channel effect

The lateral channel engineering for optimal tailoring of short channel effects such as DIBL and SCE can be investigated in detail with techniques such as solution of the pseudo-2-D Poisson's equation [53, 65], or voltage-doping transformation [66] which gives a deeper physical insight though the approach is computationally intensive. The following section illustrates some of the simplifying approximations adopted by representative compact models so that the RSCE effect can be described through a minimum number of fitting parameters.

5.5.2 BSIM3v3 Model

This assumes a step doping profile of length L_x at the edge of the source and drain and a peak volume concentration N_{pc} in the pocket, as shown in Figure 5.15(a) [92].

The average doping concentration in the channel can be written as:

$$N_{eff} = \frac{N_A(L - 2L_x) + N_{pc}2L_x}{L} \tag{5.96}$$

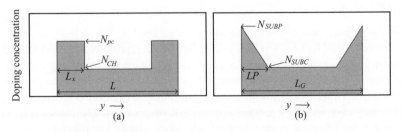

Figure 5.15 Approximations to pocket implant doping profiles

which can be put in the form:

$$N_{eff} = N_A \left(1 + \frac{N_{lx}}{L} \right) \tag{5.97}$$

where $N_{lx} = 2L_x(N_{pc} - N_A)/N_A$ and is treated as a fitting parameter. The threshold voltage in the absence of RSCE is:

$$V_T = V_{T0} + K_1 \left(\sqrt{\phi_s - V_{BS}} - \sqrt{\phi_s} \right) - K_2 V_{BS} \tag{5.98}$$

The modified threshold voltage in the presence of RSCE is given by:

$$V_T = V_{T0} + K_1 \left(\sqrt{\phi_s - V_{BS}} - \sqrt{\phi_s} \right) - K_2 V_{BS} + K_1 \left(\sqrt{1 + \frac{N_{lx}}{L}} - 1 \right) \sqrt{\phi_s} \tag{5.99}$$

where K_1 and K_2 are model parameters. The last term reflects the contribution of the pocket or pile-up induced concentration at the source/drain edges.

5.5.3 EKVv2.6 Model

In the EKV model Arora's formalism has been extended by introducing two model parameters to reflect the channel inhomogeneity [22]:

$$\Delta V_T = \frac{2\mathbf{Q_0}}{C_{ox}} \frac{1}{\left[1 + \frac{1}{2} \left(\xi + \sqrt{\xi^2 + C_\epsilon} \right) \right]^2} \tag{5.100}$$

where:

$$\xi = C_A \left(10 \cdot \frac{L}{\mathbf{LK}} - 1 \right) \tag{5.101}$$

and $C_\epsilon = 4 \times (22 \times 10^{-3})^2$ and $C_A = 28 \times 10^{-3}$. The parameters $\mathbf{Q_0}$ and $\mathbf{LK}$ physically signify the same meaning as $\mathbf{Q'_0}$ and $\mathbf{K_0}$ in Arora's model.

5.5.4 HiSIM Model

In HiSIM, RSCE is exclusively identified with a pocket implant, which has become an integral part of advanced MOS technology. The pocket implant is represented by a triangular profile, with the peak located at the source and drain edges. A linearly decaying impurity distribution along the channel is assumed. Two parameters N_{subp} and L_p denote the peak concentration and base length respectively, to describe the triangular profile of the pocket. N_{subc} is the background substrate doping over which the pocket implant is superimposed. The model is based on the perception that the effect of pocket implant can be accounted for by considering a profile closely corresponding to that of a process simulator, and taking an effective average substrate doping, N_{subeff} considering the pocket characteristics. The expression for the *averaged* doping is obtained for two cases.

Case(i). $L_{ch} \geq L_p$; There are two equal triangular profiles of base length L_p and height of impurity concentration $N_{subp} - N_{subc}$ over a uniform substrate concentration of N_{subc}. The pockets are separated and there is no superposition of the two triangular regions. Therefore, the effective impurity concentration averaged over the channel length L_c is [47]:

$$N_{subeff} = N_{subc} + 2 \left[\frac{1}{2}(N_{subp} - N_{subc})L_p \right] \frac{1}{L_c} \tag{5.102}$$

Case(ii). $L_{ch} < L_p$; In this case, the pockets superpose on each other and the base length being smaller than the channel length, each pocket profile is geometrically represented by a trapezoid with two impurity profile heights being $(N_p - N_c)$ and $(N_{subp} - N_{subc})(1 - L_p/L_c)$. With this, using a geometrical relationship in the superimposed geometrical profile, it can be shown that the effective averaged substrate doping is given by:

$$N_{subeff} = N_{subp} + (N_{subp} - N_{subc}) \left(1 - \frac{L_p}{L_c} \right) \tag{5.103}$$

Using the standard equation for threshold voltage with the effective substrate doping we have:

$$\phi_{beff} = \phi_t \ln \left(\frac{N_{subeff}}{n_i} \right) \tag{5.104}$$

$$V_{Tpi} = V_{fb} + 2\phi_{beff} + \frac{\sqrt{2qN_{subeff}\,\epsilon_{Si}(2\phi_{beff} - V_{BS})}}{C_{ox}} \tag{5.105}$$

The defining feature of the HiSIM approach to deal with length dependence of the threshold voltage in pocket implanted MOSFET structures is that inversion of both the pocket implanted and un-implanted parts of the channel need to be considered to define the threshold voltage.

The net shift in threshold voltage ΔV_{th} is composed of two components:

$$\Delta V_{TRSCE} = V_{Twpi} - V_{Tpi} \tag{5.106}$$

where V_{Twpi} is the threshold voltage of a long channel transistor without a pocket implant, and V_{Tpi} is the threshold voltage in the presence of a pocket implant.

The pocket implant introduces perturbation in the lateral field gradient predicted by the short channel model without a pocket implant. This deviation is similarly modeled using the model equation used in the normal short channel effect (Equation (5.52)) as shown below:

$$\Delta V_{TRSCE} = (V_{Twpi} - V_{Tpi}) - \frac{\epsilon_{Si}}{C_{ox}} W_d \frac{dF_{y,P}}{dy} \tag{5.107}$$

where:

$$\frac{\mathrm{d}F_{y,P}}{\mathrm{d}y} = \frac{2(\phi_{bi} - 2\phi_{beff})}{LP^2} \times \left[SCP1 + SCP2 \times V_{ds} + SCP3\frac{(2\phi_{beff} - V_{bs})}{LP} \right] \quad (5.108)$$

5.6 Carrier Mobility Reduction

The understanding of the physical mechanisms determining the inversion carrier mobility in the channel has been an object of intensive study, as it is a key parameter in the modeling of MOS transistors [28, 37]. Surface inversion carriers are subjected to various scattering mechanisms which affect their mobility. The various types of scattering mechanism identified are:

1. Coulomb scattering.
2. Phonon scattering.
3. Surface roughness scattering.

The origin of Coulomb scattering lies in the interaction of the inversion carriers with the ionized impurities, interface state charges, and fixed oxide charges. As the carriers travel along the channel they are repelled by impurity centers, as they have the same charge polarity. The scattering rate is proportional to the doping concentration at the surface where the inversion channel is located. In strong inversion, because of the screening phenomena by the mobile inversion carriers, the effect of ionized scattering gets weakened. It is, therefore, to be expected that Coulomb scattering plays a dominant role in weak and moderate inversion when the screening of the scattering centers is not strong. It may be mentioned further that if the temperature is raised, which increases the thermal energy and thereby the velocity of carriers, they encounter less interaction with the ionic charge centers. The resultant effect is an increase in the Coulomb component of mobility.

The lattice vibrations in silicon at the interface in optical and acoustic modes emit and absorb phonons while exchanging energy with the carriers resulting in lattice scattering. As lattice vibration increases with temperature, phonon scattering also increases correspondingly.[2] At room temperature phonon scattering related mobility constitutes a significant component of the total inversion layer mobility.

The discontinuity in the lattice structure at the Si–SiO$_2$ interface, where the inversion carriers are located, presents surface irregularities to the carriers moving along the channel. The surface irregularities act as scattering centers, which have a significant effect on mobility of the carriers. The amount of scattering from these centers depends upon the inversion layer thickness, as in a thin layer, a larger percentage of carriers encounter the scattering due to surface roughness.

---

[2] In high-k dielectric materials in ultra-scaled MOS devices, lattice vibrations of the optical type initiate emission and absorption of phonons which scatter carriers in the inversion layer resulting in an additional mode of remote phonon scattering mechanism.

These three scattering mechanisms outlined above can be assumed to be independent and, therefore, the resultant effective surface carrier mobility μ_s is expressed as a combination of all the mechanisms, governed by Matthiessen's rule, as:

$$\frac{1}{\mu_s} = \frac{1}{\mu_c} + \frac{1}{\mu_{ph}} + \frac{1}{\mu_{sr}} \tag{5.109}$$

where $1/\mu_c$, $1/\mu_{ph}$, and $1/\mu_{sr}$ are mobility components related to ionized impurity centers, phonon scattering, and surface scattering respectively. The primary factors determining carrier mobility in the channel are doping concentration, electric field, temperature, and crystal orientation. Though the electric field has vertical and lateral components originating from the gate and drain bias, in compact modeling, their effect on carrier mobility is treated independently. In the next section we shall discuss the mobility dependence on normal electric field (gate field) for a given temperature and crystal orientation.

5.6.1 Mobility Dependence on Gate Field

At high normal electric field, the surface roughness scattering mechanism is the dominant one determining the effective carrier mobility. It has been established experimentally that the carrier mobility in the inversion layer plotted against effective normal field at room temperature is independent of substrate doping concentration, or bias. In other words, the inversion carrier mobility shows a *universal dependence* on the effective normal surface electric field F_{eff} [29, 33, 34]. The effective normal field dependence on inversion and bulk charge under the gate is described by:

$$F_{eff} = \frac{1}{\epsilon_{Si}} \left(Q_b' + \eta Q_n' \right) \tag{5.110}$$

where η is $1/2$ for electrons and $1/3$ for holes. These values, however, generally depend on process technology. It is seen that the effective field is less than the surface field $F_s = \frac{1}{\epsilon_{Si}}(Q_b' + Q_n')$ [30], as all the electrons do not experience the same field. An average field has to be considered which can be given as:

$$F_{eff} = \frac{\int_0^\infty F(x)\, n(x)\, dx}{\int_0^\infty n(x)\, dx} \tag{5.111}$$

which coincides with Equation (5.110) for an electron.

μ_{sr} has an inverse dependence on gate electric field. This can be understood from the consideration that an increase in electric field results in a reduction of inversion layer thickness, which increases the percentage of carriers encountering surface roughness scattering. The surface scattering mobility component for an electron is represented in the form [35]:

$$\frac{1}{\mu_s} = \beta_s F_{eff}^2 \tag{5.112}$$

where β_s is a proportionality constant.

For the phonon scattering the effective field dependence of mobility is expressed empirically by [35]:

$$\frac{1}{\mu_{ph}} = \gamma F_{eff}^{1/3} + \delta \tag{5.113}$$

The Coulomb scattering, which has importance in weak inversion, is related to inversion charge density, and is modeled by the following expression:

$$\frac{1}{\mu_c} = \alpha \left(1 + \frac{Q_n'}{Q_o'} \right)^{-2} \tag{5.114}$$

where γ, δ, α, and Q_o' are dependent on technological processing parameters, and can be extracted from experimental data. The Coulomb scattering is inversely proportional to inversion charge density, as an increase in the inversion charge density reduces the Coulomb scattering due to screening. It has importance only in the subthreshold region, where the current has an exponent dependence on V_{GB}. This effect is neglected in some MOSFET models such as Philips MM11. Figure 5.16 shows the general profile of surface carrier mobility (μ_s) characteristics on effective normal field. The figure also shows how the curve can be resolved into the dominant components constituting the scattering mechanism. Substituting individual mobility components in Equation (5.109), we can write:

$$\frac{1}{\mu_s} = \delta + \gamma F_{eff}^{1/3} + \beta F_{eff}^2 + \alpha \left(1 + \frac{Q_n'}{Q_o'} \right)^{-2} \tag{5.115}$$

$$\mu_s = \frac{\mu_0}{1 + a F_{eff}^{1/3} + b F_{eff}^2 + c \left(1 + \frac{Q_n'}{Q_b'} \right)^{-2}} \tag{5.116}$$

where Q_b' is used in place of the fitting parameter Q_o', μ_0 is called the zero field mobility and a, b, and c are fitting parameters. It may be mentioned that the last term becomes negligible for substrate doping $N_A \leq 3 \times 10^{16}$ cm^{-3}.

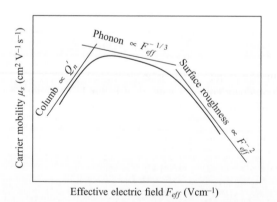

Effective electric field F_{eff} (Vcm^{-1})

Figure 5.16   Mobility dependence on effective electric field

Similarly, for holes we can write:

$$\mu_s = \frac{\mu_0}{1 + a_1 F_{eff}^{1/3} + b_1 F_{eff}^2 + c_1 \left(1 + \frac{Q_n'}{Q_b'}\right)^{-2}} \qquad (5.117)$$

It has been observed that the hole mobility is characterized by a relatively small degradation due to surface roughness scattering as compared to electrons, and further the surface roughness is initiated at lower fields than electrons [32].

The gate field dependent mobility generally centers around Equation (5.116), and has been incorporated in various compact models in different forms depending on the model framework, accuracy, and computational simplicity. The mobility model used in PSP is as given below [49]:

$$\mu_s = \frac{\text{MU0} \cdot \mu_x}{1 + (\text{MUE} \times F_{eff})^{\text{THEMU}} + \frac{\text{CS} \cdot Q_{bm}'^2}{(Q_{nm}' + Q_{bm}')^2}} \qquad (5.118)$$

where μ_x is a factor accounting for the non-universality effect and substrate inhomogeneity, and MU0 corresponds to low field mobility μ_0. The surface roughness and phonon scattering are accounted through MUE and THEMU respectively, and the last term in the denominator corresponds to Coulomb scattering which uses a model parameter CS [91].

The mobility model of Philips MM11 [35] is given as:

$$\mu_s = \frac{\mu_0}{1 + \left[(\text{THEPH} \cdot F_{eff})^{v/3} + (\text{THESR} \cdot F_{eff})^{nv}\right]^{1/v}} \qquad (5.119)$$

where n is 1 for holes and 2 for electrons. $v = 2$ at $T = 300$ k ensures a sharper transition from phonon to surface roughness scattering, as against the transition predicted by Matthiessen's rule, which in a strict sense in not valid for an electron trapped in a 2-D potential well. The model parameters THESR and THEPH correspond to surface roughness and phonon scattering respectively.

In the development of the above model, V_{DS} is kept sufficiently small (10 mV) so that the variation of F_{eff} is negligible along the channel. Under such conditions, the mobility can be assumed to be constant along the channel, and can be extracted using [81]:

$$\mu_s = \frac{g_d}{(W/L)Q_n'} \qquad (5.120)$$

where g_d is the conductance of the channel. For larger value of V_{DS} the variation is larger, and carrier mobility varies along the channel. A computationally simplified approach uses the mobility value estimated using F_{eff} evaluated at $\phi_s = \phi_m = (\phi_{sS} + \phi_{sD})/2$ as in the PSP model.

A simple universal mobility model proposed by Liang et al. [82] and Chen et al. [83] is given as:

$$\mu_s = \frac{\mu_0}{1 + \left(\frac{F_{eff}}{0.75}\right)^{1.6}} \tag{5.121}$$

where:

$$F_{eff} = \frac{V_{GS} + V_{TS}}{6T_{ox}} \tag{5.122}$$

which can be arrived at by using the approximations $Q'_n \approx C'_{ox}(V_{GS} - V_{TS})$ and $Q'_b \approx C'_{ox} V_{TS}$ in Equation (5.110). Further, in the expression for V_{TS} for a typical NMOS, we make use of the fact that $V_{fb} \approx -2\phi_f$. Variations of the above model are adopted in different earlier MOSFET models. It may however be mentioned that Equation (5.121) does not reflect the substrate bias dependence which creates inaccuracy in the estimation of F_{eff}, and thereby mobility dependence at high substrate bias.

In the EKVv2.6 and ACM models, a simple empirical model in terms of V_P is adopted, assuming that the effective normal field is independent of channel voltage. The EKVv2.6 model assumes:

$$\mu_s \approx \frac{\mu_0}{1 + \theta_F F_{eff}} \approx \frac{\mu_0}{1 + \theta V_P} \tag{5.123}$$

where the approximation $F_{eff} = \theta V_P$ is used, which is similar to that given in Equation (5.121). The ACM model uses the following model for mobility:

$$\mu_s = \frac{\mu_0}{1 + \text{THETA}\gamma'\sqrt{V_{P0} + \mathbf{PHI}}} \tag{5.124}$$

where γ' is a short channel corrected γ, and THETA is a fitting parameter.

With the use of Equation (5.121) in BSIM3v3 to represent the mobility model, the 1.6 exponent in the denominator increases the computational overhead. In order to circumvent this, three alternative mobility models have been proposed, one of which is selected by MobMod=1 which is given below:

$$\mu_s = \frac{\text{U0}}{1 + (\text{UA} + \text{UC} \cdot V_{BS})\left(\frac{V_{GS} - V_{TS}}{T_{ox}}\right) + \text{UB}\left(\frac{V_{GS} - V_{TS}}{T_{ox}}\right)^2} \tag{5.125}$$

where UA, UB, and UC are the fitted model parameters.

The HiSIM model uses the following equations for gate field dependent mobility [45], with model parameters given in uppercase:

$$\mu_c = MUE_{CB0} + MUE_{CB1} \frac{Q'_n}{q \times 10^{11}} \tag{5.126}$$

$$\mu_{ph} = \frac{MUE_{PH1}}{(T/300K)^{MUE_{TEMP}}} \times F_{eff}^{MUE_{PH0}} \tag{5.127}$$

$$\mu_{sr} = \frac{MUE_{SR1}}{F_{eff}^{MUE_{SR1}}} \tag{5.128}$$

5.6.2 Lateral Field Dependent Mobility

In the previous section the carrier mobility degradation due to a gate induced field was discussed assuming that the lateral field is small. For a long channel MOSFET with low lateral electric field F_y, the carrier velocity is proportional to field strength with to proportionality constant μ_s given by:

$$v_{dr} = \mu_s F_y \tag{5.129}$$

where v_{dr} is the carrier drift velocity. In a scaled MOSFET the lateral electric field in the channel is sufficiently high, where the low field relation $v_{dr} \propto F_y$ is no longer valid, and as the field increases, $v_{dr} = f(F_y)$ becomes nonlinear. Beyond a critical field value, the carrier velocity attains its maximum saturation value. The phenomena of velocity saturation at high electric field was already discussed in Chapter 1.

The nonlinear v_{dr}–F_y characteristic, which is fitted through an empirical relation [31], is given by:

$$v_{dr} = \mu_{eff} F_y \tag{5.130}$$

where:

$$\mu_{eff} = \frac{\mu_s}{\left[1 + \left(\frac{F_y}{F_{C1}}\right)^\beta\right]^{\frac{1}{\beta}}} \tag{5.131}$$

with $F_{C1} = v_{sat}/\mu_s$, and β a fitting parameter generally taken as 2 for electrons and 1 for holes. The above relation is shown in Figure 5.17.

In Chapter 4 the expression for drain current was obtained assuming a weak lateral channel field. Hence, a constant low field mobility value was used, which allowed the mobility term to be taken out of the integrand while evaluating the expression for the drain current. This no longer holds good in scaled MOSFETs, as the lateral field F_y is a function of position along the channel. Performing the integration for the expression of drain current with the mobility expression given in Equation (5.131) for $\beta = 2$ presents analytical difficulty. Therefore, β is taken as unity for most of the MOSFET models such as in BSIM, EKVv2.6, etc.

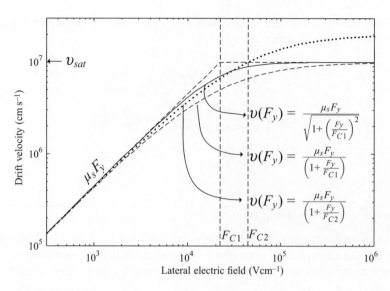

Figure 5.17 Drift velocity versus electric field characteristics for an electron

The importance of a field dependent mobility relationship can be appreciated from the fact that it not only affects the drain current in scaled MOS transistors, but also redefines the condition of drain voltage for current saturation. In long channel transistors the saturation voltage has been defined as the drain voltage at which the drain end of the channel is at weak inversion or in a pinched-off condition. In short channel MOSFETs, the carrier can attain the saturation velocity before the pinch-off condition is attained. Saturation velocity related drain saturation voltage is less than that of the value obtained assuming pinch-off phenomena.

5.6.3 Drain Current Modeling with Velocity Saturation

The drain current model with a velocity saturation effect in a short channel MOSFET operating in strong inversion has been investigated with varying degrees of accuracy. The accuracy of the model depends on the velocity versus field relation [38]. Two simpler models generally used to approximate the expression given in Equation (5.131) are:

$$v_{dr} = \frac{\mu_s}{\left[1 + \left(\frac{F_y}{F_{C1}}\right)\right]} F_y \tag{5.132}$$

and:

$$v_{dr} = \frac{\mu_s}{\left[1 + \left(\frac{F_y}{F_{C2}}\right)\right]} F_y \qquad \text{for} \quad F_y < F_{C2} \tag{5.133}$$

$$= v_{sat} \qquad\qquad\qquad \text{for} \quad F_y \geq F_{C2} \tag{5.134}$$

where $F_{C2} = 2F_{C1}$. The comparison of the above two models with that given in Equation (5.131) is shown in Figure 5.17.

The channel pinch-off is a complex 2-D phenomenon, and various approaches [84–86] for the drain current saturation condition have been proposed in the literature. One of the conditions given in Klaassen and de Groot [84] is that the channel is assumed to be pinched-off where the normal component of the gate field changes its sign. In this region because of relative weakening of the normal component of electric field, mobile inversion carriers are not confined at the Si–SiO$_2$ interface of the channel, but spread out in the direction of depth. The electric field at the pinch-off point is termed as F_P, and from numerical simulation it is found that typically $F_P > 3F_{C1}$ [84]. As the gate field loses its strength along the channel towards the drain, the validity of a GCA approximation progressively becomes weaker. It is, however, a common practice to assume that GCA is valid up to the pinch-off point $F_y = F_P$.

While the above definition of pinch-off condition can be used to arrive at the drain current relation with the mobility model given in Equation (5.132), the condition for pinch-off with the mobility model given in Equation (5.133) is relatively simple. The drain saturation voltage is defined as the voltage where the field at the drain end of the channel is F_{C2}, at which carriers move with saturation velocity. It may be mentioned that in actual terms carriers are as yet not moving with saturation velocity at $F_y = F_{C2}$, and even at $F_y = F_P$. But, according to the model given in Equation (5.131) the saturation velocity is assumed to have been reached.

The expression for drain current is written below considering field dependent mobility:

$$I_{DS} = \mu_{eff} W Q_n'(y) \frac{dV_{CS}}{dy} \tag{5.135}$$

Substituting for effective mobility from Equation (5.132) we have [84]:

$$I_{DS} = \frac{\mu_s}{1 + \frac{1}{F_{C1}} \frac{dV_{CS}}{dy}} W Q_n'(y) \frac{dV_{CS}}{dy} \tag{5.136}$$

where the relation $F_y = dV_{CS}/dy$ is used. Substituting for the inversion charge density as $Q_n'(y) = -C_{ox}'(V_{GS} - V_{TS} - nV_{CS}(y))$ in Equation (5.136), and integrating over the channel length, we have:

$$I_{DS} = \frac{\mu_s}{\left(1 + \frac{V_{DS}}{LF_{C1}}\right)} C_{ox}' \frac{W}{L} (V_{GS} - V_{TS} - (n/2)V_{DS}) V_{DS} \tag{5.137}$$

The above equation is the expression for the drain current in the linear region. When the transistor enters the saturation region, from Equation (5.136) at pinch-off point we can substitute $dV_{CS}/dy = F_P$. Thus:

$$I_{DSsat} = \frac{\mu_s}{1 + \frac{F_P}{F_{C1}}} W C_{ox}'(V_{GS} - V_{TS} - nV_{DSsat})F_P \tag{5.138}$$

Equating Equations (5.137) and (5.138) for $V_{DS} = V_{DSsat}$ and solving for V_{DSsat}, we have:

$$V_{DSsat} = \frac{LF_{C1}}{1 - \frac{F_{C1}}{F_P}} \left[1 + \frac{(V_{GS} - V_{TS})}{nLF_P}\right]$$

$$\left[\sqrt{\frac{2(V_{GS} - V_{TS})(1 - F_{C1}/F_P)}{[1 + (V_{GS} - V_{TS})/(nLF_P)]^2 nLF_{C1}} + 1} - 1\right] \quad (5.139)$$

For $F_P \gg F_{C1}$, we can have a simplified version of the above equation:

$$V_{DSsat} = LF_{C1}\left[\sqrt{\frac{2(V_{GS} - V_{TS})}{nLF_{C1}} + 1} - 1\right] \quad (5.140)$$

It can be seen from Equation (5.140) that for a long channel MOSFET such that $[2(V_{GS} - V_{TS})/(nLF_{C1})] < 1$, the Taylor series expansion of the square root term in Equation (5.140) gives:

$$V_{DSsat} = \frac{V_{GS} - V_{TS}}{n} \quad (5.141)$$

which is the expression for V_{DSsat} for a long channel MOSFET. In other words, as mentioned earlier, the velocity saturation effect reduces the drain saturation voltage with shrinking of channel length. This is shown in Figure 5.18. It is worth mentioning that gate oxide thickness also has a significant effect on drain saturation voltage. This is because velocity saturation is

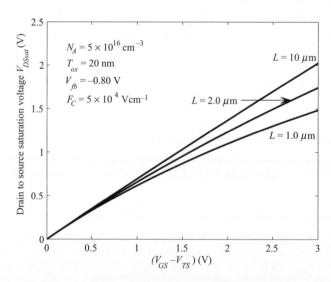

Figure 5.18   The dependence of drain saturation voltage on $(V_{GS} - V_{TS})$ in the presence of velocity saturation with channel length as a parameter

related to the gate field dependent mobility μ_s [38] which determines the critical field F_{C1}. Decreasing the oxide thickness reduces μ_s which increases F_{C1} and maintains the long channel drain saturation voltage as can be seen from Equation (5.140).

The saturated drain current can be estimated from Equation (5.138) with V_{DSsat} given either by Equation (5.139) or Equation (5.140). The effect of velocity saturation on drain current and drain saturation voltage is demonstrated in Figure 5.19. It can be seen from this figure that there is deviation of drain saturation current from the square dependence on $(V_{GS} - V_{TS})$, which is sometimes modeled as α power dependence [48] on $(V_{GS} - V_{TS})$ where $1 < \alpha < 2$.

5.6.3.1 BSIM3v3 Model

In BSIM3v3, the mobility model given in Equation (5.133) is used, where the pinch-off condition is defined by F_{C2}, at which the carrier velocity is $v_{sat} = \mu F_{C2}/2$. Therefore, the drain saturation current given in Equation (5.138) modifies to:

$$I_{DSsat} = \frac{\mu_s}{2} W C'_{ox} (V_{GS} - V_{TS} - n V_{DSsat}) F_{C2} \qquad (5.142)$$

Equate the above equation with the linear drain current given by Equation (5.137) suitably modified for the mobility model of Equation (5.133), and substitute V_{DSsat} for V_{DS}. Now solving for V_{DSsat}, we have:

$$V_{DSsat} = \frac{F_{C2} V_b L}{V_b + F_{C2} L} \qquad (5.143)$$

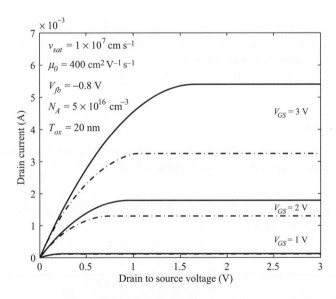

Figure 5.19 I_{DS}–V_{DS} plot using long channel (continuous) and short channel (dashed) theories

where $V_b = V_{GSeff}/n$ with[3] $V_{GSeff} = (V_{GS} - V_{TS})$. Substituting the expression for V_{DSsat} in Equation (5.142), the expression for the drain saturation current is obtained as:

$$I_{DSsat} = \frac{W\mu_s C'_{ox} V_{GSeff} F_{C2} V_b}{2(F_{C2}L + V_b)} \tag{5.145}$$

The above approach ensures the drain current continuity at $V_{DS} = V_{DSsat}$, but the continuity of the first derivative is not guaranteed. Continuity requirements dictate the use of the interpolation function as discussed in Chapter 4. The interpolation function which is asymptotic to V_{DS} in the linear region, and V_{DSsat} in saturation, used in BSIM3v3 is given below:

$$V_{DSeff} = V_{DSsat} - \frac{1}{2}\left[V_{DSsat} - V_{DS} - \delta + \sqrt{(V_{DSsat} - V_{DS} - \delta)^2 + 4\delta V_{DSsat}}\right] \tag{5.146}$$

where δ has dimensions of volts and determines the transition width between the linear and saturation regions. Its typical value is taken as 0.01 V.

5.6.3.2 EKVv2.6 Model

From Equation (5.137) we can see that the drain current with a velocity saturation effect can be written as:

$$I_{DS} = \frac{1}{1 + \frac{V_{DBeff} - V_{SB}}{F_{C1}L}} I'_{DS} \tag{5.147}$$

where I'_{DS} is the drain current without velocity saturation. V_{DBeff} is the drain-bulk bias determined using the smoothing function given in Chapter 4, which reduces to V_{DB} or V_{DBsat} depending on the linear or saturation regions respectively.

5.6.3.3 PSP Model

The symmetric linearizing technique used in the PSP model enables it to consider Caughey's expression for electron mobility given in Equation (5.131) [49].

---

[3] In BSIM3v3 the general expression for V_{GSeff} which gives a continuous expression for the Q'_{nS} valid in weak, moderate, and strong inversion is given as:

$$V_{GSeff} = \frac{2n\phi_t \ln\left[1 + e^{\left(\frac{V_{GS} - V_{TS}}{2n\phi_t}\right)}\right]}{1 + 2nC'_{ox}\sqrt{\frac{2(2\phi_f)}{q\epsilon_{Si}N_A}} e^{-\frac{V_{GS} - V_{TS} - 2V_{off}}{2n\phi_t}}} \tag{5.144}$$

The expression for the drain current for an NMOS transistor, with symmetric linearizing of the inversion charge w.r.t. surface potential outlined in Chapter 4, gives:

$$I_{DS} = W \frac{\mu_s}{\sqrt{1 + (\mu_s F_y / v_{sat})^2}} (Q'_{nm} + C'_{ox}\alpha_m \phi_t) \frac{d\phi_s}{dy} \qquad (5.148)$$

where μ_{eff} is replaced by $\frac{\mu_s}{\sqrt{1+(\mu_s F_y/v_{sat})^2}}$. Square LHS and RHS to get

$$I_{DS}^2 \left[1 + \left(\frac{\mu_s}{v_{sat}} \frac{d\phi_s}{dy}\right)^2\right] = (W\mu_s Q'^*_{nm})^2 \left(\frac{d\phi_s}{dy}\right)^2 \qquad (5.149)$$

where $Q'^*_{nm} = Q'_{nm} + C'_{ox}\alpha_m \phi_t$ and $F_y = \frac{d\phi_s}{dy}$. Rearranging Equation (5.149), we have:

$$I_{DS} = \left[(W\mu_s Q'^*_{nm})^2 - \left(\frac{\mu_s I_{DS}}{v_{sat}}\right)^2\right]^{\frac{1}{2}} \frac{d\phi_s}{dy} \qquad (5.150)$$

or:

$$I_{DS} = \mu_s W Q'^*_{nm} \left(1 - \frac{I_{DS}^2}{W^2 v_{sat}^2 Q'^{*2}_{nm}}\right)^{\frac{1}{2}} \frac{d\phi_s}{dy} \qquad (5.151)$$

For the value of v_{sat}, which is large in magnitude, the second term in brackets is less than unity. Carrying out a binomial expansion and neglecting terms beyond the second term, Equation (5.151) can be recast as:

$$I_{DS} = \mu_s W Q'^*_{nm} \left(1 - \frac{1}{2}\frac{I_{DS}^2}{W^2 v_{sat}^2 Q'^{*2}_{nm}}\right) \frac{d\phi_s}{dy} \qquad (5.152)$$

Integrating the above equation from source to drain, and using $\Delta\phi = \phi_{sS} - \phi_{sD}$, we get:

$$I_{DS} = \mu_s \frac{W}{L} \Delta\phi \left(Q'^*_{nm} - \frac{I_{DS}^2}{2W^2 v_{sat}^2 Q'^*_{nm}}\right) \qquad (5.153)$$

Solving the above quadratic equation for I_{DS}, and retaining the physically allowed root we have:

$$I_{DS} = \frac{W}{L} \frac{\mu_s Q'^*_{nm}}{\theta_{sat}^2 \Delta\phi} \left(\sqrt{1 + 2\theta^2 \Delta\phi^2} - 1\right) \qquad (5.154)$$

where $\theta_{sat} = \mu_s/(v_{sat}L)$ and:

$$I_{DS} = \frac{W}{LG_{vsat}}\mu_s Q_{nm}^{'*}\Delta\phi \tag{5.155}$$

where $G_{vsat} = \frac{1}{2}\left(1 + \sqrt{1 + 2(\theta_{sat}\Delta\phi)^2}\right)$.

For a PMOS transistor equations obtained for an n-channel transistor hold good if v_{sat} is substituted by $v_c\sqrt{1 + \mu_{eff}\Delta\phi/(Lv_c)}$, where v_c is a parameter that corresponds to the longitudinal acoustic phonon velocity [49].

In the HiSIM model the average value of the field dependent mobility is considered, which is obtained using Caughey's relation with $F_y = (\phi_{sD} - \phi_{sS})/L$. This allows us to take the μ_{eff} term out of the integration of Equation (5.135) while obtaining the expression for drain current. Such an approach, though it may appear as an oversimplification, has been found to give estimation of drain current within 7% error [44].

5.7 Velocity Overshoot

The current voltage characteristics of a short channel MOS transistor discussed so far are based on the assumption that the maximum velocity the carriers can attain is limited by v_{sat}. This is because carriers get enough time to encounter various scattering mechanisms to lose their energy which is gained from the electric field, and thereby reach a thermal equilibrium condition with the lattice. In ultra-short channel MOS transistors the electric field in the channel scales up even at low V_{DS}. Carriers are collected before they encounter an adequate number of collisions, while traversing the channel, to attain energy equilibrium. Such a condition occurs when the time the carriers spend in the channel is comparable to the energy relaxation time τ_e. The above phenomenon has been elegantly demonstrated using Monte-Carlo analysis [87, 88]. This can also be illustrated by a simpler energy–momentum relaxation model [41].

The momentum and energy equations corresponding to an average velocity v and energy E are:

$$\frac{dv}{dt} = \frac{qF}{m} - \frac{v}{\tau_m(E)} \tag{5.156}$$

$$\frac{dE}{dt} = qFv - \frac{(E - E_0)}{\tau_e(E)} \tag{5.157}$$

where $\tau_m(E)$ and $\tau_e(E)$ are the momentum and energy relaxation times respectively. These are assumed to be functions of the energy E only. Solving the momentum equation (Equation (5.156)) we get:

$$v = \frac{qF\tau_m}{m}\left(1 - e^{-\frac{t}{\tau_m}}\right) \tag{5.158}$$

From the above equation we can see that for $t \gg \tau_m$, we have:

$$v = \frac{qF}{m}\tau_m \tag{5.159}$$

Taking note of the fact that τ_m is a function of energy, the initial and final velocities of the electrons, v_i and v_f respectively, can be written as:

$$v_i = \frac{qF}{m}\tau_m(E_i) \tag{5.160}$$

$$v_f = \frac{qF}{m}\tau_m(E_f) \tag{5.161}$$

We shall consider a situation where $\tau_m \ll \tau_e$, i.e. the momentum relaxation time of the electron is much shorter than that of its energy relaxation time. Since in silicon τ_m is related to the mechanism of collision between the electrons and acoustic phonons, the higher the energy of electrons the greater will be the momentum relaxation rate, i.e. the lower will be the value of τ_m. Hence, $v_i > v_f$, indicating velocity overshoot.

For time $t \ll \tau_m$:

$$v \propto t \tag{5.162}$$

Within this time limit the velocity of the electron increases linearly with time, indicating that it is free from collision. This is called *ballistic transport*, and the carrier velocity is called *ballistic velocity*.

The velocity overshoot is a non-local phenomena in the sense that the local velocity is not determined by the local field, as is assumed in the drift-diffusion model. According to the analytical model proposed in Blakey and Joardar [39], Song *et al.* [40], and Roldan *et al.* [42], the increment in drain current is characterized by the velocity overshoot factor F_{ov}, given as:

$$F_{ov} = \frac{I_{DSov}}{I_{DS}} = 1 + \frac{v_{sat}\tau_e}{F_{sat}}\frac{dF_y}{dy} \tag{5.163}$$

where dF_y/dy can be approximated by [43]:

$$\frac{dF_y}{dy} = \frac{\eta C'_{ox}}{\epsilon_{Si}}\sqrt{\frac{qN_A}{2\epsilon_{Si}\phi_f}}\Delta V_T \tag{5.164}$$

The velocity overshoot is accounted for in HiSIM by empirically modifying the saturation velocity in the mobility model as:

$$v_{satov} = v_{sat}\left(1 + \frac{\textbf{VOVER}}{L^{\textbf{VOVERP}}}\right) \tag{5.165}$$

where **VOVER** and **VOVERP** are model parameters [45].

5.8 Channel Length Modulation: A Pseudo-2-D Analysis

Channel length modulation is discussed in Chapter 4, where a simple 1-D p–n junction theory was applied to model the decrease in channel length when a MOS transistor is operated beyond saturation. It was assumed that at the saturation point the inversion charge is zero, which was also identified as the pinch-off point at which the channel potential is V_P with respect to bulk. Although the 1-D approach for the determination of ΔL was found to be fairly good in the case of a long channel MOSFET where $\Delta L \ll L$, the accuracy of 1-D analysis is questionable in short channel MOSFETs where the ratio $\Delta L/L$ increases as the channel length is scaled.

For an accurate estimation of ΔL, a 2-D Poisson's equation is required to be solved in the pinch-off region where the space charge distribution is determined by the gate and drain fields. In fact, the channel length modulation is the outcome of addressing a 2-D phenomenon with 1-D approximation. The 2-D solution does not lead to a closed form analytical expression for ΔL which has to be used to get the effective channel length ($L - \Delta L$). A pseudo-2-D model is used to estimate the channel length reduction due to the CLM effect.

In the region between the pinched-off point and drain, the lateral field and its spatial derivative is high ($E_y > 4 \times 10^4 \, \text{V cm}^{-1}$) and this region is called the *high field* or *velocity saturation region*. In the high field region GCA breaks down, and carriers are not confined at the interface, resulting in a 2-D current flow as shown in Figure 5.20(a). The various definitions of location of the pinched-off point are discussed in Section 5.6.3. The importance of an accurate modeling of this region also arises from the fact that the field in this region is associated with high field assisted phenomena such as impact ionization, etc. These have a direct impact on hot electron generation, reliability of devices, and substrate current.

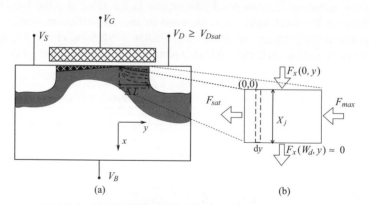

Figure 5.20   (a) Cross-section of a MOSFET under saturation conditions, and (b) Gaussian box under consideration

To derive the expression for ΔL using a pseudo-2-D model we consider a 2-D Gaussian box in the high field region of length ΔL and depth X_j, as shown in Figure 5.20(b). The potential w.r.t. source at the pinched-off point, i.e. at the distance $L - \Delta L$, for the source is V_{DSsat}. By Gauss' law, in the elemental volume element of given length $\text{d}y$ at y, with the origin taken at the pinch-off point:

$$F_y(y)WX_j - F_y(y + \text{d}y)WX_j + F_s(0, y)W\text{d}y = -\frac{Q}{\epsilon_{Si}} \qquad (5.166)$$

where we have assumed that the flux leaving the bottom of the Gaussian box is zero, and the charge enclosed in the elemental box is $Q = (qN_AX_j + Q'_{nsat})Wdy$. A Taylor series expansion of the field $F(y)$ at $y + dy$ gives

$$F_y(y + dy) = F_y + \frac{\partial F_y(y)}{\partial y}dy \tag{5.167}$$

Substituting for $F_y(y + dy)$ in Equation (5.166), and using $-dF/dy = d^2V/dy^2$, we have:

$$\frac{d^2V(y)}{dy^2} + \frac{V_{ox}(y)\epsilon_{ox}}{X_jT_{ox}\epsilon_{Si}} = -\frac{qN_AX_j + Q'_{nsat}}{X_j\epsilon_{Si}} \tag{5.168}$$

In the above expression we have assumed that the mobile charge is uniformly distributed in the Gaussian box, with the charge density Q'_{nsat} constant along y. This allows the RHS of Equation (5.168) to be estimated at the pinch-off point. At the pinch-off point, the RHS of the above equation can be written as [89]:

$$-\frac{qN_AX_j + Q'_{nsat}}{X_j\epsilon_{Si}} = \frac{1}{X_j\epsilon_{Si}}\left(C'_{ox}V_{oxP} + \frac{dF_y}{dy}\right) \tag{5.169}$$

where V_{oxP} is the voltage across the oxide at the pinch-off point, and dF_y/dy is taken at the pinch-off point which can be approximated as $F_P/(L - \Delta L)$. Equation (5.168) now becomes:

$$\frac{d^2V(y)}{dy^2} + \frac{V_{ox}(y)\epsilon_{ox}}{X_jT_{ox}\epsilon_{Si}} = \frac{1}{X_j\epsilon_{Si}}\left(C'_{ox}V_{oxP} + \frac{F_P}{L - \Delta L}\right) \tag{5.170}$$

Rearranging, we have:

$$\frac{d^2V(y)}{dy^2} = \frac{1}{l^2}(V_{oxP} - V_{ox}(y)) + \frac{F_P}{L - \Delta L} \tag{5.171}$$

where $l = [(\epsilon_{Si}/\epsilon_{ox})T_{ox}X_j]^{(1/2)}$, is the characteristic length of the field in the velocity saturation region. The term $(V_{oxP} - V_{ox}(y))$ is equivalent to $(V(y) - V_{DSsat})$, and we have:

$$\frac{d^2V(y)}{dy^2} = \frac{1}{l^2}(V(y) - V_{DSsat}) + \frac{F_p}{L - \Delta L} \tag{5.172}$$

The solution of the above linear second order differential equation is [89]:

$$V(y) = V_{DSsat} + lF_p\sinh\left(\frac{y}{l}\right) + \frac{l^2F_p}{L - \Delta L}\left[\cosh\left(\frac{y}{l} - 1\right)\right] \tag{5.173}$$

Under the assumption that GCA is valid untill the saturation point, so that $F_P/(L - \Delta L)$ is negligible, Equation (5.173) reduces to:

$$V(y) = V_{DSsat} + lF_p \sinh\left(\frac{y}{l}\right) \tag{5.174}$$

Note that for $y = 0$ at the pinch-off point, $V(y) = V_{DSsat}$. Differentiating the above equation w.r.t. y, we have:

$$F(y) = F_p \cosh\left(\frac{y}{l}\right) \tag{5.175}$$

At $y = \Delta L$, $V(y) = V_{DS}$ and the field reaches its maximum value F_m, and is given as:

$$F_m = F_p \cosh\left(\frac{\Delta L}{l}\right) \tag{5.176}$$

Now, using Equations (5.174) and (5.176), we arrive at:

$$V_{DS} = V_{DSsat} + lF_p \sinh\left(\frac{\Delta L}{l}\right) \tag{5.177}$$

$$= V_{DSsat} + lF_p \left[\exp\left(\frac{\Delta L}{l}\right) - \cosh\left(\frac{\Delta L}{l}\right)\right] \tag{5.178}$$

$$= V_{DSsat} + lF_p \left[\exp\left(\frac{\Delta L}{l}\right) - \frac{F_m}{F_p}\right] \tag{5.179}$$

Rearranging Equation (5.179), we have:

$$\Delta L = l \ln\left[\frac{(V_{DS} - V_{DSsat})}{lF_p} + \frac{F_m}{F_p}\right] \tag{5.180}$$

Now, using Equations (5.176) and (5.177), we can arrive at an alternate expression for F_m given below:

$$F_m = \sqrt{\left(\frac{V_{DS} - V_{DSsat}}{l}\right)^2 + F_p^2} = F_p \sqrt{1 + \left(\frac{V_{DS} - V_{DSsat}}{lF_p}\right)^2} \tag{5.181}$$

A simplified expression for ΔL can now be obtained by substituting for F_m from Equation (5.181) into Equation (5.180), expanding the square root term to a Taylor series, and ignoring higher order terms:

$$\Delta L = l \ln\left(1 + \frac{V_{DS} - V_{DSsat}}{lF_p}\right) \tag{5.182}$$

In the ACM and EKVv2.6 models, the above equation for ΔL is used.

5.8.1 BSIM Model

In most compact models, the ΔL calculated above is used to designate the reduced or effective channel length L_{eff} due to CLM. This, in turn, is used in the expression for the drain current, substituting L_{eff} in place for L. BSIM models CLM using the concept of *Early voltage* popularly used in bipolar transistors to describe CLM. The increased saturation drain current with a CLM effect may be given as [90]:

$$I_{DS} = I_{DS}\Big|_{V_{DS}=V_{DSsat}} + \frac{\partial I_{DS}}{\partial V_{DS}}(V_{DS} - V_{DSsat}) \tag{5.183}$$

$$I_{DS} = I_{DS}\Big|_{V_{DS}=V_{DSsat}} \left[1 + \frac{(V_{DS} - V_{DSsat})}{V_{ACLM}}\right] \tag{5.184}$$

where $V_{ACLM} = I_{DS}|_{V_{DS}=V_{DSsat}} \times (\partial I_{DS}/\partial V_{DS})^{-1}$ is a contribution to the Early voltage from the CLM effect. In this section we shall limit our discussion to voltage contributions from the CLM effect; other contributions may arise from DIBL and substrate current [92].

To derive the expression for V_{ACLM} we write:

$$1/V_{ACLM} = \frac{1}{I_{DSsat}} \left(\frac{dI_{DS}}{dV_{DSsat}}\right)\left(\frac{dV_{DSsat}}{dL}\right)\left(\frac{dL}{dV_{DS}}\right) \tag{5.185}$$

Each term in parentheses can be derived from the following expressions respectively:

$$I_{DSsat} = v_{sat}WC'_{ox}n(V_{GST} - V_{DSsat}) \tag{5.186}$$

$$V_{DSsat} = \frac{F_{C2}LV_{GST}}{(F_{C2}L + V_{GST})} \tag{5.187}$$

and:

$$\Delta L = l\ln\left(\frac{V_{DS} - V_{DSsat}}{F_{C2}l}\right) \tag{5.188}$$

where Equation (5.188) is obtained from Equation (5.182), in which we have neglected the constant 1 in the argument of the ln term, and $V_{GST} = (V_{GS} - V_{TS})$. Now substituting expressions derived from the above equations, for terms in parentheses in Equation (5.185), and with some algebraic manipulation we arrive at the following relation for V_{ACLM}:

$$V_{ACLM} = \frac{1}{\mathbf{PCLM}} \left[\frac{nF_{C2}L + (V_{GS} - V_T)}{nF_{C2}}\right] \frac{(V_{DS} - V_{DSsat})}{l} \tag{5.189}$$

where **PCLM** is a fitting parameter used to accommodate the uncertainties associated with the process dependent parameter l. It may be mentioned that the increase in saturation drain current due to CLM may be overtaken by an increase in the current due to DIBL and a hot electron effect at higher values of drain voltage, where the respective components of Early voltage are required to be considered along with that due to CLM [92].

5.8.2 HiSIM Model

In the HiSIM MOSFET model the lateral field component is assumed to be larger than the vertical component for simplicity of the model equation. Using the above assumption in the Gaussian box to neglect the vertical field gradient, and assuming a uniform charge distribution in the Gaussian box, we can write [46]:

$$-\frac{\mathrm{d}^2\phi}{\mathrm{d}y^2} = \frac{(F_P - F_D)}{\Delta L} = -\frac{1}{\epsilon_{\text{Si}}}\frac{(qN_A X_j + Q'_{nsat})}{X_j} \tag{5.190}$$

where F_D is the electric field at $y = \Delta L$. The above equation gives:

$$\Delta L = \frac{F_D - F_P}{\frac{qN_A}{\epsilon_{\text{Si}}} + \frac{Q'_{nsat}}{\epsilon_{\text{Si}}X_j}} \tag{5.191}$$

Figure 5.21 shows the variation of surface potential from source to the drain, which includes GCA and high field regions. To account for the inaccuracy due to the constant depletion width of the Gaussian box, two empirical model parameters **CLM1** and **CLM2** are introduced, as shown below:

$$\Delta L = \frac{F_D - F_P}{\mathbf{CLM1}\frac{qN_A}{\epsilon_{\text{Si}}} + \mathbf{CLM2}\frac{Q'_{nsat}}{\epsilon_{\text{Si}}X_j}} \tag{5.192}$$

In the above equation F_D is calculated from the following Poisson's equation:

$$\frac{\mathrm{d}F_y}{\mathrm{d}y} = -\frac{qN_A}{\epsilon_{\text{Si}}} \tag{5.193}$$

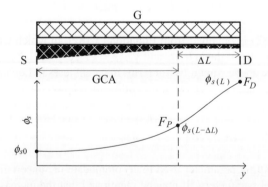

Figure 5.21 Surface potential variation along the channel

where the inversion charge contribution is ignored. Multiplying both the sides by F_y, and integrating from $y = L - \Delta L$ to $y = L$ with the substitution of $F_y \, dy = d\phi_s$, we have:

$$F_D^2 = F_P^2 + \frac{2qN_A}{\epsilon_{Si}} \left[\phi_{s(L)} - \phi_{s(L-\Delta L)} \right] \tag{5.194}$$

In the HiSIM MOSFET model the potential at $y = L$ is considered to be less than V_{DS} because a potential drop appears in transition from the p–n^+ section. Therefore, the potential at $y = L$ will be less than $\phi_{sS} + V_{DS}$, and lies between ϕ_{sS} and $\phi_{sS} + V_{DS}$. The potential at $y = L$ is expressed empirically as:

$$\phi_L = \text{CLM1}(\phi_{sS} + V_{DS}) + (1 - \text{CLM1})\phi_{s(L-\Delta L)} \tag{5.195}$$

5.9 Series Resistance Effect on Drain Current

As dimensions of the MOS transistors are scaled down, parasitic source/drain resistances become increasingly significant, and limit the device performance. With scaling of length, the channel resistance decreases. However, the source/drain parasitic resistances do not scale at the same rate, thereby restricting the exploitation of the full potential of scaling [68]. For advanced short channel structures, such as a Lightly Doped Drain (LDD) meant to reduce hot electron effects in scaled transistors, the parasitic resistance further increases due to lower levels of doping near the diffused junction forming the source and drain. The implication of parasitic resistance is reflected in device performance in terms of degradation in the drive current and transconductance.

Figure 5.22(a) shows the schematic of an LDD MOS structure identifying the physical origins of various components contributing to parasitic resistances [68, 69]. The physical origins of various constituents of parasitic source/drain resistances may be attributed to the following.

1. Contact resistance R_{co}, between metal/silicide and implanted source/drain regions, which is determined by the nature of contact and interface materials.
2. The resistance R_{sh}, due to sheet resistance of the source/drain junction regions, which is governed by the doping concentration of the implanted layer and junction depth.
3. The resistance of the electron accumulation layer R_{ac}, in the overlap region between the gate and the source at the Si–SiO$_2$ interface, due to the positive potential at the gate (for an n-channel MOSFET) in the normal operating condition.
4. The spreading resistance R_{sp}, underneath the accumulation layer and bounded by the depletion layer of the diffused region, which provides a parallel path to the current from source other than the path through the accumulation resistance R_{acc}.
5. The accumulated layer of electrons in the depleted n^--source region at the Si–SiO$_2$ interface, which provides a path to the carriers in the inversion layer to the n^+-source region, adds an additional resistance R_{dep} to the current flowing between the contact and the inversion layer.

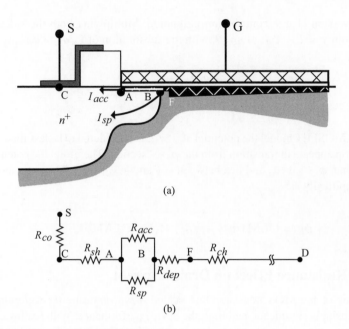

(a)

(b)

Figure 5.22   Cross-section of an LDD MOS structure showing various resistive components

As shown in Figure 5.22(b) the equivalent parasitic source resistance can be written as:

$$R_s = (R_{co} + R_{sh}) + \left[\frac{R_{acc}(V_{GS})R_{sp}}{R_{acc}(V_{GS}) + R_{sp}} \right] + R_{dep}(V_{SB}, V_{GS}) \qquad (5.196)$$

Amongst the constituents of parasitic resistance given in Equation (5.196) the first two can be assumed to be voltage independent [93], while the third term depends on the gate–source bias. The last term is dependent on gate–source and source–substrate bias. The gate–source bias determines the conductivity of the accumulation layer, and the depletion layer width determines the length of R_{dep}, which makes it dependent on source–substrate biasing.

The MOS transistor with parasitic resistances can be represented by Figure 5.23, in which S and D are the extrinsic terminals and S', D' are the intrinsic ones which are not electrically

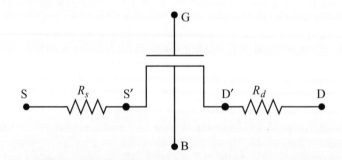

Figure 5.23   MOSFET representation including parasitic resistances

accessible. When there is a flow of current, there is a potential drop across the parasitic resistances which reduces the terminal potentials of the intrinsic device with respect to the potentials applied externally. With the source terminal as reference, V_{GS}, V_{DS}, and V_{BS} are extrinsic terminal voltages, and V'_{GS}, V'_{DS}, and V'_{BS} are the corresponding intrinsic terminal voltages as shown in Figure 5.23. Note that all the terminal voltages for the intrinsic device are reduced in magnitude due to the parasitic resistances, and can be written as:

$$V'_{GS} = V_{GS} - I_{DS}R \tag{5.197}$$

$$V'_{DS} = V_{DS} - 2I_{DS}R \tag{5.198}$$

$$V'_{BS} = V_{BS} - I_{DS}R \tag{5.199}$$

In writing the above equations we have assumed that $R_S = R_D = R$ in Figure 5.23, due to the intrinsic symmetry of the MOS transistor. Strictly speaking there can be differences in R_S and R_D due to process variation or due to their bias dependence. In the case of the bulk referred model, we have for the source and drain voltages:

$$V'_{SB} = V_{SB} + I_{DS}R \tag{5.200}$$

$$V'_{DB} = V_{DB} - I_{DS}R \tag{5.201}$$

whereas V_{GB} remains unchanged.

The effect of series resistance can be estimated either by obtaining I_{DS} with introduction of the effect of R, or by including R externally in the SPICE netlist. The latter approach introduces two additional nodes for each MOSFET, and was adopted in MOSFET models including EKVv2.6, ACM, and LEVELs 1, 2, and 3. The additional nodes increase the simulation time. Using the former approach we shall arrive at the analytical expression of the drain current in the presence of a series resistance where we shall neglect their bias dependencies.

The drain current model including the series resistance effect can be obtained by substituting the effective biases given in Equations (5.197) and (5.198) in the following equation for drain current in the linear region with lateral field dependent mobility:

$$I_{DS} = C'_{ox}\frac{W}{L}\frac{\mu}{\left(1 + \frac{V'_{DS}}{LF_{C2}}\right)}(V'_{GS} - V_{TS} - nV'_{DS})V'_{DS} \tag{5.202}$$

which results in:

$$I_{DS} = \frac{\mu}{1 + \frac{(V_{DS} - 2RI_{DS})}{LF_{C2}}}C'_{ox}\frac{W}{L}$$
$$\times [(V_{GS} - RI_{DS}) - V_{TS} - n(V_{DS} - 2RI_{DS})](V_{DS} - 2RI_{DS}) \tag{5.203}$$

The solution of the above equation, which is quadratic in I_{DS}, leads to a rather complex expression for drain current [70]. It may be noted that the effect of the series resistance on enhancing the threshold voltage due to the body effect may be neglected, as at threshold the drain current is small, and hence, the drop across the series resistance R_s may be ignored.

The BSIM3v3 [50] model follows a relatively simple approach to include the effect of series resistance using the argument that in the linear region the drain current, in the presence of a series resistance, is:

$$I_{DS} = \frac{V_{DS}}{R_{ch} + R_{sd}} \tag{5.204}$$

where $R_{sd} = R_s + R_d$, and R_{ch} is the channel resistance. Dividing the numerator and denominator by R_{ch}, and identifying $R_{ch} = V_{DS}/I_{DS0}$ where I_{DS0} is the drain current expression without an R_{sd} effect, Equation (5.204) can now be written as:

$$I_{DS} = \frac{I_{DS0}}{1 + (R_{sd}I_{DS0})/V_{DS}} \tag{5.205}$$

To arrive at the drain saturation voltage we follow the approach discussed in Section 5.6.3 for BSIM by equating Equation (5.205) to saturation current, which results in a quadratic equation in V_{DSsat} whose solution is given below:

$$V_{DSsat} = \frac{-b - \sqrt{b^2 - 4ac}}{2a} \tag{5.206}$$

where:

$$a = n^2 W R_{sd} C'_{ox} v_{sat} + \left(\frac{1}{\lambda} - 1\right) n \tag{5.207}$$

$$b = -\left[(V_{GS} - V_{TS})\left(\frac{2}{\lambda} - 1\right) + n F_{C2} L + 3n R_{ds} C'_{ox} W v_{sat}(V_{GS} - V_{TS})\right] \tag{5.208}$$

$$c = F_{C2} L(V_{GS} - V_{TS}) + 2 R_{sd} C'_{ox} W v_{sat}(V_{GS} - V_{TS})^2 \tag{5.209}$$

λ is a parameter introduced to take care of the non-saturation effect of the I–V characteristics [92].

5.9.1 PSP Model with Series Resistance Effect

The drain current equation in the PSP model, following Equation (5.155), is expressed as:

$$I_{DS} = -\frac{\mu_s}{G_{vsat}} \frac{W}{L} Q'^*_{nm} \Delta\phi' \tag{5.210}$$

In the presence of a series resistance we have $\Delta\phi' = \Delta\phi - 2I_{DS}R_s$. Substituting $\Delta\phi'$ in the above equation, we have:

$$I_{DS} = -\frac{\mu_s}{G_{vsat}} \frac{W}{L} Q'^*_{nm}(\Delta\phi - 2I_{DS}R_s) \tag{5.211}$$

where we have neglected the effect of R_s on the G_{vsat} term.

$$I_{DS}\left(1 - \frac{2\mu_s}{G_{vsat}}\frac{W}{L}Q_{nm}^{'*}R_s\right) = -\frac{\mu_s}{G_{vsat}}\frac{W}{L}Q_{nm}^{'*}\Delta\phi \qquad (5.212)$$

or:

$$I_{DS} = -\frac{\mu_s}{\left(G_{vsat} - \mu_s\frac{W}{L}Q_{nm}^{'*}R_{sd}\right)}\frac{W}{L}Q_{nm}^{'*}\Delta\phi \qquad (5.213)$$

where $R_{sd} = R_s + R_d \approx 2R_s$.

Multiplying numerator and denominator by G_s, and rearranging:

$$I_{DS} = -\frac{\mu_0}{G_s\left(1 - \frac{\mu_0}{G_sG_{vsat}}\frac{W}{L}Q_{nm}^{'*}R_{sd}\right)}\frac{W}{G_{vsat}L}Q_{nm}^{'*}\Delta\phi \qquad (5.214)$$

where $\mu_s = \mu_0/G_s$, and $G_s = 1 + (\mathbf{MUE}\cdot F_{eff})^{\mathbf{THEMU}} + \mathbf{CS}\cdot Q_{bm}^{'*2}/(Q_{bm}^{'*2} + Q_{nm}^{'*2})$.

$$I_{DS} = -\frac{\mu_0}{\left(G_s - \frac{\mu_0}{G_{vsat}}\frac{W}{L}Q_{nm}^{'*}R_{sd}\right)}\frac{W}{G_{vsat}L}Q_{nm}^{'*}\Delta\phi \qquad (5.215)$$

or:

$$I_{DS} = -\frac{\mu_0}{(G_s - G_R)}\frac{W}{G_{vsat}L}Q_{nm}^{'*}\Delta\phi \qquad (5.216)$$

where $G_R = (\mu_0/G_{vsat})(W/L)Q_{nm}^{'*}R_{sd}$.

HiSIM deals with the effect of the parasitic source–drain resistances by an iterative process which considers the voltage drop across them, and modifies the terminal voltages accordingly.

5.10 Polydepletion Effect on Drain Current

The polydepletion effect has been discussed in Section 3.4 where it is shown that due to the potential drop across the polydepletion layer in the gate, there is a reduction in the effective gate voltage. The effect can be viewed alternatively as an increase in threshold voltage thereby degrading the MOSFET drive current [102]. In BSIM this effect is modeled by replacing V_{GS} by $V_{GS} - \phi_p$ in strong inversion, where its impact is significant as discussed in Chapter 3. It may be noted that in the case of a MOS transistor operating in strong inversion in the linear region, the surface potential along the channel varies from $2\phi_f + V_{SB}$ to $V_{SB} + V_{DS} + 2\phi_f$. The amount of depletion in the polysilicon gate will vary from the source end to drain end of the gate. The polysilicon gate is more depleted near the source end compared to the drain end. The nonuniform depletion along the gate results in increasing the effective gate voltage along the channel from source to drain. The above nonuniform effective gate voltage along the channel can be modeled by linearizing the drop across the polysilicon gate as a function of channel potential. For the inversion charge density at a position y in strong inversion with the polydepletion effect, we have:

$$Q'_n(y) = -C'_{ox}\left(V_{GS} - \phi_p(y) - 2\phi_f - V_{CS}(y) - \gamma\sqrt{-V_{BS} + 2\phi_f + V_{CS}(y)}\right) \quad (5.217)$$

Charge linearization against $V_{CS}(y)$ gives:

$$Q'_n(y) = -C'_{ox}\left(V_{GS} - \phi_p(y) - V_{TS} - nV_{CS}\right) \quad (5.218)$$

where ϕ_p, as derived in Chapter 3, is:

$$\phi_p(y) = \left[\sqrt{\frac{\gamma_p^2}{4} + (V_{GB} - \phi_s(y))} - \frac{\gamma_p}{2}\right]^2 \quad (5.219)$$

Substitute $V_{GB} = V_{GS} + V_{SB}$, $\phi_p(y)$ from Equation (5.219), and the surface potential along the channel $\phi_s(y) = V_{SB} + 2\phi_f + V_{CS}(y)$ in Equation (5.217). A Taylor series expansion around V_{CS} gives:

$$Q'_n = -C'_{ox}\left(V_{GS} - V_{TS} - nV_{CS} - \phi_{p0} + n_p V_{CS}\right) \quad (5.220)$$

where:

$$\phi_{p0} = \frac{\gamma_p^2}{2}\left[1 + \frac{2}{\gamma_p^2}(V_{GS} - 2\phi_f) - \sqrt{1 + \frac{4}{\gamma_p^2}(V_{GS} - 2\phi_f)}\right] \quad (5.221)$$

and:

$$n_p = 1 - \frac{1}{\sqrt{1 + \frac{4}{\gamma_p^2}(V_{GS} - 2\phi_f)}} \quad (5.222)$$

Rearranging Equation (5.220) we get:

$$Q'_n = -C'_{ox}\left[V_{GS} - (V_{TS} + \phi_{p0}) - (n - n_p)V_{CS}\right] \quad (5.223)$$

The above equation indicates that by changing V_{TS} to $V_{TS} + \phi_{p0}$, and n to $(n - n_p)$, we can still use the I–V relation discussed in the threshold voltage based model.

BSIM3v3 accounts for the polydepletion effect by expressing the gate voltage by an effective value which is obtained from the potential balance condition $V_{GS} - V_{fb} = \phi_p + V_{ox} + \phi_s$. Substituting $\phi_p = qN_{gate}W_p^2/2\epsilon_{Si}$, the expression for V_{GSeff} takes the form:

$$V_{GSeff} = V_{fb} + \phi_s + \frac{q\epsilon_{Si}N_{gate}T_{ox}^2}{\epsilon_{ox}^2}\left[\sqrt{1 + \frac{2\epsilon_{ox}^2(V_{GS} - V_{fb} - \phi_s)}{q\epsilon_{Si}N_{gate}T_{ox}^2}} - 1\right] \quad (5.224)$$

The drain current in the linear region with a polydepletion effect can be expressed in terms of that without it through the following equation [92]:

$$I_{DS}(V_{GSeff}) = \frac{V_{GSeff} - V_T}{V_{GS} - V_T} I_{DS}(V_{GS})$$

(5.225)

EKV modifies the drain current equation by replacing the slope factor n by n_q, and the pinch-off voltage V_P suitably modified to include the polydepletion effect [95]:

$$I_{DS} = 2n_q \mu_{eff} C'_{ox} \phi_t^2 \frac{W}{L} F \left(\frac{V_P - V_{S(D)}}{\phi_t} \right)$$

(5.226)

where n_q is the charge linearization factor.

5.11 Impact Ionization in High Field Region

It was shown in Section 5.8 that the field at the drain end of the channel is very high when the transistor enters into saturation. The region between pinch-off to drain was identified as the high field or velocity saturation region, with the maximum field F_m near the drain. The length of the high field region is a function of channel length, oxide thickness, gate, and drain bias.

In spite of scaling of the supply voltage the electric field in the high field region can be strong enough ($>10^4$ V cm^{-1}). This can cause electrons to gain enough energy to create impact ionization in deep submicron MOSFETs. These electron are generally known as *hot-electrons* because of their kinetic energy, which if expressed as kT_e requires a T_e as high as 1000 K, which is much higher than the lattice temperature [100, 101].

Due to impact ionization electron–hole pairs are generated, out of which the electrons are collected by the drain increasing the drain current, and holes are pushed towards the source by the lateral field component. As holes leave the high field region they are acted upon by the vertical field from the gate directed towards the substrate, due to which holes are collected by the substrate. This results in an impact ionization induced substrate current as shown in Figure 5.24. The electric field in the high field region is as shown in Figure 5.25.

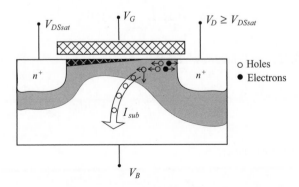

Figure 5.24 Carrier multiplication in the high field region and generation of substrate current

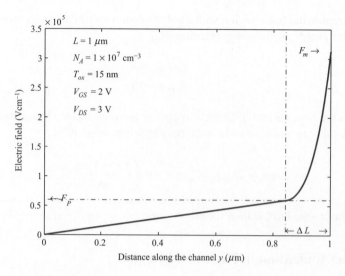

Figure 5.25   Lateral electric field variation along the channel

5.11.1 Substrate Current Model

The substrate current depends on the length of the high field region, and the impact ionization rate α_i, which depends on the local field, and will be maximum at $F = F_m$. The empirical relation describing the impact ionization rate which is mostly used in compact modeling (Equation (1.134)) is reproduced below:

$$\alpha_i(y) = A_i e^{\frac{-B_i}{F_y(y)}} \tag{5.227}$$

where α_i is the number of ionization events per unit length, and A_i and B_i are ionization constants. Therefore, the substrate current can be simply given by:

$$I_{sub} = I_{DS} \int_{L-\Delta L}^{L} \alpha_i(y)\, dy \tag{5.228}$$

From Equations (5.227) and (5.228) we have:

$$I_{sub} = A_i I_{DS} \int_{L-\Delta L}^{L} e^{\frac{-B_i}{F_y(y)}}\, dy \tag{5.229}$$

From Equation (5.175) we have:

$$\frac{dF_y}{dy} = \frac{1}{l}\sqrt{F_y^2 - F_p^2} \tag{5.230}$$

Combining Equations (5.229) and (5.230), we have:

$$I_{sub} = -A_i I_{DS} l \int_{F_P}^{F_m} e^{\frac{-B_i}{F_{y(y)}}} \frac{1}{\sqrt{F_y^2 - F_p^2}} \, dF_y \qquad (5.231)$$

Evaluation of the above integral results in

$$I_{sub} = \frac{A_i}{B_i} I_{DS} l F_m e^{\frac{-B_i}{F_m}} \qquad (5.232)$$

The accuracy of the above model depends significantly on the evaluation of the maximum electric field F_m, which can be obtained from Equation (5.181) under the assumption that $l \ll \Delta L$, giving:

$$F_m = \frac{V_{DS} - V_{DSsat}}{l} \qquad (5.233)$$

The accuracy of F_m given above can be increased by introducing fitting parameters [97] as shown below:

$$F_m = \frac{V_{DS} - \eta_i V_{DSsat}}{\beta_i l} \qquad (5.234)$$

The substrate current dependence on the gate-to-source voltage first shows an increase to reach a peak value, after which it starts dropping as shown in Figure 5.26. The peak of the

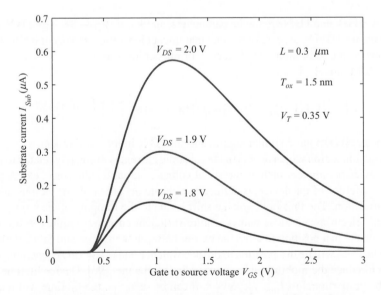

Figure 5.26   Substrate current dependence on gate and drain voltages

substrate current can be understood by observing that the substrate current is given by the product of I_{DS} and $F_m \exp(-Bi/F_m)$, where the former is an increasing function of V_{GS}, and the latter is a decreasing exponential function of V_{GS}, as $F_m \propto (V_{DS} - V_{DSsat})$. Initially, for a gate voltage less than the threshold voltage, the zero substrate current can be explained due to negligible drain current. A higher drain voltage for a given V_{GS} increases the field F_m, which increases the substrate current. From Equations (5.232) and (5.233), we can see that the substrate current goes to zero for $V_{DS} = V_{DSsat}$, at which we have no high field region, meaning thereby that for $V_{DS} < V_{DSsat}$, i.e. in the linear region, we can assume a zero substrate current. It may however be mentioned that though impact ionization is assumed to occur only in the high field region, there is a possibility that even in the GCA region near the pinch-off point, where the field is high, a certain amount of ionization may occur.

In BSIM3v3 a single equation characterizes the substrate current in all regions of operation:

$$I_{sub} = I_{DS} \left(\mathbf{ALPHA1} + \frac{\mathbf{ALPHA0}}{L} \right) (V_{DS} - V_{DSeff}) \, e^{\left(\frac{-\mathbf{BETA0}}{V_{DS} - V_{DSeff}} \right)} \qquad (5.235)$$

where **ALPHA0**, **ALPHA1**, and **BETA0** are model parameters obtained from experimental data for substrate current.

The requirement that the substrate current should go to zero in the linear region is satisfied by substituting V_{DSeff} for V_{DSsat}. **ALPHA1** represents the A_i/B_i term in Equation (5.232). The term **ALPHA0**/L is introduced to have a better fit of substrate current dependence with L.

EKV models the substrate current in the saturation region with the following equation:

$$I_{sub} = I_{DS} \frac{\mathbf{IBA}}{\mathbf{IBB}} V_{ib} e^{\left(\frac{-\mathbf{IBB} \cdot L_C}{V_{ib}} \right)} \qquad (5.236)$$

where **IBA**, **IBB**, and **IBN** are model parameters, and $V_{ib} = V_D - V_S - 2 \cdot \mathbf{IBN} V_{DSS}$. The model parameter **IBN** takes care of the fact that impact ionization can occur before pinch-off, and is known as the saturation voltage factor for impact ionization.

In the HiSIM model we have:

$$I_{sub} = I_{DS} \mathbf{SUB1} (\phi_{sS} + V_{DS} - \mathbf{SUB3} \cdot \phi_{sD}) \, e^{-\frac{\mathbf{SUB2}}{(\phi_{sS} + V_{DS} - \mathbf{SUB3} \cdot \phi_{sD})}} \qquad (5.237)$$

The parameter **SUB3** has a similar significance as **IBN** in the EKV model.

The impact ionization which results in the substrate current is intimately related to the degradation of device parameters such as threshold voltage, output conductance, etc. A parameter τ, called the lifetime of the device, is defined to indicate the effect of impact ionization on the device degradation. The threshold voltage shift is due to the trapping of electrons in the gate oxide. These electrons are those which acquire sufficient energy from the electric field near the drain, and are in a position to cross the barrier between the silicon and SiO$_2$. Further, these energetic electrons will be in a position to break covalent bonds at Si–SiO$_2$ generating interface states which reduce the mobility of the carriers in the channel [96]. Device lifetime has been shown to be proportional to $I_{sub}^{-2.9} I_{DS}^{1.9} \Delta V_T^{1.5}$. It can be seen that the lifetime will reduce for a larger drain voltage and a smaller channel length, as these increase the substrate current.

To counter the effect of impact ionization due to the high field near the drain region, as shown in Figure 5.25, the lowly doped drain structure is proposed which reduces the junction field near the drain, and thereby increases the device lifetime.

The substrate current generated as explained above may lead to a voltage drop across the substrate. This disturbs the biasing condition of the device and leads to an increase in drain current, as it reduces the threshold voltage. This increase in drain current causes an increase in current generated avalanche multiplication, and may lead to a condition where there is a sharp increase in drain current which defines the breakdown condition [98, 99].

5.12 Channel Punch-Through

In a scaled MOS transistor there is a possibility of a flow of drain current even though the gate voltage is below the threshold voltage. Such a condition sets a lower limit of scaling for the gate length. This happens for an applied drain voltage when the depletion regions for the drain–substrate/channel junction merges with that of the source–substrate/channel junction. In such a case the majority carrier electrons can thermally be injected over the source–channel barrier, and the lateral component of the electric field directed from drain–to–source causes a large amount of drain current. The transistor is said to be in a *punch-through* condition [103]. The punch-through phenomenon can be viewed from the viewpoint of a lateral field component of a 2-D field in the channel. The lateral drain induced field lowers the potential at the virtual cathode from the value of $2\phi_f$ set by the gate voltage by an amount ϕ_f, which causes high carrier injection from the source into the channel.

We shall have a rough estimate of the punch-through voltage using a one-dimensional model. The punch-through condition sets the condition that $W_{dd} + W_{ss} = L$, where W_{dd} and W_{ss} are respectively the drain and source depletion widths in the channel. Assuming an abrupt one-sided junction:

$$W_{dd} = \left[\frac{2\epsilon_{Si}(V_{PT} + \phi_{bi})}{qN_A} \right]^{\frac{1}{2}} \tag{5.238}$$

$$W_{ss} = \left[\frac{2\epsilon_{Si}(\phi_{bi})}{qN_A} \right]^{\frac{1}{2}} \tag{5.239}$$

where V_{PT} is the punch-through drain voltage.

A few select features of some of the compact models discussed in this book are tabulated in Appendix B.

5.13 Empirical Alpha Power MOSFET Model

Accurate modeling of the MOSFET drain current for circuit simulation involves complex and lengthy expressions with a large number of parameters. The complexity further increases as technology progresses to scale down the device dimensions. Short channel effects, carrier velocity saturation, quantum and polydepletion effects are the sources of complexity in model equations. Although these complex but accurate equations can be handled by circuit simulators, they are not suitable for the analytical purposes required for obtaining explicit equations for design parameters, such as delay and power dissipation, of digital CMOS subsystems for a given technology and supply voltage. If simple expressions are available they will be very handy for evaluating circuit performance. Analytical expressions for delay and power calculations are possible if simple expressions for drain current are available. Such compact and simple MOSFET models can be obtained empirically.

The drain current in the saturation region is of prime importance in digital applications. For a long channel MOSFET, Shockley's model predicts the square dependence of drain saturation current on gate overdrive ($V_{GS} - V_T$) as given below:

$$I_{Dsat} = \mu_{eff} C'_{ox} \frac{W}{L} (V_{GS} - V_T)^2 \tag{5.240}$$

For the scaled down device, where the carrier saturation effect dominates due to a high lateral field, Shockley's model fails to reproduce the $I-V$ results.

In the case of a short channel MOSFET where $L \to 0$, the lateral electric field will be very large compared to F_c, and the carriers will travel at saturation velocity. The drain current in the channel can be easily written as:

$$I_{Dsat} = -W Q'_n v_{sat} \tag{5.241}$$

where Q'_n is the inversion charge density. Using $Q'_n = -C'_{ox}(V_{GS} - V_T)$ in Equation (5.241) we have:

$$I_{Dsat} = W C'_{ox} v_{sat} (V_{GS} - V_T) \tag{5.242}$$

From the above equation it is clear that under the velocity saturation effect the drain current is less than that calculated from long channel theory. The important point to be noted is that the drain saturation current is independent of L, and varies as a linear function of gate overdrive ($V_{GS} - V_T$), as against the square dependence predicted by long channel theory [93].

From the analysis of experimental data it was proposed by Sakurai [48] that the velocity saturation effect on drain current can be taken into account by having an α-power dependence on ($V_{GS} - V_T$), where α is an empirical parameter close to unity for a short channel MOSFET. The α-power model equation for the static drain current in different operating regions is given below:

$$I_D = \begin{cases} 0 & (V_{GS} \leq V_T: \text{cutoff region}) \\ \left(\frac{I'_{D0}}{V'_{D0}}\right) V_{DS} & (V_{DS} < V'_{D0}: \text{linear region}) \\ I'_{D0} & (V_{DS} \geq V'_{D0}: \text{saturation region}) \end{cases} \tag{5.243}$$

where $I'_{D0} = B(V_{GS} - V_T)^\alpha$ and $V'_{D0} = C(V_{GS} - V_T)^{\alpha/2}$ is the drain saturation current and drain to source saturation voltage for a given V_{GS} respectively. Also, $B = I_{D0}/(V_{DD} - V_T)^\alpha$ and $C = V_{D0}/(V_{DD} - V_T)^{\alpha/2}$, in which I_{D0} is the drain saturation current, and V_{D0} is the drain saturation voltage at $V_{GS} = V_{DD}$. These equations, even though empirical, are simple with a few model parameters which can be easily extracted. They predict the drain current within tolerable error at least in the saturation region, and thus are handy in arriving at analytical expressions for critical CMOS performance factors like delay, dynamic power dissipation, noise margin, etc.

From Equation (5.243) it is seen that the α-power model consists of only four parameters, namely, V_T (threshold voltage), V_{D0} (drain saturation voltage at $V_{GS} = V_{DD}$), I_{D0} (drain saturation current $V_{GS} = V_{DD}$), and α (velocity saturation index).

The validity of this model can be justified by plotting log (I_{Dsat}), i.e. log (I'_{D0}) versus log $(V_{GS} - V_T)$. The expression for I'_{D0} indicates that such a plot should be linear, whose slope gives α, and the y-intercept gives B. The linear relationship between log (I'_{D0}) and log $(V_{GS} - V_T)$ is shown to be valid in Sakurai and Newton [48] for a device fabricated under various technologies at room temperature.

To extract the threshold voltage we can write for any three points on the line in log (I'_{D0}) versus log $(V_{GS} - V_T)$, the following equation:

$$\log \left(\frac{I_{D1}}{I_{D2}} \right) \log \left(\frac{V_{GS2} - V_T}{V_{GS3} - V_T} \right) - \log \left(\frac{I_{D2}}{I_{D1}} \right) \log \left(\frac{V_{GS1} - V_T}{V_{GS2} - V_T} \right) = 0 \qquad (5.244)$$

Solving the above equation numerically, we can easily extract the threshold voltage V_T.

In the above model the linear region characteristics are modeled by a straight line, whereas in saturation the drain current is assumed constant, i.e. the Channel Length Modulation (CLM) effect has been neglected. Plots of I_{DS} versus V_{DS} for the alpha power model are shown in Figure 5.27.

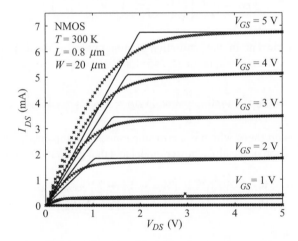

Figure 5.27 Alpha power model (V_T =0.84 V, $\alpha = 1.02$)

5.13.1 Physical Interpretation of the Alpha Power Model

The empirical nature of the α-power model limits its applicability in predicting future generations of technology. It is therefore important to understand the physical origin of the model parameter α. Though originally the α-power model was conceived purely empirically, the model parameters were linked with device physical parameters subsequently by Bowman et al. [71]

and Im *et al.* [72]. This gives a deeper insight into the origin of the empirical parameters. In the former approach the model was further refined to include subthreshold characteristics, vertical field dependent mobility, and threshold voltage roll-off due to the short channel effect. Here we shall briefly describe the procedure given in Bowman *et al.* [71].

The approximate expression for the drain current in saturation is given as:

$$I_{Dsat} \approx \mu_{eff} C'_{ox} \frac{W}{L} \left(V_{GS} - V_T - \frac{n}{2} V_{DSsat} \right) V_{DSsat} \tag{5.245}$$

where the μ_{eff} is the effective mobility taking into account the lateral and vertical field dependence [38], V_T is the threshold voltage including length dependence, $n = 1 + (1/C'_{ox})\sqrt{q\epsilon_{Si} N_A / 2(2\phi_f + V_{SB})}$, and V_{DSsat} is the drain-to-source saturation voltage taking velocity saturation into account. V_{DSsat} is given as:

$$V_{DSsat} = F_c L \left[\sqrt{1 + \frac{2}{F_c L} \left(\frac{V_{GS} - V_T}{n} \right)} \right] \tag{5.246}$$

Identifying that the α-power model parameter V_{D0} is the drain saturation voltage for $V_{GS} = V_{DD}$ from Equation (5.246), we can easily write:

$$V_{D0} = F_c L \left[\sqrt{1 + \frac{2}{F_c L} \left(\frac{V_{DD} - V_T}{n} \right)} \right] \tag{5.247}$$

Equating the drain current in the saturation region given by the α-power model Equation (5.243), and that given by Equation (5.245) for $V_{GS} = V_{DD}$, we get:

$$I_{D0} = \mu_{eff} C'_{ox} \frac{W}{L} V_{D0} \left[V_{DD} - V_T - \left(\frac{n}{2} \right) V_{D0} \right] \tag{5.248}$$

To derive an expression for α, which is a critical parameter of this model, we equate the saturation current expression of the alpha power model and that given in Equation (5.245). We get:

$$\left(\frac{V_{GS} - V_T}{V_{DD} - V_T} \right)^{\alpha} = \frac{\left(V_{GS} - V_T - \frac{n}{2} V_{DSsat} \right) V_{DSsat}}{\left(V_{DD} - V_T - \frac{n}{2} V_{D0} \right) V_{D0}} \tag{5.249}$$

Solving for α we have:

$$\alpha = \frac{\ln \left[\frac{(V_{GS} - V_T - \frac{n}{2} V_{DSsat}) V_{DSsat}}{(V_{DD} - V_T - \frac{n}{2} V_{D0}) V_{D0}} \right]}{\ln \left(\frac{V_{GS} - V_T}{V_{DD} - V_T} \right)} \tag{5.250}$$

The above equation shows that the model parameter α is a function of V_{GS}, whereas it was assumed to be constant for a given dimension and technology. However, it is found that the α is a weak function of V_{GS}, with not more than 2.5 % variation [71], which has a negligible effect in the estimation of I_{DSsat}. The typical value of α is taken for $V_{GS} = (V_{DD} + V_T)/2$. It is found that the value of α predicted by Equation (5.250) overestimates the experimentally extracted value.

The compact empirical α-power model has been used widely for analytical evaluation of basic CMOS digital circuits in view of its simplicity and the small number of parameters involved. It is a handy model to quickly evaluate the potential of a new technology for digital application. It has been found that the α-power model does not fit with the experimental data on drain current for different channel widths. A modified Sakurai–Newton model has been proposed which removes the above shortcoming [104]. The velocity saturation index, α, has been found to be temperature sensitive [105].

References

[1] R.H. Dennard, F.H. Gaennsslen, H.N. Yu, V.L. Rideout, E. Bassous and A.R. Le Blanc, Design of Ion-Implanted MOSFETs with Very Small Physical Dimensions, *IEEE J. Solid-State Circ.*, **9**(5), 256–268, 1974.

[2] J. Hayes, MOS Scaling, *Computer*, **13**(1), 8–13, 1980.

[3] B. Davani, R.H. Dennard and G.H. Shahidi, CMOS Scaling for High Performance and Low Power – The Next Ten Years, *Proc. IEEE*, **83**(4), 595–606, 1995.

[4] P.K. Bandopadhyay, Moore's Law Governs the Silicon Revolution, *Proc. IEEE*, **86**(1), 78–81, 1998.

[5] R. Woltjer, L. Tiemeijer and D. Klaassen, An Industrial View on Compact Modeling, *Solid-State Electron.*, **51**, 1572–1580, 2007.

[6] T.C. Chen, Where CMOS is Going: Trendy Hype vs. Real Technology, in *Proceedings of the IEEE International Conference on Solid-State Circuits*, pp. 1–18, 2006.

[7] E.J. Nowak, Maintaining the Benefits of CMOS Scaling When Scaling Bogs Down, *IBM J. Res. Develop.*, **46**(23), 168–180, 2002.

[8] J.R. Brews, W. Fichtner, E.H. Nicollian and S.M. Sze, Generalized Guide for MOSFET Miniaturization, *IEEE Electron Dev. Lett.*, **1**(1), 2–4, 1980.

[9] K. Kwok, S. Ng and T.D. Stanik, An improved Generalized Guide for MOSFET Scaling, *IEEE Trans. Electron Dev.*, **40**(10), 1895–1897, 1993.

[10] P.K. Chatterjee, W.R. Hunter, T.C. Holloway and Y.T. Lin, The Impact of Scaling Laws on the Choice of n-Channel or p-Channel for MOS VLSI, *IEEE Electron Dev. Lett.*, **1**(10), 220–223, 1980.

[11] G. Baccarani, M.R. Wordeman and R.H. Dennard, Generalized Scaling Theory and its Application to a $\frac{1}{4}$ micrometer MOSFET Design, *IEEE Trans. Electron Dev.*, **31**(4), 452–462, 1984.

[12] S. Wong and C.A.T. Salama, Impact of Scaling on MOS Performance, *IEEE J. Solid-State Circ.*, **18**(1), 106–114, 1983.

[13] T. Enomoto, T. Ishihara, M. Yasumoto and T. Aizawa, Design, Fabrication, and Performance of Scaled Analog ICs, *IEEE J. Solid-State Circ.*, **18**(4), 395–401, 1983.

[14] R. Singh and A.B. Bhattacharyya, Performance Oriented Scaling Laws for Mixed Analog/Digital MOS Circuits, *Solid-State Electron.*, **32**(10), 835–838, 1989.

[15] W. Liu, *MOSFET Models for SPICE Simulation, including BSIM3V3 and BSIM4*, John Wiley & Sons, Inc., New York, NY, 2001.

[16] D. Foty, *MOSFET Modeling with SPICE: Principles and Practice*, Prentice Hall PTR, Upper Saddle River, NJ, USA, 1997.

[17] H.C. Poon, L.D. Yau, R.L. Johnson and D. Beecham, DC Model for Short Channel IGFETs, *IEDM Tech. Dig.*, **19**, 156–159, 1973.

[18] L.D. Yau, A Simple Theory to Predict the Threshold Voltage of Short Channel IGFETs, *Solid-State Electron.*, **17**, 1059–1063, 1974.

[19] L. Dang, A Simple Current Model for Short-Channel IGFET and its Application to Circuit Simulation, *IEEE Trans. Electron Dev.*, **26**(4), 436–445, 1979.

[20] T.N. Nguyen and J.D. Plummer, Physical Mechanisms Responsible for Short Channel Effects in MOS Devices, *IEDM Tech. Dig.*, **27**, 596–599, 1981.

[21] T.N. Nguyen, *Small-Geometry MOS Transistors: Physics and Modeling of Surface- and Buried-Channel MOSFETs*, Ph.D. Thesis, Stanford University, Palo Alta, CA, 1984.

[22] M. Bucher, C. Lallement, C. Enz, F. Theodoloz and F. Krummenacher, *The EPFL-EKV MOSFET Model Equations for Simulation*, Technical Report, Model Version 2.6, 1998.

[23] Z.-H. Liu, C. Hu, J.-H. Huang, T.Y. Chan, M.-C. Jeng, P.K. Ko and Y.C. Cheng, Threshold Voltage Model for Deep-Submicrometer MOSFETs, *IEEE Trans. Electron Dev.*, **40**(1), 86–95, 1993.

[24] M. Orlowski, C. Mazure and F. Lau, Submicron Short Channel Effects due to Gate Reoxidation Induced Lateral Interstitial Diffusion, *IEDM Tech. Dig.*, 632–635, 1987.

[25] H.I. Hanafi, W.P. Noble, R.S. Bass, K. Varahramyan, Y. Lu and A.J. Dally, A Model for Anomalous Short Channel Behavior in Submicron MOSFETs, *IEEE Electron Dev. Lett.*, **14**(12), 575–577, 1993.

[26] C.Y. Lu and J.M. Sung, Reverse Short Channel Effects on Threshold Voltage in Submicrometer Salicided Devices, *IEEE Electron Dev. Lett.*, **10**(10), 446–448, 1989.

[27] S. Saha, Design Considerations of 25 nm MOSFET Devices, *Solid-State Electron.*, **45**(10), 1851–1857, 2001.

[28] J.R. Schrieffer, Effective Carrier Mobility in Surface–Space Charge Layers, *Phys. Rev.*, **97**(3), 641–646, 1955.

[29] A.G. Sabnis and J.T. Clemens, Characterization of the Electron Mobility in the Inverted ⟨100⟩ Si Surface, *IEDM Tech. Dig.*, **25**, 18–21, 1979.

[30] S.C. Sun and J.D. Plummer, Electron Mobility in Inversion and Accumulation Layers on Thermally Oxidized Silicon Surfaces, *IEEE Trans. Electron Dev.*, **21**(8), 1497–1508, 1980.

[31] D.M. Caughey and R.E. Thomas, Carrier Mobilities in Silicon Empirically Related to Doping and Field, *Proc. IEEE*, **55**(12), 2192–2193, 1957.

[32] S.I. Takagi, M. Iwase and A. Toriumi, On the Universality of Inversion Layer Mobility in n- and p-Channel MOSFETs, *IEDM Tech. Dig.*, 398–401, 1988.

[33] S.I. Takagi, A. Toriumi, M. Iwase and H. Tango, On the Universality of Inversion Layer Mobility in Si MOSFETs: Part I – Effects of Substrate Impurity Concentration, *IEEE Trans. Electron Dev.*, **41**(12), 2357–2362, 1994.

[34] S.I. Takagi, A. Toriumi, M. Iwase and H. Tango, On the Universality of Inversion Layer Mobility in Si MOSFETs: Part II – Effects of Surface Orientation, *IEEE Trans. Electron Dev.*, **41**(12), 2363–2368, 1994.

[35] R. van Langevelde and F.M. Klaassen, Effect of Gate Field Dependent Mobility Degradation on Distortion Analysis in MOSFETs, *IEEE Trans. Electron Dev.*, **44**(11), 2044–2052, 1997.

[36] R. van Langevelde, A.J. Scholten and D.B.M. Klaassen, *Physical Background of MOS Model 11, Level 1101, Koninklijke Philips Electronics* N. V., *Unclassified Report*, 2003.

[37] R.W. Coen and R.S. Muller, Velocity of Surface Carriers in Inversion Layers on Silicon, *Solid-State Electron.*, **23**(1), 35–40, 1980.

[38] C.G. Sodini, P.K. Ko and J.L. Moll, The Effect of High Fields on MOS Devices and Circuit Performance, *IEEE Trans. Electron Dev.*, **31**(10), 1386–1393, 1984.

[39] P.A. Blakey and K. Joardar, An Analytical Theory of the Impact of Velocity Overshoot on the Drain Characteristics of Field Effect Transistors, *IEEE Trans. Electron Dev.*, **39**(3), 740–742, 1992.

[40] J.H. Song, Y.J. Park and H.S. Min, Drain Current Enhancement due to Velocity Overshoot Effects and its Analytic Modeling, *IEEE Trans. Electron Dev.*, **43**(11), 1870–1875, 1996.

[41] S.L. Teitel and J.W. Wilkins, Ballistic Transport and Velocity Overshoot in Semiconductors: Part 1 – Uniform Field Effects, *IEEE Trans. Electron Dev.*, **30**(2), 150–153, 1983.

[42] J.B. Roldan, F. Gamiz, J.A. Lopez-Villanueva and J.E. Carceller, Modeling effects of electron-velocity overshoot in a MOSFET, *IEEE Trans. Electron Dev.*, **44**(5), 841–846, 1997.

[43] M. Miura-Mattausch and H. Jacobs, Analytical Model for Circuit Simulation with Quarter Micron Metal Oxide Semiconductor Field Effect Transistors: Subthreshold Characteristics, *Jpn. J. Appl. Phys.*, **29**(12), L2279–L2282, 1990.

[44] M. Miura-Mattausch, Analytical MOSFET Model for Quarter Micron Technologies, *IEEE Trans. Comput.-Aided Des. Integrat. Circ. Syst.*, **13**(5), 610–615, 1994.

[45] M. Miura-Mattausch, N. Sadachika, D. Navarro, G. Suzuki, Y. Takeda, M. Miyake, T. Warabino, Y. Mizukane, R. Inagaki, T. Ezaki, H.J. Mattausch, T. Ohguro, T. Iizuka, M. Taguchi, S. Kunashiro and S. Miyamoto, HiSIM2: Advanced MOSFET Model Valid for RF Circuit Simulation, *IEEE Trans. Electron Dev.*, **53**(9), 1994–2007, 2006.

[46] D. Navarro, T. Mizoguchi, M. Suetake, K. Hisamitsu, H. Ueno, M. Miura-Mattausch, H.J. Mattausch, S. Kumashiro, T. Yamaguchi, K. Yamashita and N. Nakayama, A Compact Model of the Pinch-off Region of 100 nm MOSFETs Based on the Surface-Potential, *IEICE Trans. Electron.*, **E88-C**(5), 1079–1085, 2005.

[47] H. Ueno, D. Kitamaru, K. Morikawa, M. Tanaka, M. Miura-Mattausch, H.J. Mattausch, S. Kumashiro, T. Yamaguchi, K. Yamashita and N. Nakayama, Impuity-Profile-Based Threshold-Voltage Model of Pocket-Implanted MOSFETs for Circuit Simulation, *IEEE Trans. Electron Dev.*, **49**(10), 1783–1789, 2002.

[48] T. Sakurai and A.R. Newton, Alpha-Power Law MOSFET Model and Its Application to CMOS Inverter Delay and Other Formulas, *IEEE J. Solid-State Circ.*, **25**(2), 584–594, 1990.

[49] G. Gildenblat, X. Li, W. Wu, H. Wang, A. Jha, R. van Langevelde, G.D.J. Smit, A.J. Scholten and D.B.M. Klaassen, PSP: An Advanced Surface-Potential-Based MOSFET Model for Circuit Simulation, *IEEE Trans. Electron Dev.*, **53**(9), 1979–1993, 2006.

[50] Y. Cheng, M.-C. Jeng, Z. Liu, J. Huang, M. Chan, K. Chen, P.K. Ko and C. Hu, A Physical and Scalable *I–V* Model in BSIM3v3 for Analog/Digital Circuit Simulation, *IEEE Trans. Electron Dev.*, **44**(2), 277–287, 1997.

[51] Y. Cheng, T. Sugii, K. Chen and C. Hu, Modeling of Small Size with Reverse Short Channel and Narrow Width Effects for Circuit Simulation Cases, *Solid-State Electron.*, **41**(9), 1227–1231, 1997.

[52] M. Nishida and H. Onodera, An Anomalous Increase of Threshold Voltage with Shortening the Channel Lengths for Deeply Boron-Implanted *N*-Channel MOSFETs, *IEEE Trans. Electron Dev.*, **28**(9), 1101–1103, 1981.

[53] Y.S. Pang and J.R. Brews, Analytical Subthreshold Surface Potential Model for Pocket *n*-MOSFETs, *IEEE Trans. Electron Dev.*, **49**(12), 2209–2216, 2002.

[54] J. Figueras, A.M. Brosa and A. Ferre, Test Challenges in Nanometrics CMOS Technologies, *Microelectron. Eng.*, **49**(1), 119–133, 1999.

[55] T. Sugii, Y. Momiyama and K. Goto, Continued Growth in CMOS Beyond 0.1 μm, *Solid-State Electron.*, **46**(3), 329–336, 2002.

[56] A.A. Mutlu, N.G. Gunther and M. Rehman, Analysis of Two-Dimensional Effects on Subthreshold Current in Submicron MOS Transistors, *Solid-State Electron.*, **46**(8), 1133–1137, 2002.

[57] R.R. Troutman and A.G. Fortino, Simple Model for Threshold Voltage in a Short Channel IGFET, *IEEE Trans. Electron Dev.*, **24**(10), 1266–1268, 1977.

[58] R.R. Troutman, VLSI Limitations from Drain Induced Barrier Lowering, *IEEE Trans. Solid-State Circ.*, **14**(2), 383–391, 1979.

[59] T. Skotnicki, G. Merckel and T. Pedron, The Voltage Doping Transformation: A New Approach to the Modeling of MOSFET Short-Channel Effects, *IEEE Electron Dev. Lett.*, **9**(3), 109–112, 1988.

[60] K.W. Terril, C. Hu and P.K. Ko, An Analytical Model for the Channel Electric Field in MOSFETs with Graded-Drain Structures, *IEEE Electron Dev. Lett.*, **5**(11), 440–442, 1984.

[61] H. Kurata and T. Sugii, Impact of Shallow Source/Drain on the Short Channel Characteristics of *p*MOSFETs, *IEEE Electron Dev. Lett.*, **20**(2), 95–96, 1999.

[62] K. Suzuki, Short Channel MOSFET Models Using a Universal Channel Depletion Width Parameter, *IEEE Trans. Electron Dev.*, **47**(6), 1202–1208, 2000.

[63] *HiSIM 1.1.1 User's Manual*, Hiroshima University, STARC, Hiroshima, Japan, 2002.

[64] A. Kranti, R.S. Haldar and R.S. Gupta, An Accurate Two-Dimensional CAD-Oriented Model of Retrograde Doped MOSFET for Improved Short Channel Performance, *Microelectron. Eng.*, **60**(3), 295–311, 2002.

[65] B. Yu, C.H.J. Wann, E.D. Nowak, K. Noda and C. Hu, Short Channel Effect Improved by Lateral Channel Engineering in Deep-Submicrometer MOSFETs, *IEEE Trans. Electron Dev.*, **44**(4), 627–634, 1997.

[66] R. Gwoziecki, T. Skotnicki, P. Bouillon and P. Gentil, Optimization of V_{th} Roll-off in MOSFETs with Advanced Channel Architecture – Retrograde Doping and Pockets, *IEEE Trans. Electron Dev.*, **46**(7), 1551–1561, 1999.

[67] Y. Okumura, M. Shrahate, T. Okudara, A. Hachisuka, H. Arima, T. Matsukawa and N. Tsubouchi, A Novel Source-to-Drain Nonuniformly Doped Channel (NUDC) MOSFET for High Current Derivability and Threshold Voltage Controllability, *IEDM Tech. Dig.*, 391–394, 1990.

[68] K.K. Ng and W.T. Lynch, The Impact of Intrinsic Series Resistance on MOSFET Scaling, *IEEE Trans. Electron Dev.*, **34**(3), 503–511, 1987.

[69] E. Gondro, P. Klein and F. Schuler, An Analytical Source-and-Drain Series Resistance Model of Quarter Micron MOSFETs and its Influence on Circuit Simulation, in *Proceedings of the IEEE International Symposium on Circuits and Systems*, Vol. 6, pp. 206–209, June 1999.

[70] H.-C. Chow and W.-S. Feng, An Improved Analytical Model for Short Channel MOSFETs, *IEEE Trans. Electron Dev.*, **39**(11), 2626–2629, 1992.

[71] K.A. Bowman, B.L. Austin, J.C. Eble, X. Tang and J.D. Meindl, A Physical Alpha-Power Law MOSFET Model, *IEEE Journal Solid-State Circ.*, **34**(10), 1410–1414, 1999.

[72] H. Im, M. Song, T. Hiramoto and T. Sakurai, Physical Insight into Fractional Power Dependence of Saturation Current on Gate Voltage in Advanced Short Channel MOSFETS (Alpha-Power Law Model), in *Proceedings of ISLPED '02*, pp 13–18, 2002.

[73] M.J. Deen and Z.X. Yan, DIBL in Short Channel NMOS Device at 77 K, *IEEE Trans. Electron Dev.*, **39**(4), 908–915, 1992.

[74] J.J. Ma and C.Y. Yu, A New Simplified Threshold Voltage Model for n-MOSFETs with Non-Uniformly Doped Substrate and its Application in MOSFETs Miniaturization, *IEEE Trans. Electron Dev.*, **42**(8), 1487–1494, 1995.

[75] P.M. Zeitzoff and J.E. Chung, A Perspective from the 2003 ITRS: MOSFET Scaling Trends, Challenges and Potential Solutions, *IEEE Circ. Dev. Mag.*, **21**(1), 4–15, 2005.

[76] T. Skotnicki, J.A. Hutchby, T.J. King, H.S. Philip Wong and F. Boeuf, The End of CMOS Scaling: Towards the Introduction of New Materials and Structural Changes to Improve MOSFET Performance, *IEEE Circ. Dev. Mag.*, **21**(2), 16–26, 2005.

[77] L.A. Akers and J.J. Sanches, Threshold Voltage Models for Short, Narrow and Small Geometry MOSFETs: A Review, *Solid-State Electron.*, **25**(7), 621–641, 1982.

[78] L.A. Akers, The Inverse-Width Effect, *IEEE Electron Dev. Lett.*, **7**(7), 419–421, 1986.

[79] N. Shigyo and T. Hiraoka, A Review of Narrow Channel Effects for STI MOSFETs: A difference Between Surface- and Buried-Channel Cases, *Solid-State Electron.*, **43**(11), 2061–2066, 1999.

[80] M. Bucher, C. Lallement, C. Enz, F. Theodoloz and F. Krummenacher, *The EPFL-EKV MOSFET Model Equations for Simulation*, Technical Report, Electronics Laboratories, Swiss Federal Institute of Technology (EPFL), Lausanne, Switzerland, 1998.

[81] J.R. Hauser, Extraction of Experimental Mobility Data for MOS Devices, *IEEE Trans. Electron Dev.*, **43**(11), 1981–1988, 1996.

[82] M.S. Liang, J.Y. Choi, P.K. Ko and C. Hu, Inversion Layer Capacitance and Mobility of Very Thin Gate Oxide MOSFETs, *IEEE Electron Dev. Lett.*, **33**(3), 409–413, 1986.

[83] K. Chen, H.C. Wann, J. Dunster, P.K. Ko and C. Hu, MOSFET Carrier Mobility Model Based on Gate Oxide Thickness, Threshold and Gate Voltages, *Solid-State Electron.*, **39**(10), 1515–1518, 1998.

[84] F.M. Klaassen and W.C.J. de Groot, Modeling of Scaled-Down MOS Transistors, *Solid-State Electron.*, **23**, 237–242, 1980.

[85] Y.A. El-Mansy and A.R. Boothroyd, A Simple Two Dimensional Model for IGFET Operation in the Saturation Region, *IEEE Trans. Electron Dev.*, **24**(3), 254–262, 1977.

[86] B. Hoefflinger, Output Characteristics of Short Channel Field Effect Transistors, *IEEE Electron Trans. Electron Dev.*, **28**(8), 971–976, 1981.

[87] J.G. Ruch, Electron Dynamics in Short Channel Field-Effect Transistors, *IEEE Trans. Electron Dev.*, **19**(5), 652–654, 1972.

[88] S.E. Laux and M.V. Fischetti, Monte-Carlo Simulation of Submicrometer Si n-MOSFET's at 77 and 300 K, *IEEE Electron Dev. Lett.*, **9**(9), 467–469, 1988.

[89] H. Wong and M.C. Poon, Approximation of the Length of Velocity Saturation Region in MOSFETs, *IEEE Trans. Electron Dev.*, **44**(11), 2033–2036, 1997.

[90] J.H. Huang, Z.H. Liu, M.C. Jeng, P.K. Ko and C. Hu, A Physical Model for MOSFET Output Resistance, *IEDM Tech. Dig.*, 569–572, 1992.

[91] C.L. Huang and N. Arora, Characterization and Modeling of the n- and p-Channel MOSFETs Inversion-layer Mobility in the Range 25–125 °C, *Solid-State Electron.*, **37**(1), 97–103, 1994.

[92] Y. Cheng and C. Hu, *MOSFET Modeling and BSIM3 User's Guide*, Kluwer Academic Publishers, Norwell, MA, (1999).

[93] Y. Taur and T.K. Ning, *Fundamentals of Modern VLSI Devices*, Cambridge University Press, Cambridge, UK, (1998).

[94] N.D. Arora and M.S. Sharma, Modeling the Anomalous Threshold Voltage Behaviour of Submicrometer MOSFETs, *IEEE Electron Dev. Lett.*, **13**(2), 92–94, 1992.

[95] J. Sallese, M. Bucher and C. Lallement, Improved Analytical Modeling of Polysilicon Depletion in MOSFETs for Circuit Simulation, *Solid-State Electron.*, **44**, 905–912, 2000.

[96] C. Hu, S.C. Tam, F.C. Hsu, P.K. Ko, T.Y. Chan and K.W. Terrill, Hot Electron-Induced MOSFET Degradation-Model, Monitor and Improvement, *IEEE J. Solid-State Circ.*, **20**(1), 295–305, 1985.

[97] L. Yang, Y. Hao, C. Yu and F. Han, An Improved Substrate Current Model for Ultra-Thin Gate Oxide MOSFETs, *Solid-State Electron.*, **50**, 489–495, 2006.

[98] H. Wong, A Physically-Based MOS Transistor Avalanche Breakdown, *IEEE Trans. Electron Dev.*, **42**(12), 2197–2202, 1995.

[99] H. Wong, Drain Breakdown in Submicron MOSFETs: A Review, *Microelectron. Reliabil.*, **40**, 3–15, 2000.

[100] A. Acovic, G.L. Rosa and Y.C. Sun, A Review of Hot Carrier Degradation Mechanism in MOSFETs, *Microelectron. Reliabil.*, **36**, 845–869, 1996.

[101] J. Frey, Where Do Hot Electrons Come From?, *IEEE Circ. Dev. Mag.*, **7**(6), 31–34, 1991.

[102] R. Rios, N.D. Arora and C.-L. Huang, An Analytical Polysilicon Depletion Effect Model for MOSFETs, *IEEE Electron Dev. Lett.*, **15**(4), 129–131, 1994.

[103] N. Kotani and S. Kawazaki, Computer Analysis of Punch-Through in MOSFETS, Solid-State Electronics, **22**(1), 63–70, 1979.

[104] M.M. Mansour, M.M. Mansour and A. Mehrotra, Modified Sakurai–Newton Current Model and its Application, in *Proceedings of the IEEE Computer Society Annual Symposium on VLSI [ISVLSI'03]*, pp. 62–69, 2003.

[105] A.B. Bhattacharyya, G. Dessai, C. Saha, S. Sharma, A. Singh and D.N. Singh, Temperature Dependence of Compact Alpha Power MOSFET Model Parameters, in *Proceedings of the International Workshop on Physics of Semiconductor Devices (IWPSD)*, pp. 1082–1085, 2003.

Problems

1. Using CE, CV, QCV and generalized scaling laws verify the expressions outlined in Table 5.1. Also obtain values of the drain saturation current scaling V_T as per the above rules for $V_{SB} = 0$ and $V_{SB} \neq 0$.

2. Obtain the minimum doping concentration of the substrate for the design of a MOSFET with $L = 0.1\,\mu m$, $T_{ox} = 5\,nm$, $X_j = 100\,nm$, and $V_{DD} = 2.0\,V$ that satisfies Brews' flexible scaling law.

3. In Figure 5.28, the source and drains are biased at the V_{SB} and V_{DB} voltages respectively. The triangles ADE and BCF enclose charges lost from the gate due to charge sharing by source and drain respectively. Assume that ADC forms an $\angle\alpha_1$ and BCD forms an $\angle\alpha_2$, which are assumed to remain constant for the source and drain voltage ranges considered. Distributing the charge loss due to charge sharing by source and drain uniformly, show that the reduction in the bulk depletion charge component of V_T can be expressed as a linear combination of V_{SB} and V_{DS}, where $V_{DS} = V_{DB} - V_{SB}$, and with an inverse relationship with channel length L as given below:

$$\Delta V_T = -\frac{C_1}{L}(V_{SB} + \phi_0) - \frac{C_2}{L}V_{DS} \qquad (5.251)$$

where, C_1 and C_2 are related to $\tan\alpha_1$ and $\tan\alpha_2$ respectively.

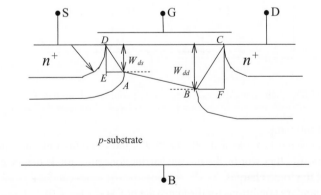

Figure 5.28   Schematic diagram representing source and drain charge sharing

4. Obtain the expression for an average electric field experienced by an inversion layer which has a thickness much less than the depletion layer under it. Assume an *arbitrary distribution of carrier density with depth* $n(x)$. For $T_{ox} = 100$ nm and $V_{fb} = 0$, obtain the gate biases necessary for obtaining the peak field at the Si–SiO$_2$ interface of a magnitude 5×10^5 V cm^{-1} for two doping concentrations $N_A = 5 \times 10^{14}$ cm^{-3} and $N_A = 10^{17}$ cm^{-3} respectively. For the above peak field, obtain the average field for the two doping concentrations. Illustrate the above through appropriate representation of the field in the F–x plane.

5. Show that the drain saturation voltage of a MOS transistor caused by velocity saturation can be given by the following simple relation:

$$V_{DSsat} = \frac{(V_{GS} - V_T)F_C L}{(V_{GS} - V_T) + F_C L} \tag{5.252}$$

where F_C is the critical field related to carrier velocity saturation by the relation $F_C = 2v_{sat}/\mu_{eff}$.

6. Using the relation $\mu_{eff} = \mu_0/(1 + \theta(V_{GS} - V_T))$, and $\theta = \frac{2}{T_{ox}} \times 10^{-7}$ V^{-1}, and given data for an NMOS transistor, $N_A = 5 \times 10^{16}$ cm^{-3} and $V_{fb} = -0.80$ V, write a program to calculate and plot the variation of the drain saturation voltage with V_{Gseff} with T_{ox} as a parameter. Use the following values of the gate oxide: 100 nm, 40 nm, 25 nm, 10 nm.

7. Figure 5.29 shows a Gaussian box in the high field region between the pinch off point and the drain where gradual channel approximation does not hold good, and the field direction between the gate and the substrate is directed from substrate to drain as against drain to source where the channel is formed. The inversion charge carriers flowing to the drain are pulled away from the surface.

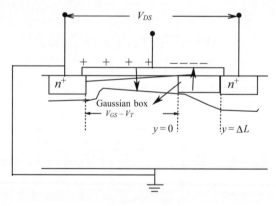

Figure 5.29   Schematic diagram representing source and drain charge sharing

Assume the following:
   (a) The Gaussian box has a depth $X_0 \approx X_j$ and contains only acceptor fixed ionic charges. In other words, the mobile carrier concentration is neglected compared to that of the ionic charges.
   (b) The boundary conditions for the Gaussian box are: $V(y = 0) = V_{DSsat}$, $F(y = 0) = F_p$, $V(y = L) = V_{DS}$, and $F(x = X_0) = 0$.

Obtain an expression for the length of the high field region ΔL in terms of F_P, V_{DS}, V_{GS}, V_{TS}, X_j, V_{DSsat}, and substrate doping concentration N_A.

8. Obtain Equation (5.23) using the expression for the spatial variation of surface potential along the length of the channel $\phi_{ss}(y)$ given in Equation (5.19).

9. Consider an LDD MOSFET structure where the length of the lowly doped n^--drain region is given to be $0.25\,\mu$m. The effective drain voltage across the channel, V_{DSeff}, is the applied drain-to-source voltage ($V_{DS} - V_{n^-}$) which can be expressed in the form: $V_{DSeff} = V_{DS}/(1 + L_{n^-}/l)$. Show the variation of threshold voltage for an LDD and the conventional non-LDD structure with V_{DS}, with channel length as a parameter, using Equation (5.23). Use the data from Figure 5.7(a).

10. For an n-channel MOS transistor, the following process parameters are given: $N_{sub} = 3.5 \times 10^{17}\,\text{cm}^{-3}$, $T_{ox} = 6\,\text{nm}$, $X_J = 0.25\,\mu$m, and $\eta = 1$. Obtain the value of the characteristic length l for a substrate bias of 2.5 V. Obtain the change in threshold voltage due to DIBL for a change in the drain bias from zero to 2.5 V. Assume $l_{eff} = 0.25\mu$m.

11. Assuming that the maximum threshold voltage shift should not exceed 20% of the long channel value, obtain the minimum channel length L_{min} permissible under quasi-constant voltage (QCV) scaling. The parameters for the long channel reference MOSFET device are: $V_{DD} = 5\,\text{V}$, $L = 5\,\mu$m, $T_{ox} = 80\,\text{nm}$, $X_j = 0.75\,\mu$m, $N_A = 10^{16}\,\text{cm}^{-3}$. Obtain the values of minimum channel length for scaling factor k_s values of 2, 4, 8, and 10.

12. Write a program for calculating and plotting the variation of the lateral electric field along the channel in a subthreshold condition for a scaling factor of 4 assuming the reference long channel value given in Problem 11 and CE scaling.

13. Briefly outline the features of the LEVEL 3, BSIM1, and BSIM3v3 models.

14. For a standard 0.5 μm process, the following is typical specification information reproduced from the MOSIS website.
    (a) Process Related Model Parameters: $C'_{ox} = 3.45\text{E} - 3$, $X_J = 0.15\text{E} - 6$.
    (b) Intrinsic Model Parameters: **KP** $= 15\text{E} - 6$, **E** $= 88.0\text{E}6$, **UCRIT** $= 4.5\text{E}6$, **DL** $= -0.05\text{E} - 6$, **DW** $= -0.02\text{E} - 6$, **LAMBDA** $= 0.23$, **LETA** $= 0.28$, **WETA** $= 0.05$, **Q0** $= 280\text{E} - 6$, **LK** $= 0.5\text{E} - 6$.
    Explain the physical significance of the parameters, and plot the $I-V$ characteristics using a SPICE circuit simulator. Given that:
    Geometry range: $W >= 0.8\,\mu$m, $L >= 0.5\,\mu$m.
    Voltage range: $|V_{GB}| < 3.3\,\text{V}$, $|V_{DB}| < 3.3\,\text{V}$, $|V_{SB}| < 2\,\text{V}$.

15. Use the data given in Problem 14 to generate $I_{DS}-V_{DS}$ characteristics for $\frac{W}{L} = 20$ and $V_{SB} = 0\,\text{V}$. Obtain the value of α of an alpha power MOSFET model for $L = 0.5\,\mu$m.

6

Quasistatic, Non-quasistatic, and Noise Models

6.1 Introduction

In Chapter 4 we discussed the operation of a MOS transistor under static bias conditions, where it was assumed that the terminal voltages remain constant with respect to time. We also discussed various modeling approaches to model the DC current–voltage characteristics. The effect of short channel and high field on DC characteristics was discussed in Chapter 5. For the DC bias conditions it was assumed that under normal operating conditions there was no flow of current through the gate and the substrate terminals.

The MOSFET, however, behaves differently under dynamic conditions where terminal voltages vary with time. To illustrate the dynamic operation we consider an n-channel MOSFET with a gate driven by a voltage pulse having a finite rise and fall time as shown in Figure 6.1(a). Initially, for $t < t_1$ we assume that the MOSFET is operating in the strong inversion with a drain current I_{DS1}. As the gate voltage increases with time for $t > t_1$ the inversion charge density also increases. The increase in the inversion charge in the channel can occur through an increase in the source current $-I_S(t)$ and a decrease in the drain current $I_D(t)$. It may be recalled that under static conditions the channel current is independent of the position y making $-I_S = I_D = I_{DS}$, which means the electrons supplied from the source are collected at the drain. But under transient conditions the requirement of build up of inversion charge makes the source and drain currents assume different values resulting in $I_D(t) \neq -I_S(t)$.

Figure 6.2 shows the input ramp excitation at the gate, and the corresponding source and drain transient currents simulated using a GSS device simulator [58].

It is also seen that with gate voltage increasing with time there is an increase in the negative depletion charge and positive gate charge. The increase in positive gate charge has to be supplied through the gate terminal constituting the gate current $I_G(t)$. The increase in the depletion of acceptor ions which have negative charges, on the other hand, is due to the outward flow of holes, which can be considered as negatives charge entering the bulk terminal constituting the bulk current $-I_B(t)$. The sign convention of various currents is shown in Figure 6.1(b). The negative charge entering through the bulk terminal in time dt equals the additional depletion

Compact MOSFET Models for VLSI Design   A.B. Bhattacharyya
© 2009 John Wiley & Sons (Asia) Pte Ltd

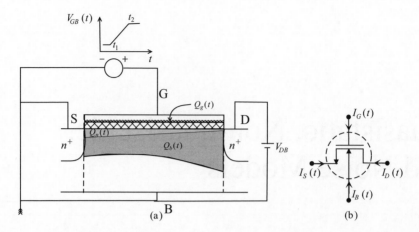

Figure 6.1 (a) Cross-section of a MOS transistor showing charge distribution under a transient state. (b) MOS transistor schematic showing conventional terminal currents

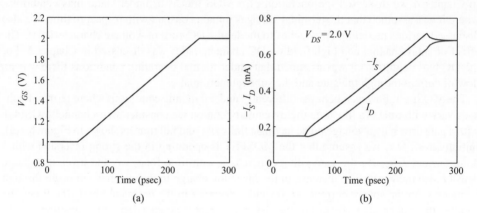

Figure 6.2 (a) Ramp voltage input to gate, and (b) transient source and drain current

charge in time dt, due to which reason the depletion charge is also called the bulk charge. Thus, the current required to increase the stored charge Q_n, Q_g, and Q_b can respectively be written as.

$$I_D(t) - I_S(t) = \frac{dQ_n(t)}{dt} \tag{6.1}$$

$$I_G(t) = \frac{dQ_g(t)}{dt} \tag{6.2}$$

$$I_B(t) = \frac{dQ_b(t)}{dt} \tag{6.3}$$

The above terminal currents are generally called charging currents. It may be mentioned that strictly speaking, the inversion charge Q_n is not a stored charge in the sense that bulk and gate charges are conceived. This may be understood from the fact that in the transient condition the inversion charges enter from the source terminal and exit through the drain terminal and, therefore, they are not in a fixed location [1].

To evaluate the source and drain terminal currents under dynamic operation, we have to consider the current continuity equation as shown below:

$$\frac{\partial Q'_n(y, t)}{\partial t} = \frac{1}{W} \frac{\partial I(y, t)}{\partial y} \tag{6.4}$$

where generation and recombination currents are neglected. Integrating Equation (6.4) from the source end of the channel to any point y in the channel we get:

$$\int_0^y \frac{\partial Q'_n(y, t)}{\partial t} \, dy = \frac{1}{W} \left[I(y, t) + I_S(t) \right] \tag{6.5}$$

where $I_S(t) = -I(0, t)$ is the current entering the source terminal. After rearranging Equation (6.5), we get:

$$I_S(t) = -I(y, t) + W \int_0^y \frac{\partial Q'_n(y, t)}{\partial t} \, dy \tag{6.6}$$

Further, integrating the above equation from 0 to L, and taking the derivative outside the integral in the second term on the RHS, we get:

$$I_S(t) = -\frac{1}{L} \int_0^L I(y, t) \, dy + W \frac{\partial}{\partial t} \int_0^L \left(1 - \frac{y}{L}\right) Q'_n \, dy \tag{6.7}$$

Similarly, the drain current $I_D(t)$:

$$I_D(t) = \frac{1}{L} \int_0^L I(y, t) \, dy + W \frac{\partial}{\partial t} \int_0^L \left(\frac{y}{L}\right) Q'_n \, dy \tag{6.8}$$

6.2 Quasistatic Approximation

Under quasistatic operation we assume that the potential and charge density at any given point in the channel follow the variation in bias voltages without any delay. It implies that $\phi_s(y, t)$ and $Q_n(y, t)$ at any instant correspond to the value calculated, assuming that the MOSFET is at the steady state at that instant. Such an approximation is found to be valid when the transition time for the voltage change is less than the transit time of the carriers from source to drain.

Assuming the quasistatic operation to be valid, we can write $I(y, t)$ in Equation (6.7) to be independent of channel position. This current, denoted as $I_T(t)$, is given by a drift–diffusion transport mechanism, and corresponds to the static drain current expression. Thus, Equation (6.7) can be written as:

$$I_S(t) = -I_T(t) + \frac{\mathrm{d}}{\mathrm{d}t} Q_s(t) \tag{6.9}$$

where $I_T(t)$ is called the transport or conduction current, and:

$$Q_s = W \int_0^L \left(1 - \frac{y}{L}\right) Q_n' \, \mathrm{d}y \tag{6.10}$$

is a charge assigned to the source terminal, which is assumed to flow through the source terminal into the channel. The second term in Equation (6.9) forms a charging current through the source terminal.

Similarly the transient drain current can be written as:

$$I_D(t) = I_T(t) + \frac{\mathrm{d}}{\mathrm{d}t} Q_d(t) \tag{6.11}$$

where:

$$Q_d = W \int_0^L \left(\frac{y}{L}\right) Q_n' \, \mathrm{d}y \tag{6.12}$$

is the charge assigned to the drain terminal, which is assumed to flow through the drain terminal into the channel. Q_s and Q_d are defined such that $Q_s + Q_d = Q_n$. Assigning the inversion charges given by Equations (6.10) and (6.12) respectively, to the source and drain terminals, is known as the Ward–Dutton partition scheme [3]. Charge conservation is ensured with this partition scheme. The variation of the integrand, as a function of y in the source and drain charge equations, is shown in Figure 6.3 for the saturation condition. It can be seen that though

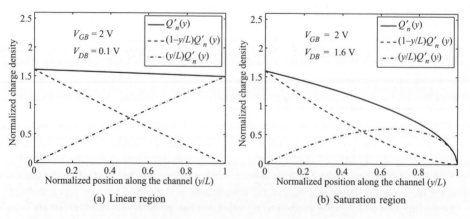

Figure 6.3   Source and drain charge distributions in (a) linear, and (b) saturation regions

the drain is isolated from the channel by the pinched-off region, and the channel charge is negligible at the drain, yet a finite amount of charge is assigned to the drain terminal which extends all the way up to the source terminal.

6.3 Terminal Charge Evaluation

6.3.1 PSP Model

The symmetric linearization of inversion charge density in terms of surface potential is given as below [42]:

$$Q'_n = Q'_{nm} + \alpha_m C'_{ox}(\phi_s - \phi_m) \tag{6.13}$$

where $\phi_m = (\phi_{sS} + \phi_{sD})/2$, and Q'_{nm} is the inversion charge density at $\phi_s = \phi_{sm}$. The inversion charge density can be integrated from source to drain to obtain the total inversion charge Q_n:

$$Q_n = W \int_0^L \left[Q_n + \alpha_m C'_{ox}(\phi_s - \phi_m) \right] dy$$

$$= W \int_{\phi_{sS}}^{\phi_{sD}} \left[Q_n + \alpha_m C'_{ox}(\phi_s - \phi_m) \right] \frac{dy}{d\phi_s} \phi_s \tag{6.14}$$

To obtain $dy/d\phi_s$, we express the drain current using CSM:

$$I_{DS} = -\mu_{eff} W \left(Q'_n \frac{d\phi_s}{dy} - \phi_t \frac{dQ'_n}{dy} \right) \tag{6.15}$$

The PSP drain current expression (Equation (4.143)) is given as:

$$I_{DS} = -\mu_{eff} \frac{W}{L} \left(Q'_{nm} - \alpha_m C'_{ox} \phi_t \right) \phi \tag{6.16}$$

where $\phi = \phi_{sD} - \phi_{sS}$. Equating expressions (6.15) and (6.16), we get:

$$\frac{\phi}{L}(Q'_{nm} - \alpha_m C'_{ox} \phi_t) \, dy = \left[Q'_{nm} + C'_{ox} \alpha_m(\phi_s - \phi_m) - \phi_t C'_{ox} \alpha_m \right] d\phi_s \tag{6.17}$$

Denoting y_m as the position in the channel where $\phi_s = \phi_m$, and integrating the above equation from $y = y_m$ to any position y in the channel, we have:

$$y = y_m + \frac{L}{\phi} \left[(\phi_s - \phi_m) - \frac{(\phi_s - \phi_m)^2}{2H} \right] \tag{6.18}$$

where $H = \phi_t - Q'_{nm}/(C'_{ox} \alpha_m)$, and $y_m = (L/2)(1 + \phi/(4H))$. Differentiating Equation (6.18) w.r.t. ϕ_s we get:

$$\frac{dy}{d\phi_s} = \frac{L}{\phi} \left[1 - \frac{(\phi_s - \phi_m)}{H} \right] \tag{6.19}$$

Substitute $dy/d\phi_s$ from Equation (6.19) in the expression for Q_n in Equation (6.14), and integrate to get [43]:

$$Q_n = WLC'_{ox} \left(\frac{Q'_{nm}}{C'_{ox}} - \frac{\alpha_m}{12H} \phi^2 \right) \tag{6.20}$$

To obtain the total bulk charge we have from symmetric linearization of the depletion charge (see Equation (4.140)):

$$Q'_b = -C'_{ox} \left[\gamma \sqrt{\phi_m - \phi_t} - (1 - \alpha_m)(\phi_s - \phi_m) \right] \tag{6.21}$$

$$Q_b = -WC'_{ox} \int_{\phi_{sS}}^{\phi_{sD}} \left[\gamma \sqrt{\phi_m - \phi_t} - (1 - \alpha_m)(\phi_s - \phi_m) \right] \frac{dy}{d\phi_s} d\phi_s \tag{6.22}$$

Substituting for $d\phi_s/dy$ from Equation (6.19) and integrating, we have:

$$Q_b = -WLC'_{ox} \left[\gamma(\phi_m - \phi_t)^{1/2} + \frac{(1 - \alpha_m)\phi^2}{12H} \right] \tag{6.23}$$

The total gate charge Q_g is given as:

$$Q_g = -(Q_n + Q_b) = WLC'_{ox} \left[-\frac{Q'_{nm}}{C'_{ox}} + \gamma(\phi_m - \phi_t)^{1/2} + \left(\frac{\phi^2}{12H} \right) \right] \tag{6.24}$$

The drain charge obtained by the Ward–Dutton partition scheme [3] is given below:

$$Q_d = \frac{W}{L} \int_{\phi_{sS}}^{\phi_{sD}} y(\phi_s) Q'_n(\phi_s) \frac{dy}{d\phi_s} d\phi_s \tag{6.25}$$

where $y(\phi_s)$ is given in Equation (6.18). On integration, the final expression for the drain charge is:

$$Q_d = \frac{1}{2} WLC'_{ox} \left[\frac{Q'_{nm}}{C'_{ox}} + \frac{\alpha_m \phi}{6} \left(1 - \frac{\phi}{2H} - \frac{\phi^2}{20H^2} \right) \right] \tag{6.26}$$

The source charge will be:

$$Q_s = Q_n - Q_d \tag{6.27}$$

Figure 6.4 shows the variation of the various normalized charges Q_g, Q_b, Q_n, Q_s, and Q_d as a function of gate voltage for two values of the drain bias. The charges are normalized with WLC'_{ox}. It can be seen from Figure 6.4(a) that as the gate voltage is increased from V_{fb}, initially there is a noticeable increase in the bulk charge until the threshold voltage, beyond which the inversion charge takes over to support the gate charge. A higher drain bias decreases

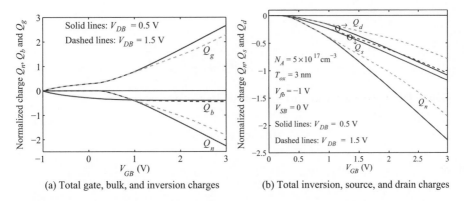

(a) Total gate, bulk, and inversion charges   (b) Total inversion, source, and drain charges

Figure 6.4   Variation of charges as a function of gate voltage, with V_{DB} as a parameter

the inversion level of the channel near the drain end, thus decreasing the total inversion charge, whereas an increased surface potential near the drain end increases the total bulk charge. Figure 6.4(b) shows the division of inversion charge into its source and drain components.

Figure 6.5 shows the variation of the various normalized charges Q_g, Q_b, Q_n, Q_s, and Q_d as a function of drain voltage for two values of the gate bias. For $V_{DS} = 0$ V, the source and drain share the inversion charge equally. The source's share of inversion charge becomes higher as compared to the drain for $V_{DS} > 0$, and saturates to 60% of the total inversion charge in the saturation region, which can be seen from Figure 6.4(b). In saturation, for $V_{GB} = 1.96$ V, the normalized $Q_n = 1$ and $Q_s = 0.6$. Due to this reason the Ward–Dutton scheme of inversion charge partition is also known as the 60–40 partition scheme.

6.3.2 EKV Model

On linearization of inversion charge as a function of ϕ_s (i.e. $\mathrm{d}Q_n = -n_q C'_{ox}\, \mathrm{d}\phi_s$) [51], Equation (6.15) for the drain current, assuming the charge sheet approximation, can be expressed

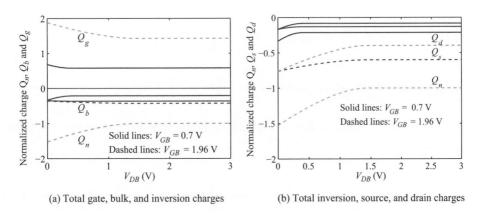

(a) Total gate, bulk, and inversion charges   (b) Total inversion, source, and drain charges

Figure 6.5   Variation of charges as a function of gate voltage, with V_{GB} as a parameter

as:

$$I_{DS} = \mu_{eff} W \left(-\frac{Q'_n}{n_q C'_{ox}} + \phi_t \right) \frac{dQ'_n}{dy} \tag{6.28}$$

Integrating the above expression from source to drain, we have:

$$I_{DS} = I_S(i_f - i_r) \tag{6.29}$$

where:

$$i_{f(r)} = \left(\frac{Q'_{nS(D)}}{2n_q C'_{ox} \phi_t} \right)^2 - \frac{Q'_{nS(D)}}{2n_q C'_{ox} \phi_t} \tag{6.30}$$

and $I_S = 2\mu_{eff} \frac{W}{L} n_q C'_{ox} \phi_t^2$; $i_{f(r)}$ is a function of $(V_P - V_{S(D)})/\phi_t$ given by Equation (4.123). For a channel potential V_C where the inversion charge density is Q'_n, we can write the following expression from Equation (6.30):

$$i = \left(\frac{Q'_n}{2n_q C'_{ox} \phi_t} \right)^2 - \frac{Q'_n}{2n_q C'_{ox} \phi_t} \tag{6.31}$$

From Equation (6.31), the inversion charge density is expressed as [52, 53]:

$$Q'_n = -2n_q \phi_t C'_{ox} \left(\sqrt{\frac{1}{4} + i} - \frac{1}{2} \right) \tag{6.32}$$

The total inversion charge Q_n is obtained by integrating the above equation over the entire channel length, as given below:

$$Q_n = -2n_q \phi_t C'_{ox} W \int_0^L \left(\sqrt{\frac{1}{4} + i} - \frac{1}{2} \right) dy \tag{6.33}$$

From Equation (6.28) dy can be written as:

$$dy = \frac{\mu_{eff} W}{I_{DS}} \left(-\frac{Q'_n}{n_q C'_{ox}} dQ'_n + \phi_t dQ'_n \right) \tag{6.34}$$

which can be expressed as:

$$dy = -\frac{L \, di}{i_f - i_r} \tag{6.35}$$

Substituting dy from Equation (6.35) into Equation (6.33), and changing the limits of integration, we get:

$$Q_n = 2n_q\phi_t C'_{ox} W \int_{i_f}^{i_r} \left(\sqrt{\frac{1}{4}+i} - \frac{1}{2}\right) \frac{L di}{(i_f - i_r)} \tag{6.36}$$

On integration we get:

$$Q_n = -2n_q\phi_t C_{ox} \left[\frac{2}{3}\frac{(x_f^2 + x_r x_f + x_r^2)}{x_f + x_r} - \frac{1}{2}\right] \tag{6.37}$$

where $x_f = \sqrt{\frac{1}{4}+i_f}$, $x_r = \sqrt{\frac{1}{4}+i_r}$, and $C_{ox} = WLC'_{ox}$.

In order to obtain the total bulk charge we start with the bulk charge density expression:

$$Q'_b = -C'_{ox}\gamma\sqrt{\phi_s} \tag{6.38}$$

Linearizing around $Q'_n = 0$, i.e. at $\phi_s = \phi_{po} \approx (V_P + \phi_o)$, we have:

$$Q'_b = -C'_{ox}\gamma \left[\sqrt{\phi_{po}} - \frac{1}{2\sqrt{\phi_{po}}}(\phi_s - \phi_{po})\right]$$

$$= -C'_{ox}\gamma\sqrt{\phi_{po}} + C'_{ox}(n_q - 1)(\phi_s - \phi_{po}) \tag{6.39}$$

where $n_q = 1 + \gamma/2\sqrt{\phi_{po}}$. We have from Equation (2.176), with n_q replacing n assuming the polydepletion effect, $Q'_n = -n_q C'_{ox}(\phi_{po} - \phi_s)$. Substituting this value of Q'_n in Equation (6.39), and integrating from source to drain, gives the total bulk charge:

$$Q_b = -C_{ox}\gamma\sqrt{V_P + \phi_o} - \left(\frac{n_q - 1}{n_q}\right)Q_n \tag{6.40}$$

The total gate charge, Q_g, will be:

$$Q_g = -(Q_b + Q_n) = C_{ox}\gamma\sqrt{V_P + \phi_o} + 2\phi_t C_{ox} \left[\frac{2}{3}\frac{(x_f^2 + x_r x_f + x_r^2)}{x_f + x_r} - \frac{1}{2}\right] \tag{6.41}$$

The drain charge using the Ward–Dutton partition scheme [3] is:

$$Q_d = \frac{W}{L} \int_0^L y Q'_n \, dy \tag{6.42}$$

Substituting $dy = -L di/(i_f - i_r)$ and $y = L(i_f - i)/(i_f - i_r)$, and integrating we have:

$$Q_d = -n_q C_{ox} \left[\frac{4}{15} \frac{3x_r^3 + 6x_r^2 x_f + 4x_r x_f^2 + 2x_f^2}{(x_f + x_r)^2} - \frac{1}{2} \right] \tag{6.43}$$

with $Q_g = -(Q_n + Q_b)$ and $Q_s = Q_n - Q_d$.

6.3.3 BSIM Model

6.3.3.1 Strong Inversion

To begin with we evaluate the total inversion charge given as [49, 54]:

$$Q_n = W \int_0^L Q_n'(y) \, dy \tag{6.44}$$

where $Q_n'(y) = -C_{ox}'(V_{gs} - V_{TS} - n V_C(y))$. To evaluate the above integral we make a change of variable dy to dV_C from the expression for drain current:

$$I_{DS} = -W \mu_{eff} Q_n' \frac{dV_{CS}}{dy} \tag{6.45}$$

We can now write:

$$dy = -\frac{W \mu_{eff}}{I_{DS}} Q_n' \, dV_{CS} \tag{6.46}$$

where, in terms of terminal voltage, I_{DS} for the linear region can be written as:

$$I_{DS} = \mu_{eff} C_{ox}' \frac{W}{L} \left[(V_{GS} - V_{TS}) V_{DS} - n \frac{V_{DS}^2}{2} \right] \tag{6.47}$$

Substituting dy from Equation (6.46) into Equation (6.44), we have:

$$Q_n = -\frac{W L C_{ox}'}{(V_{GS} - V_{TS}) V_{DS} - n \frac{V_{DS}^2}{2}} \int_0^{V_{DS}} (V_{GS} - V_{TS} - n V_C)^2 \, dV_C \tag{6.48}$$

which gives:

$$Q_n = -W L C_{ox}' \left[V_{GS} - V_{TS} - \frac{n V_{DS}}{2} + \frac{1}{12} \frac{n^2 V_{DS}^2}{(V_{GS} - V_{TS} - n V_{DS}/2)} \right] \tag{6.49}$$

The total bulk charge is given as:

$$Q_b = W \int_0^L Q_b'(y)\,dy = -WC_{ox}' \left[V_{TS} - V_{fb} - 2\phi_f - (1-n)V_C\right]\,dy \qquad (6.50)$$

Making a change of variable, as we had done for the inversion charge calculation, we can write:

$$Q_b = -\frac{WLC_{ox}'}{\left[(V_{GS} - V_{TS})V_{DS} - nV_{DS}^2/2\right]}$$

$$\times \int_0^{V_{DS}} \left[V_{TS} - V_{fb} - 2\phi_f - (1-n)V_C\right](V_{GS} - V_{TS} - nV_C)\,dV_C \qquad (6.51)$$

which gives:

$$Q_b = WLC_{ox}' \left[V_{fb} - V_{TS} + 2\phi_f + (1-n)\frac{V_{DS}}{2} - \frac{n(1-n)V_{DS}^2}{12(V_{GS} - V_{TS} - nV_{DS}/2)}\right] \qquad (6.52)$$

The total gate charge is nothing but $-(Q_n + Q_b)$, which is given as:

$$Q_g = WLC_{ox}' \left[V_{GS} - V_{fb} - 2\phi_f - \frac{V_{DS}}{2} + \frac{nV_{DS}^2}{12(V_{GS} - V_{TS} - nV_{DS}/2)}\right] \qquad (6.53)$$

The BSIM model has various options for the selection of charge partition between source and drain, controlled by the model parameter **X$_{PART}$**. The simplest model to partition the channel charge is to assign 50% of the inversion charge to source, and the remaining 50% to drain (**X$_{PART}$** = 0.5). Thus, we can write:

$$Q_s = Q_d = 0.5 Q_n \qquad (6.54)$$

A more rigorous approach is to solve the Ward–Dutton integral equation for drain charge given by the Ward–Dutton partition scheme (**X$_{PART}$** < 0.5), and write the source charge as $Q_s = Q_n - Q_d$. Thus, the drain charge is:

$$Q_d = \frac{W}{L} \int_0^L y Q_n'\,dy \qquad (6.55)$$

Making a change of variable, and writing $y = \int dy$ and integrating, we have:

$$Q_d = -WLC'_{ox} \left[\frac{V_{GS} - V_{TS}}{2} - \frac{nV_{DS}}{2} \right.$$

$$\left. + \frac{nV_{DS} \left(\frac{(V_{GS}-V_{TS})^2}{6} - \frac{nV_{DS}(V_{GS}-V_{TS})}{8} + \frac{n^2 V_{DS}^2}{40} \right)}{(V_{GS} - V_{TS} - nV_{DS}/2)^2} \right] \tag{6.56}$$

The above treatment assumes the transistor to be in the linear region. When the transistor is in the saturation region the channel potential at the drain end of the channel is $V_{DSsat} = (V_{GS} - V_{TS})/n$. Substituting $V_{DS} = V_{DSsat}$ into Equation (6.53), we have the expression for Q_g in the saturation region as:

$$Q_g = WLC'_{ox} \left(V_{GS} - V_{fb} - 2\phi_f - \frac{V_{DSsat}}{3} \right) \tag{6.57}$$

and for Q_b, we get:

$$Q_b = WLC'_{ox} \left[V_{fb} + 2\phi_f - V_{TS} + \frac{(1-n)V_{DSsat}}{3} \right] \tag{6.58}$$

and the total inversion charge:

$$Q_n = -\frac{2}{3} WLC'_{ox}(V_{GS} - V_{TS}) \tag{6.59}$$

From the 50–50 partition scheme, we have:

$$Q_s = Q_d = Q_n/2 = -\frac{1}{3} WLC'_{ox}(V_{GS} - V_{TS}) \tag{6.60}$$

The Ward–Dutton partitioned drain charge in the saturation condition, derived from Equation (6.56), is given as:

$$Q_d = -\frac{4}{15} WLC'_{ox}(V_{GS} - V_{TS}) \tag{6.61}$$

which comes out to be 40% of the total inversion charge given in Equation (6.49).

6.3.3.2 Weak Inversion

In weak inversion the inversion charge is assumed to be zero, and the gate charge is equated to the bulk charge which is given as:

$$Q_g = -Q_b = \gamma C_{ox} \sqrt{\phi_s} \tag{6.62}$$

The surface potential ϕ_s in weak inversion, as derived in Equation (2.60), is reproduced below:

$$\phi_s = \left(\sqrt{V_{GB} - V_{fb} + \frac{\gamma^2}{4}} - \frac{\gamma}{2} \right)^2 \tag{6.63}$$

Substituting for ϕ_s in Equation (6.62), we get:

$$Q_b = -\gamma C_{ox} \left(\sqrt{V_{GB} - V_{fb} + \frac{\gamma^2}{4}} - \frac{\gamma}{2} \right) \tag{6.64}$$

6.3.4 Terminal Charge in Accumulation Region

The previous section discusses the terminal charges, and their evaluation under weak and strong inversion conditions. In the accumulation region the space charge is mainly due to the accumulated holes, which come from the bulk. The source and drain terminal charges are zero, due to absence of an inversion channel. Thus, we can write:

$$Q_n = Q_s = Q_d = 0 \tag{6.65}$$

and the charge balance gives:

$$Q_g = -Q_b = -WLC'_{ox}V_{ox} = -WLC'_{ox}(V_{GB} - V_{fb} - \phi_s) \tag{6.66}$$

The BSIM model expression for gate charge is given as:

$$Q_g = -WLC'_{ox}(V_{GS} - V_{fb} - V_{BS}) \tag{6.67}$$

6.4 Quasistatic Intrinsic Small Signal Model

For analog circuit design and analysis small signal dynamic models for MOS transistors are required. It has been recognized that small signal modeling is more complex, as not only the values of the operating current, but also the derivatives are important [5].

To obtain the small signal model we assume that the MOS transistor is biased at a certain operating point around which the AC small signal is superimposed. It may be mentioned that the circuit simulator can obtain the small signal parameters from a large signal model. Hand calculations for AC analysis and analytical design, however, require compact expressions for small signal parameters.

6.4.1 Small Signal Conductances

For a MOSFET operating at low frequency, only the transport current from drain to source is significant. The gate and bulk transport currents are assumed to be zero under normal operating bias conditions. The drain current I_{DS} in a MOS transistor is a function of gate, drain, and source voltages which is described by:

$$I_{DS} = f(V_{GB}, V_{DB}, V_{SB}) \tag{6.68}$$

where the function f for long and short channel MOSFETs has been discussed in Chapters 4 and 5. For small variations in terminal voltages around the bias point, we can write, taking bulk as the reference terminal, the change in drain current as:

$$i_{ds} = \left[\frac{\partial I_{DS}}{\partial V_{GB}}\right]_{V_{SB},V_{DB}} v_{gb} + \left[\frac{\partial I_{DS}}{\partial V_{DB}}\right]_{V_{GB},V_{SB}} v_{db} + \left[\frac{\partial I_{DS}}{\partial V_{SB}}\right]_{V_{GB},V_{DB}} v_{sb} \tag{6.69}$$

where v_{gb}, v_{db}, and v_{sb} are the incremental changes in the gate, drain, and source voltages respectively. Using the small signal notation we write:

$$i_{ds} = g_{mg}v_{gb} + g_{md}v_{db} + g_{ms}v_{sb} \tag{6.70}$$

where:

$$g_{mg} = \left[\frac{\partial I_{DS}}{\partial V_{GB}}\right]_{V_{SB},V_{DB}=\text{const.}} \qquad \text{gate transconductance} \tag{6.71}$$

$$g_{md} = \left[\frac{\partial I_{DS}}{\partial V_{DB}}\right]_{V_{GB},V_{SB}=\text{const.}} \qquad \text{drain transconductance} \tag{6.72}$$

$$g_{ms} = \left[\frac{\partial I_{DS}}{\partial V_{SB}}\right]_{V_{GB},V_{DB}=\text{const.}} \qquad \text{source transconductance} \tag{6.73}$$

If we take the source terminal as the reference terminal, Equation (6.70) takes the form:

$$i_{ds} = g_m v_{gs} + g_{mb}v_{bs} + g_{ds}v_{ds} \tag{6.74}$$

where

$$g_m = \frac{\partial I_{DS}}{\partial V_{GS}}\bigg|_{V_{DS},V_{BS}=\text{const.}} \qquad \text{gate transconductance} \tag{6.75}$$

$$g_{mb} = \frac{\partial I_{DS}}{\partial V_{BS}}\bigg|_{V_{GS},V_{DS}=\text{const.}} \qquad \text{bulk transconductance} \tag{6.76}$$

$$g_{ds} = \frac{\partial I_{DS}}{\partial V_{DS}}\bigg|_{V_{GS},V_{BS}=\text{const.}} \qquad \text{output conductance} \tag{6.77}$$

The small signal current to voltage ratio g_{ds} is known as the *output conductance*, as it gives the change in current flowing from drain to source for change in drain-to-source voltage. The bulk and source referred small signal equivalent circuits are shown in Figure 6.6.

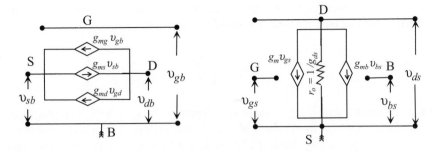

Figure 6.6 Bulk and source referred low frequency equivalent circuits for a MOS transistor

The small signal parameters with bulk as the reference can be easily related to the parameters with source as the reference as shown below:

$$g_m = g_{mg} \tag{6.78}$$

$$g_{mb} = g_{ms} - g_{mg} - g_{md} \tag{6.79}$$

$$g_{ds} = g_{md} \tag{6.80}$$

Since the transconductance g_{md} is equal to the output conductance g_{ds}, we shall henceforth refer to g_{md} as the output conductance.

We shall now obtain the expressions for the above small signal parameters with bulk as the reference for the linear and saturation modes of operation for a long channel MOS transistor.

6.4.2 Transconductances (g_{mg} and g_{ms})

The Memelink–Wallinga drain current model has been used for illustration, which is valid for a strong inversion and a long channel, to obtain the small signal parameters. Equation (4.90), is recalled for the appropriate expression for the linear region of operation, and is given below:

$$I_{DS} = \frac{\beta}{2n} \left[(V_{GB} - V_{TB0} - nV_{SB})^2 - (V_{GB} - V_{TB0} - nV_{DB})^2 \right] \tag{6.81}$$

where $\beta = \mu_{neff} C'_{ox} W/L$. On partial differentiation of the above equation w.r.t. V_{GB}, we get an expression for the gate transconductance g_{mg} as:

$$g_{mg} = \beta(V_{DB} - V_{SB}) \tag{6.82}$$

If the drain-to-source voltage is increased such that the device enters saturation, the transconductance also reaches a maximum value, and is given as:

$$g_{mgMAX} = \beta(V_P - V_{SB}) \tag{6.83}$$

The maximum transconductance for a given transistor dimension is an important parameter in analog design, and can be expressed in terms of the drain current as:

$$g_{mgMAX} = \sqrt{\frac{2\beta}{n} I_{DSsat}} \tag{6.84}$$

The gate transconductance g_{mg} is found to be independent of V_{GB} in the linear region as seen from Equation (6.82), and independent of V_{DB} in the saturation region of operation being dependent only on V_P. It may, however, be noted that these conclusions are valid under the assumption that the carrier mobility is independent of gate and drain voltages, and the threshold voltage does not vary with bias.

The source transconductance can similarly be arrived at by differentiating the drain current w.r.t. V_{SB}. In the linear as well as in the saturation regions, it is given by:

$$g_{ms} = -\beta(V_{GB} - V_{TB0} - nV_{SB}) \tag{6.85}$$

The negative sign is attributed to the fact that an increase in the source voltage results in a decrease in drain current. From Equations (6.83) and (6.85), we can see that in saturation g_{ms} is related to g_{mg} as:

$$g_{ms} = ng_{mgMAX} \tag{6.86}$$

and from Equation (6.78), with source as reference:

$$g_{mb} = (n - 1)g_{mMAX} \tag{6.87}$$

6.4.3 Drain Conductance g_{md}

The drain conductance is defined as the incremental change in drain current for an incremental change in the drain voltage V_{DB}. From the output $I-V$ characteristics it can be seen that the drain current has a strong dependence on V_{DB} in the linear region, and a weak dependence in the saturation region. The weak dependence in saturation is attributed to the phenomenon called channel length modulation, by which the channel length varies as function of V_{DB}. In the linear region the drain conductance is obtained from Equations (6.81) and (6.74) as:

$$g_{md} = \beta \frac{W}{L}(V_{GB} - V_{TB0} - nV_{DB}) \tag{6.88}$$

In saturation, the drain current is expressed as [29]:

$$I_{DS} = I_{DSsat}\left[1 + \frac{\zeta}{L}\sqrt{(V_{DB} - V_P)}\right] \tag{6.89}$$

where I_{DSsat} is the drain saturation current without channel length modulation, and $\zeta = \sqrt{2\epsilon_{Si}/qN_A}$. Differentiating Equation (6.89) w.r.t. V_{DB}, we get the expression for g_{md} as:

$$g_{md} = \frac{\zeta}{2L} \frac{I_{DSsat}}{\sqrt{V_{DB} - V_P}} \tag{6.90}$$

From the LEVEL 1 model of a MOSFET the output conductance in saturation is expressed as:

$$g_{ds} = \lambda I_{DSsat} \tag{6.91}$$

6.4.4 Small Signal Capacitances

The charging current at the jth terminal of a MOS transistor, in terms of the charge Q_j associated with the terminal j, in expressed as:

$$i_{cg} = \frac{dQ_g}{dt} \tag{6.92}$$

where the subscript c indicates charging current. The rate of change of charge Q_g in Equation (6.92) is due to variation of potential at any terminal of the MOSFET at the instant t, which can be expressed in terms of partial derivatives as follows:

$$i_{cg} = \frac{dQ_g}{dt} = \frac{\partial Q_g}{\partial V_G} \frac{dV_G}{dt} + \frac{\partial Q_g}{\partial V_S} \frac{dV_S}{dt} + \frac{\partial Q_g}{\partial V_D} \frac{dV_D}{dt} + \frac{\partial Q_g}{\partial V_B} \frac{dV_B}{dt} \tag{6.93}$$

Similarly, we can write the expression for the charging current for the remaining three nodes, i.e. source, drain, and bulk.

Defining:

$$C_{jk} = \pm \frac{\partial Q_j}{\partial V_k} \tag{6.94}$$

where the negative sign has to be taken for $j \neq k$ so that the capacitive elements are positive.[1] Now the set of four equation describing the charging currents at the respective nodes can be given in matrix form as:

$$\begin{bmatrix} i_{cg} \\ i_{cd} \\ i_{cs} \\ i_{cb} \end{bmatrix} = \begin{bmatrix} C_{gg} & -C_{gd} & -C_{gs} & -C_{gb} \\ -C_{dg} & C_{dd} & -C_{ds} & -C_{db} \\ -C_{sg} & -C_{sd} & C_{ss} & -C_{sb} \\ -C_{bg} & -C_{bd} & -C_{bs} & C_{bb} \end{bmatrix} \begin{bmatrix} V_G' \\ V_D' \\ V_S' \\ V_B' \end{bmatrix} \tag{6.95}$$

where $V_{G,S,D,B}'$ stands for the time derivative of $V_{G,S,D,B}$.

---

[1] The two capacitive elements C_{ds} and C_{sd} are exceptions.

It needs to be noted that in general $C_{jk} \neq C_{kj}$, which implies that the capacitance elements are *non-reciprocal*. Physically, this means that change in the charge at terminal j for a change in the voltage at terminal k need not be equal to the change in the charge at terminal k for the same change in voltage at terminal j. As an illustration, when a MOS transistor is in saturation a change in drain voltage does not increase the gate charge because the gate node is isolated by the pinch-off region. However, a change in the gate voltage can cause a change in the drain charge by changing the inversion charge density, and hence, a change in the drain current implying $C_{gd} \ll C_{dg}$ in saturation. In general, the origin of non-reciprocity can be attributed to the nonlinear dependence of node charges on more than one node voltages, which cannot be superposed linearly. Since the transcapacitances are non-reciprocal, they cannot be represented by conventional two terminal capacitors in an equivalent circuit representation. This concept is similar to the fact that transconductance cannot be represented by a conventional resistance in the equivalent circuit model.

In Equation (6.95) we have sixteen capacitive elements to describe the charging currents, in which only nine capacitances are independent, which follows from the application of charge conservation. In order to satisfy charge conservation, the sum of the elements of each row and column in the capacitive matrix in Equation (6.95) must be zero. For example, the sum of the elements of the first column in the capacitive matrix is:

$$C_{gg} - C_{dg} - C_{sg} - C_{bg} = \left[\frac{\partial Q_g + \partial Q_d + \partial Q_s + \partial Q_b}{\partial V_G} \right]_{V_S, V_D, V_B} \tag{6.96}$$

For charge to be conserved, the numerator of Equation (6.96) should be zero. Similar arguments can be applied to other columns as well. On account of the above consideration, Equation (6.95) can be written with:

$$C_{bg} = C_{gg} - C_{dg} - C_{sg} \tag{6.97a}$$

$$C_{bd} = C_{dd} - C_{gd} - C_{sd} \tag{6.97b}$$

$$C_{bs} = C_{ss} - C_{gs} - C_{ds} \tag{6.97c}$$

$$C_{bb} = C_{gb} + C_{db} + C_{sb} \tag{6.97d}$$

Now, the sum of the row elements in the capacitive matrix in Equation (6.96) can be written as:

$$C_{gg} - C_{gd} - C_{gs} - C_{gb} = \frac{\partial Q_g}{\partial V_G} + \frac{\partial Q_g}{\partial V_D} + \frac{\partial Q_g}{\partial V_S} + \frac{\partial Q_g}{\partial V_B} \tag{6.98}$$

The increase in the gate charge due to an increase in the gate voltage should be the same as the decrease in gate charge by increasing the voltages at other terminals individually and holding the gate potential constant, which implies that the summation in Equation (6.98) should be zero. The same argument holds for other rows as well. The last column in the capacitive matrix in Equation (6.95) can be replaced by the following column matrix:

$$
\begin{bmatrix}
-(C_{gg} - C_{gd} - C_{gs}) \\
-(C_{dd} - C_{dg} - C_{ds}) \\
-(C_{ss} - C_{sg} - C_{sd}) \\
C_{bb}
\end{bmatrix}
\tag{6.99}
$$

where $C_{bb} = (C_{gg} - C_{dg} - C_{sg}) + (C_{dd} - C_{gd} - C_{sd}) + (C_{ss} - C_{gs} - C_{ds})$.

From Equations (6.96), (6.97), and (6.99) it can be seen that we require only nine independent capacitances from which other capacitances can be derived. These nine capacitances are given in matrix form as:

$$
\begin{bmatrix}
C_{gg} & C_{gd} & C_{gs} \\
C_{dd} & C_{dg} & C_{ds} \\
C_{ss} & C_{sg} & C_{sd}
\end{bmatrix}
\tag{6.100}
$$

6.4.5 Equivalent Circuit Model

To have an equivalent circuit model for small signal variation, we refer to Figure 6.7, in which the potentials at gate, drain, and source are applied w.r.t. bulk. In this configuration the small signal variation at the gate, drain, and source terminal can be written as:

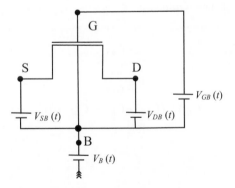

Figure 6.7 Circuit diagram showing gate, drain, and source potentials w.r.t. bulk

$$
dV_G = dV_{GB} + dV_B
\tag{6.101a}
$$

$$
dV_D = dV_{DB} + dV_B
\tag{6.101b}
$$

$$
dV_S = dV_{SB} + dV_B
\tag{6.101c}
$$

Substituting Equation (6.101) into Equation (6.93), we have:

$$i_{cg} = \frac{dQ_g}{dt} = C_{gg}\left(\frac{dV_{GB}}{dt} + \frac{dV_B}{dt}\right) - C_{gs}\left(\frac{dV_{SB}}{dt} + \frac{dV_B}{dt}\right)$$

$$- C_{gd}\left(\frac{dV_{DB}}{dt} + \frac{dV_B}{dt}\right) - C_{gb}\frac{dV_B}{dt} \qquad (6.102)$$

We can obtain similar equations for i_{cd}, i_{cs}, and i_{cb} giving:

$$\begin{bmatrix} i_{cg} \\ i_{cd} \\ i_{cs} \\ i_{cb} \end{bmatrix} = \begin{bmatrix} C_{gg} & -C_{gd} & -C_{gs} & C_{gg}-C_{gd}-C_{gs}-C_{gb} \\ -C_{dg} & C_{dd} & -C_{ds} & -C_{dg}+C_{dd}-C_{ds}-C_{db} \\ -C_{sg} & -C_{sd} & C_{ss} & -C_{sg}-C_{sd}+C_{ss}-C_{sb} \\ -C_{bg} & -C_{bd} & -C_{bs} & -C_{bg}-C_{bd}-C_{bs}+C_{bb} \end{bmatrix} \begin{bmatrix} V'_{GB} \\ V'_{DB} \\ V'_{SB} \\ V'_B \end{bmatrix} \qquad (6.103)$$

The elements in the right-most column in the above capacitance matrix will sum to zero, and therefore we have:

$$\begin{bmatrix} i_{cg} \\ i_{cd} \\ i_{cs} \\ i_{cb} \end{bmatrix} = \begin{bmatrix} C_{gg} & -C_{gd} & -C_{gs} & 0 \\ -C_{dg} & C_{dd} & -C_{ds} & 0 \\ -C_{sg} & -C_{sd} & C_{ss} & 0 \\ -C_{bg} & -C_{bd} & -C_{bs} & 0 \end{bmatrix} \begin{bmatrix} V'_{GB} \\ V'_{DB} \\ V'_{SB} \\ V'_B \end{bmatrix} \qquad (6.104)$$

The above equation can also be written as:

$$\begin{bmatrix} i_{cg} \\ i_{cd} \\ i_{cs} \\ i_{cb} \end{bmatrix} = \begin{bmatrix} C_{gg} & -C_{gd} & -C_{gs} & 0 \\ -C_{dg} & C_{dd} & -C_{ds} & 0 \\ -C_{sg} & -C_{sd} & C_{ss} & 0 \\ -(C_{gg}-C_{dg}-C_{sg}) & -(C_{dd}-C_{gd}-C_{sd}) & -(C_{ss}-C_{gs}-C_{ds}) & 0 \end{bmatrix}$$

$$\times \begin{bmatrix} V'_{GB} \\ V'_{DB} \\ V'_{SB} \\ V'_B \end{bmatrix} \qquad (6.105)$$

The above equation can be realized with the small signal equivalent model as shown in Figure 6.8. Substituting $V_{DB} = V_{DG} + V_{GB}$, and $V_{SB} = V_{SG} + V_{GB}$ in Equation (6.93), we can write:

$$i_{cg} = -C_{gs}\frac{dV_{SG}}{dt} - C_{gs}\frac{dV_{GB}}{dt} + C_{gg}\frac{dV_{GB}}{dt} - C_{gd}\frac{dV_{DG}}{dt} - C_{gd}\frac{V_{GB}}{dt} \qquad (6.106)$$

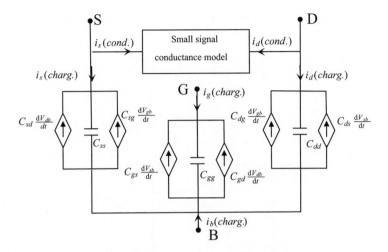

Figure 6.8 High frequency small signal MOSFET equivalent circuit

Rearranging:

$$i_{cg} = C_{gs}\frac{dV_{GS}}{dt} + \left(C_{gg} - C_{gs} - C_{gd}\right)\frac{dV_{GB}}{dt} + C_{gd}\frac{dV_{GD}}{dt} \tag{6.107}$$

As $C_{gb} = C_{gg} - C_{gs} - C_{gd}$, we have:

$$i_{cg} = C_{gs}\frac{dV_{GS}}{dt} + C_{gb}\frac{dV_{GB}}{dt} + C_{gd}\frac{dV_{GD}}{dt} \tag{6.108}$$

Similar transformations give:

$$i_{cd} = C_{gd}\frac{dV_{DG}}{dt} + C_{sd}\frac{dV_{DS}}{dt} + C_{bd}\frac{dV_{DB}}{dt} - C_m\frac{dV_{GS}}{dt} - C_{mb}\frac{dV_{BS}}{dt} \tag{6.109a}$$

$$i_{cb} = C_{bd}\frac{dV_{BD}}{dt} + C_{gb}\frac{dV_{BG}}{dt} - C_{mx}\frac{dV_{GB}}{dt} + C_{bs}\frac{dV_{BS}}{dt} \tag{6.109b}$$

where:

$$C_m = C_{dg} - C_{gd} \tag{6.110}$$

$$C_{mb} = C_{db} - C_{bd} \tag{6.111}$$

$$C_{mx} = C_{bg} - C_{gb} \tag{6.112}$$

Alternatively, the matrix in Equation (6.105) can also be realized in the circuit form shown in Figure 6.9.

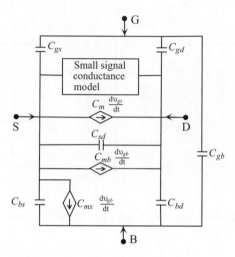

Figure 6.9   High frequency small signal MOSFET alternative equivalent circuit

6.4.6 Medium Frequency Capacitance Model

In the previous section we derived the generalized high frequency small signal equivalent circuit within the range of validity of quasistatic approximation. The general equivalent circuit can be simplified significantly for low and medium frequency applications under valid approximations.

Medium frequency operation will be defined by the condition that the transport current between the drain and source dominates the charging current, as the rate of change of voltage with time is relatively small. It implies that in Figure 6.9 we can introduce the following approximations:

$$g_m v_{gs} \gg C_m \frac{dV_{GS}}{dt} \tag{6.113}$$

$$g_{sd} v_{sd} \gg C_{sd} \frac{dV_{SD}}{dt} \tag{6.114}$$

$$g_{gmb} v_{bs} \gg C_{mb} \frac{dV_{BS}}{dt} \tag{6.115}$$

The above simplification leads to a simplified small signal model for the mid frequency range shown in Figure 6.10. This simplification, which ignores $C_m dV_{gs}/dt$, C_{sd}, and $C_{mb} dV_{bs}/dt$, does not lead to any error in computation of $i_{\delta g}$ and $i_{\delta b}$, as they are primarily governed by C_{gs}, C_{gd}, C_{gb}, and C_{bs}, C_{gb}, and C_{bd} respectively. The errors introduced in $i_{\delta s}$ and $i_{\delta d}$ are within acceptable limits in the medium frequency range.

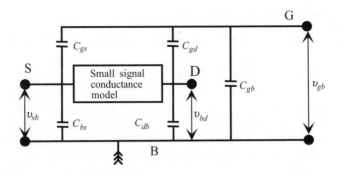

Figure 6.10 Medium frequency small signal MOS transistor equivalent circuit

6.4.7 EKV Capacitance Model

A compact expression for the capacitance in medium frequency operation can be obtained in terms of terminal voltages by differentiation of the terminal charge. The gate-to-bulk capacitances will be:

$$C_{gs} = \frac{2}{3}\left[1 - \frac{x_r^2 + x_r + 0.5x_f}{(x_f + x_r)^2}\right] \tag{6.116a}$$

$$C_{gd} = \frac{2}{3}\left[1 - \frac{x_f^2 + x_f + 0.5x_r}{(x_f + x_r)^2}\right] \tag{6.116b}$$

$$C_{gb} = \left(\frac{n_q - 1}{n_q}\right)(1 - C_{gs} - C_{gd}) \tag{6.116c}$$

$$C_{sb} = (n_q - 1)C_{gs} \tag{6.116d}$$

$$C_{db} = (n_q - 1)C_{gd} \tag{6.116e}$$

Figure 6.11 shows the variation of small signal capacitances with V_{GB} and V_{DB}.

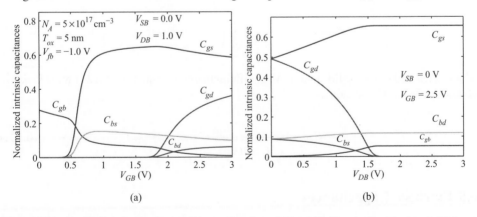

(a)   (b)

Figure 6.11 Small signal capacitances (a) as a function of V_{GB} with V_{DB} as a parameter, and (b) as a function of V_{DB} with V_{GB} as a parameter

6.4.8 Meyer's Simplified Capacitance Model

The inversion charge density in the strong inversion region in a simple form is:

$$Q'_n = -C'_{ox}(V_{GS} - V_{TS} - V_C) \tag{6.117}$$

and the drain current is $I_{DS} = -\mu_{eff}C_{ox}(W/L)(V_{GS} - V_{TS} - V_C)\,dV_C/dy$. This allows us to express dy as $dy = -\mu_{eff}C'_{ox}W(V_{GS} - V_{TS} - V_C)\,dV/I_{DS}$. The simplified expression for total inversion charge density in strong inversion is obtained by integrating Equation (6.117) from the source to drain, with a change of variable from y to V_C. Thus:

$$Q_n = -\frac{2}{3}C_{ox}\frac{(V_{GS} - V_{TS})^3 - (V_{GD} - V_{TS})^3}{(V_{GS} - V_{TS})^2 - (V_{GD} - V_T)^2} \tag{6.118}$$

In Meyer's model [2, 4] the gate charge is assumed to be equal to the inversion charge in strong inversion ($Q_g + Q_n = 0$), and the bulk depletion charge is neglected ($Q_b = 0$). This condition also implies ($C_{gb} = 0$).

Hence, in strong inversion, in the linear region, the capacitances are given by:

$$C_{gs} = \frac{2}{3}C_{ox}\left\{1 - \left[\frac{V_{GS} - V_{TS} - V_{DS}}{2(V_{GS} - V_{TS}) - V_{DS}}\right]^2\right\} \tag{6.119}$$

$$C_{gd} = \frac{2}{3}C_{ox}\left\{1 - \left[\frac{V_{GS} - V_{TS}}{2(V_{GS} - V_{TS}) - V_{DS}}\right]^2\right\} \tag{6.120}$$

In saturation, $V_{DS} = V_{GS} - V_{TS}$, which gives $C_{gs} = \frac{2}{3}C_{ox}$, and $C_{gd} = 0$.

In weak inversion we have $Q_g = -Q_b$ and $Q_n = 0$, which gives $C_{gs} = C_{gd} = 0$, and:

$$Q_b = -\gamma C_{ox}\sqrt{\phi_s} = -\gamma C_{ox}\left(\sqrt{V_{GB} - V_{fb} + \frac{\gamma^2}{4}} - \frac{\gamma}{2}\right) \tag{6.121}$$

which is derived in the BSIM model. After differentiation of Equation (6.121) w.r.t. V_{GB}, we obtain C_{gb} in weak inversion as:

$$C_{gb} = \frac{C_{ox}}{\sqrt{1 + 4(V_{GB} - V_{fb})/\gamma^2}} \tag{6.122}$$

6.5 Extrinsic Capacitances

The small signal capacitance model given in the previous section is incomplete unless the extrinsic components, invariably associated with the real device, are included. The extrinsic capacitance components of the MOS transistor equivalent circuit are primarily constituted of the source-to-bulk and drain-to-bulk reverse bias capacitances, overlap and fringe capacitances. These capacitances, at a given operating bias condition, are required to be estimated.

6.5.1 Junction Capacitance

In a MOS transistor the highly doped source and drain regions form a one-sided abrupt junction with the substrate, due to which it can be assumed that the depletion width between source/drain and the substrate predominantly penetrates in the substrate region. These junctions are usually reverse biased in normal MOSFET operation. Such a reverse biased *p–n* junction has a capacitance due to a stored depletion charge. This junction capacitance defined by the depletion width is bias dependent. As the source/drain forms a junction with the substrate in three spatial directions, namely, length, width, and depth, for estimating the value of source and drain junction capacitances their sidewall and bottom components have to be estimated. Figure 6.12 shows the sidewall and bottom junctions with level of doping across the junction.

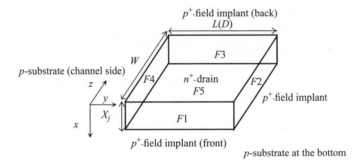

Figure 6.12   Sidewall and bottom drain–substrate junctions

It is usual practice to take an average constant doping concentration for the heavily doped source and drain diffused regions. In reality, there is a concentration gradient from the silicon surface to the bottom part of the junction. Further, on the substrate side, the bottom and peripheral parts of the junction may see different concentrations because the sidewall junctions interface with the field implanted doping in the substrate. The field implant is not seen by the channel side of the sidewall junction and the bottom planar junction. Table 6.1 presents a consolidated view of the constituents of the source/drain junction capacitances along with doping environments across the faces of the junction.

In Table 6.1, p_{fi} denotes the field implanted region of the substrate, which has higher doping compared to the substrate doping, and serves the purpose of increasing the threshold voltage under field oxide (FOX). The expressions for the bottom and sidewall components of source and drain can be written by using the simple *p–n* junction theory as discussed in Chapter 1. The bottom capacitance is expressed as:

$$C_{jb(S,D)} = A\frac{C_{jb0}}{\left(1 + \frac{V_{(S,D)B}}{\phi_{bi}}\right)^{m_b}} \tag{6.123}$$

Table 6.1 Bottom and sidewall capacitances of source–drain junctions

Capacitance type	Face label	Junction type	Capacitance label
Sidewall	$F1$	n^+-p_{fi}	C_{jp1}
Sidewall	$F2$	n^+-p_{fi}	C_{jp2}
Sidewall	$F3$	n^+-p_{fi}	C_{jp3}
Sidewall	$F4$	n^+-p	C_{jp4}
Bottom	$F5$	n^+-p	C_{jb}

where $A = WL_d$ is the bottom area labelled $F5$, m_b is the grading coefficient character-izing the nature of transition of the n^+-p layer at the bottom part of the junction, ϕ_{bi} is the zero bias junction potential, and C_{jb0} is the zero bias junction capacitance per unit area.

As the junction depth X_j is constant for a given technology it is convenient to express the sidewall capacitance in terms of per unit length, which is given as

$$C_{jsw(S,D)} = P \frac{C_{jsw0}}{\left(1 + \frac{V_{(S,D)B}}{\Phi_{bsw}}\right)^{m_{sw}}} \tag{6.124}$$

where C_{jsw0} is the zero bias sidewall capacitance per unit length, m_{sw} is the gradient coefficient for the sidewall, and $P = 2L_y + 2W$ is the perimeter of the junction.

The dependence of junction capacitance on substrate doping concentration is shown in Figure 6.13 with junction bias as a parameter. The junction capacitance is plotted per unit area

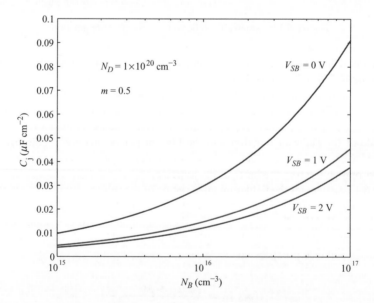

Figure 6.13 Junction capacitance variation as a function of N_B with V_{SB} as a parameter

for an abrupt junction ($m = 0.5$). The plot indicates that for a scaled transistor the supply voltage being small, and doping concentration large, the junction capacitance increases compared to a long transistor.

6.5.2 Overlap Capacitances

A MOS transistor is fabricated with self-aligned technology where the polysilicon gate defines the edges of the source/drain junction defining the channel. High temperature thermal cycles involved during the fabrication of the device invariably introduce overlaps of the polysilicon gate over source/drain in the direction of channel length. The penetration of source and drain under the gate, L_D, is called the lateral diffusion source/drain extension under the gate, as shown in Figure 6.14.

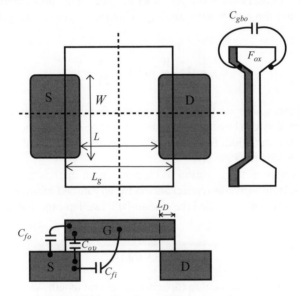

Figure 6.14   Overlap and fringe capacitances

The overlap capacitance C_{ov} is obtained using the classical formula of a parallel plate capacitor, and is given by:

$$C_{ov} = \frac{\epsilon_{ox}}{T_{ox}}(WL_D) \tag{6.125}$$

The above equation is valid for V_{GD} such that the drain/source overlapped region is in accumulation. In the case of the gate-to-drain bias, for which there is a depletion region in the drain, the overlap capacitance will be bias dependent [56]. The vertical side of the polysilicon gate acts as a conducting plane, and the source as another conducting plane, which are perpendicular to each other with oxide separating them. This results in an additional outer *fringe capacitance.* Such a capacitance is calculated using a conformal mapping technique with appropriate boundary conditions [55], and is given as:

$$C_{fo} = \epsilon_{ox} \frac{W}{\theta} \ln \left(1 + \frac{T_p}{T_{ox}} \right) \qquad (6.126)$$

where $\theta = \frac{\pi}{2}$ for the vertical edge of the polysilicon gate, and T_p is the thickness of the polysilicon gate. Similar to the outer fringe, there exists an inner fringe capacitance as well between the bottom conducting plane of the polysilicon gate and the vertical sides of the source and drain junctions adjacent to the channel, with the depleted channel as the insulating medium between the electrodes in the absence of inversion. The internal fringe capacitances between the gate and source/drain terminals are also shown in Figure 6.14, and exist only in the absence of an inversion layer or accumulation layer under the gate, in the presence of which shielding takes place. Applying the conformal mapping technique as before, we get the inner fringe capacitance as:

$$C_{fi} = 2 \frac{\epsilon_{si} W}{\pi} \ln \left(1 + \frac{X_j}{2T_{ox}} \right) \qquad (6.127)$$

The above capacitances, as seen from the respective formulae, are directly proportional to width, and therefore, they can be lumped together and expressed as capacitance per unit width. These capacitance per unit width form the model parameters C_{GSO} and C_{GDO} for the source and drain overlap regions respectively.

Similar to the overlap in the direction of channel length there is an overlap in the width direction, which results in an overlap capacitance between gate and substrate as can be seen from Figure 6.14. This capacitance is called the gate-to-bulk overlap capacitance, and is expressed as:

$$C_{gbo} = C_{GBO} L \qquad (6.128)$$

where C_{GBO} is the gate-to-bulk overlap capacitance per unit length. Various types of extrinsic capacitive components and their physical origins are summarized in Table 6.2.

Table 6.2 Various extrinsic capacitive components

Capacitive component	Physical origin
C_{gso} and C_{gdo}	Gate-to-source and gate-to-drain overlap capacitances because of lateral diffusion of the source and drain boundaries under the gate during device fabrication
C_{gbo}	Gate-to-bulk overlap capacitance due to extension of the gate over the field oxide
C_{fo} and C_{fi}	Outer and inner fringe capacitances between gate and source/drain
$C_{jsb}, C_{jdb} = (C_{jb} + C_{jsw})_{(S,D)}$	Source/drain-to-substrate capacitances comprising aerial and peripheral components

6.6 Non-quasistatic (NQS) Models

6.6.1 Non-quasistatic Large Signal Models

If the MOSFET is driven by an input signal with a very short rise time (less than the transit time of carriers through the channel), or very high frequency, the assumption that the channel charge is only a function of terminal voltages does not hold good. The channel charge is to be expressed in the following form to account for the non-quasistatic carrier build up in the channel of the MOS transistor [8]:

$$Q_n(t) = Q[t, V_j(t)] \qquad (6.129)$$

where $j = S, D, G$, and B. The current transport and continuity equations are rewritten for the quasistatic condition for ready reference:

$$I_c(y, t) = W\mu_n Q_n(y, t)\frac{\partial V_C(y, t)}{\partial y} \qquad (6.130)$$

$$W\frac{\partial Q_n(y, t)}{\partial t} = \frac{\partial I_c(y, t)}{\partial y} \qquad (6.131)$$

The non-quasistatic effect for circuit simulation may further be classified into two categories, namely (i) large signal and (ii) small signal, depending on the application domain.

The NQS effect has been adapted in compact MOSFET models in a manner consistent with the framework of the respective model. We shall briefly give an overview of the representative categories.

6.6.1.1 BSIM3v3

The MOSFET is divided into n sub-MOSFETs of smaller channel length. The RC network representing the distributed channel is now replaced by an *Elmore lumped equivalent circuit that preserves the lowest frequency pole of the distributed channel* [7]. The Elmore resistance is given by:

$$R_{Elmore} = \frac{L_{eff}}{E_{ELM}\epsilon_f \mu_{eff} W_{eff} Q_{ch}} \qquad (6.132)$$

where ϵ_f is a fitting parameter. Here, $Q_{ch} = C'_{ox}(V_{gs} - V_{th})$, is the inversion charge density.

Figure 6.15 illustrates the above approach. The distributed RC representation of the channel, where the original MOSFET has been replaced by a series of sub-MOSFETs of smaller channel length, is such that for the reduced length the quasistatic condition may be assumed to hold good at the highest frequency of operation. The value of R_{ELM} is obtained by matching the simulated small signal and time response of the fast rise time of the distributed and lumped structures of the MOSFET. This approach has a price in terms of computational effort because

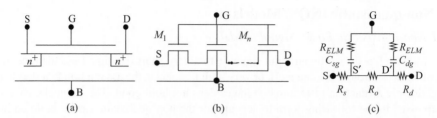

Figure 6.15 (a) Primary MOSFET, (b) segmented and distributed sub-MOSFETs, and (c) lumped equivalent circuit with an Elmore resistance

it introduces additional nodes in the netlist describing the MOSFET. Therefore, an alternative computationally simpler approach is adopted in BSIM3v3 which is as follows:

- As already described, the total transient currents, $I_X(t)$, are constituted of transport component, $I_{TX}(t)$, and charging component $X_{xpart}dQ_{ch}(t)/dt$. X stands for the terminal designation G, D, or S, and X_{xpart} denotes the charge partition ratio assigned to terminal X as per the partition scheme. The following relationships hold good:

$$D_{xpart} + S_{xpart} = 1 \tag{6.133}$$

$$G_{xpart} = -1 \tag{6.134}$$

- Whereas the channel potential responds to the terminal voltages instantaneously, the charge built up takes a finite time. Hence, $Q_{cheq}(t) \neq Q_{ch}(t)$. In other words, with reference to the quasistatic value of the charge, at any given instant, in the non-quasistatic condition there is a deficit of channel charge. We can call it as the *differential charge*, $\Delta Q_{ch} = Q_{cheq} - Q_{ch}(t)$. The time dependence of the differential charge may be described by the following equation:

$$\frac{d\Delta Q_{ch}}{dt} = \frac{dQ_{cheq}}{dt} - \frac{dQ_{ch}}{dt} \tag{6.135}$$

In BSIM3v3 the differential charge ΔQ_{ch} is labelled as Q_{def}.
- The charging current dQ_{ch}/dt can then be equated to $\Delta Q_{ch}/dt$. Thus, Equation (6.135) takes the form:

$$\frac{d\Delta Q_{ch}}{dt} = \frac{dQ_{cheq}}{dt} - \frac{\Delta Q_{ch}}{dt} \tag{6.136}$$

The term $dQ_{ch}(t)/dt$, representing the rate of change of the mobile charge in the channel during the transient state, may be attributed to the delay in charging the channel through the distributed resistance–capacitance network constituting the channel. If τ is the time constant associated with the delay, the charging current dQ_{ch}/dt can then be equated to $\Delta Q_{ch}/\tau$. Thus, Equation (6.135) takes the form:

$$\frac{d\Delta Q_{ch}}{dt} = \frac{dQ_{cheq}}{dt} - \frac{\Delta Q_{ch}}{\tau} \tag{6.137}$$

- The relaxation time, τ, for a unified expression valid for both strong and weak inversion modes of operation, is given as a combination of the drift and diffusion mechanism:

$$\frac{1}{\tau} = \frac{1}{\tau_{diff}} + \frac{1}{\tau_{dr}} \tag{6.138}$$

$$\tau_{diff} = \frac{\left(\frac{L_{eff}}{4}\right)^2}{\phi_t \mu} \tag{6.139}$$

$$\tau_{dr} = \frac{1}{2} R_{ELM} C'_{ox} W_{eff} L_{eff} \tag{6.140}$$

In BSIM3v3, the solution of Equation (6.137) representing the non-quasistatic current is obtained with the aid of a sub-circuit, shown in Figure 6.16, which is then added to the DC current. The NQS model is activated with the command **nqsMod =1**.

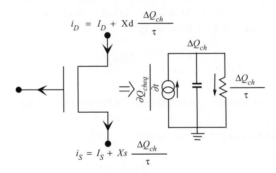

Figure 6.16   BSIM sub-circuit for NQS transient simulation

6.6.1.2 HiSIM Model

The NQS model in HiSIM is based on the physical mechanism of build up of carriers in the channel of a MOSFET considering the carrier transit delay [9, 10]. It is assumed that the potential is established instantaneously, unlike the channel charge, which requires a finite time to establish itself. The NQS charging mechanism can be visualized through the process illustrated through Figure 6.17 (adapted from Figure 4 of [10]):

- Assume that $Q_c(t_n)$ is the total steady state and *quasistatic* inversion charge in the channel between source and drain corresponding to the instant t_n. The channel charge is represented by the area ABDE. The quasistatic charge distribution profile along the channel length for the voltage $V(t_n)$, $Q_n(y, V(t_n))$, assumed linear, is shown in Figure 6.17(a). Due to the delay in charging the channel, the total inversion charge at the instant t_n, will not be able to attain the quasistatic value $Q_c(t_n)$ and has a value $q_c(t_n)$, shown in Figure 6.17(d) represented by the area ABFG. The corresponding charge distribution is represented by $q_n(y, t_n)$.
- The dynamic stored charge at the instant t_{n-1} is $q_c(t_{n-1})$, and is represented by the area ABC in Figure 6.17(b).

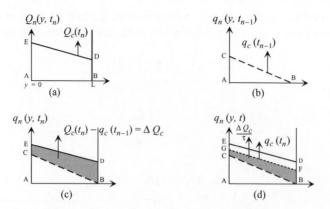

Figure 6.17   HiSIM transit delay based NQS model

- During the time interval $\Delta t = (t_n - t_{n-1})$, the charge build up takes place at the rate $(Q_c(t_n) - q_c(t_{n-1}))/\tau$. Thus, in an interval Δt, the channel charge at the instant t_n with respect to that at the instant t_{n-1} is related by:

$$q_c(t_n) = q_c(t_{n-1}) + \frac{\Delta t}{\tau} \left[Q_c(t_n) - q_c(t_{n-1}) \right] \qquad (6.141)$$

The charge build up is represented by the area CBFG which is added to the area ABC to give the charge at t_n equal to the area ABFG shown in Figure 6.17(d). The dynamic charging equation can be written in the following form:

$$\frac{dq_c}{dt} = \frac{(Q_c - q_c)}{\tau} \qquad (6.142)$$

In other words, the rate of change of channel charge in the non-quasistatic process is directly proportional to the charge deficit between the quasistatic value and instantaneous value, and is inversely related to the carrier transit delay mechanism determining τ. In HiSIM two delay mechanisms are assumed to be involved:

1. The delay due to the diffusion transit-delay mechanism τ_{diff} that determines the propagation time of the channel charge wavefront from source to drain to form the channel. The expression for diffusion delay is:

$$\tau_{diff} = \frac{L^2}{\phi_t} \mu \qquad (6.143)$$

2. The conduction transit-delay, τ_{cond}, which is the transit time of the carriers when the channel is formed as described in Chapter 4 (Equation (4.113)).

$$\tau_{cond} = \frac{|Q_n|}{I_{DS}} \qquad (6.144)$$

The total delay is obtained from:

$$\tau^{-1} = \tau_{diff}^{-1} + \tau_{cond}^{-1} \qquad (6.145)$$

The computational process involves storage of the channel charge at the instant t_{n-1} and addition of the second term in Equation (6.141) to this stored value to obtain the charge at the instant t_n. The process is believed to be computationally more efficient than simulation through sub-circuits.

6.6.1.3 PSP Model

The PSP model is based on the solution of the continuity equation, which is the partial differential equation describing the transient process of current build up by replacing it with a set of ordinary differential equations using spline collocation method. Using a sub-circuit approach both large and small signal NQS problems have been solved [6].

Substituting the drift diffusion current expression with the charge sheet approximation in the continuity equation (Equation (6.131)), we have:

$$\frac{\partial q_n}{\partial t} + \frac{\partial}{\partial y}\left[\mu_{eff}\left(\frac{q_n}{dQ_n/d\phi_s} - \phi_t\right)\frac{\partial q_i}{\partial y}\right] = 0 \qquad (6.146)$$

where $q_n = Q'_n/C'_{ox}$. In the PSP model Equation (6.146) is reduced to several ordinary differential equations using the method of collocation version of weighted residuals. The resulting set of ordinary differential equations can be solved directly by a built-in numerical algorithm in the SPICE circuit simulator or by the sub-circuits approach. Solution of these ordinary differential equations gives the approximate inversion charge density at the collocation points within the channel. The accuracy of this method can be increased by increasing the number of collocation points. It has been shown that three collocation points are sufficient for most practical purposes [6, 43]. Using the cubic spline function the inversion charges at the collocation points are interpolated to give an approximate and smooth distribution of inversion charge density $Q'_n(y, t)$. The two point spline collocation based method is discussed in Appendix C.

6.6.2 Non-quasistatic Small Signal Models

As the frequency of the applied signal increases, there is a delay in the channel charge response to the applied signal. The small signal equivalent circuit based on the quasistatic approach is no longer valid if the inversion charge does not follow the applied signal instantaneously. By studying the amplitude and phase of the inversion charge variation with respect to the low frequency value, the criteria for onset of a non-quasistatic frequency of operation can be established [11].

6.6.2.1 EKV Model

A first order small signal non-quasistatic model is built through the following procedure [29, 30]:

- The MOSFET channel is considered as a distributed non-uniform transmission line as shown in Figure 6.18. Figure 6.19 represents a first order EKV non-quasistatic small signal equivalent circuit of a MOS transistor with frequency dependent conductances.

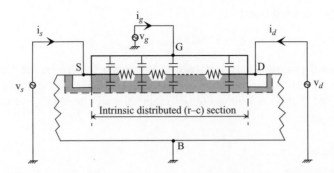

Figure 6.18 Distributed transmission line circuit representation of a MOSFET channel at high frequency

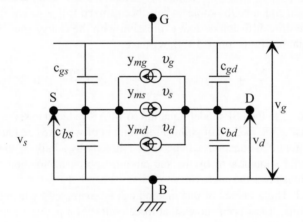

Figure 6.19 EKV non-quasistatic small signal transadmittance model

- Divide the MOSFET into *elemental* MOS transistors with small, elemental channel lengths for which the quasistatic approximation holds good. The gates of elemental transistors are connected together to constitute the original transistor. This enables the use of a quasistatic equivalent circuit valid for all regions of operation for the elemental transistors.
- The first order non-quasistatic small signal circuit can now be constructed directly from the mid-frequency equivalent circuit by replacing the transadmittance terms with terms having low pass frequency dependence, which are given as:

$$Y_{mg} = \left. \frac{i_d}{v_g} \right|_{(v_d, v_s = 0)} = \frac{g_{mg}}{1 + s\tau} \qquad (6.147)$$

$$Y_{ms} = \left. \frac{i_d}{v_s} \right|_{(v_g, v_d = 0)} = \frac{g_{ms}}{1 + s\tau} \qquad (6.148)$$

$$Y_{md} = \left. \frac{i_d}{v_d} \right|_{(v_s, v_d = 0)} = \frac{g_{md}}{1 + s\tau} \qquad (6.149)$$

The time constant corresponding to the elemental section is:

$$\tau = \int_0^L R(x)C''(x)\,dx \tag{6.150}$$

where $R(x)$ corresponds to the value of unit elemental resistance and $C''(x)$ stands for the total capacitance of the elemental section w.r.t. ground. $C''(x)$ is position dependent along the channel, as the distributed line is bias dependent, and it has the unit of capacitance per unit length.

For weak inversion:

$$\tau_0 = \frac{L_{eff}^2}{2\mu_n\phi_t} \tag{6.151}$$

For strong inversion:

$$\tau = \frac{4}{15}\frac{\tau_0}{\sqrt{i_f}} \qquad\qquad \text{saturation} \tag{6.152}$$

$$\tau = \frac{1}{6}\frac{\tau_0}{\sqrt{i_f}} \qquad\qquad \text{linear region} \tag{6.153}$$

6.6.2.2 BSIM Model

For small signal quasistatic simulation, the Elmore equivalent circuit can as well be adapted by transforming the time dependent differential charge expression ΔQ_{cheq} given in Equation (6.135) to the frequency domain as given below [12]:

$$\Delta Q_{ch}(t) = \frac{\Delta Q_{cheq}(t)}{(1+jw\tau)} \tag{6.154}$$

The transformation results in rendering the transcapacitance and conductance terms into complex quantities, such as:

$$G_m = \frac{G_{m0}}{(1+\omega^2\tau^2)} - j\frac{(\omega G_{m0})\tau}{(1+\omega^2\tau^2)} \tag{6.155}$$

and:

$$C_{XY} = \frac{C_{XY0}}{(1+\omega^2\tau^2)} - j\frac{(C_{XY0})\tau}{(1+\omega^2\tau^2)} \tag{6.156}$$

where G_{m0} and C_{XY0} are respectively, DC transconductance and transcapacitance values.

Recognizing that in RF circuit analysis, gate and substrate resistance play important parts in determining the NQS characteristics, the BSIM4 RF circuit model provides options for model selection. The gate resistance has two components (i) R_{geltd} which is bias independent and defined by parameters such as R_{SHG}, N_{GCON}, X_{GW}, X_{GL}, and N_F corresponding respectively to gate sheet resistance, number of gate contacts, separation between gate contact and channel edge, lithography variability related gate length offset, number of finger pairs in the gate, and (ii) intrinsic-input resistance, R_{ii}, which is bias dependent.

1. **rgateMod = 0**:   represents a conventional MOSFET without any gate resistance.
2. **rgateMod = 1**:   represents the model with a constant, bias independent resistance, R_{geltd} at the gate input.
3. **rgateMod = 2**:   represents the sum of two components of gate resistances ($R_{ii} + R_{geltd}$), which, however, does not add any additional node.
4. **rgateMod = 3**:   represents a gate network from the junction nodes of R_{geltd} and R_{ii}, C_{gso} and C_{gdo} which are connected to the source and drain terminals respectively. This means that from the junction node between the two resistors the charging current gets divided between the resistive and capacitive paths, as shown in Figure 6.20(a).

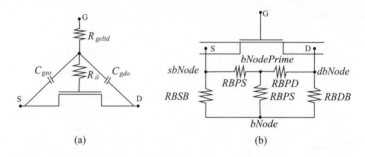

Figure 6.20   RF lumped gate and substrate resistor based equivalent circuit

Similarly, for the substrate resistance, BSIM4 provides options for the substrate resistance model selection through the **rbodyMod** model selector.

1. **rbodyMod = 0**:   default option with no substrate resistance.
2. **rbodyMod = 1**:   The option selects five resistor network RBPSs with parameters such as RBPS, RBPD, RBSB, RBDB, and RBPS. The network is illustrated in Figure 6.20(b).

6.7 Noise Models

Noise in a MOS transistor is caused by small fluctuations in current and voltages, which are attributed to phenomena generated within the device. The noise sets the lower limit of the signal that can be amplified by the device. Noise signal is not periodic but random. Therefore, noise is measured in terms of average power. *Power Spectral Density*, PSD, expressed by the term $S_x(f)$, indicates the average power delivered by the signal, x, within 1 Hz bandwidth around the frequency f and, therefore, is the metric of the power that the noise signal carries at a given frequency [14].

For MOSFETs the noise is broadly classified into the following categories [13, 15]:

1. **Thermal noise.** The noise is attributed to the thermal motion of charge carriers when the inversion layer is formed. Considering the channel in a MOSFET as a sheet of resistor, the noise signal is expressed in voltage and current form as:

$$\overline{v^2} = 4kTR\Delta f \tag{6.157}$$

$$\overline{i^2} = \frac{4kT\Delta f}{R} \tag{6.158}$$

The expression for thermal noise Power Spectral Density (PSD) for a MOS transistor is obtained through the following steps:

(a) Elementary noise voltage source. Consider an infinitesimally small section of the noiseless channel of the MOS transistor of length dy. Let the resistance of this elemental section be dR. Assume that the resistance produces an incremental thermal noise voltage ΔV_{ch} across it. For a channel current I_{DS}, the thermal noise voltage dV_{ch} is:

$$dV_C = I_{DS}\,dR = W\mu_n(-Q_n')\frac{dV_C}{dy}\,dR \tag{6.159}$$

or:

$$dR = \frac{dy}{W\mu_n(-Q_n')} \tag{6.160}$$

The power spectral density for the elemental noise voltage is given by:

$$dS_{v_C} = 4kT\,dR \tag{6.161}$$

(b) Elemental noise current PSD. From the elemental noise voltage source the corresponding elemental noise current PSD is obtained using the following relation:

$$dS_{i_D} = g_C^2\,dS_{v_C} \tag{6.162}$$

(c) Computation of conductance for the elementary channel segment:

$$g_C = \frac{dI_{DS}}{dV_C} = \frac{d}{dV_C}\left[\mu_n C_{ox}'\frac{W}{L}\int \frac{-Q_n'(V_C)}{C_{ox}'}\,dV_C\right] \tag{6.163}$$

$$= \mu_n\frac{W}{L}(-Q_n') \tag{6.164}$$

Substitute g_C using Equation (6.164), and dS_{v_C} from Equation (6.161), into Equation (6.162) to get:

$$dS_{i_D} = 4kT\frac{\mu_n}{L^2}W(-Q_n')\,dy \tag{6.165}$$

Integrating dS_{iD}, the elementary thermal noise source, along the channel from source to drain gives the total noise current spectral density, S_{iD} for the drain current:

$$S_{iD} = 4kT \frac{\mu_n W}{L^2} \int_0^L -Q'_n \, dy \qquad (6.166)$$

The integral term in Equation (6.166) multiplied by W represents the total inversion charge under the gate, Q_n. Therefore, the final expression for the current spectral density is as follows:

$$S_{iD} = 4kT \frac{\mu_n}{L^2} Q_n \qquad (6.167)$$

The thermal noise power spectral density is also expressed in the following alternative form that is commonly used for MOSFET thermal noise modeling.

$$S_{iD} = \frac{4kT}{L_{eff}^2 I_{DS}} \int g^2(V_C) \, dV_C \qquad (6.168)$$

where, $g(V_C)$ denotes the *local specific channel conductance*. The above expression is generic and is known as the Klaassen–Prins equation for thermal noise [37]. It may, however, be mentioned that the above equation is valid only for long channel MOSFETs where the mobility dependence on channel lateral electric field need not be taken into consideration [38].

2. **Flicker or $1/f$ noise.** This is associated with current flow in the device, and the spectral density has the following characteristic form:

$$\overline{i^2} = K \frac{I_{DS}^a}{f^b} \Delta f \qquad (6.169)$$

The symbols have the following meaning: I_{DS} is drain–source current, a, b, and K are parameters and Δf is a small bandwidth around the frequency f.

In order to understand the origin of flicker noise let us have a look at the basic expression for conductivity due to drift current:

$$\frac{j}{F} = \sigma = qn\mu \qquad (6.170)$$

where j is the current density, F is the electric field and μ is the carrier mobility. It is obvious from this equation that a fluctuation in the current density will be caused by perturbations in the carrier concentration and/or mobility of carriers. Therefore, as will be discussed later, $1/f$ noise has been associated with number fluctuation ΔN (ΔN noise) or/and mobility fluctuation $\Delta \mu$ ($\Delta \mu$ noise) models.

At low frequencies, $1/f$ noise is the dominant source of noise in MOS devices, which has a spectrum with a slope that varies between -0.7 to -1.3 on a double log-plot. Due to 'up conversion', $1/f$ noise has a serious impact on RF CMOS circuits, where it causes a significant increase in the phase noise. Therefore, a good $1/f$ noise model is an important ingredient of an RF design kit. Many different theories have been proposed to explain the physical origin of $1/f$ noise in MOSFETs [18, 24]. These can be categorized into three major types.

According to the carrier number fluctuation (ΔN) model of McWorther, $1/f$ noise is attributed to charge carrier trapping and a detrapping mechanism through traps located within the gate oxide. Every single trap leads to a Lorentzian noise power spectrum. In the case of a uniform spatial trap distribution, the spectrum is to be integrated to give the resultant $1/f$ spectrum. The carrier number fluctuation theory has been observed to successfully account for a $1/f$ noise spectrum in n-channel devices. The characteristic feature is that input-referred $1/f$ noise, defined by:

$$S_{vG} = \frac{S_{iD}}{g_m^2} \tag{6.171}$$

is almost independent of V_{GS}. The symbol S_{iD} is the drain current noise spectral density, and g_m, the transconductance at the bias point [18].

The mobility fluctuation ($\Delta\mu$) theory, on the other hand, considers the flicker noise as a result of the fluctuation in bulk mobility. It is based on Hooge's empirical relation for the spectral density of flicker noise in a homogeneous sample:

$$\frac{S_I}{I^2} = \frac{\alpha_H}{fN_{total}} \tag{6.172}$$

where I is the mean current flowing through the sample, S_I is the spectral density of the noise in the current, N_{total} is the total number of free carriers in the sample, and α_H, known as Hooge's parameter, is an empirical constant with a value of about 2×10^{-3} [1]. When the above expression is directly applied to MOSFETs, it predicts an input referred noise power, which is proportional to $(V_{GS} - V_T)C_{ox}^{-1}$ [19]. The Hooge model, attributes the $1/f$ noise to bulk mobility fluctuations caused by phonon scattering. In contrast to the carrier number fluctuation theory, the Hooge model is more successful in describing the observed $1/f$ noise in p-channel devices, where the input-referred $1/f$ noise is found to be strongly dependent on V_{GS} [18].

The unified $1/f$ noise model [19, 20] combines the above two models to give a unified $1/f$ model. The mobility fluctuations are induced by the Coulomb scattering of free carriers from trapped interface charges in the oxide as against those attributed to phonon scattering assumed in the earlier mobility fluctuation models. Both fluctuations in transporting carrier number and perturbation in mobility have their physical origins attributed to trapped charges in the oxide. Thus, noise generated due to the two mechanisms are *correlated*. The unified model shows good fittings with experimental $1/f$ noise results [22].

The unified noise model starts with the following equation for the drain current in the linear region:

$$I_{DS} = W\mu_{eff}qn_nF_y \tag{6.173}$$

where W is the channel width, μ_{eff} the effective mobility, q the electron charge, n_n the inversion carrier density, and F_y the electrical field along the y-axis between the source and drain. Fluctuations in the local drain current are given by [19]:

$$\frac{\delta I_{DS}}{I_{DS}} = -\left(\frac{1}{\Delta N} \frac{\delta \Delta N}{\delta \Delta N_t} \pm \frac{1}{\mu_{neff}} \frac{\delta \mu_{neff}}{\delta \Delta N_t} \right) \delta \Delta N_t \qquad (6.174)$$

where N is the number of carriers in the channel per unit area, and N_t is the number of occupied traps per unit area. The total current noise power spectral density, on integrating from source to drain, becomes:

$$S_{iD} = \frac{kTI_{DS}^2}{\gamma fWL} \left(\frac{1}{N_n} + \frac{\mu_{neff}}{\mu_{co}\sqrt{N_n}} \right)^2 N_t(E_{F_n}) \qquad (6.175)$$

where f is the frequency, T the absolute temperature, k the Boltzmann's constant, γ the attenuation coefficient of the electron wave function in the oxide (in cm), N_n the inversion carrier density, μ_{co} the Coulomb scattering related constant (which has the dimension of mobility), and $N_t(E_{F_n})$ the effective oxide trap density at the quasi-Fermi level (in cm). However, in the SPICE implementation of the BSIM3v3 unified $1/f$ noise model, three additional parameters are introduced in the expression of $N_t(E_{F_n})$ to fit the measurement results [23].

3. **Generation–recombination noise.** The generation–recombination noise originates due to trapping of carriers from the channel into the traps in the oxide closely located near the interface, and subsequent detrapping of them due to thermal energy. The noise spectral density model is based on the assumption that (i) traps are neutral when empty, (ii) capture and emission processes are mutually exclusive events, and (iii) adjacent traps are electrostatically isolated. The capture–emission process results in the fluctuation in the number of carriers constituting the current from source-to-drain, which is known as the *Random Telegraph Signal (RTS) noise* [21]. A schematic of RTS noise is shown in Figure 6.21. The

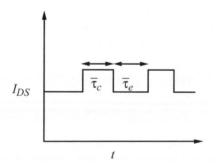

Figure 6.21   RTS noise schematic diagram

capture of electrons from the inversion channel by the traps in the oxide results in a decrease in drain current, and emission of electrons increases the current. The capture and emission are stochastic processes, and therefore, they are characterized by average values of $\overline{\tau}_{(c,e)}$, the mean time of capture and emission respectively. The RTS amplitude may be expressed by the relative change in the drain current which is expressed by [17]:

$$\frac{\delta I_{DS}}{I_{DS}} = \alpha \frac{g_m}{I_{DS}} \frac{q}{WLC'_{ox}} \left(1 - \frac{x_{tr}}{T_{ox}}\right) \qquad (6.176)$$

The symbols have the following meanings: x_{tr} is the trap distance in the oxide referred to the Si–SiO$_2$ interface, and α is a fitting parameter. It is observed that the RTS amplitude is bias dependent through the conductance term which depends on the operating condition. It is also seen that the further the trap location is from the interface, the weaker is its influence on noise amplitude.

The expression for the PSD of RTS noise shows a frequency dependence, and has the following form [16]:

$$S_{iD}(f) = \frac{4(\Delta I_{DS})^2}{(\overline{\tau}_c + \overline{\tau}_e)\left[(\frac{1}{\overline{\tau}_c} + \frac{1}{\overline{\tau}_e})^2 + (2\pi f)^2\right]} \qquad (6.177)$$

The 'corner frequency' is given by:

$$f_c = \frac{1}{2\pi}\left(\frac{1}{\overline{\tau}_c} + \frac{1}{\overline{\tau}_e}\right) \qquad (6.178)$$

4. **Gate resistance noise.** The gate electrode, commonly made of polysilicon, introduces a finite value of resistance with which thermal noise is associated, which may not be negligible. The resistance value is reduced (i) through technological means by using metal silicides which gives low sheet resistivity of the gate layer, and (ii) through layout design by effectively paralleling the fingers constituting the gate [49, 50].

5. **Substrate resistance noise.** The inversion channel in a MOS transistor is controlled on the top side by the gate electrode, and on the bottom side by the bulk substrate. The bulk substrate across source-to-drain can be replaced by a resistance, the noise voltage of which is coupled to the channel through the bulk transconductance term g_{mb}. Though realistically the substrate resistance is distributed, a lumped value R_{sb} contributes to the drain current noise through the following relation:

$$S_{iD} = 4kTR_{sb}\, g_{mb}^2 \qquad (6.179)$$

Though bulk transconductance is weaker than gate transconductance, nevertheless, under certain circumstances the noise contribution may not be insignificant. High doping of the substrate and reverse biasing the substrate reduces the bulk noise contribution [15].

6. **Gate Tunneling Current (GTC) Shot Noise.** Scaled MOS transistors with a gate oxide thickness of a few nanometers have a finite amount of gate current, as has been mentioned in Chapter 3, and will be discussed in detail in Chapter 7. The channel carriers are separated from the gate by a potential barrier which, if thin, can be penetrated by the tunneling carriers.

The carriers cross the barrier randomly and constitute shot noise. The PSD of the gate noise current is given by:

$$S_{iG} = 2qI_G \tag{6.180}$$

The above does not include the edge tunneling current. The frequency dependence of the gate noise current has been shown to be inversely related to frequency [27]:

$$S_{iG} \propto \frac{I_G^2}{f} \tag{6.181}$$

Amongst the noise sources mentioned above, the channel noise is composed of both thermal and flicker noise sources, and the gate current noise includes induced gate noise and shot noise due to the GTC in the case of leaky or thin gate oxides. Since the output current noise of the MOSFET is mainly due to the channel noise, i_{ds}, gate current noise i_g, and the gate resistance noise, v_{Rg} only these three noise sources are discussed.

6.7.1 Analytical Channel Thermal Noise Model for MOSFETs with Velocity Saturation

We have given a brief outline of the origin of channel thermal noise of MOSFETs in the previous section. Modern MOS transistors being of short channel, the carriers get velocity saturated during transport from source to drain. An analytical channel thermal noise model considering velocity saturation, which would be useful for understanding the relation of noise with device parameters, is briefly discussed below [27].

A common feature of short-channel models is to segment the MOSFETS channel into (i) GCA (low lateral field) and (ii) velocity saturation (high field between velocity saturation and drain) regions as shown in Figure 5.20. Early short-channel models, e.g. [31–34], tried to explain the increased noise by modeling the noise of the velocity saturated region, and adding it to the noise from the linear region. However, it was successfully shown in Chen and Deen [36] that the contribution of the velocity saturation region to the output noise current is negligible as the carriers in that region travel at their saturation velocity v_{sat}, and they do not respond to the fluctuations of the electric field caused by voltage noise in that region. The analytical models presented in Chen and Deen [36], Scholten et al., [41] and Han et al. [44] are, however, not suitable for hand calculations for circuit design. A simpler analytical approach that relates noise spectral density to channel length, given in Asgaran et al. [28], is discussed below.

- Start with the equation for channel current, i.e. Equation (5.136) given in Chapter 5.
- Integrating the above expression from source $y = 0$ to a point y, obtain the expression for the channel potential $V_C(y)$.
- Obtain the expression for the lateral electric field along the channel $F(y)$ by differentiating $V_C(y)$ w.r.t. y.
- Equate $F(y)$ to F_C, the critical field at $y = L_C$, in the channel where velocity saturation onset occurs and the high field region starts. The condition provides a relation between L_C and V_0:

$$V_0 = \frac{I_{DS}}{WC'_{ox}v_{sat}} = V_{GT} - nV_{DSsat} \tag{6.182}$$

where:

$$V_{GT} = V_{GS} - V_T \tag{6.183}$$

The expression for L_C is:

$$L_C = \frac{V_{GT}(V_{GT} - V_0)}{n F_C V_0} = \frac{V_{GT} \cdot V_{DSsat}}{F_C V_0} \tag{6.184}$$

- Substituting the expression for $F(y)$ in Equation (5.136), and integrating with the limits $(0, L_C)$, provides the total inversion charge Q_n.
- Using the expression for Q_n in Equation (6.167), the expression for drain current noise spectral density is given as:

$$S_{iD} = 4kT \frac{4V_{GT}^2 + V_0^2 - 2V_0 V_{GT}}{3V_{GT}^2(V_{GT} - V_0)} n I_{DS} \tag{6.185}$$

At the point of velocity saturation in the channel, i.e. at the start of the high field region, the drain current is given by:

$$I_{DS} = W C_{ox} (V_{GT} - n V_{DSsat}) v_{sat} \tag{6.186}$$

Substituting expressions for I_{DS} from Equation (6.186), V_0 from Equation (6.182), and V_{GT} from Equation (6.183) into Equation (6.185), a compact expression for the thermal noise spectral density for a short channel MOSFET with velocity saturation is obtained:

$$S_{iD} = 4kT I_{DS} \left(\frac{1}{V_{DSsat}} + \frac{n^2 V_{DSsat}}{3V_{GT}^2} \right) \tag{6.187}$$

In order to correlate noise spectral density with channel length, a simplified expression for S_{iD} can be obtained through the substitution of Equations (6.182) and (6.184) into Equation (6.185), and using the approximation that for a small channel length $L_c \to 0$, the second term in the square bracket can be neglected compared to the first term. The resulting simplified expression of the drain current PSD is:

$$S_{iD} = 4kT \left(\frac{W \mu_{eff} V_{GT} C'_{ox}}{2} \right) \frac{1}{L_e} \tag{6.188}$$

It is seen that the for scaled MOSFETs, the thermal noise spectral density is inversely proportional to channel length. This implies that the smaller the channel length the greater is the thermal noise.

The simplified Equation (6.185) provides the channel thermal noise of MOSFETs close to the value obtained from the relatively involved expression reported in Chen and Deen [36], Scholten et al. [40], and Han et al. [44]. It indicates the suitability of the expression for accurately estimating the channel thermal noise down to a 10 nm gate length.

It may be mentioned that the Klaassen–Prins expression has been shown to be applicable for scaled MOSFETs as well, with the change that in Equation (6.168) real values of local channel specific conductivity and channel lengths are to be replaced with the corrected values g_c and L_{vsat} respectively [38].

$$S_{iD} = \frac{4kT}{L_{vsat}^2 I_{DS}} \int g_c^2(V_C)\, \mathrm{d}V_C \tag{6.189}$$

6.7.2 Compact Thermal and Flicker Noise Models

Thermal noise spectral density expressions as reported in some of the compact models are given below. A survey of the status of various noise sources in compact MOSFET models is given in [59].

6.7.2.1 BSIM3v3

For the thermal noise spectral density expression in BSIM3v3, Equation (6.167) is to be used with appropriate values of Q_n, where Q_n is the *total inversion charge* which is computed from the capacitance model options available in BSIM3v3. There are four capacitance model options available, corresponding to which there are expressions for inversion charge in the channel. The options available are (i) **capMod = 0**, (ii) **capMod = 1**, (iii) **capMod = 2**, and (iv) **capMod = 3**. The corresponding expressions are given in [48].

BSIM3v3 also includes the thermal noise due to the drain/source extrinsic resistances, R_d and R_s. These resistances are incorporated via the model parameter **R_{SH}**. The PSDs due to R_d and R_s are given by the following expressions:

$$S_{iS}(f) = \frac{4kT}{R_s} \tag{6.190}$$

$$S_{iD}(f) = \frac{4kT}{R_d} \tag{6.191}$$

R_s and R_d can be obtained following the procedure to extract parasitic resistances:

$$R_{s,d} = \mathbf{R_{SH}} \times \mathbf{N_{RS,D}} \tag{6.192}$$

$\mathbf{N_{RS}}$ and $\mathbf{N_{RD}}$ are the number of squares in the source and drain diffusion regions, respectively.

BSIM3v3 offers two options to users for the flicker noise.

noiMod = 1.   The model is simple though not considered to be very accurate. The expression for the $1/f$ noise power spectral density S_{id} is given by:

$$S_{iD}(f) = \frac{\mathbf{K_F} I_{DS}^{\mathbf{A_F}}}{C_{ox} L_{eff}^2 f^{\mathbf{E_F}}} \tag{6.193}$$

where $\mathbf{K_F}$ is a multiplying factor in the noise expression, $\mathbf{A_F}$ is an exponent associated with the drain current, and $\mathbf{E_F}$ is another coefficient associated with the frequency. The bias dependence of $1/f$ PSD is primarily reflected through the drain current term I_{DS}. The noise exponent $\mathbf{A_F}$ is reported to have values in the range of 0.5 to 2. The coefficient $\mathbf{K_F}$ is normally taken as constant though there are some results to show that it is dependent on bias. The frequency exponent $\mathbf{E_F}$ varies within the range 0.7–1.2, and also happens to be bias dependent [17].

noiMod $= 2$. The model is based on the unified approach discussed earlier. The flicker noise model is available for different modes of operation: (i) strong inversion (linear), (ii) strong inversion (saturation), and (iii) subthreshold operation. The expressions for the respective regions are given in Cheng and Hu [48].

The flicker noise model adopted in BSIM4 follows the framework of BSIM3v3 with improvements due to smoothing functions and incorporation of the bulk charge effect [57].

6.7.2.2 EKV

Equation (6.167) can be rewritten in terms of thermal noise conductance G_{nth} through the following expression:

$$S_{i_D} = 4kTG_{nth} \tag{6.194}$$

where the thermal noise conductance is defined as:

$$G_{nth} = \frac{\mu_n}{L^2} Q_n \tag{6.195}$$

The above relation is valid for all regions of operation of the MOSFET, and appropriate values of total inversion charge for strong and weak inversion can be expressed in terms of normalized current.

The EKV model uses the following relation for flicker noise which is based on the experimental results which show that the noise PSD varies inversely proportional to (i) gate area WL of the transistor, and (ii) frequency f [29]:

$$S_{i_D}(f) = \frac{\mathbf{KF}\, g_{mg}^2}{WLC_{ox}'} \times \frac{1}{f} \tag{6.196}$$

In the above equation $\mathbf{KF}$ is a constant which is strongly process sensitive and weakly bias dependent. The value of $\mathbf{KF}$ is higher for an n-channel transistor than for p-channel devices.

6.7.2.3 PSP

The PSP flicker noise model is also based on consideration of both mobility and carrier fluctuation with adaptation of symmetric linearization as required by the model framework [39]:

$$S_{iD} = \frac{q^2 kT\mu I_{DS}}{\gamma_{FN}\alpha_m L_e^2 C_{ox}' f^{NEF}} \left[\frac{C}{2}\left(N_0^2 - N_L^2\right) + (B - CN_*)(N_0 - N_L) \right.$$
$$\left. \times \left(A - BN_* + CN_*^2\right)\ln\left(\frac{N_0 + N_*}{N_L + N_*}\right) \right]$$

$$\tag{6.197}$$

where N_0 and N_L are the number of carriers at the source and drain sides of the channel respectively, assuming GCA is valid for the inverted channel region. A, B, C, and NEF are technology related parameters, α_m is the linearization coefficient at the symmetric potential midpoint of the channel, γ_{FN} is a physical constant, and N_* is defined by the following equation:

$$N_* = \frac{C'_{ox}\phi_t}{q}\left(1 + \frac{\gamma}{\sqrt{\phi_m + 10^{-6}\phi_t}}\right) \tag{6.198}$$

6.7.2.4 HiSIM

Another approach, based on surface potential and its derivatives, has been adopted in HiSIM, and the compact expression of spectral density of thermal noise in MOSFETs is given as [26]:

$$S_{iD} = 4kT\frac{W_{eff}C'_{ox}V_G\phi_t\mu}{(L_{eff} - \Delta L)}$$
$$\times \frac{(1 + 3\eta + 6\eta^2)\mu_d^2 + (3 + 4\eta + 3\eta^2)\mu_d\mu_s + (6 + 3\eta + \eta^2)\mu_s}{15(1+\eta)\mu_{av}^2} \tag{6.199}$$

$$\eta = 1 - \frac{(\phi_{SL} - \phi_{S0}) + \chi(\phi_{SL} + \phi_{S0})}{V_G\phi_t} \tag{6.200}$$

$$\chi = 2\frac{const0}{C'_{ox}}\left(\left\{\frac{2}{3}\phi_t\frac{[\frac{1}{\phi_t}(\phi_{SL} - V_{BS}) - 1]^{\frac{3}{2}} - [\frac{1}{\phi_t}(\phi_{S0} - V_{BS}) - 1]^{\frac{3}{2}}}{\phi_{SL} - \phi_{S0}}\right\}\right.$$
$$\left. - \sqrt{\frac{1}{\phi_t}(\phi_{S0} - V_{BS}) - 1}\right) \tag{6.201}$$

$$const0 = \sqrt{2\epsilon_{Si}qN_A\phi_t} \tag{6.202}$$

It may be noted that the thermal noise spectral density has been expressed in terms of surface potentials at the ends of the channel, ϕ_{S0} and ϕ_{SL}.

The $1/f$ noise spectral density in HiSIM is based on the combined effect of carrier and mobility fluctuation [25], and is given by:

$$S_{iD} = \frac{\phi_t I_{DS}^2\mathbf{NFTRP}}{f(L_{eff} - \Delta L)W_{eff}}N_{fluct} \tag{6.203}$$

$$N_{fluct} = \frac{1}{(N_0 + N^*)(N_L + N^*)}$$
$$+ \frac{2\mu E_g\mathbf{NFALP}}{N_L - N_0}\ln\left(\frac{N_L + N^*}{N_0 + N^*}\right) + (\mu E_g\mathbf{NFALP})^2 \tag{6.204}$$

where **NFTRP** is the trap density to attenuation coefficient ratio, and **NFALP** is the mobility fluctuation contribution. N_0 and N_L are the carrier densities at the source and drain, respectively.

6.7.3 Induced Gate Noise

At high frequency, the thermal noise is coupled to the gate through the gate capacitance producing an *induced gate noise*. The approach to induced gate noise modeling is similar to that adopted for channel thermal noise. The channel under the gate is partitioned into GCA and velocity saturation regions. As the contribution of the velocity saturation, or high field, region contribution to thermal noise is negligible, only the GCA region needs to be considered for induced gate noise. The spectral density of the induced gate noise current [47] is given as:

$$
S_{iGind} = \frac{4kT\omega^2 W^4 C_{ox}'^4 \mu_{eff}^2}{I_{DS}^3}
$$
$$
\times \left[V_{GT}^2 V_{as}^2 V_{DS} - V_{GT} V_{as} (V_{GT} + V_{as}) V_{DS}^2 \right.
$$
$$
+ \frac{V_{GT}^2 + 4 V_{GT} V_{as} + V_{as}^2}{3} V_{DS}^3
$$
$$
\left. - \frac{V_{GT} + V_{as}}{2} V_{DS}^4 + \frac{V_{DS}^5}{5} \right] \tag{6.205}
$$

where:

$$
V_{as} = V_{DS} - \frac{\frac{V_{GT} V_{DS}}{2} - \frac{V_{DS}^2}{6}}{V_{GT} - \frac{V_{DS}}{2}} \tag{6.206}
$$

Induced gate noise is correlated to channel thermal noise [35, 45, 46]. The correlation coefficients of the channel noise and induced gate noises have been observed to be proportional to channel length. The induced gate noise current is represented by a current source connected in parallel with the gate–source capacitance.

References

[1] Y. Tsividis, *Operation and Modeling of the MOS Transistor*, McGraw Hill, New York, NY, 1999.

[2] M.A. Cirit, The Meyer Model Revisited: Why is Charge Not Conserved, *IEEE Trans. Comput.-Aided Des.*, **8**(10), 1033–1037, 1989.

[3] D.E. Ward and R.W. Dutton, A Charge-Oriented Model for MOS Transistor Capacitances, *IEEE J. Solid-State Cir.*, **13**(5), 703–708, 1978.

[4] J.E. Meyer, MOS Models and Circuit Simulation, *RCA Rev.*, **32**(3), 42–63, 1971.

[5] Y.P. Tsividis and K. Suyama, MOSFET Modeling for Analog Circuit CAD: Problems and Prospects, *IEEE J. Solid-State Circ.*, **29**(3), 210–216, 1994.

[6] H. Wang, X. Li, W. Wu, G. Gildenblat, R. van Langevelde, G.D.J. Smit, A.J. Scholten and D.B.M. Klaassen, A Unified Nonquasi-Static MOSFET Model for Large-Signal and Small-Signal Simulations, *IEEE Trans. Electron Dev.*, **53**(9), 2035–2043, 2006.

[7] M. Chan, K. Hui, C. Hu and P. K. Ko, A Robust and Physical BSIM3 Non-Quasi-Static Transient and AC Small-Signal Model for Circuit Simulation, *IEEE Trans. Electron Dev.*, **45**(4), 834–841, 1998.

[8] A.S. Roy, J.M. Vasi and M.B. Patil, A New Approach to Model Nonquasistatic (NQS) Effects for MOSFETs– Part1: Large-Signal Analysis, *IEEE Trans. Electron Dev.*, **50**(12), 2393–2400, 2003.

[9] N. Nakayama, D. Navarro, M. Tanaka, H. Ueno, M. Miura-Mattausch, H.J. Mattausch, T. Ohguro, S. Kumashiro, M. Taguchi, T. Kage and S. Miyamoto, Non-quasi-static Model for MOSFET Based on Carrier-transit Delay, *Electron. Lett.*, **40**(4), 276–277, 2004.

[10] D. Navarro, Y. Takeda, M. Miyake, N. Nakayama, K. Machida, T. Ezaki, H.J. Mattausch and M. Miura-Mattausch, A Carrier-Transit-Delay-Based Nonquasi-Static MOSFET Model for Circuit Simulation and its Application to Harmonic Distortion Analysis, *IEEE Trans. Electron Dev.*, **53**(9), 2025–2034, 2006.

[11] A.F.-L. Ng, P.K. Ko and M. Chan, Determining the Onset Frequency of Non-quasistatic Effects of the MOSFET in AC Simulation, *IEEE Electron Dev. Lett.*, **23**(1), 37–39, 2002.

[12] T. Ytterdal, Y. Cheng and T.A. Fjeldly, *Device Modeling for Analog and RF CMOS Circuit Design*, John Wiley & Sons Ltd., Chichester, UK, 2003.

[13] P.R. Gray and R.G. Meyer, *Analysis and Design of Analog Integrated Circuits*, 3rd Edition, John Wiley & Sons, Inc., New York, NY, 1993.

[14] R. Razavi, *Design of Analog CMOS Integrated Circuits*, McGraw Hill, New York, NY, 2001.

[15] R. Jindal, Compact Noise Model for MOSFETs, *IEEE Trans. Electron Dev.*, **53**(9), 2051–2061, 2006.

[16] J.S. Kolhatkar, *Steady-State and Cyclo-Stationary RTS Noise in MOSFETs*, Ph.D thesis, University of Twente, The Netherlands, 2005.

[17] N.H. Hamid, A.F. Murray and S. Roy, Time-Domain Modeling of Low Frequency Noise in Deep-Submicrometer MOSFET, *IEEE Trans. Circ. Syst.*, **55**(1), 245–257, 2008.

[18] E. Simoen and C. Claeys, On the Flicker Noise in Submicron Silicon MOSFETs, *Solid-State Electron.*, **43**(5), 865–882, 1999.

[19] K.K. Hung, P.K. Ko, C. Hu and Y.C. Cheng, A Unified Model for the Flicker Noise in Metal-Oxide-Semiconductor Field Effect Transistors, *IEEE Trans. Electron Dev.*, **37**(3), 654–665, 1990.

[20] K.K. Hung, P.K.Ko, C. Hu and Y.C. Cheng, A Physics-Based MOSFET Noise Model for Circuit Simulators, *IEEE Trans. Electron Dev.*, **37**(5), 1323–1333, 1990.

[21] G. Ghibaudo and T. Boutchacha, Electrical Noise and RTS Fluctuation in Advanced CMOS Devices, *Microelectron. Reliabil.*, **42**(4), 573–582, 2002.

[22] M.V. Heijningen, E.P. Vandamme, L. Deferm and L.K.J. Vandamme, Modeling the $1/f$ Noise and Extraction of the SPICE Noise Parameters Using a New Extraction Procedure, in *Proceedings of the 28th European Solid-State Device Research Conference, (ESSDERC)*, pp. 468–471, 1998.

[23] E.P. Vandamme and L.K.J. Vandamme, Critical Discussion on Unified $1/f$ Noise Models for MOSFETs, *IEEE Trans. Electron Dev.*, **47**(11), 2146–2152, 2000.

[24] L.K.J. Vandamme, X. Li and D. Rigaud, $1/f$ Noise in MOS Devices, Mobility or Number Fluctuations?, *IEEE Trans. Electron Dev.*, **41**(11), 1936–1945, 1994.

[25] S. Matsumoto, H. Ueno, S. Hosokawa, T. Kitamura, M. Miura-Mattausch, H.J. Mattausch, T. Ohguro, S. Kumashiro, T. Yamaguchi, K. Yamashita and N. Nakayama, $1/f$-Noise Characteristics in 100 nm-MOSFETs and its Modeling in Circuit Simulation, *IEICE Trans. Electron.*, **E88-C**(2), 247–254, 2005.

[26] M. Miura-Mattausch, N. Sadachika, D. Navarro, G. Suzuki, Y. Takeda, M. Miyake, T. Warabino, Y. Mizukane, R. Inagaki, T. Ezaki, H.J. Mattausch, T. Ohguro, T. Iizuka, M. Taguchi, S. Kumashiro and S. Miyamoto, HiSIM2: Advanced MOSFET Model Valid in RF Circuit Simulation, *IEEE Trans. Electron Dev.*, **53**(9), 1994–2007, 2006.

[27] M.J. Deen, C.H. Chen, S. Asgaran, G.A. Rezvani, J. Tao and Y. Kiyota, High-Frequency Noise of Modern MOSFETs: Compact Modeling and Measurement Issues, *IEEE Trans. Electron Dev.*, **53**(9), 2062–2081, 2006.

[28] S. Asgaran, M.J. Deen and C.-H. Chen, Analytical Modeling of MOSFETs Channel Noise, and Noise Parameters, *IEEE Trans. Electron Dev.*, **51**(12), 2109–2114, 2004.

[29] C.C. Enz, F. Krummenacher and E.A. Vittoz, An Analytical MOS Transistor Model Valid in All Regions of Operation and Dedicated to Low-Voltage and Low-Current Applications, *Analog Integr. Circ. Signal Proc.*, **8**, 83–114, 1995.

[30] C.C. Enz, *High Precision CMOS Micropower Amplifiers*, Ph.D. Thesis, Ecole Polytechnique Federal de Lausanne, Lausanae Switzerland, 1989.

[31] R.P. Jindal, Hot-Electron Effect on Channel Thermal Noise in Fine-Line NMOS Field Effect Transistor, *IEEE Trans. Electron Dev.*, **ED-33**(9), 1395–1397, 1986.

[32] P. Klein, An Analytical Thermal Noise Model of Deep Submicron MOSFETs, *IEEE Electron Dev. Lett.*, **20**(8), 399–401, 1999.

[33] C.H. Park, and Y.J. Park, Modeling of Thermal Noise in Short-Channel MOSFETs at Saturation, *Solid-State Electron*, **44**(11), 2053–2057, 2000.

[34] D.P. Triantis, A.N. Birbas and D. Kondis, Thermal Noise Modeling of Short-Channel MOSFETs, *IEEE Trans. Electron Dev.*, **43**(11), 1950–1955, 1996.

[35] D.P. Triantis, A.N. Birbas and S.E. Plevridis, Induced Gate Noise in MOSFETs Revised: The Submicron Case, *Solid-State Electron.*, **41**(12), 1937–1942, 1997.

[36] C.H. Chen and M.J. Deen, Channel Noise Modeling of Deep Submicron MOSFETs, *IEEE Trans. Electron Dev.*, **49**(8), 1484–1487, 2002.

[37] F.M. Klaassen and J. Prins, Thermal Noise of MOS Transistors, *Philips Res. Rep.*, **22**, 505–514, 1967.

[38] J.C.J. Paasschens, A.J. Scholten and R. van Langevelde, Generalizations of the Klaassen–Prins Equation for Calculating the Noise of Semiconductor Devices, *IEEE Trans. Electron Dev.*, **52**(11), 2463–2472, 2005.

[39] G. Gildenblat, H. Wang, T.-L. Chen, X. Gu and X. Cai, SP: An Advanced Surface-Potential-Based Compact MOSFET Model, *IEEE J. Solid-State Circ.*, **39**(9), 1394–1406, 2004.

[40] A.J. Scholten, H.J. Tromp, L.F. Tiemeijer, R.V. Langevelde, R.J. Havens, P.W.H. De Vreede, R.F.M. Roes, P.H. Woerlee, A.H. Montree and D.B.M. Klaassen, Accurate Thermal Noise Model for Deep-Submicron CMOS, *IEDM Tech. Dig.*, 155–158, 1999.

[41] A.J. Scholten, L.F. Tiemeijer, R.V. Langevelde, R.J. Havens, A.T.A. Zegeres-van Duijnhoven and V.C. Venezia, Noise Modeling for RF CMOS Circuit Simulation, *IEEE Trans. Electron Dev.*, **50**(3), 618–632, 2003.

[42] T.L. Chen and G. Gildenblat, Symmetric Bulk Charge Linearization on Charge-Sheet MOSFET Model, *Electron Lett.*, **37**(12), 791–793, 2001.

[43] H. Wang, T.L. Chen and G. Gildenblat, Quasi-static and Nonquasi-static Compact MOSFET Models Based on Symmetric Linearization of the Bulk and Inversion Charges, *IEEE Trans. Electron Dev.*, **50**(11), 2262–2272, 2003.

[44] K. Han, H. Shin and K. Lee, Analytical Drain Thermal Noise Current Model Valid for Deep Submicron MOSFETs, *IEEE Trans. Electron Dev.*, **51**(2), 261–269, 2004.

[45] M. Shoji, Analysis of High-frequency Thermal Noise of Enchancement Mode MOS Field-effect Transistors, *IEEE Trans. Electron Dev.*, **13**(6), 520–524, 1966.

[46] A.V.D. Ziel, Gate Noise in Field Effect Transistors at Moderately High Frequencies, *Proc. Inst. Elect. Engg.*, **51**(3), 461–467, 1963.

[47] Y. Cheng, M.J. Deen and C.H. Chen, MOSFET Modeling for RF IC Design, *IEEE Trans. Electron Dev.*, **52**(7), 1286–1303, 2005.

[48] Y. Cheng and C. Hu, *MOSFET Modeling and BSIM3 User's Guide*, Kluwer Academic Publishers, Boston, MA, 1999.

[49] W. Liu, *MOSFET Models for SPICE Simulation, including BSIM3v3 and BSIM4*, John Wiley & Sons, Inc., New York, NY, 2001.

[50] Z.Y. Chang, *Low-Noise Wide Band Amplifiers in Bipolar and CMOS Technologies*, Ph.D. Thesis, University of Leuven, Belgium, 1990.

[51] A.I.A. Cunha, M.C. Schneider and C.G-Montoro, An Explicit Physical Model for the Long-Channel MOS Transistor Including Small-Signal Parameters, *Solid-State Electron.*, **38**(11), 1945–1952, 1995.

[52] M. Bucher, C. Lallement, C.C. Enz, F. Theodoloz and F. Krummenacher, The EPFL-EKV MOSFET Model Equations for Simulation, Technical Report), 1–18, 1998.

[53] M. Bucher, J.-M. Sallese, C. Lallement, W. Grabinski, C.C. Enz and F. Krummenacher, Extended Charges Modeling for Deep Submicron CMOS, In *Proceedings of the International Semiconductor Device Research Symposium (ISDRS'99)*, 397–400, 1999.

[54] B.J. Sheu, W.-J. Hsu and P.K. Ko, An MOS Transistor Charge Model for VLSI Design, *IEEE Trans. Comput.-Aided Des.*, **7**(4), 520–527, 1988.

[55] R. Shrivastava and K. Fitzpatrick, A Simple Model for the Overlap Capaciatance of a VLSI MOS Device, *IEEE Trans. Electron Dev.*, **29**(12), 1870–1875, 1982.

[56] N. Wakita and N. Shigyo, Verification of Overlap and Fringing Capacitance Models for MOSFETs, *Solid-State Electron.*, **44**(6), 1105–1109, 2000.

[57] T. Noulis, S. Siskos and G. Sarrabayrouse, Comparison Between BSIM4.X and HSPICE Flicker Noise Models in NMOS and PMOS Transistors in all Operating Regions, *Microelectron. Reliabil.*, **47**(8), 1222–1227, 2007.
[58] G. Ding, *General-purpose Semiconductor Simulator* (http://gss-tcad.sourceforge.net/) 2008.
[59] R.P. Jindal, Compact Noise Models for MOSFETs, *IEEE Trans. Electron Dev.*, **53**(9), 2051–2061, 2006.

Problems

1. Derive Equations (6.7) and (6.8), which are used for source and drain charge evaluation in the quasistatic model.
2. Using the PSP model formulation show that the coordinate of the surface potential mid-point where $\phi_s = \phi_m$ is given by $y_m = (L/2)(1 + \phi/(4H))$, where $H = \phi_t - Q'_{nm}/(C'_{ox}\alpha_m)$.
3. Establish the relation between the bulk referred small signal conductance parameters g_{mgb}, g_{msb}, and g_{mdb}, and the source referred parameters g_m, g_{mb}, and g_{ds} of a MOSFET.
4. Show that Equations (6.109a) and (6.109b) for i_{cd} and i_{cb} respectively, are equivalent to those given by the matrix Equation (6.95).
5. Obtain the expression for unity gain frequency from the equivalent circuit of a small signal model of a MOSFET with and without a source terminal connected to bulk.
6. Derive the expressions for EKVv2.6 model in strong inversion for the small signal capacitances C_{gs}, C_{gd}, C_{gb}, C_{sb}, and C_{db}.
7. For a representative 0.8 micrometer technology node the following are given:
$N_A = 10^{15}\,\text{cm}^{-3}$, $T_{ox} = 18.0\,\text{nm}$, $\mu_n = 486\ \text{cm}^2\,\text{V}^{-1}\text{s}^{-1}$, $L_D = 0.1\,\mu\text{m}$, $C_{j0} = 2.5 \times 10^{-8}\,\text{F cm}^{-2}$, $C_{jsw0} = 0.2 \times 10^{-11}\,\text{F cm}^{-1}$, $X_j = 0.15\,\mu\text{m}$, $\phi_{bi} = 0.9\,\text{V}$, $L_g = 1.6\,\mu\text{m}$, $W = 3.2\,\mu\text{m}$, $I_{DS} = 10\,\mu\text{A}$, $V_{DS} = 2.5\,\text{V}$, $V_{SB} = 0\,\text{V}$, $\phi_{gs} = -0.6\,\text{V}$ and $Q'_f = q \times 10^{11}\,\text{C cm}^{-2}$.

    Obtain the complete small signal model of the NMOS transistor. For lengths of the source/drain regions, assume each of them has a length as 5 LAMBDA as per the design rule.
8. Assume that the potential variation in the channel of an MOS transistor is given by:

$$V_{CS}(y) = \frac{V_{GS} - V_{TS}}{n}\left[1 - \sqrt{\left(1 - \frac{y}{L}\right)}\right] \qquad (6.207)$$

    Obtain the expressions for source and drain charges as per the Ward–Dutton partitioning approach discussed in Section 6.2.
9. Construct a complete EKV noise model for a MOSFET in saturation with total noise referred to gate, and obtain the corner frequency.
10. Equation (6.187), discussed in Section 6.7.1, has been given without showing intermediate steps. The intermediate equations have been indicated but not derived. Do a complete derivation of the above expression.
11. Discuss critically the bias dependence of fringe capacitance in a MOSFET structure in the framework of the BSIM model.

7

Quantum Phenomena in MOS Transistors

7.1 Introduction

In modern VLSI chips the device density is increased by scaling down the device dimensions. An increase in device density increases the logic function implemented on the chip. The reduction in device dimensions introduces a number of undesirable second order effects on the I–V characteristics of the device. As discussed in previous chapters the scaling theory proposes that these effects can be reduced by an increase in the substrate doping ($> 5 \times 10^{17}$ cm^{-3}), and a decrease in the oxide thickness (< 50 Å). Such measures, however, increase the normal field in the gate oxide, as well as at the Si–SiO$_2$ interface ($F > 10^5$ V cm^{-1}) in silicon where the inversion layer is formed. The energy band diagram of a MOS structure under such a condition is shown in Figure 7.1, where the slope of the conduction band is proportional to the electric field component in the x-direction. When the electric field is high at the interface, the well in which the electrons are trapped is narrow, and its width is comparable to the De Broglie wavelength of the confined electrons. From quantum theory we know that in such a case the electron energy will be quantized in the conduction band, and the classical treatment does not hold good. The first two quantized energy levels are also shown in Figure 7.1.

Quantum confinement of the carriers leads to an increase in threshold voltage from that of the classical value. The gate capacitance also gets degraded in strong inversion. The above two parameters, in turn, directly affect the drain current.

Furthermore, scaling down of the oxide thickness reduces the width of the SiO$_2$ potential barrier which separates the carriers in the semiconductor from the gate terminal. The reduced width of the barrier enhances the quantum tunneling probability of carriers through the barrier. The assumption that the gate oxide acts as an insulating material no longer holds good in the presence of tunneling. The carrier tunneling phenomenon leads to a gate current, which has been neglected so far in our discussion. The presence of this gate current increases the static power of MOSFET based circuits, which is a matter of serious concern.

In this chapter we shall discuss the quantum mechanical theory related to the carrier confinement of mobile carriers in the potential well at the Si–SiO$_2$ interface, which leads to

Compact MOSFET Models for VLSI Design   A.B. Bhattacharyya
© 2009 John Wiley & Sons (Asia) Pte Ltd

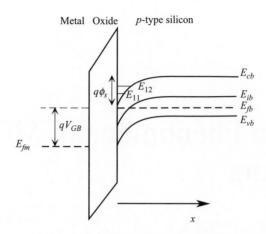

Figure 7.1   Carrier energy quantization at the Si–SiO$_2$ interface in a MOS capacitor

quantization of energy levels. Its implication on threshold voltage, inversion charge density, C–V characteristics of MOS capacitors, etc. will be discussed. The basic theory of quantum tunneling has also been discussed with compact model perspectives in view.

7.2 Carrier Energy Quantization in MOS Capacitor

The increase in substrate doping and decrease in gate oxide thickness for reducing short channel effects cause an enhancement of the field strength at the interface. The high field results in the formation of a narrow potential well as shown in Figure 7.1. The inverted carriers find themselves in a potential well whose wavelengths are comparable to the well width. In such a potential well the carrier behavior is governed by quantum mechanics. From quantum mechanics it is known that the allowed energies are quantized in the direction of confinement (x-direction). The dimensions of the devices are assumed such that in the y–z plane the carrier can have a continuum of energies.

As the potential well becomes narrower, the difference between the quantized energy levels increases, and the motion of an electron in a direction normal to the interface gets confined. Thus, the transport properties of an electron in the inversion layer is more appropriately described by a two-dimensional electron gas (2DEG) as against classical 3DEG where an electron can be considered to move freely in all three spatial directions. The behavior of a trapped electron in a triangular-like potential well is described by Schrödinger's and Poisson's equations, which have to be solved self-consistently.

To solve the above problem for electron energy and their distribution, we start with the three-dimensional time independent Schrödinger's equation:

$$\left(-\frac{\hbar^2}{2m_n}\nabla^2 + V(\mathbf{R})\right)\Psi(\mathbf{R}) = E\Psi(\mathbf{R}) \qquad (7.1)$$

where m_n represents the mass of the electron.

At the atomic scale, the potential in the bulk semiconductor varies periodically in all of the three dimensions, and cannot be determined easily. The variation in potential can, however, be ignored under effective mass approximation. The applied gate bias creates a potential well in the x-direction. Taking the above points into consideration we can write $V(\mathbf{R}) = V(x)$.

The electron wave function in a direction parallel to the interface can be assumed to be a plane wave, as it is free to move along this plane. The wave function describing the electron behaviour in a two-dimensional confined well can be expressed as a product of a plane wave parallel to the interface and the envelop function $\xi(x)$ normal to the interface [2]. This envelop function determines the quantized energy in the direction normal to the interface. Let these energy levels be denoted by E_n. That is, using the effective mass approximation, the electron wave function can be written as the product of the Bloch function at the bottom of the conduction band $e^{ik_y y + ik_z z} e^{i\theta x}$, and an envelop function $\xi(x)$:

$$\Psi_n(x, y, z) = e^{ik_y y + ik_z z} e^{i\theta x} \xi_n(x) \tag{7.2}$$

Here, the y- and z-directions are along and across the channel respectively, and the x-direction is normal to the Si–SiO$_2$ interface. The quantity θ depends on k_y and k_z, and represents the z-dependence of the Bloch function; $k_y = \frac{2\pi l}{L}$, $k_z = \frac{2\pi r}{W}$, l and r are quantum numbers which can be positive or negative integers.

Substituting the above solution in the Equation (7.1), we have:

$$\left[-\frac{\hbar^2}{2m_n^*} \left(\frac{\partial}{\partial x^2} + \frac{\partial}{\partial y^2} + \frac{\partial}{\partial z^2} \right) + V(x) \right] e^{ik_y y + ik_z z} \xi_n(x) = E e^{ik_y y + ik_z z} \xi_n(x) \tag{7.3}$$

which can be written as:

$$\left(\frac{\hbar^2 k_y^2}{2m_n^*} + \frac{\hbar^2 k_z^2}{2m_n^*} - \frac{\hbar^2}{2m_n^*} \frac{d^2}{dx^2} + V(x) \right) \xi_n(x) = E_n \xi_n(x) \tag{7.4}$$

The first two terms denote the energy of the plane waves along the y- and z-direction respectively. Taking the first two terms to the RHS, we have:

$$\left(-\frac{\hbar^2}{2m_n^*} \frac{d^2}{dx^2} + V(x) \right) \xi_n(x) = \left(E_n - \frac{\hbar^2 k_y^2}{2m_n^*} - \frac{\hbar^2 k_z^2}{2m_n^*} \right) \xi_n(x) \tag{7.5}$$

The above equation can be written as:

$$\left(-\frac{\hbar^2}{2m_n^*} \frac{d^2}{dx^2} + V(x) \right) \xi_n(x) = E_n' \xi_n(x) \tag{7.6}$$

which is a one-dimensional Schrödinger equation that requires to be solved for $\xi(x)$ and E_n'. The electron energy will, therefore, be given by:

$$E_n(\mathbf{k}) = E_n' + \frac{\hbar^2 k_y^2}{2m_n^*} + \frac{\hbar^2 k_z^2}{2m_n^*} \tag{7.7}$$

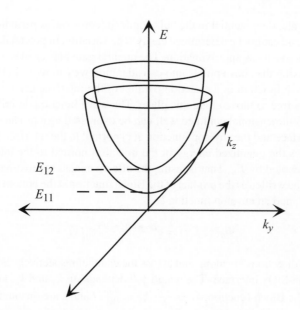

Figure 7.2   Bell-shaped paraboloids showing the total energies including transverse kinetic energies for the first two sub-bands

The E–k plots look like bell-shaped concentric paraboloids for motion constrained in the k_y–k_z plane, with the bottoms separated by E'_n called sub-bands, as shown in Figure 7.2. The solution of the 1-D Schrödinger equation for a given potential well shape gives the electron eigenfunctions ξ_n and eigenvalues E'_n, from which we can determine the electron distribution $n(x)$ as a function of x. This is given as:

$$n(x) = \sum_{n=1}^{\infty} N_n |\xi_n(x)|^2 \tag{7.8}$$

where N_n is the density of electrons ($\#/\mathrm{cm}^2$) occupying the nth sub-band energy level. N_n can be obtained from a knowledge of the density of states and distribution function, which will be discussed later.

The distribution $n(x)$ has to satisfy Poisson's equation, which is given by:

$$\frac{\mathrm{d}^2 V(x)}{\mathrm{d}x^2} = -\frac{\rho(x)}{\epsilon_{\mathrm{Si}}} \tag{7.9}$$

where

$$\rho(x) = q[N_A(x) + n(x) + p(x)] \tag{7.10}$$

is the charge density constituted by acceptor ions, electrons, and holes.[1]

---

[1] N_D is assumed to be zero.

We desire a solution for $V(x)$ which satisfies both Schrödinger's and Poisson's equations. Solving Equation (7.9), with ρ calculated from the $n(x)$ obtained from the solution of Schrödinger's equation and distribution function (Equation (7.10)), may result in a $V(x)$ which differs from the $V(x)$ used originally to solve Schrödinger's equation. The effect of the altered $V(x)$ on the eigenfunction and eigenvalue is taken into consideration by solving Schrödinger's equation again for a new estimate of $n(x)$, which is then used in solving Poisson's equation again. This procedure has to be carried out iteratively until a self-consistent solution has been reached with acceptable errors where further iteration has little effect on $n(x)$ or $V(x)$.

In solving Schrödinger's equation, the appropriate value of electron conduction mass, which depends on band structure and the direction of motion of electrons, has to be used. For silicon, the E versus k curve is different in different directions, and there are six constant energy ellipsoids located symmetrically located on the k_x, k_y, and k_z planes for a $\langle 100 \rangle$ orientation of the silicon surface, as discussed in Chapter 1. So the degeneracy factor will be six, and the conduction electron mass of the electrons in these ellipsoids will depend upon the directions in which they move. We need the effective conduction mass along the x-direction for solving Schrödinger's equation.

For electrons in the constant energy ellipsoid along the x-direction (valley-1), the conduction effective mass m_{x1} will be equal to the longitudinal mass m_l, whereas for electrons which are in the constant energy ellipsoid along the y- and z-directions (valley-2), the conduction effective mass m_{x2} will be equal to the transverse mass[2] m_t.

The solution of Schrödinger's equation gives two sets of eigenfunctions, viz. $\xi_n(m_{x1})$ and $\xi_n(m_{x2})$ with eigenvalues $E_{n1}(m_{x1})$ and $E_{n2}(m_{x2})$ respectively.[3] We shall call the energy levels $E_{n1}(m_{x1})$ as ladder 1 and the energy levels $E_{n2}(m_{x2})$ as ladder 2. Qualitatively we can see that as $m_{x1} > m_{x2}$, we will have $E_{11} < E_{12}$ as shown in Figure 7.1.

The effective density of state mass to be used for calculating the density of states in the following section will also be different for electrons in different valleys, based on the direction in which they move. For electrons moving in the y–z plane, and considering a constant energy ellipsoid along the x-direction, the masses along the y- and z-directions have the same value. Its value is $m_{yz_1} = m_t = 0.196m_0$ [2]. For the electron in a constant energy ellipsoid along the y- or z-directions, the masses will be different depending upon the ellipsoid and direction. The effective value is taken as the geometric mean, and is denoted as $m_{yz_2} = \sqrt{m_t m_l} = 0.42m_0$. From the above discussion we can see that electrons with an effective density of state mass m_{yz_1} have a degeneracy of two, and electrons with an effective density of state mass m_{yz_2} have a degeneracy of four.

7.3 2-D Density of States

The density of states (DOS) in a three-dimensional system is the number of available electronic states per unit volume per unit energy at an energy E. In the case of a two-dimensional system, the DOS is the number of available energy states per unit area per unit energy interval at the energy value E, and will be denoted by $g(E)$.

---

[2] $m_l = 0.916m_0$ and $m_t = 0.196m_0$ are masses along the longitudinal and transverse directions if we consider a single constant energy ellipsoid.

[3] The first subscript denotes the energy level, and the second is for the ladder (valley) number corresponding to mass m_{x1} and m_{x2} respectively [19].

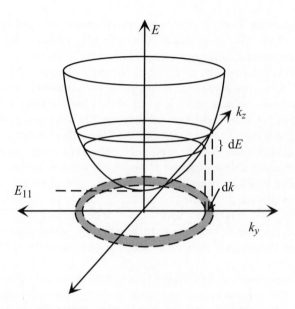

Figure 7.3 E–k parabola along the y–z plane

The DOS for the two-dimensional electron gas can be calculated from the energy (E) versus wave vector (**k**) relation [18, 20] which is given in Equation (7.7). From Figure 7.3 we can see that the energies E and $E + dE$ represent the strip of the paraboloid between k and $k + dk$. For the sake of illustration, we consider electrons in valley-1 and assume the first and only allowed energy level for motion along the x-direction to be E_{11}, and the density of states mass is m_{yz1}. The area of the strip can be calculated as:

$$A = \pi(k + dk)^2 - \pi k^2 \tag{7.11}$$

In the limit $dk \to 0$, we can write:

$$A = \pi 2k\,dk = 2\pi k\,dk \tag{7.12}$$

The area per state in k space is $\left(\frac{2\pi}{L}\right)^2$. Therefore, the number of states G between k and $k + dk$ is:

$$G = 2\left[\frac{2\pi k\,dk}{\left(\frac{2\pi}{L}\right)^2}\right] \tag{7.13}$$

where the multiplication factor of 2 takes care of the direction of spin of the electrons.

$$G = \frac{L^2 k\,dk}{\pi} \tag{7.14}$$

Dividing the above equation by L^2 (area), we have $g(k)\,dk$ as:

$$g(k)\,dk = \frac{k\,dk}{\pi} \qquad (7.15)$$

Energy is related to k as:

$$E = \frac{\hbar^2 k^2}{2m_{yz1}} \qquad (7.16)$$

Differentiating w.r.t. k we have:

$$k\,dk = \frac{m_{yz1}}{\hbar^2}\,dE \qquad (7.17)$$

Substituting for $k\,dk$ in Equation (7.15), we have:

$$g(E)\,dE = \frac{m_{yz1}}{\pi \hbar^2}\,dE \qquad (7.18)$$

That is, the density of states $g(E)$ can be written as:

$$g(E) = \frac{m_{yz1}}{\pi \hbar^2} \qquad (7.19)$$

From the above derivation we conclude that the density of states for a given energy level is constant. Taking into account the degeneracy of 2 corresponding to the mass m_{yz1}:

$$g(E) = 2\frac{m_{yz1}}{\pi \hbar^2} \qquad (7.20)$$

On considering all the sub-band energies corresponding to mass m_{yz1}, we have:

$$g(E) = 0 \qquad\qquad\qquad \text{for} \quad E < E_{11} \qquad (7.21)$$

$$g(E) = 2\left(\frac{m_{yz1}}{\pi \hbar^2}\right) \qquad \text{for} \quad E_{11} < E < E_{21} \qquad (7.22)$$

$$g(E) = 4\left(\frac{m_{yz1}}{\pi \hbar^2}\right) \qquad \text{for} \quad E_{21} < E < E_{31} \qquad (7.23)$$

Figure 7.4 shows the variation of density of states as a function of energy. We can see that below E_{11} the density of states is zero with step increase.

It may be reminded that in the above analysis we have not taken into account the density of states available from the other four valleys with a density of states mass m_{yz2} and degeneracy factor of four. The actual step increase in the DOS function is non-uniform, and is due to the difference in m_{yz1} and m_{yz2}.

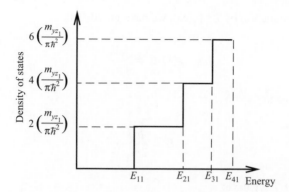

Figure 7.4   Step variation of density of states in valley-1 as a function of energy

7.4 Electron Concentration Distribution

The solution of Schrödinger's equation gives the eigenfunctions ξ_{n1} and eigenvalues E_{n1} for the effective mass corresponding to m_{x1}. Similarly, for the effective mass m_{x2}, the corresponding eigenfunctions and eigenvalues are ξ_{n2} and E_{n2} respectively. Let the density of electrons per unit area in the nth sub-band of ladder 1 be N_{n1}, and in the nth sub-band of ladder 2 be N_{n2}. With the above definition, and recognizing that $|\xi(x)|^2$ gives the probability density per unit length, we can write the volume density of electrons $n(x)$ as:[4]

$$n(x) = \sum_{n=1}^{\infty} N_{n1}|\xi_{n1}|^2 + \sum_{n=1}^{\infty} N_{n2}|\xi_{n2}|^2 \tag{7.24}$$

The density of electrons N_{n1} in the nth sub-band can be written as the integral of the product of density of states and the probability of occupancy given by the Fermi–Dirac distribution function, as given below:

$$N_{n1} = \int_{E_{n1}}^{\infty} N(E) f(E) \, dE \tag{7.25}$$

$$N_{n1} = \int_{E_{n1}}^{\infty} \frac{2m_{yz_1}}{\pi \hbar^2} \left(e^{\frac{E-E_F}{kT}} + 1 \right)^{-1} dE \tag{7.26}$$

Let $p = e^{-(E-E_F)/kT}$, such that $dE = -\frac{kT}{p} dp$. As $E \to E_{n1}$, we have $p \to e^{-(E_{n1}-E_F)/kT}$, and as $E \to \infty$, we have $p \to 0$. Thus:

$$N_{n1} = \frac{2m_{yz_1}}{\pi \hbar^2} kT \int_{0}^{e^{\frac{-(E_{n1}-E_F)}{kT}}} \frac{1}{1+p} \, dp \tag{7.27}$$

---

[4] ξ should be normalized before using in Equation (7.24).

Therefore, we have:

$$N_{n1} = \frac{2m_{yz_1}kT}{\pi\hbar^2} \ln\left(1 + e^{\frac{E_F - E_{n1}}{kT}}\right) \tag{7.28}$$

Similarly:

$$N_{n2} = \frac{4m_{yz_2}kT}{\pi\hbar^2} \ln\left(1 + e^{\frac{E_F - E_{n2}}{kT}}\right) \tag{7.29}$$

Therefore, we have:

$$n(x) = \sum_{n=1}^{\infty} \frac{2m_{yz_1}kT}{\pi\hbar^2} \ln\left(1 + e^{\frac{E_F - E_{n1}}{kT}}\right) |\xi_{n1}|^2 + \sum_{n=1}^{\infty} \frac{4m_{yz_2}kT}{\pi\hbar^2} \ln\left(1 + e^{\frac{E_F - E_{n2}}{kT}}\right) |\xi_{n2}|^2 \tag{7.30}$$

We can also write the expression for Q_n', the inversion surface charge density, as:

$$Q_n' = -q\left(\sum_{n=1}^{\infty} N_{n1} + \sum_{n=1}^{\infty} N_{n2}\right) \tag{7.31}$$

which is nothing but the summation of surface charge density contribution from each quantized level.

7.4.1 Self-consistent Results

In this section, the results obtained from SCHRED [34] and UTQUANT [35] simulators, which solve Schrödinger's and Poisson's equations self-consistently, are discussed. The quantized sub-band energy levels obtained from SCHRED are shown in Figure 7.5. The energies are measured from the bottom of the conduction band E_{cb0} at the Si–SiO$_2$ interface. The simulation shows that as the gate voltage increases, the sub-band energy level shifts away from the conduction band E_{cb0}. This is because of narrowing of the potential well width at a higher gate voltage. As we approach inversion the energy difference between the Fermi level and the sub-bands decreases, which indicates an increase in electron population in these sub-bands. The share of increased electron population in the sub-band E_{11} also increases with gate voltage, as the Fermi level moves closer to E_{11}, and the other sub-band moves away from it. Even though the sub-bands E_{21} and E_{12} have almost the same energy value, the electron population in E_{12} is higher than at E_{21} because of the higher density of states available in E_{12}, as has been discussed in the previous section. Figure 7.6 shows the distribution of electron concentration in the inversion region. It can be seen that the inversion charge density is zero at the Si–SiO$_2$ interface, as against the maximum value predicted by the classical theory. This can be explained by considering the continuity requirement of the wave function at the Si–SiO$_2$ interface. At the Si–SiO$_2$ interface, the carrier electrons encounter a high potential barrier, which reduces

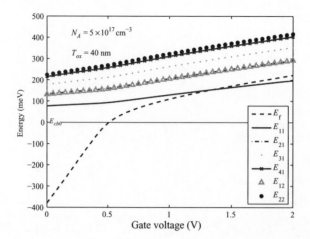

Figure 7.5   Sub-Band energies as a function of gate voltage numerically obtained from SCHRED

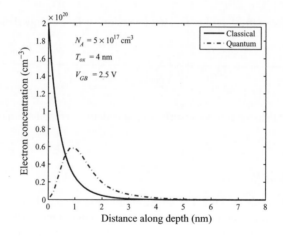

Figure 7.6   Classical and quantum mechanical carrier distributions at the silicon-silicon dioxide interface

the probability of their occurrence in the SiO_2 to zero. To meet the continuity requirement, the electron wave function at the interface is thus zero.

It can also be seen from this figure that the peak of the inversion layer is shifted away from the interface towards the bulk. Due to this shift there is an increase in inversion charge thickness, as shown in Figure 7.7. The increase in inversion charge thickness has serious implications on surface potential, inversion charge density, and capacitance.

Figure 7.8 shows the variation of surface potential as a function of gate-to-bulk bias, as obtained from both classical and quantum calculations. It is observed that in strong inversion for a given gate bias the quantum theory gives a higher surface potential than that predicted by classical theory. This can be attributed to two factors: (i) spreading of inversion charge which results in more electric field lines penetrating deep inside the semiconductor, thus increasing

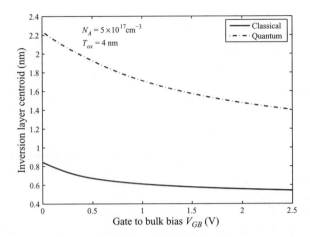

Figure 7.7 Variation of the inversion charge centroid as a function of gate voltage

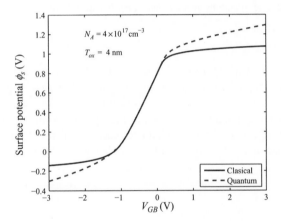

Figure 7.8 Classical and quantum mechanical surface potentials simulated using UTQUANT

the surface potential, and (ii) additional band bending to accommodate inversion charge in the sub-bands which are away from the bottom of the conduction band.

The increase in surface potential from that of the classically estimated value decreases the potential drop across the oxide, which results in a decrease in the inversion charge density, as shown in Figure 7.9. It is also seen from this figure that there is an increase in threshold voltage due to the quantum effect.

The gate-to-bulk capacitance of a MOS capacitor is determined by the series combination of oxide capacitance C'_{ox} and space charge capacitance C'_{sc}. The space charge capacitance in normal operation is viewed as a parallel combination of depletion capacitance C'_b and inversion capacitance C'_n, of which the inversion capacitance is the dominant term in the inversion condition. Classically, the inversion layer thickness is negligible, resulting in a large value of the corresponding capacitance. Therefore, in strong inversion the gate-to-bulk capacitance

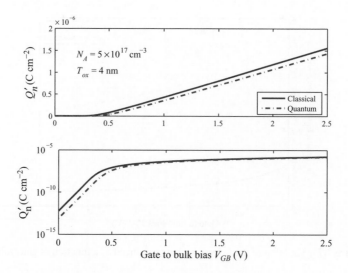

Figure 7.9 Classical and quantum mechanical carrier distributions at the silicon-silicon dioxide interface

approximates to C'_{ox}. But quantum mechanically, the increase in inversion layer thickness reduces the inversion capacitance, and hence the value of the corresponding capacitance. The total gate-to-bulk capacitance, taking the quantum effect into account, will thus be less than the value predicted classically, as shown in Figure 7.10. Similarly, in the accumulation region, the quantum effect on the C–V characteristics can also be observed as a reduction in accumulation capacitance. The quantization of hole energy accounts for the reduced value in capacitance.

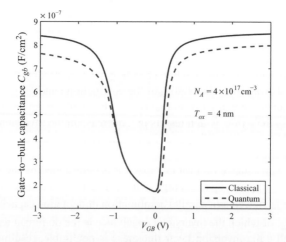

Figure 7.10 Classical and quantum mechanical carrier gate capacitances

7.5 Approximate Methods

The self-consistent calculation demands numerical computation, which is not a convenient option for circuits simulators. The two commonly used approximations are the triangular well approximation and the variational approach discussed below. Most of the compact MOSFET models use the outcome of these approximations to account for the quantum effect on surface potential, threshold voltage, etc.

7.5.1 Triangular Well Approximation Method

The quantized energy levels and carrier distributions in the substrate can be approximately calculated by assuming the potential well to be triangular, which decouples the Schrödinger's and Poisson's equations. The Triangular Well Approximation (TWA) is valid only in weak inversion where the inversion charge density is small as compared to the depletion charge density. For an electron in a triangular potential well, Schrödinger's equation has a closed form solution involving Airy functions, given as:

$$\xi_{in}(x) = \mathrm{Ai}\left[\left(\frac{2m_{xn}qF_s}{\hbar^2}\right)^{\frac{1}{3}}\left(x - \frac{E_{in}}{qF_s}\right)\right] \tag{7.32}$$

Here:

$$E_{in} = \left(\frac{\hbar^2}{2m_{xn}}\right)^{\frac{1}{3}}\left[\frac{3}{2}\pi qF_s\left(i - \frac{1}{4}\right)\right]^{\frac{2}{3}} \tag{7.33}$$

denote eigenvalues corresponding to mass m_{xn}, F_s denotes the electric field at the semiconductor surface and $i = 1, 2, 3, ...,$ is an integer. The triangular well approximation provides good accuracy for quantized energies only in weak inversion. In strong inversion, the presence of the inversion charge greatly perturbs the shape of the potential well for the triangular approximation to be valid. In spite of its inaccuracy the results of the TWA are extended in strong inversion with empirical fitting parameters.

7.5.2 Variational Method

Another commonly used approximation method in the analysis of the quantum confinement effect in a MOS transistor is the variational approach. In the variational method, we can have an approximate estimation of the lowest allowed energy state of the quantum well. A guess of the lowest energy wave function satisfying the boundary condition is made with the least number of parameters. For example, the ground state wave function for a quantum well in a MOS structure as proposed by Howard [16] is:

$$\xi_1(x) = \left(\frac{1}{2}b^3\right)^{\frac{1}{2}} x e^{-bx/2} \tag{7.34}$$

The parameter b has to be found, either analytically or numerically, such that it minimizes the energy of the system. The analytical expression for b given in [1] is:

$$b = \left[\frac{12m_{x1}q^2}{\epsilon_{Si}\hbar^2} \left(N_b' + \frac{11}{32}N_n' \right) \right]^{\frac{1}{3}} - \frac{4}{3} \frac{N_A}{\left(N_b' + \frac{11}{32}N_n' \right)} \tag{7.35}$$

where N_b' and N_n' are the number of depletion and inversion charges per unit area respectively. The second term in Equation (7.35) is negligible compared to the first term and is usually neglected in compact model development. The eigenvalue corresponding to the wave function given in Equation (7.34) is:

$$E_{11} = 2 \left(\frac{q^2\hbar}{\sqrt{m_{x1}\epsilon_{Si}}} \right)^{\frac{2}{3}} \left(N_b' + \frac{55}{96}N_n' \right) \left(N_b' + \frac{11}{32}N_n' \right)^{-\frac{1}{3}} \tag{7.36}$$

As we have seen from the self-consistent solution, with an increase in inversion carrier concentration, the potential well narrows at the surface resulting in an increase in spacing between the energy levels. Therefore, in strong inversion more than 80% of the carriers are located in the first allowed or ground energy state [23], as the probability of occupancy given by Fermi–Dirac statistics decreases exponentially for higher sub-bands. The assumption that most of the carriers are in the first allowed energy state is called the *First Sub-band Approximation* (FSA). The energy level estimates obtained from the variational method are more accurate than the values obtained from the triangular well approximation, especially in strong inversion. The average inversion layer depth or centroid of the inversion carrier distribution is found to be $W_i = 3/b$ [2].

7.6 Quantization Correction in Compact MOSFET Models

Though the above approximations have reduced the computational complexity as compared to the self-consistent approach, they cannot be used to explicitly obtain the quantum effect on surface potential, $C\text{–}V$, and $I\text{–}V$ characteristics.

7.6.1 Threshold Voltage Model

Due to the quantization effect, the first allowed energy level is not available at the bottom of the conduction band, but is displaced upwards. This effectively widens the band gap. Hence, an increase in surface potential is required to attain the inversion charge density required at the threshold. This implies that an additional band bending is required at the surface to create the requisite threshold inversion charge density Q_{nT}'. This results in an increase in the threshold voltage. Further, the quantum mechanical distribution of carriers *effectively* increases the oxide thickness, which further necessitates an additional gate voltage to attain the required band bending.

The increase in threshold voltage can be expressed as a function of the additional bend bending $q\Delta\phi_s$ as given by Dort [11]:

$$\Delta V_T = \Delta\phi_s + \Delta V_{ox} = \Delta\phi_s \left(1 + \frac{1}{2C'_{ox}}\sqrt{\frac{q\epsilon_{Si}N_A}{2\phi_f}}\right) \tag{7.37}$$

In the above equation, $\Delta\phi_s$ is the additional surface potential increase over the classical value $(2\phi_f)$ required to maintain the same amount of carrier density at threshold. Thus:

$$\Delta\phi_s = \phi_{sT} - 2\phi_f \tag{7.38}$$

ϕ_{sT} is calculated following the approach given in [21], which makes use of (i) the triangular well approximation for the potential well, and (ii) the fact that at inversion, most of the inversion charges are located in sub-bands E_{11} and E_{12}. From Equations (7.28) and (7.29), and noting that at threshold $(E_{1n} - E_f) \gg 3kT$, we can write:

$$|Q'_{nT}| = qN_{C1}\exp\left(\frac{E_F - E_{11}}{kT}\right) + qN_{C2}\exp\left(\frac{E_F - E_{12}}{kT}\right) \tag{7.39}$$

where $N_{C1} = 2m_{yz1}kT/\pi\hbar^2$ and $N_{C2} = 4m_{yz2}kT/\pi\hbar^2$. The energy level E_{11} is calculated from the triangular well approximation as discussed in Section 7.5.1:

$$E_{11} = \left(\frac{\hbar^2}{2m_{x1}}\right)^{1/3}\left(\frac{9}{8}\pi qF_s\right)^{2/3} \tag{7.40}$$

Similarly, the energy level $E_{12} = 1.70E_{11}$, where E_{12} is obtained by substituting m_{x2} for m_{x1} in Equation (7.40). From Figure (7.1) it can be seen that E_{1i} (where $i = 1, 2$) can be expressed as:

$$E_{1i} - E_F = q\phi_f + \frac{E_g}{2} - q\phi_s + E_{1i} \tag{7.41}$$

Substituting for $E_{1i} - E_F$ in Equation (7.39), and solving for ϕ_s, at threshold we have:

$$\phi_{sT} = \phi_t \ln\left\{\frac{|Q'_{nT}|}{qN_{C1}P\exp\left(\frac{-E_{11}}{kT}\right)\left[1 + 4.42\exp\left(\frac{-0.70E_{11}}{kT}\right)\right]}\right\} \tag{7.42}$$

where $P = \exp\left[-(q\phi_f + E_g/2)/kT\right]$.

Substitution of $\Delta\phi_{sT} = \phi_s - 2\phi_f$ in Equation (7.37) gives the shift in threshold voltage. From the plots given in Figure 7.11 it is seen that ΔV_T, given by Equation (7.37), is in close agreement with the numerical simulation results obtained with SCHRED. As seen from this figure, there is an increase in threshold voltage shift with doping. A higher substrate doping increases the surface field F_s, raising sub-band energy away from the bottom of the conduction band. This requires more band bending which increases the threshold surface potential and also the potential drop across the oxide through an increase in depletion charge density.

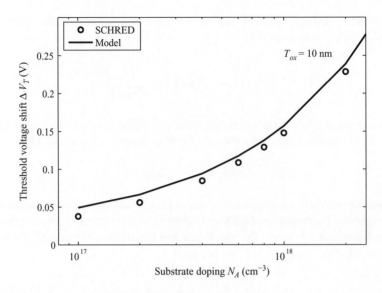

Figure 7.11   Shift in threshold voltage due to quantum effect as a function of substrate doping

7.6.2 Effective Oxide Thickness

For the compact modeling purpose, the displacement of the inversion layer peak away from the interface is modeled as an effective increase in the oxide layer thickness T_{oxQM} [14]. The expression for the effective oxide thickness can be obtained as follows. The inversion charge is given by:

$$Q_n' = -C_{ox}' \left(V_{GB} - V_{fb} - \phi_s + \frac{Q_b'}{C_{ox}'} \right) \tag{7.43}$$

where ϕ_s is the surface potential, which can be written as:

$$\phi_s = \phi_{dep} + \frac{1}{\epsilon_{Si}} Q_n' W_i \tag{7.44}$$

In the above, ϕ_{dep} is the contribution of the depletion charge to the surface potential. It is equal to ϕ_s under the charge sheet approximation with $W_i = 0$, where W_i is the average inversion layer thickness. From Equation (7.44) it can be interpreted that once the centroid of the charge distribution is known, the inversion charge contribution to surface potential is $\epsilon_{Si}^{-1} Q_n' W_i$. Substituting for ϕ_s in Equation (7.43), we can write:

$$Q_n' = -\frac{C_{ox}'}{1 + C_{ox}' \frac{W_i}{\epsilon_{Si}}} \left(V_{GB} - V_{fb} - \phi_{dep} + \frac{Q_b'}{C_{ox}'} \right) \tag{7.45}$$

The term ϵ_{Si}/W_i is known as the *inversion capacitance per unit area*. Then, the expression $C'_{ox}/(1 + C'_{ox}W_i/\epsilon_{Si})$ represents the series combination of oxide capacitance and inversion capacitance. Thus, the total gate capacitance value is reduced. Alternatively, it can be considered that the effective oxide thickness is increased. It follows that:

$$T_{oxQM} = T_{ox} + \frac{\epsilon_{ox}}{\epsilon_{Si}} W_i \tag{7.46}$$

With the inversion layer centroid given by the variational method in Equation (7.35), the effective oxide thickness is given as:

$$T_{oxQM} = T_{ox} + C_1 \left(Q'_b + \frac{11}{32} Q'_n \right)^{-\frac{1}{3}} \tag{7.47}$$

where $C_1 = (12 m_{x1} q/\epsilon_{Si} \hbar^2)^{-1/3}$ and can be used as a fitting parameter.

7.6.3 HiSIM Model

The effective oxide thickness obtained in Equation (7.47) cannot be used to explicitly evaluate Q_n or Q_b as they depend on the oxide thickness itself. In the HiSIM model the increase in oxide thickness is modeled by empirical fitting as a function of gate-to-source voltage V_{GS}, given as:

$$\Delta T_{ox} = \text{QME1} \cdot (V_{GS} - V_T - \text{QME2})^2 + \text{QME3} \tag{7.48}$$

where QME1, QME2, and QME3 are empirical model fitting parameters.

7.6.4 Philips MM11 and PSP Model

In the Philips MM11 and PSP models, the increase in band gap and effective oxide thickness due to the quantum confinement effect are considered. The increase in band gap is included as a correction to the bulk potential. The effective oxide thickness is derived as follows [25]. The effective inversion layer thickness given as $3/b$ can be transformed using Equations (7.35) and (7.36) to:

$$W_i = \frac{2}{3} \frac{E_{11}}{q F_s} \tag{7.49}$$

and E_{11} is approximated as:

$$E_{11} = \frac{3}{2} \left(\frac{3q\hbar}{2\epsilon_{Si}\sqrt{m_{x1}}} \right)^{2/3} \frac{Q'_b + \frac{55}{96} Q'_n}{\left(Q'_b + \frac{11}{32} Q'_n \right)^{1/3}} \tag{7.50}$$

or:

$$E_{11} \approx \frac{3}{2} \left(\frac{3q\hbar F_{eff}}{2\sqrt{m_{x1}}} \right)^{2/3} = \frac{3}{5} qQM(\epsilon_{Si} F_{eff})^{2/3} \tag{7.51}$$

where $QM = \frac{5}{2}q^{-1/3}(3\hbar/2\sqrt{m_{x1}}\epsilon_{Si})^{2/3}$ is a physical constant, which differs for electrons and holes due to their effective masses, and F_{eff} is the effective normal field. The increased oxide thickness can then be written as:

$$T_{oxQM} = T_{ox} + \frac{\epsilon_{ox}}{\epsilon_{Si}} W_i = T_{ox} + \frac{\epsilon_{ox}}{\epsilon_{Si}} \frac{2}{3} \frac{q^{\frac{3}{5}} QM(\epsilon_{Si} F_{eff})^{2/3}}{qF_s} \tag{7.52}$$

Replacing F_s by F_{eff} we have:

$$T_{oxQM} = T_{ox} + T_{ox} QM_{tox} \left(\frac{C'_{ox}}{\epsilon_{Si} F_{eff}} \right)^{1/3} \tag{7.53}$$

where $QM_{tox} = \frac{2}{5} QMC_{ox}^{\prime(2/3)}$. The oxide capacitance can now be written as:

$$C'_{oxQM} = \frac{C'_{ox}}{1 + QM_{tox} \left(\frac{C'_{ox}}{\epsilon_{Si} F_{eff}} \right)^{1/3}} \tag{7.54}$$

In the PSP model the quantum correction introduced into the oxide capacitance in depletion and inversion conditions are given below [32]:

$$C'_{oxQM} = \frac{C'_{ox}}{1 + q_q/[(q_{bm} + n_\mu q_{im})^2 + q_{lim}^2]^{1/6}} \tag{7.55}$$

where $q_{lim} = 10\phi_t, q_q = 0.4 \cdot QMC \cdot QM \cdot C_{ox}^{\prime 2/3}$, and QMC is a model parameter. The band gap widening is modeled as an increase in the bulk potential ϕ_f as given below:

$$\phi_{bQM} = \phi_f + 0.75 q_q q_{b0}^{2/3} \tag{7.56}$$

where $q_{b0} = \gamma \sqrt{\phi_f}$.

7.6.5 EKV Model

The EKV model takes into account the quantization effect by writing the surface potential in the form [13]:

$$\phi_s = \phi_1 + \Delta\phi_s \tag{7.57}$$

where $q\phi_1$ is the band bending up to the first allowed energy level, and $q\Delta\phi_s$ is the first allowed energy level from the bottom of the conduction band at the Si–SiO$_2$ interface. From the triangular approximation of the potential well, we can write:

$$\Delta\phi_s = \frac{E_{11}}{q} = C_1 F_s^{2/3} \tag{7.58}$$

where C_1 is a constant. In Equation (7.58) using $F_s = -(Q_n' + Q_b')/\epsilon_{Si}$ and assuming (100) silicon plane, $\Delta\phi_s = -3.53(Q_n' + Q_b')^{2/3}/q$. In the EKV model this is approximated by a second order Taylor series for implementation convenience in the core model framework, and is given as:

$$\Delta\phi_s = a_0 + a_1(Q_n' + Q_b') + a_2(Q_n' + Q_b')^2 \tag{7.59}$$

The above equation is substituted in the potential balance equation, which is reproduced below:

$$V_{GB} = V_{fb} + (\phi_1 + \Delta\phi_s) - \frac{Q_n'}{C_{ox}'} - \frac{Q_b'}{C_{ox}'} + \frac{1}{\gamma_p} \left(\frac{Q_n' + Q_b'}{C_{ox}'} \right)^2 \tag{7.60}$$

The last term in the above equation is the potential drop across the polydepletion layer. Substituting $\Delta\phi_s$ into the above equation, and rearranging, we have:

$$V_{GB} = V_{fb}^q + \phi_1 - \frac{Q_n' + Q_b'}{C_{ox}'^q} + a_2(Q_n' + Q_b')^2 + \frac{1}{\gamma_p^q} \left(\frac{Q_n' + Q_b'}{C_{ox}'^q} \right)^2 \tag{7.61}$$

where $V_{fb}^q = V_{fb} + a_0$, $C_{ox}'^q = C_{ox}'/(1 - a_1 C_{ox}')$, and $(1/\gamma_p^q)^2 = (1/\gamma_p^2 + a_2 C_{ox}'^2)/(1 - a_1 C_{ox}')^2$. Approximating $Q_b' = -C_{ox}'\gamma_s\sqrt{\phi_1}$, and letting $\gamma_s^q = \gamma_s(1 - a_1 C_{ox}')$, we write:

$$Q_b' = -C_{ox}'^q \gamma_s^q \sqrt{\phi_1} \tag{7.62}$$

The new slope factor $n_q^q = (1/C_{ox}'^q) \partial Q_n'/\partial\phi_1$ at $\phi_1 = \phi_0 + V_P/2$ is:

$$n_q^q = \frac{\gamma_s}{2\sqrt{\phi_0 + V_P/2}} + \frac{|\gamma_p^2|}{\sqrt{(\gamma_p^2 + 2\gamma_s\sqrt{\phi_0 + V_P})^2 + 2\gamma_p^2 V_P}} \tag{7.63}$$

It may be noted that ϕ_1 now plays the role of ϕ_s in the calculation of current with new values of C_{ox}', V_{fb}, γ_s, and γ_p as discussed above. In strong inversion, where the quantum effect is important, ϕ_1 can be written as:

$$\phi_1 = \phi_0 + V_C \tag{7.64}$$

Using the modified parameters as discussed above, the quantum effect is empirically included in the EKV model.

7.7 Quantum Tunneling

In the previous section we have calculated quantized energy levels in the inversion layer. In deriving the expression for energy levels, we had assumed that electrons are trapped in a potential well where the Si–SiO$_2$ barrier height was considered to be infinitely high. Consequently, the wave functions associated with electrons are assumed to be zero at the interface.

In reality, for the SiO$_2$ gate oxide the barrier height is about 3.1 eV. For such a barrier height the assumption that the electron wave function does not penetrate in the gate oxide is not valid. There is always a finite probability that the wave function can penetrate into the oxide due to the requirement of the continuity of wave function at the interface [30].

With the scaling of MOS technology there is a progressive reduction in the thickness of the gate oxide to improve the drive current, and to overcome short channel effects. The projection of ITRS predicts that the scaling of gate oxide is expected to be in the range of 15 Å for a 65 nm node. In this range of gate oxide thicknesses, which defines the width of the potential barrier for carriers, the wave function can extend to the other side of the barrier leading to parasitic gate currents.

The tunneling current can be classified into two categories depending on the type of the barrier. If the tunneling path is through the band gap of the oxide then it is called *Direct Tunneling* (DT). On the other hand, if while tunneling the carrier finds itself in the oxide conduction band then it is called *Fowler–Nordheim* (FN) tunneling. The two different tunneling processes are shown in Figure 7.12, where the shape of the potential barrier is trapezoidal and triangular for DT and FN tunneling respectively. From this figure it is seen that for FN tunneling to occur, the field in the oxide should be sufficiently large so as to bend the conduction band of the oxide more than 3 eV. In modern low-power CMOS technology the direct tunneling, however, is the dominant tunneling mechanism. It was shown in Chapter 1 that the tunneling current depends inversely on the exponent of both the barrier thickness and square root of barrier height.

Tunneling of carriers through the gate constitutes a significant component of leakage current in nanoscale MOSFET circuits, contributing to static power dissipation which limits the

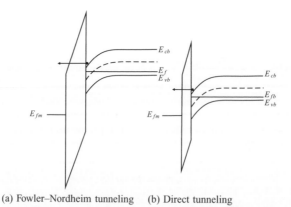

(a) Fowler–Nordheim tunneling   (b) Direct tunneling

Figure 7.12 Tunneling mechanisms though a gate oxide

maximum device count in a silicon chip [31]. Therefore, a proper understanding of the various mechanisms involved is an imperative necessity.

7.8 Gate Current Density

The gate current can be further classified depending upon the type of tunneling carrier, and whether the original state of the tunneling carrier lies in the valence or conduction band. Thus, there can be electron tunneling from the conduction band (ECB), hole tunneling from the valence band (HVB), and electron tunneling from the valence band (EVB). The current densities corresponding to ECB, HVB, and EVB are denoted as J_N, J_P, and J_{VE} respectively. The possible conditions for each of the tunneling current components are shown in Figure 7.13 for a p-substrate and an n^+-gate MOS capacitor.

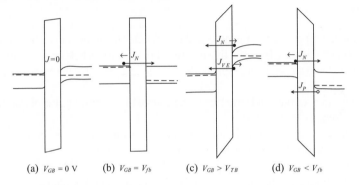

(a) $V_{GB} = 0$ V   (b) $V_{GB} = V_{fb}$ (c) $V_{GB} > V_{TB}$ (d) $V_{GB} < V_{fb}$

Figure 7.13   Carrier types constituting the gate tunneling current

In the accumulation region the gate current is due to electron tunneling from gate-to-bulk and hole tunneling from bulk, and is denoted as I_{GB}. Under the inversion condition the electron tunneling from channel-to-gate is supplied from the source and drain terminal, and the gate-to-channel current I_{GC} gets divided into the gate-to-source current I_{GS} and the gate-to-drain current I_{GD}, as shown in Figure 7.14.

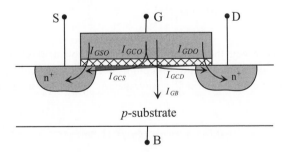

Figure 7.14   Gate current components

In a MOS transistor the gate current due to tunneling can also occur from the gate and the source/drain extension overlap region, as shown in Figure 7.14. The tunneling from these overlap regions is known as Edge Direct Tunneling (EDT). The current in the source and drain overlap region is denoted as I_{GSO} and I_{GDO} respectively. Depending on the relative alignment of the energy band diagrams of the gate and the substrate, and gate and source/drain material, due to the biasing condition, dominance of the respective tunneling components (ECB, EVB, HVB) may differ in the channel and overlap regions. The current through the overlap region is significantly important in modern MOS devices as the overlap area does not scale in the same proportion as the channel area.

7.8.1 The Tsu–Esaki Model

In this section we shall derive the expression for the gate current density due to the tunneling of carriers through the barrier [7, 27]. We shall make the following approximations for deriving the gate current density on account of direct tunneling in the presence of a trapezoidal potential barrier.

- Effective mass approximation is assumed to be valid in silicon and even in oxide of a few atomic layers thickness. The effective masses of the electrons belonging to different valleys of the silicon band structure are represented by a single effective mass m_{Si}, and that in SiO_2 by m_{ox}.
- Though the nature of the band structure in the gate oxide is complex, the parabolic dispersion relation is used.
- The transverse wave vector k_t in a plane parallel to direction of tunneling is assumed to remain unaltered during the tunneling process.

The wave vector $\mathbf{k}$ and kinetic energy E can be decomposed into components along x and the transverse directions as given below:

$$\mathbf{k}^2 = k_x^2 + k_{||}^2 \quad \text{and} \quad E = E_x + E_{||} \tag{7.65}$$

For electrons in the conduction band of silicon, the E–$\mathbf{k}$ relation has the form:

$$E(\mathbf{k}) = E_{cb} + \frac{\hbar^2 k_{||}^2}{2m_{Si}} + \frac{\hbar^2 k_x^2}{2m_{Si}} \tag{7.66}$$

where $k_{||}^2 = k_y^2 + k_z^2$. Now with the assumptions mentioned above, the elemental current density from gate to bulk, $dJ_{G \to B}$, due to the tunneling electrons from bulk to gate as shown in Figure 7.15 with the x-component of the wave vector within the interval dk_x around k_x is written as;

$$dJ_{G \to B} = qv_x(k_x)TC(k_x)n(k_x) \tag{7.67}$$

where v_x is the velocity of the tunneling electrons which can be written as $v_x = \hbar k_x / m_{Si}$, TC is the transmission coefficient and $TCn(k_x)$ is the density of the electrons which can tunnel with a wave vector within the interval dk_x around k_x.

For electrons to tunnel from bulk-to-gate with a total energy $E_t = E_c + E$, it is necessary that the energy state with energy E_t in the bulk should be filled, and that in the gate should be

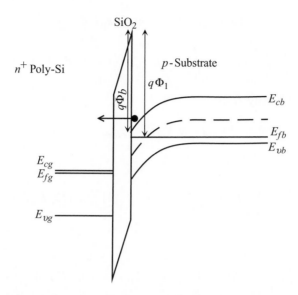

Figure 7.15   Electron tunneling from substrate to gate (ECB)

empty. The probability distribution of filled and vacant states is given by the Fermi function $f_b(E_t, E_{fb})$ and $[1 - f_g(E_t, E_{fg})]$, where E_{fb} and E_{fg} are the Fermi levels in bulk and gate respectively. Thus, with $g(k_x)$ density of k_x states, $n(k_x)$ can be written as:

$$n(k_x) = g(k_x) f_b(E_t, E_{fb}) \left[1 - f_g(E_t, E_{fg})\right] dk_x \qquad (7.68)$$

which can be also expressed by:

$$n(k_x) = \int_0^\infty \int_0^\infty g(k) f_b(E_t, E_{fb}) \left[1 - f_g(E_t, E_{fg})\right] dk_y \, dk_z \qquad (7.69)$$

where $g(k)$ is the 3-D k-space density of states including electron spin, which is given by $1/4\pi^3$:

$$n(k_x) = \frac{1}{4\pi^3} \int_0^\infty \int_0^\infty f_b(E_t, E_{fb}) \left[1 - f_g(E_t, E_{fg})\right] dk_y \, dk_z \qquad (7.70)$$

Substituting for v_x and $n(k_x)$ in Equation (7.67) we have:

$$dJ_{G \to B} = \frac{q}{4\pi^3} TC(k_x) \frac{\hbar k_x}{m_{Si}} \int_0^\infty \int_0^\infty f_b(E_t, E_{fb}) \left[1 - f_g(E_t, E_{fg})\right] dk_y \, dk_z \qquad (7.71)$$

If we make the change of variables $k_y = k_{||} \cos \theta$ and $k_z = k_{||} \sin \theta$, such that the elemental area $dy dz$ will be $k_{||} dk_{||} d\theta$ in polar coordinates, then Equation (7.71) can be written as:

$$J_{G \to B} = \frac{m_{Si} q}{4\pi^3 \hbar^3} \int_0^\infty TC(k_x) \left[\int_0^{2\pi} \int_0^\infty f_b(E_t, E_{fb}) \left[1 - f_g(E_t, E_{fg})\right] dE_{||} d\theta\right] dE_x \qquad (7.72)$$

where we have made use of the relation $dk_\| = (m_{Si}/\hbar k_\|)dE_\|$.

$$J_{G \to B} = \frac{m_{Si}q}{2\pi^2\hbar^3} \int_0^\infty TC(k_x) \left[\int_0^\infty f_b(E_t, E_{fb}) \left[1 - f_g(E_t, E_{fg}) \right] dE_\| \right] dE_x \qquad (7.73)$$

The net gate current density J_G is:

$$J_G = J_{G \to B} - J_{B \to G} \qquad (7.74)$$

where:

$$J_{B \to G} = \frac{m_{Si}q}{2\pi^2\hbar^3} \int_{-qV_{ox}}^\infty TC(k_x) \left[\int_0^\infty f_g(E_t, E_{fg}) \left[1 - f_b(E_t, E_{fb}) \right] dE_\| \right] dE_x \qquad (7.75)$$

Therefore we have,

$$J_G = \frac{m_{Si}q}{2\pi^2\hbar^3} \int_0^\infty TC(k_x) \left[\int_0^\infty [f_b(E_t, E_{fb}) - f_g(E_t, E_{fg})] dE_\| \right] dE_x \qquad (7.76)$$

The above equation can be written as:

$$J_G = \frac{m_{Si}q}{2\pi^2\hbar^3} \int_0^\infty TC(E_x) S(E_x) \, dE_x \qquad (7.77)$$

where:

$$S(E_x) = \int_0^\infty [f_b(E_t, E_{fb}) - f_g(E_t, E_{fg})] dE_\| \qquad (7.78)$$

is called the *supply function*. Equation (7.77) is known as the Tsu–Esaki relation. Using the Fermi–Dirac distribution function, we have the following relation for the supply function:

$$S(E_x) = kT \ln \left(\frac{1 + \exp\left\{ \frac{[E_{fb} - (E_{cb0} + E_x)]}{kT} \right\}}{1 + \exp\left\{ \frac{[E_{fg} - (E_{cg} + E_x + qV_{ox})]}{kT} \right\}} \right) \qquad (7.79)$$

whereas the Maxwell–Boltzmann distribution gives:

$$S(E_x) = kT \left(\exp\left\{ \frac{[E_{fb} - (E_{cb0} + E_x)]}{kT} \right\} - \exp\left\{ \frac{[E_{fg} - (E_{cg} + E_x + qV_{ox})]}{kT} \right\} \right)$$

$$(7.80)$$

where E_{cb0} is the conduction band minimum at the Si–SiO$_2$ interface.

The simulated gate current density for a MOS capacitor obtained from the UTQUANT simulator is shown in Figure 7.16. Part (a) shows the gate current density for an n^+–SiO$_2$–p-sub-MOS capacitor, while part (b) shows the gate current for an n^+–SiO$_2$–n^+-sub-MOS which

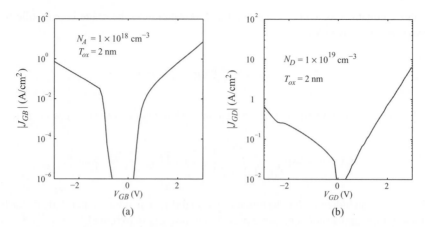

Figure 7.16 Gate tunneling current versus gate voltage: (a) gate-to-bulk tunneling current density in the channel region, and (b) gate-to-drain tunneling current density in the overlap region

resembles the drain overlap region of the n-channel MOS transistor. More detailed analysis of the various components of the gate current, and their dependences on bias, is available in Cai and Sah [7] and Yang et al. [24].

7.8.2 Tunneling Probability

Evaluating the tunneling probability/coefficient for an arbitrary potential barrier requires a numerical approach. In the transfer matrix method the tunneling barrier of trapezoidal shape is divided into small regions, where in every region the energy barrier height can be assumed to be constant. This allows the analytical solution of Schrödinger's equation in that region. The transfer matrix of the whole barrier is the product of all the matrices of each region, which is obtained by maintaining the continuity in wave function and its derivative at the boundaries of each region. Such a method requires numerical techniques and is not suitable for a compact model development of gate current.

The Wentzel–Kramers–Brillouin (WKB) approximation can be used to get an analytical expression for the tunneling probability. From WKB theory, the direct tunneling probability TC for a slowly varying potential barrier can be approximated as:[5]

$$TC(E_x) = \exp\left[-2\int_{-T_{ox}}^{0} k(x)\,dx\right] \tag{7.81}$$

---

[5] A modified expression of TC appropriate for an abrupt barrier is given in [7].

In the above equation we will assume that the E–k relation in the oxide is parabolic with an effective mass m_{ox} given as:

$$k(x) = \sqrt{\frac{2m_{ox}}{\hbar^2}[E_b(x) - E_x]} \tag{7.82}$$

where $E_b(x)$ is the spatial variation of the conduction band edge of the oxide w.r.t. E_{cb0}. Thus:

$$TC(E_x) = \exp\left(-\frac{2}{\hbar}\int_{-T_{ox}}^{0}\left\{\sqrt{2m_{ox}[E_b(x) - E_x]}\right\}dx\right) \tag{7.83}$$

The above equation indicates that the tunneling probability depends exponentially on the barrier height and width, and a more energetic particle is more likely to tunnel.

In the case of MOS structures, $E_b(x)$, neglecting the image potential, will be:

$$E_b(x) = q\Phi_b + \frac{qV_{ox}}{T_{ox}}x \tag{7.84}$$

where Φ_b is the barrier height above E_c at the Si–SiO$_2$ interface.

Integration of the equation for tunneling probability with $E_b(x)$ given in Equation (7.84), gives:

$$TC(E_x) = \exp\left[-\frac{2^{3/2}\sqrt{m_{ox}}}{\hbar}\int_{-T_{ox}}^{0}\left(\sqrt{q\Phi_b + \frac{qV_{ox}}{T_{ox}}x - E_x}\right)dx\right] \tag{7.85}$$

Integrating:

$$TC(E_x) = \exp\left\{-\frac{2^{3/2}\sqrt{m_{ox}}}{\hbar}\frac{2}{3}\frac{T_{ox}}{qV_{ox}}\left[(q\Phi_b - E_x)^{3/2} - (q\Phi_b - qV_{ox} - E_x)^{3/2}\right]\right\} \tag{7.86}$$

7.8.3 Gate Current and its Partition

In the previous section we have derived the expression for current density in terms of transmission probability and supply function. In the case of a MOS transistor with inversion channel, the current density varies along the channel from source to drain in the presence of tunneling for an applied V_{DS} and the total gate-to-channel current can be written as:

$$I_{GC} = W\int_{0}^{L} J_{GC}(y)\,dy \tag{7.87}$$

As shown in Figure 7.14 this gate-to-channel current will be divided into its source and drain current components, denoted as J_{GS} and J_{GD} respectively. From the current continuity equation it can be shown that the partition of J_{GC} into J_{GS} and J_{GD} is similar to the inversion charge partition [26] discussed in Chapter 6, and is written as:

$$I_{GS} = W \int_0^L \left(1 - \frac{y}{L}\right) J_{GC}(y)\,\mathrm{d}y \tag{7.88}$$

and:

$$I_{GD} = W \int_0^L \frac{y}{L} J_{GC}(y)\,\mathrm{d}y \tag{7.89}$$

7.9 Compact Gate Current Models

To obtain the gate current model we have to evaluate the integral given in Equation (7.77), with the tunneling probability given by the WKB approximation, and the supply function given by the Fermi–Dirac or Maxwell–Boltzmann statistics. In order to be able to integrate the expression for gate current we make the following approximation for the supply function.

Assume that the channel is strongly inverted (where EVB is significant, $E_f < E_{cb0}$), and the Fermi–Dirac distribution is approximated by a step function. Now, electrons with energy $E_t > E_{cb0}$ and $E_t < E_{fb}$ will be capable of tunneling.

$$S(E_x) = \int_0^{E_{fb} - E_{cb0} - E_x} [f_b(E_t, E_{fb}) - f_g(E_t, E_{fg})]\,\mathrm{d}E_{\|} \tag{7.90}$$

Under the above-stated conditions, we have:

$$S(E_x) = \int_0^{E_{fb} - E_{cb0} - E_x} [1 - 0]\,\mathrm{d}E_{\|} = E_{fb} - E_{cb0} - E_x \tag{7.91}$$

Substituting $S(E_x)$ from Equation (7.91), and TC from Equation (7.86), into Equation (7.77) gives:

$$J_G = a \int_0^{E_{fb} - E_{cb0}} \exp\left\{d\left[(q\Phi_b - E_x)^{3/2} - (q\Phi_b - qV_{ox} - E_x)^{3/2}\right]\right\}$$
$$\times (E_{fb} - E_{cb0} - E_x)\,\mathrm{d}E_x \tag{7.92}$$

where $a = m_{Si}q/2\pi^2\hbar^3$, and $d = -(2^{3/2}\sqrt{m_{ox}}/\hbar) \times \frac{2}{3} \times (T_{ox}/qV_{ox})$.

Now substituting for $q\Phi_b = q\Phi_1 + E_{fb} - E_{cb0}$, where $q\Phi_1$ is the energy difference between the oxide conduction band and the Fermi level at the Si–SiO$_2$ interface, gives:

$$J_G = a \int_0^{E_{fb} - E_{cb0}} \exp\left(d\left\{[q\Phi_1 - (E_{cb0} - E_{fb} + E_x)]^{3/2}\right.\right.$$
$$\left.\left. - [q\Phi_1 - qV_{ox} - (E_{cb0} - E_{fb} + E_x)]^{3/2}\right\}\right)(E_{fb} - E_{cb0} - E_x)\,\mathrm{d}E_x \tag{7.93}$$

To simplify further, consider the Taylor series expansions of the sub-expressions in the exponent term at $q\Phi_1$ and $(q\Phi_1 - qV_{ox})$ respectively. We get:

$$[q\Phi_1 - (E_{cb0} - E_{fb} + E_x)]^{3/2} = (q\Phi_1)^{3/2} - \frac{3}{2}(E_{cb0} - E_{fb} + E_x)(q\Phi_1)^{1/2} \qquad (7.94)$$

and:

$$[q\Phi_1 - qV_{ox} - (E_{cb0} - E_{fb} + E_x)]^{3/2} = (q\Phi_1 - qV_{ox})^{3/2}$$
$$- \frac{3}{2}(E_{cb0} - E_{fb} + E_x)(q\Phi_1 - qV_{ox})^{1/2} \quad (7.95)$$

Substituting into Equation (7.93), we have:

$$\exp\left(d\left\{(q\Phi_1)^{3/2} - \frac{3}{2}(E_{cb0} - E_{fb} + E_x)(q\Phi_1)^{1/2}\right.\right.$$
$$\left.\left. - \left[(q\Phi_1 - qV_{ox})^{3/2} - \frac{3}{2}(E_{cb0} - E_{fb} + E_x)(q\Phi_1 - qV_{ox})^{1/2}\right]\right\}\right) \quad (7.96)$$

$$\exp\left\{d\left[(q\Phi_1)^{3/2} - (q\Phi_1 - qV_{ox})^{3/2}\right] - \frac{3}{2}d\left[(q\Phi_1)^{1/2} - (q\Phi_1 - qV_{ox})^{1/2}\right]\right.$$
$$\left. \times (E_{cb0} - E_{fb} + E_x)\right\} \qquad (7.97)$$

Let $b = d\left[(q\Phi_1)^{3/2} - (q\Phi_1 - qV_{ox})^{3/2}\right]$. The term now transforms to:

$$\exp(b)\exp\left\{-\frac{3}{2}d\left[(q\Phi_1)^{1/2} - (q\Phi_1 - qV_{ox})^{1/2}\right](E_{cb0} - E_{fb} + E_x)\right\} \qquad (7.98)$$

Substituting back into Equation (7.93), we have:

$$J_G = a\exp(b)\int_0^{E_{fb} - E_{cb0}} \exp\left\{-\frac{3}{2}d\left[(q\Phi_1)^{1/2} - (q\Phi_1 - qV_{ox})^{1/2}\right]\right.$$
$$\left. \times (E_{cb0} - E_{fb} + E_x)\right\}(E_{fb} - E_{cb0} - E_x)\,dE_x \quad (7.99)$$

or:

$$J_G = a\exp(b)\int_0^{E_{fb} - E_{cb0}} (E_{fb} - E_{cb0} - E_x)\exp\left[c(E_{cb0} - E_{fb} + E_x)\right]\,dE_x \qquad (7.100)$$

where $c = -\frac{3}{2}d\left[(q\Phi_1)^{1/2} - (q\Phi_1 - qV_{ox})^{1/2}\right]$.

Now make a change of variables, i.e. $(E_{cb0} - E_{fb} + E_x) = y$. For the new limits, at $E_x = 0$ we have $y = (E_{cb0} - E_{fb})$, and at $E_x = E_{fb} - E_{cb0}$ we have $y = 0$. Thus:

$$JG = a \exp(b) \int_{(E_{cb0} - E_{fb})}^{0} -y \exp(cy) \, dy \tag{7.101}$$

$$JG = a \exp(b) \int_{0}^{(E_{cb0} - E_{fb})} y \exp(cy) \, dy \tag{7.102}$$

$$JG = a \exp(b) \left[\frac{y}{c} \exp(cy) - \frac{1}{c^2} \exp(cy) \Big|_{0}^{|(E_c - E_{fb})|} \right] \tag{7.103}$$

$$JG = a \exp(b) \left\{ \frac{1}{c^2} + \frac{(E_{cb0} - E_{fb})}{c} \exp[c(E_{cb0} - E_{fb})] - \frac{1}{c^2} \exp[c(E_{cb0} - E_{fb})] \right\} \tag{7.104}$$

which is approximately $(a/c^2) \exp(b)$ for $E_{fb} > E_{cb0} + 2kT$. Substituting the values for a, b, and c, the resulting approximate expression for JG (when $E_{fb} \gg E_{cb0}$) works out to be:

$$JG = \frac{q^2}{16\pi^2 \hbar \Phi_b} \frac{1}{\left[1 - \sqrt{1 - \frac{V_{ox}}{\Phi_b}} \right]^2} F_{ox}^2 \exp \left\{ -\frac{4}{3} \frac{\sqrt{2m_{ox}}(q\Phi_b)^{3/2}}{\hbar q F_{ox}} \left[1 - \left(1 - \frac{V_{ox}}{\Phi_b} \right)^{3/2} \right] \right\} \tag{7.105}$$

which is known as the Schuegraf model for direct tunneling [9, 10]. Note that Φ_b has been substituted for Φ_1, as they are considered to be approximately equal. An average barrier height approximation is sometimes made as in Stern [6] to arrive at an analytical expression for gate current.

7.9.1 BSIM4 Model

BSIM4 adopts the model developed by Lee and Hu [12], which introduces a semi-empirical term in the pre-exponential factor of Equation (7.105), that can predict all the significant tunneling components such as *ECB*, *EVB*, and *HVB*:

$$JT = \frac{q^2}{8\pi h \Phi_b \epsilon_{ox}} C(V_G, V_{ox}, T_{ox}, \Phi_b) \exp \left\{ \frac{-8\pi \sqrt{2m_{ox}}(q\Phi_b)^{3/2} \left[1 - \left(1 - \frac{|V_{ox}|}{\Phi_b} \right)^{3/2} \right]}{3hq|F_{ox}|} \right\} \tag{7.106}$$

where:

$$C(V_G, V_{ox}, T_{ox}, \Phi_b) = \exp \left[\frac{20}{\Phi_b} \left(\frac{|V_{ox}| - \Phi_b}{\Phi_{bo}} + 1 \right)^{\alpha} \left(1 - \frac{|V_{ox}|}{\Phi_b} \right) \right] \left(\frac{V_G}{T_{ox}} \right) N \tag{7.107}$$

α is a process dependent fitting parameter, Φ_{bo} is the Si–SiO$_2$ barrier height, which is 3.1 eV for electrons and 4.5 eV for holes, and Φ_b is the actual tunneling barrier height corresponding to the category of tunneling mechanism such as ECB, EVB, or HVB mechanism. The term N represents the carrier population at the injection end, and for ECB and HVB tunneling mechanism it is given as:

$$N = \frac{\epsilon_{ox}}{T_{ox}} \left\{ n_{inv}\phi_t \ln\left[1 + e^{\left(\frac{V_{GE}-V_{TS}}{n_{inv}\phi_t}\right)}\right] + n_{acc}\phi_t \ln\left[1 + e^{\left(-\frac{V_G-V_{fb}}{n_{acc}\phi_t}\right)}\right] \right\} \tag{7.108}$$

where n_{inv} and n_{acc} are fitting parameters. V_{GE} is $V_G - \phi_p$. For the EVB tunneling mechanism N is given by:

$$N = \frac{\epsilon_{ox}}{T_{ox}} \left\{ n_{EVB}\phi_t \ln\left[1 + e^{\left(\frac{|V_{ox}|-E_g/q}{n_{EVB}\phi_t}\right)}\right] \right\} \tag{7.109}$$

where n_{EVB} is a fitting parameter.

The BSIM4 model equation for the tunneling current is:

$$J_T = A \left(\frac{T_{oxref}}{T_{ox}P}\right)^{ntox} \frac{V_{aux}V_{appl}}{(T_{ox}P)^2} e^{[-B(\alpha-\beta|V_{ox}|)(1+\gamma|V_{ox}|)T_{ox}P]} \tag{7.110}$$

where $A = q^2/(8\pi h \Phi_b)$, $B = 8\pi\sqrt{2qm_{ox}}(q\Phi_b)^{3/2}/(3h)$, T_{oxref} is the reference oxide thickness for parameter extraction, $ntox$ is a fitting parameter, and V_{aux} is an auxiliary function modeling the density of tunneling carriers and available states in various regions of operation, as given in Table 7.1 [36]. The model parameters α, β, γ, and P have different values depending

Table 7.1 Expressions for V_{aux} and V_{appl} used in Equation (7.110) for the evaluation of the tunneling current components in different regions of operation

Current component	Region of operation	V_{aux}, V_{appl}
J_{GB}	Acc.	$V_{aux} = NIGBACC\phi_t \ln\left[1 + e^{\left(-\frac{V_{GB}-V_{fbb}}{NIGBACC\phi_t}\right)}\right]$
		$V_{appl} = V_{GB}$
	Inv./Dep.	$V_{aux} = NIGBINV\phi_t \ln\left[1 + e^{\left(-\frac{V_{oxdepinv}-EIGBINV}{NIGBINV\phi_t}\right)}\right]$
		$V_{appl} = V_{GB}$
J_{GC}	Inv./Dep.	$V_{aux} = NIGC\phi_t \ln\left[1 + e^{\left(-\frac{V_{GSE}-V_{THO}}{NIGC\phi_t}\right)}\right]$
		$V_{appl} = V_{GSE}$
J_{GS}	All	$V_{aux} = |V_{GS} - V_{fbsd}|$
		$V_{appl} = V_{GS}$
J_{GD}	All	$V_{aux} = |V_{GD} - V_{fbsd}|$
		$V_{appl} = V_{GD}$

on the tunneling mechanism, regions of operation, and current components described in Figure 7.14. They are $AIGACC$, $AIGBINV$, $AIGC$, and $AIGSD$ for α, $BIGACC$, $BIGBINV$, $BIGC$, and $BIGSD$ for β, and $CIGACC$, $CIGBINV$, $CIGC$, and $CIGSD$ for γ respectively in the current component J_{GB} in accumulation, J_{GB} in inversion/depletion, J_{GC}, and $J_{(GS,D)}$.

To obtain a gate current partition we first obtain the perturbation on the electron quasi-Fermi $V_{CS}(y)$ due to the presence of gate current. The $V_{CS}(y)$ must satisfy the following current continuity equation:

$$\frac{1}{W}\frac{\mathrm{d}I(y)}{\mathrm{d}y} + J_G(y) = 0 \tag{7.111}$$

After substituting the current expression in terms of the channel potential given in Chapter 4, the above equation can be written as:

$$\frac{\mathrm{d}[(V_{GST} - nV_{CS})(\mathrm{d}V_{CS}/\mathrm{d}y)]}{\mathrm{d}y} + \frac{J_G(y)}{\mu_{eff}C'_{ox}} = 0 \tag{7.112}$$

The complementary solution of the above equation is the channel potential $V_{CS}(y)$ in the absence of a gate current, and can be obtained following the procedure discussed in Section (4.7). Assuming $n = 1$:

$$V_{CS0}(y) = V_{GST} - \sqrt{V_{GST}^2 - 2\,(V_{GST} - V_{DS}/2)\,V_{DS}(y/L)} \tag{7.113}$$

where $V_{GST} = V_{GS} - V_{TS}$. This equation is rearranged in a different form to allow a suitable approximation as shown later.

$$V_{CS0}(y) = V_{GST} - V_{GST}\left[\sqrt{1 - \frac{2\,(V_{GST} - V_{DS}/2)\,V_{DS}(y/L)}{V_{GST}^2}}\right] \tag{7.114}$$

$V_{CS0}(y)$ can now be further approximated as a linear function of y, by ignoring higher order terms in the binomial expansion of the RHS in Equation (7.114), to give:

$$V_{CS0}(y) = Ky \tag{7.115}$$

where $K = (V_{GST} - V_{DS}/2)V_{DS}/LV_{GST}$. The above approximation is not valid as we go towards the saturation or pinch-off point where $V_{CS0} = V_{DSsat}$. However, this loss of accuracy is not of prime concern as for the channel near this point, the gate current is insignificant.

In the presence of a gate current there is a perturbation of channel potential given as:

$$V_{CS}(y) = V_{CS0}(y) + V_{CS1}(y) \tag{7.116}$$

where $V_{CS1}(y)$ is the perturbation introduced. It is assumed that $V_{CS0}(y) \gg V_{CS1}(y)$ is valid. Using this assumption, Equation (7.112) can be written as:

$$(V_{GST} - V_{CS0})\frac{\mathrm{d}V_{CS1}^2}{\mathrm{d}y^2} - V_{CS1}\frac{\mathrm{d}^2 V_{CS0}}{\mathrm{d}y^2} - 2\frac{\mathrm{d}V_{CS0}}{\mathrm{d}y}\frac{\mathrm{d}V_{CS1}}{\mathrm{d}y} + \frac{J_G(y)}{\mu_{eff}C_{ox}'} = 0 \qquad (7.117)$$

The above differential equation can be solved with the boundary condition that $V_{CS1}(0) = V_{CS1}(L) = 0$. The solution is given as:

$$V_{CS1}(y) = -\frac{J_{G0}\left[(e^{-B^*Ky} - 1) + \frac{y}{L}\left(1 - e^{-B^*KL}\right)\right]}{\mu_{eff}C_{ox}'B^{*2}K^2(V_{GST} - Ky)} \qquad (7.118)$$

In obtaining the above solution the gate current $J_G(y)$ is approximated as:

$$J_G(y) \approx A F_{ox}^2(y)e^{-B/F_{ox}(y)} \approx A F_{oxs}^2 e^{-BT_{ox}/(V_{oxs}-V_{CS})} \approx J_{G0}e^{-B^*V} \qquad (7.119)$$

where J_{G0} is the gate tunneling current density with $V_{DS} = 0$, $B^* = P_{igcd}BT_{ox}/V_{oxs}^2$, $V_{oxs} \approx V_{gs}$ is the voltage across the gate oxide at $y = 0$, and P_{igcd} is the fitting parameter.

The partitioned gate currents I_{GCS} and I_{GCD} can be obtained as the gradient of the quasi-Fermi levels of electrons w.r.t. y at the source and drain ends respectively:

$$I_{GCS} = \mu_{eff}C_{ox}'V_{GST}W \left.\frac{\mathrm{d}V_{CS1}}{\mathrm{d}y}\right|_{y=0} \qquad (7.120)$$

$$= \frac{J_{G0}WL(B^*V_{DSeff} + e^{-B^*V_{DSeff}} - 1)}{B^{*2}V_{DSeff}^2} \qquad (7.121)$$

and:

$$I_{GCD} = -\mu_{eff}C_{ox}'(V_{GST} - V_{DSeff})W \left.\frac{\mathrm{d}V_{CS1}}{\mathrm{d}y}\right|_{y=L} \qquad (7.122)$$

$$= -\frac{J_{G0}WL(B^*V_{DSeff}e^{-B^*V_{DSeff}} + e^{-B^*V_{DSeff}} - 1)}{B^{*2}V_{DSeff}^2} \qquad (7.123)$$

where $V_{DSeff} = KL$. The gate-to-channel current as a function of the drain bias is:

$$I_{GC} = \frac{J_{G0}WL(1 - e^{-B^*V_{DSeff}})}{B^*V_{DSeff}} \qquad (7.124)$$

where the model parameter B^* is named as PIGCD.

7.9.2 PSP Model

In the PSP model it is assumed that the electrons that tunnel do so with a kinetic energy of $E_T = q\psi_t$ [28, 29], as shown in Figure 7.17. Under this assumption we can write:

$$J_G(y) = J_0 TC(y) S_P(y) \qquad (7.125)$$

where $J_0 = qm_{Si}k^2T^2/2\pi^2\hbar^3$ and $S_P(y) = S(y)/kT$. The transmission coefficient $TC(y)$ has been derived in Equation (7.86) which, after a second order Taylor series expansion, can be written as:

$$TC(y) = \exp\left\{-B_0\left[G_1 + G_2\left(\frac{V_{ox}}{\Phi_b}\right) + G_3\left(\frac{V_{ox}}{\Phi_b}\right)^2\right]\right\} \qquad (7.126)$$

In the above expression $B_0 = 4T_{ox}\sqrt{m_{ox}q\Phi_B}/3\hbar$. The coefficients G_1, G_2, and G_3, though they result from the expansion, are preferred as fitting parameters in order to account for the inaccuracies resulting from various assumptions in deriving the tunneling probability. Referring to Equation (7.79) and Figure 7.17, the supply function can be written as:

$$S(y) = kT \ln\left[\frac{1 + e^{\frac{(\phi_s - V_{CB} - \alpha_b - \psi_t)}{\phi_t}}}{1 + e^{\frac{(\phi_s - V_{GB} - \alpha_b - \psi_t)}{\phi_t}}}\right] \qquad (7.127)$$

The term ψ_t in Equation (7.127) is taken to be zero for the inversion condition, when $V_{ox} > 0$ and electrons tunnel from channel to gate. On the other hand, in the accumulation mode of operation, when $V_{ox} < 0$ and electrons tunnel from gate to channel, $\psi_t = -V_{ox} + G_0\phi_t$, as shown in Figure 7.17. G_0 acts as a fitting parameter to accommodate for a possible conduction band offset difference of the Si–SiO$_2$ and polysilicon–SiO$_2$ boundaries. $q\alpha_b$, shown in Figure 7.17, represents the energy difference between quasi-Fermi level and bottom of the conduction band in the bulk.

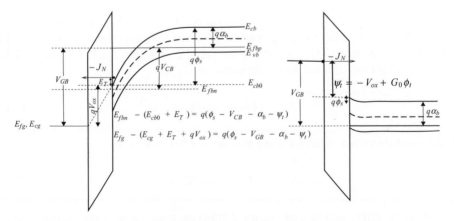

Figure 7.17   Band diagram showing electron tunneling in inversion and accumulation

The gate-to-channel current is:

$$I_{GC} = W \int_0^L J_G(y)\, dy = W \int_{-\phi/2}^{\phi/2} J_G(u) \frac{dy}{du}\, du \tag{7.128}$$

where $\phi = \phi_{sD} - \phi_{sS}$ and $u = \phi_s - \phi_{sm}$. From the symmetric linearization scheme of the PSP model, we can write:

$$\frac{dy}{du} = \frac{L(H - u)}{H\phi} \tag{7.129}$$

where $H = q_{im}/\alpha_m + \phi_t$ and q_{im} denotes the normalised inversion charge density at $\phi_s = \phi_m$. Substituting Equation (7.129) into Equation (7.128), we have:

$$I_{GC} = \frac{WL}{H\phi} \int_{-\phi/2}^{\phi/2} J_G(u)(H - u)\, du \tag{7.130}$$

Integration of the above expression, with J_G derived in Equation (7.125), still requires a numerical technique. Taylor series expansions of the power term in the expressions for TC (Equation (7.126)) and S_P (Equation (7.127)) around ϕ_{sm} give:

$$TC(u) = TC_m \exp\left(\frac{-u}{u_0}\right) \tag{7.131}$$

and:

$$S_P(u) = S_{Pm} + \frac{u}{u_F} \tag{7.132}$$

where $TC_m = \exp\left\{-B_0\left[G_1 + G_2\left(V_{oxm}/\Phi_b\right) + G_3\left(V_{oxm}/\Phi_b\right)^2\right]\right\}$, $u_0 = \Phi_b/(G_2 + 2G_3 |V_{oxm}/\Phi_b|)$, and $u_F = (\partial S_P/\partial \phi_s)_{\phi_s = \phi_m}$ is given in Gu $et\ al.$ [28]. Substituting Equations (7.131) and (7.132) into Equation (7.125), and then substituting the value for $J_G(u)$ in Equation (7.128) gives, after integration, a closed form expression for I_{GC} as:

$$I_{GC} = I_{GINV} S_{Pm} TC_m p_{gc} \tag{7.133}$$

The gate-to-drain current evaluated using Equation (7.89) is given as:

$$I_{GCD} = I_{GINV} S_{Pm} TC_m p_{gd} \tag{7.134}$$

where $I_{GINV} = J_0 WL$ is taken as an empirical fitting parameter, and p_{gc} and p_{gd} are functions of ϕ_m, ϕ_{sS}, and ϕ_{sD} as given in the PSP 102.1 User Manual [32].

7.9.3 HiSIM Model

Based on the physical mechanism of direct tunneling as modeled by Kane [4], and from the experimental observation of the gate current dependence on the entire channel length, the following empirical model is proposed in HiSIM involving three model parameters, surface potentials at source and drain ends ϕ_{sS} and ϕ_{sD} respectively, and the voltage drop across the gate oxide:

$$I_G = qGLEAK1.\frac{F^2}{E_g^{\frac{1}{2}}} \exp\left(-GLEAK2.\frac{E_g^{\frac{3}{2}}}{F}\right) WL \tag{7.135}$$

where $F = (V_G' - \phi_S)/T_{ox}$, $\phi_S = (\phi_{sS} + \phi_{sD})/GLEAK3$ [33], and $V_G' = V_{GS} - V_{fb} + \Delta V_{th}$.

7.10 Gate Induced Drain Leakage (GIDL)

In a MOS transistor the gate, substrate, and drain form a three-terminal gated diode structure. For the normal 'off' condition, with drain at the positive potential and the gate grounded, the drain portion under the gate overlapped region is in the depletion condition as shown in Figure 7.18.

The physical processes involved under such a biasing condition are summarized as under:

- The depletion region at the gate overlapped Si–SiO$_2$ interface of the n^+-drain region is primarily due to the vertical field along the x-direction originating from the gate. This field is strong due to the high doping concentration in the drain region, and the thin gate oxide in scaled MOSFETs. There is a corresponding bending of the energy band over the depletion region, which is maximum at the interface ($x = 0$). The band bending is a function of drain doping concentration, oxide thickness, and drain-to-gate potential difference V_{DG}.
- When the band bending is sufficiently high, and exceeds the band gap of silicon, 1.2 eV, the valence electrons near the band edge see empty energy levels in the conduction band across the barrier separating the valence and conduction bands. The barrier seen by the

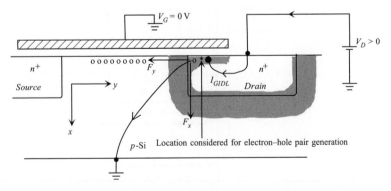

Figure 7.18 Schematic diagram showing the generation of electron–hole pairs under the gate–drain overlap region, and the flow of drain leakage current

valence electron E_g is sufficiently thin due to the very small depletion width, arising from the high doping concentration associated with the drain. The condition is appropriate for band-to-band tunneling. Electrons tunneling from the valence band to the conduction band create holes in the valence band, which is equivalent to generation of electron–hole pairs under the gate overlapped region by the gate–drain bias V_{DG}.

- The electrons are collected by the drain, and holes are pushed towards the channel by the lateral electric field directed towards the channel originating from the drain-to-substrate voltage. Thus, the region under discussion is maintained in the depletion condition, using which the amount of band bending can be calculated under the gate overlapped region. In reality the doping concentration decreases laterally in the gate overlapped region, but to a first order approximation, uniform doping may be assumed. The gate induced drain current formulation involves two steps: (1) calculation of electric field at the surface where tunneling is confined, and (2) estimation of tunneling probability.

7.10.1 Energy Band Bending at the Si–SiO$_2$ Interface

A simple model proposed by Chen *et al.* [15] assumes that for band-to-band tunneling to occur, the band bending at the interface under the gate in the drain region due to the applied gate–drain voltage V_{DG} must be maintained at 1.2 eV, which fixes the surface potential value at $\phi_s = 1.2$ V. Thus, the potential drop across the oxide $V_{ox} = V_{DG} - 1.2$. Figure 7.19 shows the band diagram for this model. From the continuity of electric displacement at the Si–SiO$_2$ interface we get:

$$\epsilon_{Si} F_{Si}(0) = \epsilon_{ox} F_{ox} \tag{7.136}$$

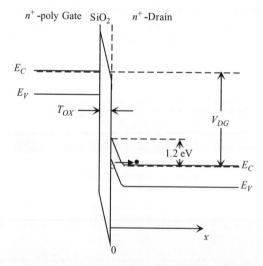

Figure 7.19 A simple model for gate induced energy band bending under the gate

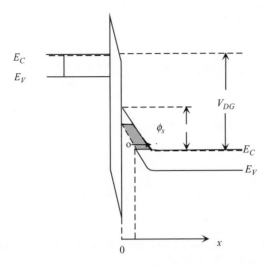

Figure 7.20 Energy band diagram showing the range of the band due to gate induced band bending in the drain for tunneling

which gives:

$$F_{Si}(0) = \frac{\epsilon_{ox}}{\epsilon_{Si}} \frac{V_{ox}}{T_{ox}} = \frac{(V_{DG} - 1.2)}{3T_{ox}} \qquad (7.137)$$

The above model is oversimplified, and does not reflect the role of the drain doping concentration. A more realistic model is outlined below. The corresponding band diagram is shown in Figure 7.20.

Using the depletion approximation, the spatial dependence of the vertical electric field in silicon under the gate in the overlap region is given as:

$$F_{Si}(x) = F_{Si}(0)\left(1 - \frac{x}{W_d}\right) \qquad (7.138)$$

where $F_{Si}(0)$ is the maximum electric field at the interface, and W_d the depletion width corresponding to the voltage V_{DG}.

The field at the surface, $(x = 0)$, in the gate overlapped region can be given by the following expression, assuming depletion approximation:

$$F_{Si}(0) = -\frac{Q'_{bd}}{\epsilon_{Si}} = \frac{\sqrt{2q\epsilon_{Si}N_D\phi_s}}{\epsilon_{Si}} \qquad (7.139)$$

where $Q'_{bd} = -\sqrt{2q\epsilon_{Si}N_D}\sqrt{\phi_s}$ is the drain depletion (bulk) charge density created by V_{DG}. Using Gauss' condition of continuity of electric displacement at the interface of oxide and drain in the overlap region, we have:

$$\epsilon_{Si}\frac{\sqrt{2qN_D\epsilon_{Si}}\sqrt{\phi_s}}{\epsilon_{Si}}=\epsilon_{ox}\frac{V_{DG}-V_{fb}-\phi_s}{T_{ox}} \tag{7.140}$$

Rearranging the above equation, we get:

$$(\sqrt{\phi_s})^2-\sqrt{2qN_D\epsilon_{Si}}\frac{T_{ox}}{\epsilon_{ox}}\sqrt{\phi_s}-(V_{DG}-V_{fb})=0 \tag{7.141}$$

Solving Equation (7.141) for ϕ_s, and considering the physically acceptable root, the surface potential can be expressed in the following form:

$$\phi_s=(V_{DG}-V_{fb})+\frac{qN_D\epsilon_{Si}T_{ox}^2}{\epsilon_{ox}^2}$$

$$-\sqrt{\left(\frac{qN_D\epsilon_{Si}T_{ox}^2}{\epsilon_{ox}^2}\right)^2+2\left(\frac{qN_D\epsilon_{Si}T_{ox}^2}{\epsilon_{ox}^2}\right)(V_{DG}-V_{fb})} \tag{7.142}$$

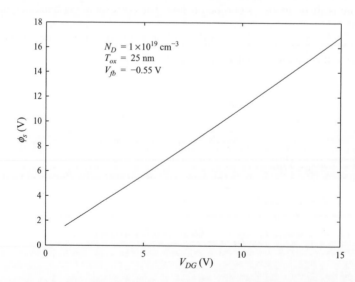

Figure 7.21   Variation of energy band bending in the gate–drain overlap area with potential difference between the gate and drain

$$\phi_s = (V_{DG} - V_{fb}) + \frac{qN_D\epsilon_{Si}T_{ox}^2}{\epsilon_{ox}^2}$$

$$-\sqrt{\left[\left(\frac{qN_D\epsilon_{Si}T_{ox}^2}{\epsilon_{ox}^2}\right) + (V_{DG} - V_{fb})\right]^2 - (V_{DG} - V_{fb})^2} \qquad (7.143)$$

Figure 7.21 shows the variation of ϕ_s with V_{DG} at the interface of the overlap region using Equation (7.143).

7.10.2 Analytical Expression for GIDL Current

The GIDL current has been found to have the following dependencies:

1. Band-to-band tunneling rate.
2. The amount of band bending defining the value of the peak field at the Si–SiO$_2$ interface where tunneling is assumed to be confined to a first approximation.
3. Traps and their distribution at the interface. The influence of traps on the tunneling probability is required to be formulated from the information on the characteristics of the traps, which is process dependent.
4. The lateral electric field which affects the energy of carriers generated, and thereby the tunneling probability.
5. Temperature, by virtue of its influence on the band gap of silicon, which influences the tunneling probability.

In the simple theory described below, the effect of traps, lateral field, and temperature will not be considered. The expression for the GIDL current is obtained from evaluation of the following integral [17]:

$$I_{GIDL} = W.L_{ov} \int_0^\infty qP\left(F_{Si}(x)\right)dx \qquad (7.144)$$

where P is the Kane's tunneling probability rate, which is a function of electric field [3], W is the effective width of the MOS transistor, and L_{ov} is the gate overlap length.

In Kane's model, the tunneling rate, $P(F_{Si})$, has an exponential dependence on the electric field as given below:

$$P(F_{Si}(x)) = AF_{Si}^2(x)\exp\left(-\frac{F_k}{F_{Si}(x)}\right) \qquad (7.145)$$

The pre-exponential multiplying factor A and F_k are given as:

$$A = \frac{q^2 m_r^{\frac{1}{2}}}{18\pi\hbar^2 E_g^{\frac{1}{2}}} \tag{7.146}$$

$$F_k = \frac{\pi m_r^{\frac{1}{2}} E_g^{\frac{3}{2}}}{2q\hbar} \tag{7.147}$$

The quantity m_r is the reduced electron mass of the electron.

The value of I_{GIDL} given by Equation (7.144) is reduced to a simple analytical form by substituting for $F_{Si}^2(x)$ from Equation (7.138), and neglecting the quadratic term involving $(x/W_d)^2$ before integrating. This simplifying approximation leads to:

$$I_{GIDL} = \frac{A W L_{ov} \epsilon_{Si} F_{Si}^4(0)}{F_k N_D} \exp\left(-\frac{F_k}{F_{Si}(0)}\right) \tag{7.148}$$

Figure 7.22 shows the variation of I_{GIDL} with V_{GD}.

From Equation (7.148) the following qualitative conclusions can be drawn:

- Higher negative gate voltage for a given positive drain voltage increases the GIDL current, because of the increasing surface field induced by an increased V_{GD}.
- For the same reason, scaling down of oxide thickness for a given V_{GD}, increases the GIDL current.
- GIDL current is strongly temperature dependent because of the variation of energy band gap of silicon on temperature. The band gap of silicon decreases with temperature as given by Equation (1.28). Hence, the GIDL current increases with temperature.

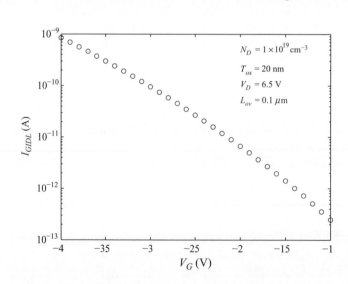

Figure 7.22 Variation of gate induced drain leakage current with gate–drain potential difference

BSIM4 accounts for the GIDL current through the following expression [22]:

$$I_{GIDL} = NF \times AGIDL.W_{eff} \left(\frac{V_{DS} - V_{GSeff} - EGIDL}{3TOXE} \right)$$

$$\times \exp \left(\frac{-3TOXE.BGIDL}{V_{DS} - V_{GSeff} - EGIDL} \right) \frac{V_{DB}^3}{CGIDL + V_{DB}^3} \quad (7.149)$$

where AGIDL (the pre-exponential factor), BGIDL (the exponential coefficient), CGIDL (bulk bias related), and EGIDL (the minimum band bending) are model parameters.

The HiSIM model describes the drain and gate voltage dependence of the GIDL current through the following equation:

$$I_{GIDL} = qGIDL1 \frac{F^2}{E_g^{\frac{1}{2}}} \exp \left(-GIDL2 \frac{E_g^{\frac{3}{2}}}{F} \right) W_{eff} \quad (7.150)$$

where $F = (GIDL3 \times V_{DS} - V_G')/T_{ox}$, and GIDL1, GIDL2, and GIDL3 are model parameters. E_g is the band gap energy of silicon, and $V_G' = V_{GS} - V_{fb} + \Delta V_{th}$.

In the PSP and Philips MM11 models, the GIDL model equation is expressed in the following form [25, 32]:

$$I_{GIDL} = A_{GIDL} V_{DB} V_{tovL}^2 \exp \left(\frac{-B_{GIDL}}{V_{tovL}} \right) \quad (7.151)$$

where, $V_{tovL} = (V_{ovL}^2 + (C_{GIDL} V_{DB})^2)^{1/2}$, and A_{GIDL} and B_{GIDL} are fitting parameters.

Similar expressions for the gate induced source leakage current (GISL) can also be obtained.

References

[1] F. Stern and W. E. Howard, Properties of Semiconductor Surface Inversion Layers in the Electric Quantum Limit, *Phys. Rev.* 163, 816–835, 1967.

[2] F. Stern, Self-consistent Results for *n*-type Si Inversion Layers, *Phys. Rev., B*, 5(12), 4891–4899, 1972.

[3] E.O. Kane, Zener Tunneling in Semiconductors, *J. Phys. Chem. Solids*, 12, 181–188, 1959.

[4] E.O. Kane, Theory of Tunneling, *J. Appl. Phys.*, 32(1), 83–91, 1961.

[5] L.F. Register, E. Rosenbaum, and K. Yang, Analytic Model for Direct Tunneling Current in Polycrystalline Silicon–Gate Metal-Oxide–Semiconductor Devices, *Appl. Phys. Lett.*, 74(3), 457–459, 1999.

[6] L. Wang, *Quantum Mechanical Effects on MOSFET Scaling Limit*, Ph.D. Thesis, Georgia Institute of Technology, Atlanta, GA, 2006.

[7] J. Cai and C.T. Sah, Gate Tunneling Currents in Ultrathin Oxide Metal-Oxide–Silicon Transistors, *J. Appl. Phys.*, 89(4), 2272–2285, 2001.

[8] M. Depas, B. Vermeire, P.W. Mertens, R.L. Van Meirhaeghe and M.M. Heyns, Determination of Tunneling Parameters in Ultra-thin Oxide Layer Poly-Si/SiO2/Si Structures, *Solid-State Electron.*, 38(8), 1465–1471, 1995.

[9] K.F. Schuegraf, C.C. King and C. Hu, Ultra-thin Silicon-Dioxide Leakage Current and Scaling Limit, in *Proceedings of the Symposium on VLSI Technology Digest of Technical Papers*, 18–19, 1992.

[10] K.F. Schuegraf and C. Hu, Hole Injection SiO2 Breakdown Model for Very Low Voltage Lifetime Extrapolation, *IEEE Trans. Electron Dev.*, 41(5), 761–767, 1994.

[11] M.J. van Dort, P.H. Worlee, A.J. Walker, C.A.H. Juffermans and H. Lifka, Influence of High Substrate Doping Levels on the Threshold Voltage and the Mobility of Deep-Submicrometer MOSFETS, *IEEE Trans. Electron Dev.*, **39**(4), 932–938, 1992.

[12] W.C. Lee and C. Hu, Modeling CMOS Tunneling Currents Through Ultrathin Gate Oxide Due to Conduction- and Valence-Band Electron and Hole Tunneling, *IEEE Trans. Electron Dev.*, **48**(7), 1366–1373, 2001.

[13] C. Lallement, J. Sallese, M. Bucher, W. Grabinski and P. C. Fazan, Accounting for Quantum Effects and Polysili- con Depletion From Weak to Strong Inversion in a Charge-Based Design-Oriented MOSFET Model, *IEEE Trans. Electron Dev.*, **50**(2), 406–417, 2003.

[14] J.A. Lopez-Villanueva, P. Cartujo-Casinello, J. Banqueri, F. Gamiz and S. Rodriguez, Effects of the Inversion Layer Centroid on MOSFET Behavior, *IEEE Trans. Electron Dev.*, **44**(11), 1915–1922, 1997.

[15] J. Chen, T.Y. Chan, I.C. Chen, P.K. Ko and C. Hu, Subbreakdown Drain Leakage Current in MOSFET, *IEEE Electron Dev. Lett.*, **8**(11), 515–517, 1987.

[16] F. F. Fang and W. E. Howard, Negative Field-Effect Mobility on (100) Si Surfaces, *Phys. Rev. Lett.*, **16**(18), 797– 799, 1966.

[17] A. Bouhdada, S. Bakkali and A. Touhami, Modelling of Gate-Induced Drain Leakage in Relation to Technological Parameters and Temperature, *Microelectron. Reliabil.*, **37**(4), 649–652, 1997.

[18] A.P. Gnädinger and H.E. Talley, Quantum Mechanical Calculation of the Carrier Distribution and the Thickness of the Inversion Layer of a MOS Field-Effect Transistor, *Solid-State Electron.*, **13**(9), 1301–1309, 1970.

[19] J.H. Davies, *The Physics of Low-Dimensional Semiconductors: An Introduction*, Cambridge University Press, Cambridge, UK, 1998.

[20] G. Dessai, Quantum Effects on Density of State and Carrier Distribution in MOS Capacitor, *JIITU, NOIDA, Internal Project Memo*, 2006.

[21] G. Dessai and A.B. Bhattacharyya, On Threshold Voltage Model of MOS Capacitor in the Presence of Quantiza- tion Effect, in *Proceedings of the IEEE Mini-Colloquia on Compact MOSFET Modeling of Advance MOSFET Structures and Mixed Model Application*, University of Delhi South Campus, New Delhi, India, 5–6, January 2008.

[22] W. Liu, *MOSFET Models for SPICE Simulation, Including BSIM3V3 and BSIM4*, John Wiley & Sons, Inc., New York, NY, 2001.

[23] N. Yang, W.K. Henson, J.R. Hauser and J.J. Wortman, Modeling Study of Ultrathin Gate Oxides using Direct Tun- neling Current and Capacitance Voltage Measurements in MOS Dev., *IEEE Trans. Electron Devices*, **46**(7), 1464– 1471, 1999.

[24] K.N. Yang, H.T. Huang, M.J. Chen, Y.M. Lin, M.C. Yu and S.N. Jang, D.C. H. Yu and M.S. Liang, Characterization and Modeling of Edge Direct Tunneling (EDT) Leakage in Ultrathin Gate Oxide MOSFETs, *IEEE Trans. Electron Dev.*, **48**(6), 1159–1164, 2001.

[25] R. Langevelde, A.J. Scholten and D.B.M. Klaassen, *Physical Background of MOS Model 11, Level 1101*, Net. Lab. Unclassified Report 2003/00239 Koninklijke Philips Electronics N.V., 2003.

[26] Y. Shih, R. Rios, P. Packman, K. Mistry and T. Abbott, A General Partition Scheme for Gate Leakage Current Suitable for MOSFET Compact Models, *IEDM Tech. Dig.*, 293–296, 2001.

[27] A. Gehring, *Simulation of Tunneling in Semiconductor Devices*, Ph.D. Thesis, Technische Universität Wien, Wien, Austria, 2003.

[28] X. Gu, T.L. Chen, G. Gildenblat, G.O. Workman, S. Veeraraghavan, S. Shapira and K. Stiles, A Surface Potential- Based Compact Model of *n*-MOSFET Gate-Tunneling Current, *IEEE Trans. Electron Dev.*, **51**(1), 127–135, 2004.

[29] G. Gildenblat, X. Li, W. Wu, H. Wang, A. Jha, R. van Langevelde, G.D.J. Smit, A.J. Scholten and D.B.M. Klaassen, PSP: An Advanced Surface-Potential-Based MOSFET Model for Circuit Simulation, *IEEE Trans. Electron Dev.*, **53**(9), 1979–1993, 2006.

[30] F. Li, P. Mudarai, Y.-Y. Fan, L.F. Register and S.K. Banerjee, Physically Based Quantum-Mechanical Compact Model of MOS Devices Substrate-Injected Tunneling Current Through Ultrathin (EOT $\sim$ 1 nm) SiO$_2$ and High-κ Gate Stacks, *IEEE Trans. Electron Dev.*, **53**(5), 1096–1106, 2006.

[31] K. Roy, S. Mukhopadhyay and H. M. Meimand, Leakage Current Mechanisms and Leakage Reduction Techniques in Deep-Submicrometer CMOS Circuits, *Proc. IEEE*, **91**(2), 305–327, 2003.

[32] *PSP 102.1 Manual*. (http://pspmodel.asu.edu/documentation.html), 2007.

[33] *HiSIM 1.1.1 User's Manual*, Hiroshima University, STARC, Hiroshima, Japan, 2002.

[34] D. Vasileska, S. S. Ahmed, M. Mannino, A. Matsudaira and G. Klimeck, *SCHRED* (https://www.nanohub.org/tools/schred/), 2006.

[35] W.K. Shih, S. Jallepalli and G. Chindalore, *UTQUANT2.0 User's Guide*, University of Texas at Austin, Austin, TX, 1997.

[36] J.C. Ranuarez, M.J. Deen and C-H. Chen, A Review of Gate Tunneling Current in MOS Devices, *Microelectronics Reliability*, **46**, 1939–1956, 2006.

Problems

1. Using the variational method show that the centroid of the inversion layer is given by $\frac{3}{b}$, assuming the trial function as:

$$\psi(x) = \left(\frac{b^3}{2}\right)^{\frac{1}{2}} x \exp\left(-\frac{b}{2}x\right) \tag{7.152}$$

2. Write a program for solving Schrodinger's equation in a triangular well, and obtain the eigenvalues and eigenfunctions for electrons corresponding to valley-1. Plot the wave function for typical values of a MOS capacitor such that the quantization condition holds good. Assume the capacitor to be in the $\langle 100 \rangle$ plane.

3. For a MOS capacitor obtain the conduction band bending using simulation by SCHRED with the following data: $T_{ox} = 5$ nm, $N_A \geq 10^{17}$ cm^{-3}, $V_{GB} = 3$ V, $T = 300$ K. The capacitor is fabricated on the (100) plane.

4. Using the data given in Problem 3 obtain the energy levels for various sub-bands using SCHRED. Interpret the result.

5. Derive the expression for ϕ_{sT} given in Equation (7.42).

6. Obtain the expression for W_i given in Equation (7.49).

7. Show the variation of I_{GIDL} as a function of gate oxide thickness for the data given in Figure 7.22 with oxide thickness varying from 10 nm to 30 nm, with the gate voltage V_G at -3 V and $V_{fb} = 0$ V.

8. Derive the expression for change in the threshold voltage due to the quantum effect given in Equation (7.37).

9. Derive the expression for the barrier transmission coefficient TC in the PSP model from Equation (7.86) using a Taylor's series expansion.

10. For an n^+-polysilicon–SiO$_2$–n-Si MOS structure assume that the boundary between FN and direct tunneling is given by the following conditions:

$$V_{ox} > \Phi_b \rightarrow \text{FN tunneling;}$$

$$V_{ox} < \Phi_b \rightarrow \text{Direct tunneling.}$$

For a gate oxide thickness $T_{ox} = 6.2$ nm, obtain the tunnel current densities at 2.0 V and 6.0 V, Given that Φ_b for an n-type silicon substrate is 3.2 eV.

Hint: Use the expressions for FN and direct tunneling given in Depas *et al.* [8].

8

Non-classical MOSFET Structures

8.1 Introduction

In spite of decades of lithography driven *classical* scaling of bulk MOSFET device structure to meet higher packing density and superior performance, the basic structure remained invariant except for innovations mostly in process technology. The modifications in mainstream bulk MOSFETs revolve around drain and substrate engineering leading to the following illustrative structures [1–3]:

1. Lightly Doped Drain (LDD) structure. This architecture requires drain engineering for a creation of a lightly doped drain by ion implantation. This reduces the electric field near the drain, reducing hot electron related reliability problems in MOS devices.
2. Retrograde doping profile in the substrate. This architecture needs channel engineering to reduce bulk punchthrough and resulting leakage current.
3. Ground plane structure with a thin layer of silicon film. This structure realizes the property of a retrograde substrate structure by substitution of implant technology through CVD based atomic layer technology of silicon on the bulk substrate.
4. Pocket or Halo implant. The DIBL effect is minimized through creation of highly doped pockets near the source and drain in the channel. This creates inhomogeneities in the channel adjacent to the source and drain, which are responsible for an increased leakage current due to band-to-band tunneling between the source and pocket regions.

The performance of a nanoscale MOS device, bordering near the atomic limit, has surfaced many emerging problem areas such as tunneling, quantization, leakage current, parasitic resistances, etc. which are identified as major roadblocks in the path of continuation of scaling. We shall briefly outline below the physical basis of emerging non-classical MOS structures where the limits of scaling of classical devices can be extended [4, 5, 10].

The guiding principle driving scaling to the limits set by the ITRS roadmap has to ensure that the short channel effect and drain induced barrier lowering, which get aggravated in the

Compact MOSFET Models for VLSI Design   A.B. Bhattacharyya
© 2009 John Wiley & Sons (Asia) Pte Ltd

scaling process, are overcome in the new approach [6]. In a MOSFET structure the conducting channel is under the influence of the following fields: (i) the gate induced vertical field acting in the direction normal to the channel, (ii) the field induced by the bulk terminal acting also normal to the channel but in a direction opposite to that of the gate countering thereby the effect of the gate, and (iii) the drain/source induced depletion charges contributing to the lateral field, reducing the control of the gate on the conducting channel. If the scaling has to be pushed to the limits of ultimate scalability, the structure has to be such that the conducting channel is dominantly under gate control, and penetration of the lateral field is minimized. This minimization can be accomplished by increased substrate doping. However, an increase in substrate doping leads to thinning of the depletion layer under the channel, and strong coupling of the bulk to the channel. This reduces inversion charge density, and weakens the control of the gate.

The gate gets strongly coupled to the channel through scaling down the thickness of the oxide. With thinning of the gate oxide, the supply voltage is scaled down, and the threshold voltage needs to be controlled more precisely. Higher bulk doping results in more variation in threshold voltage due to dopant fluctuation. Therefore, the bulk doping concentration needs to be reduced, which is in conflict with the condition of the tighter bulk control requirement mentioned above. The limits of a scaled bulk MOS structure depend on the extent that the conflicting requirements can be traded off.

An approach to minimize the impact of bulk doping concentration is to reduce the bulk material. A Silicon-on-Insulator (SOI) MOSFET structure for radiation hardened applications, with a limited amount of bulk material on an insulating substrate, is an example of this approach. In an SOI structure, an ultrathin single crystal silicon layer is grown for MOSFET fabrication. Obviously, the bulk material being grown on an insulator, direct bulk contact has not been natural, and the bulk remains floating with attendant problems. Nevertheless, partially depleted SOI MOSFETs have been successfully integrated in a number of applications. With the advancement of technology, the silicon film could be thinned to realize fully depleted SOI MOSFETs [2].

Evolution of technology has led to innovative variations in SOI MOSFET structures. The bulk control can be handed over to the insulator substrate in the form of an additional back gate. A highly doped conducting silicon layer below the back gate oxide provides the gate terminal for a double gate MOSFET (DG MOSFET) structure.

In Chapter 5 we outlined several laws which form the basis of scaling of classical MOSFET devices. These laws are not applicable for scaling of DG MOSFET based structures. Figure 8.1 is a cross-section of a DG MOSFET structure, showing a thin layer of silicon of thickness t_{Si} controlled by gate terminals which are normal to the direction of the source drain terminals. For a symmetrical structure, the front (top) and back (bottom) gate oxides have the same thickness.

The potential distribution within the device structure can be obtained by solving the 2-D or 3-D Poisson's equations with appropriate boundary conditions relevant to the specific structure. There are a number of approaches available, both 2-D and 1-D, with intrinsic and finite doping of the silicon film having a large width. The 1-D solution with an intrinsic silicon film is more common. The boundary conditions are related to (i) potential, potential gradients at the silicon–oxide interfaces, and (ii) the vertical field at the center $x = \frac{t_{Si}}{2}$, which is zero by virtue of operational and structural symmetry. Solutions for the potential, specific to the problem addressed, will be obtained in later sections. In this section we shall discuss some general conclusions related to scaling based on reported results. For devices which are wide, the 2-D Poisson's equation is used. Assuming that the film is fully depleted, which is

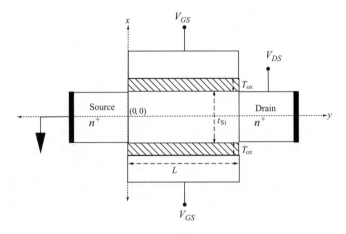

Figure 8.1   Symmetric double gate MOSFET structure

considered to be the most promising from performance viewpoint, the potential distribution in the x-direction is obtained as [10]:

$$\phi(x, y) = C_0(y) + C_1(y)x + C_2(y)x^2 \tag{8.1}$$

The constants C_0, C_1, and C_2 are to be evaluated from the boundary conditions provided by the structure under consideration.

A parameter designated as the *natural scale length,* λ_1, given by:

$$\lambda_1 = \sqrt{\frac{\epsilon_{Si}}{\epsilon_{ox}} t_{Si} T_{ox}} \tag{8.2}$$

relates the short channel effect of the structure with two key design variables T_{ox} and t_{Si}. A small value of λ_1 physically implies superior immunity to short channel effect. [1, 17]. The scaling theory of the SOI double gate structure provides a guideline for the device design, where the channel length can be reduced without degrading the subthreshold slope performance relating the channel length with the natural scale length through the following equation:

$$\alpha_1 = \frac{L_g}{2\lambda_1} \tag{8.3}$$

Equation (8.3) shows that after optimizing the scale length through t_{Si} and T_{ox} to a small value, the gate length can be scaled down to a limit lower than bulk structures. An improved relation [18] gives the expression for natural scale length as follows:

$$\lambda_2 = \sqrt{\frac{\epsilon_{Si}}{2\epsilon_{ox}} \left(t_{Si} T_{ox} + \frac{\epsilon_{ox}}{4\epsilon_{Si}} t_{Si}^2 \right)} \tag{8.4}$$

The difference in the expressions for the natural scale lengths λ_1 and λ_2 lies in the fact that the latter gives the same scale factor for various combination of device parameters [18].

The potential of the DG MOSFET structure and its variations need to be estimated from the viewpoint of its immunity to short channel effects. A parameter known as *Electrostatic Integrity (EI)* is such a metric, which is mathematically expressed by [7]:

$$EI = \left(1 + \frac{X_J^2}{L^2}\right) \frac{T_{OXE}}{L} \times \frac{W_{dep}}{L} \tag{8.5}$$

In the above expression L denotes the distance between the source–drain junctions, and T_{OXE} is the electric oxide thickness. When applied to evaluation of non-classical structures, X_J is replaced by t_{Si} and W_{dep} by $t_{Si}/2$. It turns out that with a double gate structure, the source-to-drain distance defining channel length can be scaled to a much lower value while maintaining high electrical integrity for the structure. This extends the range of scalability through the non-classical structure route.

8.2 Non-classical MOSFET Structures

The above relations pertain to non-bulk Silicon-on-Insulator (SOI) MOSFET structures, which we shall classify as belonging to the non-classical MOSFET family. Most of the structures in the family of non-classical structures are variations around the basic DG MOSFET. We shall briefly introduce below the variations in *non-classical structures* that are identified as potential candidates to possibly replace the existing classical bulk CMOS structures currently in production. The important structures are [7, 8]:

- Single Gate Ultra Thin Body (UTB) MOSFETs.
- Double Gate/Multigate UTB MOSFETs.
- Strained-Silicon MOSFETs.

8.2.1 Single Gate UTB MOSFET on SOI

The cross-section of a UTB MOSFET structure is shown in Figure 8.2. The device is operated in the fully depleted condition in the ON state. The active thin silicon film is grown on an oxide layer which is on top of a silicon substrate. The silicon film is lowly doped compared to bulk MOS structures, and its thickness is optimized to ensure that the gate voltage which switches on the device is able to deplete it completely. The obvious advantages are low junction capacitance and less voltage variation due to dopant fluctuation. The high cost of SOI wafers and less control of short channel effects in scaled structures are limitations.

8.2.2 Double Gate MOSFETs

The double gate structure, an extension of the single gate UTB MOSFET device described above, is currently considered to be the most promising architecture in view of the potential of scalability as per the DC MOSFET scaling law. The essential features of the structure are demonstrated in Figure 8.3. The structure belongs to the ultra-thin body family of SOIs with an extra degree of freedom to add an additional gate on the back side. Thus, the planar DG MOSFET has two gates designated as *front gate* and *back gate* within which the ultra-thin

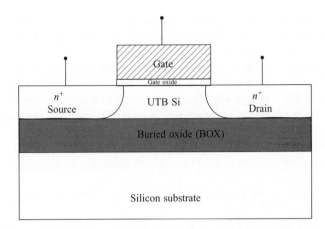

Figure 8.2 Cross-section of a single gate ultra-thin body SOI MOSFET structure

silicon layer is sandwiched. The features of the planar DG MOSFET structure, shown in Figure 8.3, are:

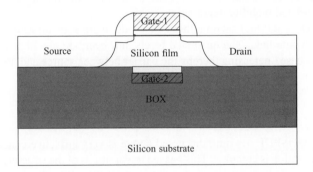

Figure 8.3 Cross-sectional view of a planar double gate MOSFET structure

- The 3-D planar MOSFET structure requires a number of specialized processes for the formation of the back gate oxide, the back gate contact, control of silicon film thickness sandwiched between the front and back gate, symmetric operation, etc.
- Being planar, the structure is attractive and is compatible to the standardized bulk CMOS process.
- The most common mode of operation is to switch both the gates simultaneously. An alternative possible variation of operation of the device is to apply a back gate bias to create a conducting plane, and switch the top gate which provides an additional dimension of flexibility in circuit design.
- As both the gates contribute to inversion carriers, it has a higher drive capability.
- For CMOS operation a single midgap material can be used to control the threshold voltage for both NMOS and PMOS transistors. For better V_T control, a dual metal gate process may be needed, adding to the process complexity.

- Generally, the structure has less parasitic capacitances compared to bulk structures. However, alignment of both the front and back gate is a demanding condition. Misalignment contributes to parasitic capacitances.
- Amongst the quantities which determine the variability in device parameters are T_{ox}, the gate oxide thickness, and ϕ_{gs}, the work function difference between the gate material and silicon substrate. They are dependent on thermal cycles in process technology, which are of magnitudes which are amenable to precise control. Gate length L and film thickness T_{ox}, controlled by etching and lithography, are more susceptible to variations.
- The semiconductor material determining the device characteristics is a thin layer of undoped silicon. As scattering due to impurity or coulomb scattering is absent, the carrier mobility is high.
- The gates are close to the channel, which gives them dominant control on the channel electrostatics.
- In bulk devices, in the subthreshold region, the leakage current originates due to surface or subsurface conduction leading to the need of complex channel engineering. However, in a DG MOSFET the thin silicon substrate is under a nearly volume inversion condition, due to which the conduction of carriers is across the entire volume of the material.
- The substrate film being undoped or lowly doped, the normal gate field is smaller in magnitude for a given gate oxide thickness compared to that in bulk structures. This reduces the gate field related mobility degradation.
- In the scaled DG MOSFETs with a thin gate oxide, there is quantization of energy levels due to the high gate field. The quantized levels, however, are at a higher value compared to the classical case, increasing thereby the barrier between source and drain, and reducing leakage current [22].

8.2.3 Nonplanar MOSFET Structures

The double gate MOSFET structure discussed above is very attractive conceptually, but the technological realization is complex. This has led to variations of the structure which allow the exploitation of the stated advantages with technological flexibility. The nonplanar structures belong to this category.

The quasi-planar FinFET structure shown in Figure 8.4 is a popular version of the DG MOSFET structure where a thin silicon fin of height W and thickness t_{Si} is realized on a BOX. The silicon fin is wrapped by the polysilicon gate as shown in the figure. The gate planes

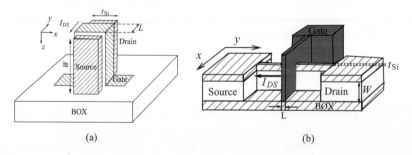

(a)   (b)

Figure 8.4   Quasi-planar FinFET structure: (a) 3-D view, and (b) alternate perspective view

being vertical to the surface in the y–z plane can be simultaneously oxidized ensuring equal thickness. The polysilicon wraps the fin, providing a tied gate structure, as the same voltage is applied to both the gates. The source/drain connects the ends of the fin. The current flow is in the y-direction and is perpendicular to the direction of the gate field, which is in the x-direction. The film thickness t_{Si} is realized through lithographic and etching processes. The device terminals are easily accessible for connection compared to the planar DG structure. The structure has the following characteristics [9]:

- It is a nonplanar structure, and therefore a special process needs to be developed.
- The MOSFET is built perpendicular to the plane of the silicon wafer.
- There are two planes of gate, i.e. front and back gate. The width of the gate (i.e. fin height) is normal to the wafer plane, thereby providing a high packing density as usually the width is larger than the channel length.
- All the device terminals are accessible for connection.
- The current flow direction is normal to the gate-to-gate direction.

Figure 8.5 gives an overview of the vertical SOI MOSFET structure where the gate induced field is in the y-direction, and the current flow is normal to it in the x-direction. The structure provides the possibility of putting up gates in the z-direction, enabling the silicon film to be controlled by four gates. Such an SOI MOSFET structure belongs to the *Surrounding Gate* or *Gate All Around (GAA)* family. The structure is classified into two broad categories: (i) square section, and (ii) circular cross-section. The GAA structure has a superior scalability potential and the scale length for the square type is obtained as:

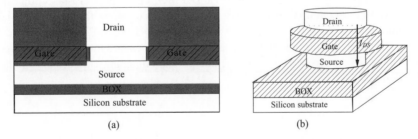

Figure 8.5   (a) Vertical DG MOSFET structure, and (b) Gate All Around SOI MOSFET structure

$$\lambda_3 \approx \sqrt{\frac{\epsilon_{Si}}{4\epsilon_{ox}} t_{Si} T_{ox}} \tag{8.6}$$

8.2.4 Strained-Silicon Structure

SiGe strained-silicon has been used to enhance mobility of carriers in the classical bulk MOSFET structures. It has become an integral part of the CMOS process, especially for chips intended for RF applications. There have been two categories of strain engineering: (i) local strain, and (ii) global strain [16]. In the case of global strain structures, the strain is engineered by depositing an epitaxial SeGe buffer layer of suitable mix on the silicon substrate. The physical principle involved in increasing the carrier mobility has been discussed in Chapter 1.

The strain introduced in this process is biaxial, which increases the mobility of both holes and electrons. Strained-silicon technology has been incorporated in SOI structures as well to take advantage of improved transport characteristics of strained-silicon [14, 15]. Figure 8.6 is an illustrative overview of a strain engineered DG MOSFET. In the local strain engineered variety, the source and drain regions are substituted by an SiGe composition. The lattice constant of the SiGe alloy is large compared to that of silicon. Thus, the silicon channel, sandwiched between the SiGe S/D regions, is put under uniaxial stress which is compressive. The process has been demonstrated to have improved hole mobility.

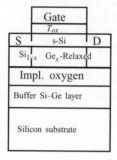

Figure 8.6   Globally strained DG MOSFET structure

8.3 Double Gate MOSFET Models

Consider an undoped (lightly doped) symmetric DG MOSFET whose cross-section is shown in Figure 8.1. The direction of current flow is assumed to be along the y-axis and the gate field acts normal to it along the x-axis. The width of the DG MOSFET is taken to be in the z-direction. The center of the silicon film is assumed to be located symmetrically between the gates, and thereby taken to be at $x = 0$. The energy band diagram without bias condition is shown in Figure 8.7.

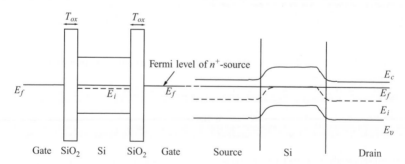

Figure 8.7   Band diagram of a symmetric DG MOSFET without applying any gate bias

At zero gate voltage, the positions of the silicon bands are determined by the gate work function. As we consider a undoped body symmetric DG MOSFET with no work function difference, the bands remain flat as shown. Since no direct contact to the body is available, *the energy levels are referenced to the quasi-Fermi level of the n^+-source.*

When the gate voltage is applied ($V_{GS} > 0$), below threshold, the Fermi level of the gate shifts down, and along with it the energy bands of the silicon body, as there is negligible depletion charge. At this time volume inversion takes place. When the gate voltage approaches close to threshold, sufficient mobile charges (electrons in case of $V_{GS} > 0$) build up at the surface. More bending of the bands is seen at the surface than at the center, as shown in Figure 8.8.

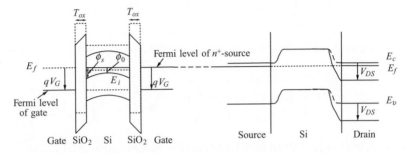

Figure 8.8   Band diagram of a symmetric DG MOSFET with an applied gate bias

8.3.1 Electrostatic Potential

We shall consider the model of a long channel DG MOSFET with an undoped silicon layer. Since the variation of electric field along the channel direction is much less compared to the variation in electric field in the direction perpendicular to the channel, only the perpendicular field is considered, resulting in gradual channel approximation (GCA). Thus, a 1-D Poisson's equation is written, considering the mobile charge (electron) density [11, 12]:

$$\frac{d^2\phi}{dx^2} = \frac{q}{\epsilon_{Si}} n_i e^{\frac{(\phi - V_C)}{\phi_t}} \tag{8.7}$$

Here ϕ is the potential in the channel, q is the electronic charge, ϵ_{Si} is the silicon permittivity, ϕ_t is the thermal voltage, n_i is intrinsic concentration, and V_C is the electron quasi-Fermi potential of the channel.

Referring to Figure 8.1, the boundary conditions are:

* $\frac{d\phi}{dx} = 0$ at $x = 0$;
* $\phi(x = 0) = \phi_0$, the potential at the center of the film.

It is assumed that the work function difference between the gate and the semiconductor body is zero.

From the potential balance equation for the structure under consideration:

$$V_{GS} - V_{fb} = \phi \bigg|_{x = \frac{t_{Si}}{2}} + \frac{\epsilon_{Si}}{C'_{ox}} \frac{d\phi}{dx} \bigg|_{x = \frac{t_{Si}}{2}} \tag{8.8}$$

where $\mathrm{d}\phi/\mathrm{d}x|_{x=t_{\mathrm{Si}}/2}$ represents the field in the silicon at the interface, and using continuity of displacement vector, the potential drop across the oxide is represented by the second term.

Integrating Equation (8.7) with the above boundary conditions gives the electric field $F = -\frac{\mathrm{d}\phi}{\mathrm{d}x}$:

$$\frac{\mathrm{d}\phi}{\mathrm{d}x} = \sqrt{\frac{2kTn_i}{\epsilon_{\mathrm{Si}}} \left[e^{\frac{(\phi - V_C)}{\phi_t}} - e^{\frac{(\phi_0 - V_C)}{\phi_t}} \right]} \tag{8.9}$$

Integrating Equation (8.9) gives the electrostatic potential at any point x in the body:

$$\phi(x) = \phi_0 - 2\phi_t \ln \left[\cos \left(e^{\frac{\phi_0 - V_C}{2\phi_t}} \sqrt{\frac{q^2 n_i}{2kT\epsilon_{\mathrm{Si}}}} x \right) \right] \tag{8.10}$$

At $x = t_{\mathrm{Si}}/2$, $\phi(x) = \phi_s$. Substituting Equation (8.9) into Equation (8.8) yields:

$$V_{GS} - V_{fb} = \phi_s + \frac{\sqrt{2kTn_i\epsilon_{\mathrm{Si}}}}{C'_{ox}} e^{\frac{(\phi_s - V_C)}{2\phi_t}} \sqrt{\left(1 - e^{\frac{\phi_0 - \phi_s}{\phi_t}} \right)} \tag{8.11}$$

8.3.1.1 Ortiz-Conde *et al.* Model

An explicit analytical solution for the surface potential applicable to symmetric DG MOSFET structures has been obtained by Ortiz-Conde et al. [24, 25]. The potential at the surface is obtained by substituting $x = \frac{t_{\mathrm{Si}}}{2}$ in Equation (8.10):

$$\phi_s = \phi_0 - 2\phi_t \ln \left\{ \cos \left[\sqrt{\frac{q^2 n_i}{2kT\epsilon_{\mathrm{Si}}}} e^{\frac{(\phi_0 - V_C)}{2\phi_t}} \frac{t_{\mathrm{Si}}}{2} \right] \right\} \tag{8.12}$$

Rearranging the above equation, we can write:

$$e^{\left(\frac{\phi_0 - \phi_s}{\phi_t} \right)} = \cos^2(\zeta) \tag{8.13}$$

where:

$$\zeta = \sqrt{\frac{q^2 n_i}{2kT\epsilon_{\mathrm{Si}}}} e^{\frac{(\phi_0 - V_C)}{2\phi_t}} \frac{t_{\mathrm{Si}}}{2} \tag{8.14}$$

Combining Equation (8.13) and Equation (8.11), we now have:

$$V_{GS} - V_{fb} = \phi_s + \frac{\sqrt{2kTn_i\epsilon_{\mathrm{Si}}}}{C'_{ox}} e^{\frac{(\phi_s - V_C)}{2\phi_t}} \sin(\zeta) \tag{8.15}$$

The above equation results in an implicit expression for ϕ_s, which would need to be solved numerically. To get an explicit closed form for ϕ_s, the above equation can be expressed in the form of $z = xe^x$ so that the Lambert W-function can be applied. To recast the equation, we first divide both sides by $2\phi_t$:

$$\frac{V_{GS} - V_{fb}}{2\phi_t} = \frac{\phi_s}{2\phi_t} + \frac{\sqrt{2kTn_i\epsilon_{Si}}}{2\phi_t C'_{ox}} e^{\frac{(\phi_s - V_C)}{2\phi_t}} \sin(\zeta) \tag{8.16}$$

Now, considering both sides to be exponents we get:

$$e^{\frac{V_{GS} - V_{fb}}{2\phi_t}} = e^{\frac{\phi_s}{2\phi_t}} \times e^{\frac{\sqrt{2kTn_i\epsilon_{Si}}}{2\phi_t C'_{ox}} e^{\frac{(\phi_s - V_C)}{2\phi_t}} \sin(\zeta)} \tag{8.17}$$

$$= e^{\left[\frac{\phi_s}{2\phi_t} + \frac{\sqrt{2kTn_i\epsilon_{Si}}}{2\phi_t C'_{ox}} e^{\frac{(\phi_s - V_C)}{2\phi_t}} \sin(\zeta) \right]} \tag{8.18}$$

Finally, multiply both sides of Equation (8.17) by $(\sqrt{2kTn_i\epsilon_{Si}}/2\phi_t C'_{ox}) e^{-V_C/2\phi_t} \sin(\zeta)$ to get:

$$\frac{\sqrt{2kTn_i\epsilon_{Si}}}{2\phi_t C'_{ox}} e^{\frac{-V_C}{2\phi_t}} \sin(\zeta) e^{\frac{V_{GS} - V_{fb}}{2\phi_t}} = \frac{\sqrt{2kTn_i\epsilon_{Si}}}{2\phi_t C'_{ox}} e^{\frac{(\phi_s - V_C)}{2\phi_t}} \sin(\zeta)$$

$$\times e^{\frac{\sqrt{2kTn_i\epsilon_{Si}}}{2\phi_t C'_{ox}} e^{\frac{(\phi_s - V_C)}{2\phi_t}} \sin(\zeta)} \tag{8.19}$$

Using the Lambert W-function ($W(z) = x$):

$$W\left(\frac{\sqrt{2kTn_i\epsilon_{Si}}}{2\phi_t C'_{ox}} e^{\frac{-V_C}{2\phi_t}} \sin(\zeta) e^{\frac{(V_{GS} - V_{fb})}{2\phi_t}} \right) = \frac{\sqrt{2kTn_i\epsilon_{Si}}}{2\phi_t C'_{ox}} \sin(\zeta) e^{\frac{\phi_s - V_C}{2\phi_t}} \tag{8.20}$$

Substituting the RHS of Equation (8.20) into Equation (8.15), we get:

$$\phi_s = V_{GS} - V_{fb} - 2\phi_t W\left(\frac{\sqrt{2kTn_i\epsilon_{Si}}}{2\phi_t C'_{ox}} e^{\frac{-V_C}{2\phi_t}} \sin(\zeta) e^{\frac{V_{GS} - V_{fb}}{2\phi_t}} \right) \tag{8.21}$$

Therefore, to evaluate the surface potential, we need to know ζ, which depends on ϕ_0, as indicated by Equation (8.14). It is observed that Equation (8.21) is an explicit solution of ϕ_s in terms of ζ and the terminal voltages.

8.3.1.2 Potential at the Center, ϕ_0

A bound on the maximum value of ϕ_0 (ϕ_{0max}) can be obtained from Equation (8.10). The value of the cosine term is bounded in the range $[0, 1]$ for the possible values that the argument terms of the cosine function can take. This results in a range of $[-\infty, 0]$ for the natural logarithm of the cosine term. Hence, the value of $\phi(x)$ is really a sum of two terms. For the bound, consider the argument of the cosine term which will maximize the value of the second term for $\phi(x)$. This will happen for an argument of $\frac{\pi}{2}$. Thus:

$$\left[e^{\frac{(\phi_{0max} - V_C)}{2\phi_t}} \sqrt{\frac{q^2 n_i}{2kT\epsilon_{Si}}} \frac{t_{Si}}{2} \right] = \frac{\pi}{2} \tag{8.22}$$

which on rearrangement gives:

$$\phi_{0max} = V_C + \phi_t \ln \left(\frac{2\epsilon_{Si}kT}{q^2 n_i} \frac{\pi^2}{t_{Si}^2} \right) \tag{8.23}$$

The behavior of ϕ_0 as a function of gate voltage is represented by Ortiz-Conde et al. [25] using a smoothing function, given as:

$$\phi_0 = U - \sqrt{U^2 - (V_{GS} - V_{fb}) \phi_{0max}} \tag{8.24}$$

where:

$$U = \frac{1}{2} \left[(V_{GS} - V_{fb}) + (1 + r) \phi_{0max} \right] \tag{8.25}$$

r is a smoothing parameter weakly dependent on T_{ox}, t_{Si}, and V_C, which is determined from the equation given by:

$$r = (A T_{ox} + B) \left(\frac{C}{t_{Si}} + D \right) e^{-E V_C} \tag{8.26}$$

Here $A = 0.0267\,\text{nm}^{-1}$, $B = 0.0270$, $C = 0.4526\,\text{nm}$, $D = 0.0650$, and $E = 3.2823\text{V}^{-1}$. These values are appropriate for device dimensions of $T_{ox} < 10\,\text{nm}$ and $t_{Si} > 5\,\text{nm}$.

Figure 8.9 is obtained from Equation (8.24) and Equation (8.21) for different values of gate voltages. To study ϕ_0, it is important to understand its behaviour in two regions, i.e. below threshold and above threshold. Below threshold the mobile charge density is low, and volume inversion takes place. As a consequence both ϕ_s and ϕ_0 closely follow the gate voltage. In other words, there is negligible potential drop across the oxide. Above threshold voltage, mobile charges at the surface screen the charge present at the center of the silicon film, thereby limiting its potential to a maximum value, i.e. ϕ_{0max}. Surface potential increases with increasing gate voltage as more and more electrons accumulate near the surface, but the potential at the center remains constant which can be seen in the given plot.

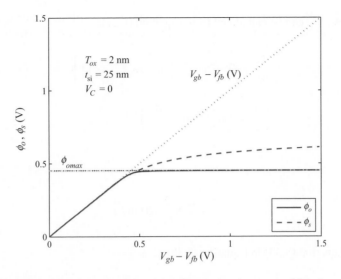

Figure 8.9   Channel potential at the surface and at the center of the channel

8.3.1.3 Taur's Model

In this model, 1-D Poisson's equation is solved for the DG MOSFET structure considering the gradual channel approximation, for a long channel DG MOSFET [12]. The solution is parameterized with a variable β whose value varies from source to drain, as it is a function of V_C. Equation (8.10) is re-written in the following format, in terms of the parameter β:

$$\phi(x) = V_C - 2\phi_t \ln \left[\frac{t_{Si}}{2\beta} \sqrt{\frac{qn_i}{2\epsilon_{Si}\phi_t}} \cos \left(\frac{2\beta x}{t_{Si}} \right) \right] \tag{8.27}$$

where:

$$\beta = \frac{t_{Si}}{2} \sqrt{\frac{qn_i}{2\epsilon_{Si}\phi_t}} \, e^{\frac{(\phi_0 - V_C)}{2\phi_t}} \tag{8.28}$$

The constant β has to be found out using Equation (8.8):
Substituting Equation (8.27) into Equation (8.8) leads to:

$$2\phi_t \frac{\epsilon_{ox}}{T_{ox}} \left[\frac{(V_{GS} - V_{fb} - V_C)}{2\phi_t} + \ln \left(\frac{t_{Si}}{2\beta} \sqrt{\frac{qn_i}{2\epsilon_{Si}\phi_t}} \cos \beta \right) \right] = \pm \epsilon_{Si} \frac{d\phi}{dx} \bigg|_{x = \pm \frac{t_{Si}}{2}} \tag{8.29}$$

$$2\phi_t \frac{\epsilon_{ox}}{T_{ox}} \left[\frac{(V_{GS} - V_{fb} - V_C)}{2\phi_t} + \ln \left(\frac{1}{\beta} \right) + \ln \left(\frac{t_{Si}}{2} \sqrt{\frac{qn_i}{2\epsilon_{Si}\phi_t}} \right) + \ln(\cos \beta) \right] =$$
$$\pm \epsilon_{Si} \frac{d\phi}{dx} \bigg|_{x = \pm \frac{t_{Si}}{2}} \tag{8.30}$$

For the above, $\frac{d\phi}{dx}$ can be calculated from Equation (8.27):

$$\frac{d\phi}{dx} = 0 - 2\phi_t \frac{d}{dx}\left\{ \ln\left[\cos\left(\frac{2\beta x}{t_{Si}}\right)\right]\right\} \tag{8.31}$$

$$\frac{d\phi}{dx} = 2\phi_t \frac{2\beta}{t_{Si}} \tan\left(\frac{2\beta x}{t_{Si}}\right) \tag{8.32}$$

At $x = \frac{t_{Si}}{2}$:

$$\frac{d\phi}{dx} = 2\phi_t \frac{2}{t_{Si}} (\beta \tan \beta) \tag{8.33}$$

Substituting Equation (8.33) in Equation (8.30) gives:

$$\left[\frac{(V_{GS} - V_{fb} - V_C)}{2\phi_t} + \ln\left(\frac{1}{\beta}\right) + \ln\left(\frac{t_{Si}}{2}\sqrt{\frac{qn_i}{2\epsilon_{Si}\phi_t}}\right) + \ln(\cos\beta)\right] =$$
$$\frac{2\epsilon_{Si}T_{ox}}{\epsilon_{ox}t_{Si}}(\beta\tan\beta) \tag{8.34}$$

Thus, we obtain an equation, which on solving gives the value of β at different values of V_C:

$$\frac{(V_{GS} - V_{fb} - V_C)}{2\phi_t} - \ln\left(\frac{2}{t_{Si}}\sqrt{\frac{2\epsilon_{Si}\phi_t}{qn_i}}\right) = \ln\beta - \ln(\cos\beta) + \frac{2\epsilon_{Si}T_{ox}}{\epsilon_{ox}t_{Si}}(\beta\tan\beta) \tag{8.35}$$

For $V_C = 0$ and V_{DS}, we obtain β_s at source and β_d at drain respectively: β can be computed numerically or analytically, the details of which are presented in Yu *et al.* [13].

8.3.2 Drain Current Models

Drain current has been computed by various approaches [26]. A few of these approaches are discussed below.

8.3.2.1 Taur's Model

I_{DS} is given by Equation (4.12), which is reproduced below for convenience:

$$I_{DS} = -W\mu_{neff}Q'_n(y)\frac{dV_C(y)}{dy} \tag{8.36}$$

Integrating $I_{DS}\, dy$ from source to drain over length L gives:

$$\int_0^L I_{DS}\, dy = -W\mu_{neff} \int_0^y Q'_n(y)\, dV_C(y) \tag{8.37}$$

$$I_{DS} = -\mu_{neff}\frac{W}{L}\int_0^{V_{ds}} Q'_n(V_C)\, dV_C = -\mu_{neff}\frac{W}{L}\int_{\beta_s}^{\beta_d} Q'_n(\beta)\frac{dV_C}{d\beta}\, d\beta \tag{8.38}$$

The total mobile charge per unit gate area is given by:

$$Q'_n = 2\epsilon_{Si}\left(\frac{d\phi}{dx}\right)\Bigg|_{x=\frac{t_{Si}}{2}} \tag{8.39}$$

$$Q'_n = 2\epsilon_{Si}\left(-\frac{2kT}{q}\right)\left(\frac{2}{t_{Si}}\right)\beta\tan\beta \tag{8.40}$$

where the factor of 2 is due to incorporation of the effect of both the gates.
In the above, $dV_C/d\beta$ is obtained by differentiating Equation (8.35):

$$-\frac{q}{2kT}\frac{dV_C}{d\beta} = \frac{1}{\beta} + \tan\beta + \frac{2\epsilon_{Si}T_{ox}}{\epsilon_{Si}t_{Si}}\frac{d}{d\beta}(\beta\tan\beta) \tag{8.41}$$

Thus, I_{DS} is obtained as:

$$I_{DS} = \mu_{neff}\frac{W}{L}\frac{4\epsilon_{Si}}{t_{Si}}(2\phi_t)^2$$
$$\times \int_{\beta_d}^{\beta_s}\left[\tan\beta + \beta\tan^2\beta + \frac{2\epsilon_{Si}T_{ox}}{\epsilon_{ox}t_{Si}}\beta\tan\beta\frac{d}{d\beta}(\beta\tan\beta)\right]d\beta \tag{8.42}$$

The expression for the drain current is:

$$I_{DS} = \mu_{neff}\frac{W}{L}\frac{4\epsilon_{Si}}{t_{Si}}(2\phi_t)^2 \times \left[\beta\tan\beta - \frac{\beta^2}{2} + \frac{\epsilon_{Si}T_{ox}}{\epsilon_{ox}t_{Si}}\beta^2\tan^2\beta\right]_{\beta_d}^{\beta_s} \tag{8.43}$$

where:

$$r_1 = \frac{\epsilon_{Si}T_{ox}}{\epsilon_{ox}t_{Si}} \tag{8.44}$$

Let:

$$g_{r_1}(\beta) = \beta \tan \beta - \frac{\beta^2}{2} + r_1 \beta^2 \tan^2 \beta \tag{8.45}$$

We have:

$$I_{DS} = \mu_{neff} \frac{W}{L} \frac{4\epsilon_{Si}}{t_{Si}} (2\phi_t)^2 \times \left[g_{r_1}(\beta_s) - g_{r_1}(\beta_d) \right] \tag{8.46}$$

As explained in the previous section, Equation (8.35) is evaluated to obtain the value of β_s and β_d; the details of the computation are presented in Yu *et al.* [13].

8.3.2.2 Ortiz-Conde *et al.* Model

According to the Pao–Sah current model, both drift and diffusion contribute to the current in the silicon film. Assuming mobility to be constant, current can be evaluated from the following expression:

$$I_{ds} = -\mu_{neff} \frac{W}{L} \int_0^{V_{DS}} Q_n' \, dV \tag{8.47}$$

where I_{ds} represents current for a single gate. Using the relation $F = -\frac{d\phi}{dx}$, the inversion charge density (Q_n') inside the silicon film at a given location for a gate can be expressed as:

$$Q_n' = -2q \int_{\phi_0}^{\phi_s} \frac{(n - n_i)}{F} \, d\phi \tag{8.48}$$

Since $n \gg n_i$, Equations (8.47) and (8.48) can be combined to give:

$$I_{ds} = 2\mu_{neff} \frac{W}{L} \int_0^{V_{DS}} \int_{\phi_0}^{\phi_s} \frac{qn}{F} \, d\phi \, dV \tag{8.49}$$

Rearranging Equation (8.9), the electric field, as calculated from Poisson's equation, is shown below:

$$F = -\sqrt{\frac{2kTn_i}{\epsilon_{Si}} e^{\frac{\phi_s - V_C}{\phi_t}} + \alpha} \tag{8.50}$$

where:

$$\alpha = \frac{2kTn_i}{\epsilon_{Si}} e^{\frac{(\phi_0 - V_C)}{\phi_t}} \tag{8.51}$$

and:

$$n = n_i e^{\frac{\phi_s - V_C}{\phi_t}} \tag{8.52}$$

Differentiating Equation (8.50) with respect to channel voltage following the procedure of Pierret and Shields [23]:

$$\frac{\partial F}{\partial V_C} = \frac{1}{2F} \frac{d\alpha}{dV_C} - \frac{qn}{F\epsilon_{Si}} \tag{8.53}$$

After rearrangement:

$$\frac{qn}{F} = \epsilon_{Si} \left(\frac{1}{2F} \frac{d\alpha}{dV_C} - \frac{\partial F}{\partial V_C} \right) \tag{8.54}$$

Substituting Equation (8.54) into Equation (8.49):

$$I_{DS} = 2\mu_{neff} \frac{W}{L} \epsilon_{Si} \int_0^{V_{DS}} \int_{\phi_0}^{\phi_s} \left(\frac{1}{2F} \frac{d\alpha}{dV_C} - \frac{\partial F}{\partial V_C} \right) d\phi \, dV_C \tag{8.55}$$

Following the method of Ortiz-Conde et al. [19], the drain current can be written as the difference of two currents, i.e. I_1 and I_2:

$$I_{DS} = I_1 - I_2 \tag{8.56}$$

where:

$$I_1 = 2\mu_{neff} \frac{W}{L} \epsilon_{Si} \int_0^{V_{DS}} \int_{\phi_0}^{\phi_s} \frac{1}{2F} \frac{d\alpha}{dV_C} d\phi \, dV_C \tag{8.57}$$

Using the relation $\int_0^{t_{Si}/2} dx = \int_{\phi_0}^{\phi_s} \frac{d\phi}{F}$, the final expression for I_1 is expressed as:

$$I_1 = \mu_{neff} \frac{W}{L} \epsilon_{Si} \int_0^{V_{DS}} \frac{t_{Si}}{2} d\alpha = \mu_{neff} \frac{W}{L} \epsilon_{Si} \frac{t_{Si}}{2} (\alpha_S - \alpha_D) \tag{8.58}$$

where α_S and α_D are evaluated at source and drain respectively. The second integral I_2 is given by:

$$I_2 = 2\mu_{neff} \frac{W}{L} \epsilon_{Si} \int_0^{V_{DS}} \int_{\phi_0}^{\phi_s} \frac{\partial F}{\partial V_C} d\phi \, dV_C \tag{8.59}$$

Figure 8.10 shows the dependence of the surface potential and potential at the center of the film on the channel voltage for a specific gate voltage. The double integrals mentioned in Equation (8.55) evaluate the area ABCD shown on the plot. The drain current can be easily

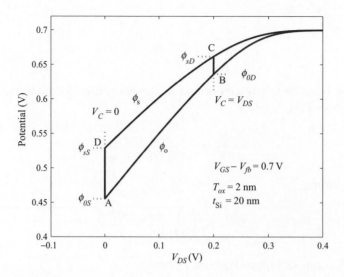

Figure 8.10 Electrostatic potential ϕ_s at the silicon–gate oxide interface, and ϕ_0 at the center of a silicon film as a function of drain voltage

computed by splitting it into four single integrals. $V = V_1$ when $\phi = \phi_s$, and $V = V_2$ when $\phi = \phi_0$. Therefore:

$$\int_0^{V_{DS}} \int_{\phi_0}^{\phi_s} \frac{\partial F}{\partial V_C}\, d\phi\, dV_C = \int_{\phi_{0D}}^{\phi_{sD}} F(\phi, V = V_{DS})\, d\phi$$

$$- \int_{\phi_{0S}}^{\phi_{sS}} F(\phi, V = 0)\, d\phi$$

$$- \int_{\phi_{sS}}^{\phi_{sD}} F(\phi, V = V_1)\, d\phi$$

$$+ \int_{\phi_{0S}}^{\phi_{0D}} F(\phi, V = V_2)\, d\phi \qquad (8.60)$$

After integration of the above integrals, and combining Equation (8.58) and Equation (8.59), we get the final expression for drain current.[1]

$$I_{DS} = \mu_{neff} \frac{W}{L} \left\{ 2C'_{ox} \left[V_{GS}(\phi_{sD} - \phi_{sS}) - \frac{1}{2}(\phi_{sD}^2 - \phi_{sS}^2) \right] \right.$$

$$\left. + 4\frac{kT}{q} C'_{ox}(\phi_{sD} - \phi_{sS}) + t_{Si} kT n_i \left(e^{\frac{\phi_{0D} - V_{DS}}{\phi_t}} - e^{\frac{\phi_{0S}}{\phi_t}} \right) \right\} \qquad (8.61)$$

---

[1] For a detailed description refer to Ortiz-Conde *et al.* [19].

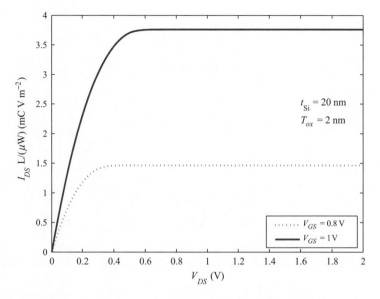

Figure 8.11 Drain current as a function of drain voltage for two different values of gate voltages using the Ortiz-Conde *et al.* model

Figure 8.11 shows the simulated static drain current versus drain voltage characteristics for two gate bias voltages based on the Ortiz-Conde *et al.* model.

8.3.2.3 He *et al.* Model

Poisson's equation is solved in terms of the electron concentration:

$$\frac{d^2\phi}{dx^2} = \frac{qn(x)}{\epsilon_{Si}} \tag{8.62}$$

According to Boltzmann statistics, electron concentration can be expressed as:

$$n(x) = n_i \exp\frac{q(\phi - V_C)}{kT} \tag{8.63}$$

Also, at the center of the film ($x = 0$), $n = n_0$, and $\phi = \phi_0$, which gives:

$$n_0 = n_i \exp\frac{q(\phi_0 - V_C)}{kT} \tag{8.64}$$

Differentiating Equation (8.63), we get:

$$\frac{dn}{dx} = \frac{qn}{kT}\frac{d\phi}{dx} \tag{8.65}$$

Rearranging:

$$\frac{d\phi}{dx} = \frac{kT}{qn}\frac{dn}{dx} \tag{8.66}$$

$$\frac{d^2\phi}{dx^2} = \frac{kT}{qn}\frac{d^2n}{dx^2} - \frac{kT}{qn^2}\left(\frac{dn}{dx}\right)^2 \tag{8.67}$$

Substituting Equation (8.67) into Equation (8.62):

$$\frac{d^2n}{dx^2} = \frac{1}{n}\left(\frac{dn}{dx}\right)^2 + \frac{q^2n^2}{\epsilon_{Si}kT} \tag{8.68}$$

The differential equation can be solved assuming the boundary condition $x = 0$ and $n(x) = n_0$:

$$n(x) = \frac{n_0}{\cos^2\left[\left(\frac{q^2n_0}{2\epsilon_{Si}kT}\right)^{\frac{1}{2}}x\right]} \tag{8.69}$$

Substituting $n(x)$ from Equation (8.69) into Equation (8.62), and integrating with the boundary condition $F(x = 0) = 0$, gives the potential at the surface, ϕ_s ($x = t_{Si}/2$):

$$\phi_s - \phi_0 = \frac{kT}{q}\ln\left\{\cos^{-2}\left[\left(\frac{q^2n_0}{2\epsilon_{Si}kT}\right)^{\frac{1}{2}}x\right]\right\} \tag{8.70}$$

The value of ϕ_0 can be obtained from Equation (8.64):

$$\phi_s = V_C + \frac{kT}{q}\ln\left\{\frac{n_0}{n_i}\cos^{-2}\left[\left(\frac{q^2n_0}{2\epsilon_{Si}kT}\right)^{\frac{1}{2}}\frac{t_{Si}}{2}\right]\right\} \tag{8.71}$$

Now, the total inversion charge can be obtained from:

$$Q'_n = -2q\int_0^{\frac{t_{Si}}{2}} n(x)\,dx \tag{8.72}$$

On substituting for $n(x)$ from Equation (8.69):

$$Q'_n = -2qn_0\int_0^{\frac{t_{Si}}{2}} \sec^2\left[\left(\frac{q^2n_0}{2\epsilon_{Si}kT}\right)^{\frac{1}{2}}x\right]\,dx \tag{8.73}$$

$$Q_n' = -2\sqrt{2\epsilon_{Si}n_0kT} \, \tan\left[\left(\frac{q^2 n_0}{2\epsilon_{Si}kT}\right)^{\frac{1}{2}}\frac{t_{Si}}{2}\right] \tag{8.74}$$

The above equation can be expressed in an alternative form so that the surface potential can be expressed in terms of inversion charge density. On squaring both sides of the Equation (8.74), we get:

$$Q_n'^2 = 8\epsilon_{Si}n_0kT\frac{\sin^2\left[\left(\frac{q^2 n_0}{2\epsilon_{Si}kT}\right)^{\frac{1}{2}}\frac{t_{Si}}{2}\right]}{\cos^2\left[\left(\frac{q^2 n_0}{2\epsilon_{Si}kT}\right)^{\frac{1}{2}}\frac{t_{Si}}{2}\right]} \tag{8.75}$$

According to the approximation used by He *et al.* [20]:

$$\sin^2\left[\left(\frac{q^2 n_0}{2\epsilon_{Si}kT}\right)^{\frac{1}{2}}\frac{t_{Si}}{2}\right] = -\frac{qQ_n'}{2\epsilon_{Si}kT}\frac{t_{Si}}{4}\exp(f) \tag{8.76}$$

Substituting the above approximation into Equation (8.75), the inversion charge density becomes:

$$Q_n' = -\frac{qn_0 t_{Si}}{\cos^2\left[\left(\frac{q^2 n_0}{2\epsilon_{Si}kT}\right)^{\frac{1}{2}}\frac{t_{Si}}{2}\right]}\exp(f) \tag{8.77}$$

where f is a dimensionless correction factor. Therefore, the surface potential can be expressed in terms of inversion charge by substituting Equation (8.77) into Equation (8.71):

$$\phi_s = V_C + \frac{kT}{q}\left[\ln(-Q_n') - f\ln(qn_i t_{Si})\right] \tag{8.78}$$

According to Gauss's law, (potential balance at the surface):

$$V_{GS} - V_{fb} = \phi_s - \frac{Q_n'}{2\epsilon_{ox}}T_{ox} \tag{8.79}$$

Substituting for ϕ_s from Equation (8.78), we get:

$$V_{GS} - V_{fb} - V_C = \frac{kT}{q}\left[\ln(-Q_n') - f\ln(qn_i t_{Si})\right] - \frac{Q_n'}{2\epsilon_{ox}}T_{ox} \tag{8.80}$$

Therefore:

$$dV_C = -\frac{kT}{q}\frac{dQ_n'}{Q_n'} + \frac{dQ_n'}{2C_{ox}'} \tag{8.81}$$

After normalizing the biasing by $\phi_t = \frac{kT}{q}$, we get:

$$dv_c = -\frac{dQ'_n}{Q'_n} + \frac{dQ'_n}{2\phi_t C'_{ox}} \tag{8.82}$$

Current, including diffusion and drift component, can be written as:

$$I_{ds} = -\mu W \left(\frac{kT}{q}\right)(Q'_n)\frac{dv_c}{dy} \tag{8.83}$$

Therefore after integrating:

$$I_{DS} = \frac{\mu W}{L}\left[\left(\frac{kT}{q}\right)(Q'_{nD} - Q'_{nS}) - \frac{Q'^2_{nD} - Q'^2_{nS}}{4C'_{ox}}\right] \tag{8.84}$$

8.3.2.4 Sallese *et al.* Model

Sallese and coworkers obtained the following expression for drain current [21]:

$$I_{DS} = \mu\frac{W}{L}\left\{\frac{2kT}{q}(Q'_{nD} - Q'_{nS}) - \frac{(Q'^2_{nD} - Q'^2_{nS})}{4C_{ox}}\right.$$
$$\left. +8\left(\frac{kT}{q}\right)^2\frac{\epsilon_{Si}}{t_{Si}}\ln\left[1 - \frac{qt_{Si}(Q'_{nD} - Q'_{nS})}{8\epsilon_{Si}kT}\right]\right\} \tag{8.85}$$

where Q'_{nS} and Q'_{nD} can be evaluated using the surface potential at the source and drain ends.

$$Q'_n = -2C_{ox}\left(V_{GB} - V_{fb} - \phi_s\right) \tag{8.86}$$

The above drain current equation can be expressed in the normalized form conforming to the EKV formalism by expressing $q = Q/Q_o$ and $i = I/I_s$ where $Q_o = 4C'_{ox}\phi_t$ and $I_s = 4C'_{ox}\phi_t^2 W/L$. The potential is also normalized w.r.t. ϕ_t.

Figure 8.12 shows the simulated drain current as a function of gate voltage for a given drain voltage, i.e. $V_{DS} = 1$ mV, for the various drain current models discussed above. It is observed that the results are in close agreement with each other.

8.3.3 Short Channel Effect (SCE)

Like bulk channel MOSFETs, the short channel effect is a matter of concern for scaled DG MOSFETs because of increase of leakage current in the subthreshold region of operation. In bulk MOSFETs, the enhanced conduction is attributed to barrier lowering between source and drain at the surface, and due to bulk punchthrough. For a long channel DG MOSFET, in the subthreshold region of operation, the bulk potential of the entire film is raised. The magnitude

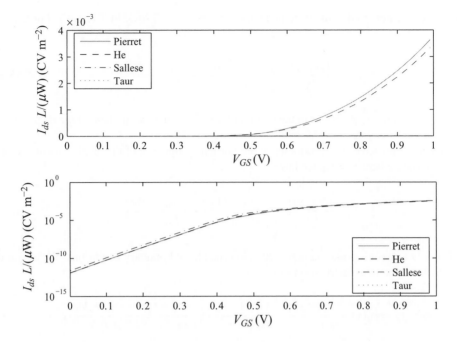

Figure 8.12 Simulated drain current versus V_{GS} for various models

of this potential is dependent on gate voltage and technological parameters, leading to so-called *volume inversion*. Therefore, the entire bulk provides the path for leakage current from source to drain. When the channel length is scaled, the bulk potential of the film in the subthreshold region is further increased due to the influence of drain potential. The potential barrier between source and the bulk material is now reduced compared to the long channel case, leading to a higher level of volume inversion and leakage current. For short channel double gate MOSFETs the electric field lines penetrate the silicon film from the gate vertically and drain laterally in the planar structure under consideration. Now both the gates and the drain compete for the control of charge in the film, and thus GCA fails. Thus, we need to solve the 2-D Poisson's equation in the bulk material of a double gate MOSFET.

We shall present the approach of Diagne *et al.* [29] for the derivation of an SCE in a DG MOSFET. The 2-D Poisson's equation is written taking the channel doping concentration N_A. In this formulation, the origin is taken at the SiO$_2$–front gate interface, and boundary conditions have to be set accordingly:

$$\frac{d^2\phi(x, y)}{dx^2} + \frac{d^2\phi(x, y)}{dy^2} = \frac{qN_A}{\epsilon_{Si}} \tag{8.87}$$

Following Young [17], the potential $\phi(x, y)$ can be described in the vertical x-direction by a second order parabolic function:

$$\phi(x, y) = c_0(y) + c_1(y)x + c_2(y)x^2 \tag{8.88}$$

This gives the potential at the front and back gates of the DG MOSFET along y. Further,

$$\left. \frac{d\phi(x, y)}{dx} \right|_{x=0} = \frac{\epsilon_{ox}}{\epsilon_{Si}} \frac{\phi_s(y) - (V_{GS} - V_{fb})}{T_{ox}} \tag{8.89}$$

The above equation follows from the continuity of the normal component of the displacement vector at the front gate, $x = 0$.

As a consequence of the continuity of the normal component of the displacement vector at the back gate, where $x = t_{Si}$, we have:

$$\left. \frac{d\phi(x, y)}{dx} \right|_{x=t_{Si}} = \frac{\epsilon_{ox}}{\epsilon_{Si}} \frac{(V_{GS} - V_{fb}) - \phi_s(y)}{T_{ox}} \tag{8.90}$$

The 2-D Poisson's equation is solved using the above boundary conditions. We obtain the 2-D channel potential distribution $\phi(x, y)$ as:

$$\phi(x, y) = \phi_s(y) + \frac{\epsilon_{ox}}{\epsilon_{Si}} \frac{\phi_s(y) - V_{GS} + V_{fb}}{T_{ox}} x - \frac{\epsilon_{ox}}{\epsilon_{Si}} \frac{\phi_s(y) - V_{GS} + V_{fb}}{T_{ox} t_{Si}} x^2 \tag{8.91}$$

It is observed that for a short channel DG MOSFET, the potential at the center becomes more than the potential at the surface, unlike the long channel case, where the surface potential is more than the potential at the center. This is due to the encroachment of the lateral electric field, which results in more charge at the center of the film, and thus current starts flowing through the center. Thus, the potential at the center is evaluated using the above equation, and it is substituted in the 2-D Poisson's equation, neglecting the channel doping:

$$\frac{d^2\phi_c(y)}{dy^2} - \frac{1}{l^2}\phi_c(y) = -\frac{1}{l^2}\left(V_{GS} - V_{fb}\right) \tag{8.92}$$

Here:

$$l = \sqrt{\frac{\epsilon_{Si} t_{Si} T_{ox}}{2\epsilon_{ox}} \left(1 + \frac{\epsilon_{ox} t_{Si}}{4\epsilon_{Si} T_{ox}}\right)} \tag{8.93}$$

Equation (8.92) is solved using the following boundary conditions:

$$\phi_c(x, 0) = \phi_{bi} - \phi_f = \frac{kT}{q} \ln\left(\frac{N_D}{n_i}\right) \tag{8.94}$$

$$\phi_c(x, L) = \phi_{bi} + V_{DS} - \phi_f \tag{8.95}$$

where $\phi_{bi} = (kT/q)\ln(N_D N_A/n_i^2)$ and $\phi_f = (kT/q)\ln(N_A/n_i)$ is the Fermi potential.

Thus:

$$\phi_c(y) = (V_{GS} - V_{fb}) + \left[(\phi_{bi} - \phi_f) - (V_{GS} - V_{fb}) + V_{DS}\right] \frac{\sinh\left(\frac{y}{l}\right)}{\sinh\left(\frac{L}{l}\right)}$$

$$+ \left[(\phi_{bi} - \phi_f) - (V_{GS} - V_{fb})\right] \frac{\sinh\left(\frac{L-y}{l}\right)}{\sinh\left(\frac{L}{l}\right)} \quad (8.96)$$

The minimum channel potential value is evaluated as ϕ_{cmin} at the virtual cathode position, and is evaluated as:

$$\phi_{cmin} = \phi_c(L/2) \quad (8.97)$$

Thus ϕ_{cmin} is obtained as:

$$\phi_{cmin} = (V_{GS} - V_{fb}) + \left\{2\left[(\phi_{bi} - \phi_f) - (V_{GS} - V_{fb})\right] + V_{DS}\right\} \frac{\sinh\left(\frac{L}{2l}\right)}{\sinh\left(\frac{L}{l}\right)} \quad (8.98)$$

Let $V'_{GS} = V_{GS} - V_{fb}$. We obtain from Equation (8.98):

$$V'_{GS} = \frac{\phi_{cmin} - \left[2(\phi_{bi} - \phi_f) + V_{DS}\right] \frac{\sinh\left(\frac{L}{2l}\right)}{\sinh\left(\frac{L}{l}\right)}}{1 - 2\frac{\sinh\left(\frac{L}{2l}\right)}{\sinh\left(\frac{L}{l}\right)}} \quad (8.99)$$

The short channel threshold voltage is defined as the voltage at which ϕ_{cmin} becomes equal to the threshold voltage for the long channel V_{TL}. Thus, the threshold voltage for a short channel DG MOSFET model becomes:

$$V_T = \frac{V_{TL} - \left[2(\phi_{bi} - \phi_f) + V_{DS}\right] \frac{\sinh\left(\frac{L}{2l}\right)}{\sinh\left(\frac{L}{l}\right)}}{1 - 2\frac{\sinh\left(\frac{L}{2l}\right)}{\sinh\left(\frac{L}{l}\right)}} \quad (8.100)$$

where V_{TL} is given by:

$$V_{TL} = \Delta\phi_{ms} - \phi_t \ln\left(\frac{q n_i t_{Si}}{4 C'_{ox} \phi_t}\right) \quad (8.101)$$

In the above, $\Delta\phi_{ms}$ is the gate-intrinsic silicon work function difference. The threshold voltage roll-off ΔV_T is the difference of the long channel threshold voltage and short channel threshold voltage. To calculate ΔV_T substitute the value of V_T from Equation (8.100), and use the identity $\sinh 2x = 2 \sinh x \cosh x$ to transform the equation to get:

$$\Delta V_T = V_{TL} - V_T = \frac{2\left(\phi_{bi} - \phi_f - V_{TL}\right) + V_{DS}}{2\cosh\left(\frac{L}{2l}\right) - 2} \quad (8.102)$$

Equation (8.102) is approximated by considering $l \ll L$:

$$\Delta V_T = \gamma_1 \left[2 \left(\phi_{bi} - \phi_f - V_{TL} \right) + V_{DS} \right] \tag{8.103}$$

where $\gamma_1 = e^{-\sigma L/2l}$: σ is a fitting parameter, given by $\sigma = 0.98$. For a low drain to source voltage, Equation (8.103) is evaluated for the threshold roll-off as $\Delta V_{T_{roll}} = \Delta V_T|_{V_{DS}=0}$.

$$\Delta V_T = 2\gamma_1 \left(\phi_{bi} - \phi_f - V_{TL} \right) \tag{8.104}$$

DIBL is defined as $d\Delta V_T/dV_{DS}$:

$$\Delta V_{T_{DIBL}} = \gamma_1 V_{DS} \tag{8.105}$$

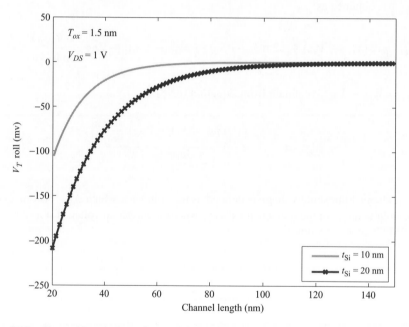

Figure 8.13   Channel length dependence of threshold voltage roll-off in a DG MOSFET structure

Figure 8.13 shows the threshold voltage roll-off with channel length for two values of film thickness. It is noted that the roll-off starts earlier for the thicker film. In other words, the SCE can be minimized by making the film thinner [30].

Figure 8.14 shows that the DIBL effect is more pronounced, both in terms of onset and magnitude, for a thicker film.

8.3.4 Quantum Confinement Effect

It has been established that the thiner the bulk film in DG MOSFETs the better is the SCE behavior of the device, as the gate has more effective control on the channel. When the film

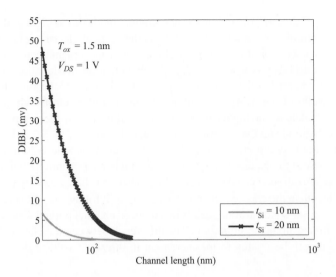

Figure 8.14 Channel length dependence of DIBL in a DG MOSFET structure

thickness t_{Si} approaches the size of the de Broglie wavelength, however, quantization of energy levels take place [32]. The quantization may take place due to two factors: (a) Structural Confinement (SC) of carriers within the ultrathin film, and (b) Electrical Confinement (EC) of carriers within the potential barrier induced by the gate field [33, 35]. Figure 8.15(a) shows the energy band diagrams of DG MOSFETs where the film thickness induced energy level discretization is illustrated. Figure 8.15 displays the gate field induced energy level discretization. In general terms, the energy level quantization is known as the Quantum Confinement Effect (QCE). Using the variational approach the ground state energy level, E_0, can be obtained by solving Schrödinger's equation [37]:

$$E_0 \approx \frac{\hbar^2}{2m_x} \left(\left(\frac{\pi}{t_{Si}} \right)^2 + b_0^2 \left\{ 3 - \frac{4}{3} \frac{1}{\left[\left(\frac{b_0 t_{Si}}{\pi} \right)^2 + 1 \right]} \right\} \right) \qquad (8.106)$$

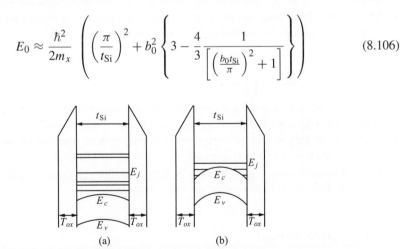

Figure 8.15 Energy quantization in DG MOSFETs: (a) structural confinement, and (b) electrical confinement

The term, $b_0 = (\frac{3}{4} \times 2m_x q F_x / \hbar^2)^{1/3}$, in the above equation owes its origin to the gate induced electric field creating the potential well, and F_x is the vertical electric field in the film. The electric field is obtained from the continuity of electric displacement vector at the film oxide interface, $\epsilon_{Si} F_x = \epsilon_{ox}(V_{GB} - V_{fb} - \phi_s)/T_{ox}$. An examination of Equation (8.106) clearly shows that the first term is related to the physical dimension, i.e. the film thickness which determines the width of the well, and confines the motion of electrons within the structure. The smaller the thickness, the higher is the ground level energy of the sub-band. The second term has a contribution from the electric field through the term b_0.

In the subthreshold condition, the normal field strength in the silicon film being small, the carriers can be viewed to be confined within a potential barrier of infinite height of width t_{Si}. Due to energy quantization, it can be shown that the inversion charge density $Q'_{n,qm}$ for a given gate voltage is less than the classical charge density $Q'_{n,cl}$. It follows from Gauss' theorem that the reduction in carrier concentration results in a reduced channel potential as compared to the classical value in the subthreshold region. The difference between the classical and quantum channel potentials can be related to the increment in threshold voltage due to the quantum effect [22, 35].

8.3.5 Compact DG MOSFET Models

8.3.5.1 BSIM-MG Model

BSIM-MG is a surface potential based model generalized for a finite doped silicon film [27, 28, 37].

- In the 1-D Poisson's equation, both bulk and inversion charges are included. A perturbation scheme is used to solve the Poisson's equation by first ignoring the doping related term. The solution of the Poisson's equation, ϕ_1, thus obtained corresponds to the inversion carrier term.
- The effect of finite body doping is represented through the superposition of an additional perturbation term ϕ_2, providing the total potential as:

$$\phi(x) = \phi_1(x) + \phi_2(x) \tag{8.107}$$

- The corrected surface potential term has a unique feature that a single expression is valid both for partially depleted and fully depleted modes of operation.
- BSIM-MG is able to characterize various scaling related phenomena for DG MOSFET structures such as: V_T roll-off due to charge sharing, the DIBL effect, sub-threshold swing, mobility degradation, capacitances including parasitics, quantization of carriers due to structural and field related confinement of carriers, etc.

8.3.5.2 HiSIM-SOI Model

The characteristic features of the HiSIM-SOI model, for an SOI structure with a front and back gate on the top and bottom plane of a thin silicon film, is summarized below [31]:

- Three core potentials are computed: (i) ϕ_{sf}, the potential at the front or top gate–silicon body interface, (ii) ϕ_{sb}, the potential at the silicon body–back or bottom gate interface, and (iii) ϕ_{sc}, the potential at the back gate–silicon substrate interface.
- The potential balance is described for the structure using the potentials mentioned above by computing the potential drop across the front gate and back gate oxides with the charges in the substrate and the film:

$$V_{GS} - V_{fb} - \Delta V_T = V_{ox1} + (\phi_{sf} - \phi_{sb}) + V_{ox2} + \phi_{sc} \qquad (8.108)$$

where $V_{ox1} = -(Q'_{depf} + Q'_n)/T_{ox1}$, and ΔV_T corresponds to the threshold voltage shift due to lateral field penetration from source/drain, and accounts for the short channel effect.
- The short channel effect is formulated by evaluating the divergence of the electric field around the boundary enclosing the silicon field, and equating it to the enclosed charge, assuming that the vertical field strength is constant. A relation thereby can be established between the gate voltage and the charge in the silicon film. For a short channel device, the change in the threshold voltage compared to long channel value is given by:

$$\Delta V_T = \frac{\epsilon_{Si} t_{Si}}{C'_{ox1}} \frac{\mathrm{d} F_y}{\mathrm{d} y} \qquad (8.109)$$

Important charge related equations are:

$$Q'_{depf} = -q N_A t_{Si} \qquad (8.110)$$

$$Q'_{bulk} = \sqrt{2 q \phi_t N_B \epsilon_{Si}} \sqrt{\frac{\phi_{sc}}{\phi_t} - 1} \qquad (8.111)$$

- The parameters required to be extracted are:
  (i) Back gate thickness T_{ox2}, (ii) thickness of the film, (t_{Si}), and (iii) substrate doping concentration, N_B.

8.3.5.3 UFDG Model

The university of Florida DG MOSFET compact model, which is a generic physics and process based approach, has been in use for SOI based circuit simulation [36, 38]. It is generic in the sense that it can handle single gate, double gate, symmetric, asymmetric structures, and updates include structures such as FinFET. It addresses nanoscale dimensions by including quantum effects for which the S–P equations are solved self-consistently in the framework of DG structures. The updates provide for estimation of parasitic calculations for nonplanar structures. It relates device characteristics with device process parameters such as body and gate oxide thicknesses and device dimensions.

8.3.5.4 PSP Model

PSP-based model for FinFET is respresentative of the attributes of the surface potential driven approach. It obtains expression for drain current similar to that by Taur [12]. It has capabilities to address higher order issues such as SCE, mobility degradation, quantum effect, etc. [39].

References

[1] R.-H. Yan, A. Ourmazd and K. F. Lee, Scaling the Si MOSFET: From Bulk to SOI to Bulk, *IEEE Trans. Electron Dev.*, **39**(7), 1704–1710, 1992.

[2] C.H. Wann, K. Noda, T. Tanaka, M. Yoshida and C. Hu, A Comparative Study of Advanced MOSFET Concepts, *IEEE Trans. Electron Dev.*, **43**(10), 1742–1753, 1996.

[3] W. Grabiński, M. Bucher, J.-M. Sallese and F. Krummenacher, Advanced Compact Modeling of the Deep Submicron Technologies, *J. Telecom. IT*, **3–4**, 31–42, 2000.

[4] E.J. Nowak, Maintaining the Benefits of CMOS Scaling when Scaling Bogs Down, *IBM Res. Dev.*, **46**(2/3), 169–179, 2002.

[5] W. Haensch, E.J. Nowak, R.H. Dennard, R.M. Solomon, A. Bryant, O.H. Dokumaci, A. Kumar, X. Wang, J.B. Johnson and M.V. Fiscetti, Silicon CMOS devices, Beyond Scaling, *IBM J. Res. Dev.*, **50**(4/5), 339–361, 2006.

[6] Q. Chen, K.A. Bowman, E.M. Harrell and J. D. Meindl, Double Jeopardy in the Nanoscale Court?, *IEEE Circ. Dev. Mag.*, **19**(1), 28–34, 2003.

[7] T. Skotnicki, J.A. Hutcby, T.-J. King, H.-S.P. Wong and F. Boeuf, The End of CMOS Scaling toward the Introduction of New Materials and Structural Changes to Improve MOSFET Performance, *IEEE Circ. Dev. Mag.*, **21**(1), 16–26, 2005.

[8] P.M. Zeitzoff, J.A. Hutchby, G. Bersuker and H.R. Huff, Integrated Circuit Technologies: From Conventional CMOS to the Nanoscale Era, in *Nano and Giga Challenges in Microelectronics,* J. Greer, A. Korkin and J. Labanoski (Eds), Elsevier Press, Amsterdam, The Netherlands, 2003.

[9] H.P. Wong, D.J. Frank, P.M. Solomon, C.H.J. Wann and J.J. Welser, Nanoscale CMOS, *Proc. IEEE*, **87**(4), 537–569, 1999.

[10] J.-P. Colinge, Multiple-gate SOI MOSFETs, *Solid-State Electron.*, **48**(6), 897–905, 2004.

[11] Y. Taur, An Analytical Solution to a Double-Gate MOSFET with Undoped Body, *IEEE Electron Dev. Lett.*, **21**(5), 245–247, 2000.

[12] Y. Taur, X. Liang, W. Wang and H. Lu, A Continuous, Analytic Drain–Current Model for DG MOSFETs, *IEEE Electron Dev. Lett.*, **25**(2), 107–109, 2004.

[13] B. Yu, H. Lu, M. Liu and Y. Taur, Explicit Continuous Models for Double-Gate and Surrounding-Gate MOSFETs, *IEEE Trans. Electron Dev.*, **54**(10), 2715–2722, 2007.

[14] S. Takagi, N. Sugiyama, T. Mizuno, T. Tezuka and A. Kurobe, Device Structure and Electrical Characteristics on Strained-Si-on-Insulator (strained-SOI) MOSFETs, *Mater. Sci. Eng., B, Solid-State Adv. Technol.*, **89**(1), 426–434, 2002.

[15] V. Venkataraman, S. Nawal, and M.J. Kumar, Compact Analytical Threshold-Voltage Model of Nanoscale Fully Depleted Silicon-Germanium-on-Insulator (SGOI) MOSFETs, *IEEE Trans. Electron Dev.*, **54**(3), 554–562, 2007.

[16] E. Parton and P. Verheyen, Strained Silicon–the Key to Sub-45 nm CMOS, *Adv. Semiconductor Mag.*, **19**(3), 28–31, 2006.

[17] K.K. Young, Short-Channel Effect in Fully Depleted SOI MOSFETs, *IEEE Trans. Electron Dev.*, **36**(2), 399–402, 1989.

[18] K. Suzuki, T. Tanaka, Y. Tosaka, H. Horie and Y. Arimoto, Scaling Theory of Double-Gate SOI MOSFETs, *IEEE Trans. Electron Dev.*, **40**(12), 2326–2329, 1993.

[19] A. Ortiz-Conde, F.J. Garcia Sanchez and J. Muci, Rigorous Analytical Solution for Drain Current of Undoped Symmetric Dual-Gate MOSFETs, *Solid-State Electron.*, **49**(4), 640–647, 2005.

[20] J. He, X. Xuemei, M. Chan, C-H. Lin, A. Niknejad and C. Hu, A Non-Charge-Sheet Based Analytical Model of Undoped Symmetric Double-Gate MOSFETs Using SPP Approach, in *Proceedings of the 5th International Symposium on Quality Electronic Design*, pp. 45–50, 2004.

[21] J.M. Sallese, F. Krummenacher, F. Pregaldiny, C. Lallement, A.Roy and C. Enz, A Design-Oriented Charged-Based Current Model for Symmetric DG MOSFET and its Correlation with the EKV Formalism, *Solid-State Electron.*, **49**(3), 485–489, 2005.

[22] H. Ananthan and K. Roy, A Compact Physical Model for Yield Under Gate Length and Body Thickness Variations in Nanoscale Double-Gate CMOS, *IEEE Trans. Electron Dev.*, **53**(9), 2151–2159, 2006.

[23] R.F. Pierret and J.A. Shields, Simplified Long-Channel MOSFET Theory, *Solid-State Electron.*, **26**(2), 143–147, 1983.

[24] S. Malobabic, A. Ortiz-Conde and F.J.G. Sanchez, Modeling the Undoped-Body Symmetric Dual-Gate MOSFET, in *Proceedings of the 5th IEEE International Caracas Conference on Devices, Circuits and Systems*, Vol. 1, 19–25, 2004.

[25] A. Ortiz-Conde, F.J.G. Sanchez and S. Malobabic, Analytical Solution of the Channel Potential in Undoped Symmetric Dual-Gate MOSFETs, *IEEE Trans. Electron Dev.*, **52**(7), 1669–1672, 2005.

[26] A. Ortiz-Conde, F.J. Garcia-Sanchez, J. Muci, S. Malobabic and J.J. Liou, A Review of Core Compact Models for Undoped Double-Gate SOI MOSFETs, *IEEE Trans. Electron Dev.*, **54**(1), 131–140, 2007.

[27] M.V. Dunga, C.-H. Lin, X. Xi, D.D. Lu, A.M. Niknejad and C. Hu, Modeling Advanced FET Technology in a Compact Model, *IEEE Trans. Electron Dev.*, **53**(9), 1971–1978, 2006.

[28] M.V. Dunga, C.-H. Lin, D.D. Lu, W. Xiong, C.R. Cleavelin, P. Patruno, J.-R. Hwang, F.-L. Yang, A.M. Niknejad and C. Hu, BSIM-MG: A Versatile Multi-Gate FET Model for Mixed Signal Design, in *Proceedings of the IEEE Symposium on VLSI Technology*, pp. 60–61, 2007.

[29] B. Diagne, F. Pregaldiny, C. Lallement, J.-M. Sallese and F. Krummenacher, Explicit Compact Model for Double-Gate MOSFETs Including Solutions for Small-Geometry Effects, *Solid-State Electron.*, **52**, 99–106, 2008.

[30] H.A.E. Hamid, J.R. Guitart and B. Iñíguez, Two-Dimensional Analytical Threshold Voltage and Subthreshold Swing Models of Undoped Symmetric Double-Gate MOSFETs, *IEEE Trans. Electron Dev.*, **54**(6), 1402–1408, 2007.

[31] N. Sadachika, D. Kitamaru, Y. Uetsuji, D. Navarro, M.M. Yusoff, T. Ezaki, H.J. Mattausch and M.-M. Mattausch, Completely Surface-Potential-Based Compact Model of the Fully Depleted SOI-MOSFET Including Short-Channel Effects, *IEEE Trans. Electron Dev.*, **53**(9), 2017–2024, 2006.

[32] G. Baccarani and S. Reggiani, A Compact Double-Gate MOSFET Model Comprising Quantum Mechanical and Non-static Effects, *IEEE Trans. Electron Dev.*, **46**(8), 1656–1666, 1999.

[33] L. Ge and J.G. Fossum, Analytical Modeling of Quantization and Volume Inversion in Thin Si-film Double-Gate MOSFETs, *IEEE Trans. Electron Dev.*, **49**(2), 287–294, 2002.

[34] V. Trivedi and J.G. Fossum, Scaling Fully Depleted SOI CMOS, *IEEE Trans. Electron Dev.*, **50**(10), 2095–2013, 2003.

[35] V.P. Trivedi and J.G. Fossum, Quantum Mechanical Effects on the Threshold Voltage of Undoped Double-Gate MOSFETs, *IEEE Trans. Electron Dev.*, **26**(8), 579–582, 2005.

[36] J.G. Fossum, L. Ge, H.-M. Chaing, V.P. Trivedi, M.M. Chowdhury, L. Mathew, G.O. Workman, and B.-Y.Nguyen, A Process/Physics-based Compact Model for Nonclassical CMOS Device and Circuit Design, *Solid-State Electron.*, **48**(6), 919–926, 2004.

[37] C.-H. Lin, *Compact Modeling of Nanoscale CMOS*, Ph.D. Thesis (Technical Report No. UCB/EECS-2007-169), University of California at Berkeley, CA, 2007.

[38] *UFDG MOSFET Model User's Guide* (Version 3.5), SOI Group, University of Florida, Gainsville, Fl, 2005.

[39] G.D.J Smit, A.J. Scholten, G. Curatola, R. van Langevelde, G. Gildenblat and D.B.M. Klaassen, PSP-based Scalable Compact FinFET Model, *NSTI-Nanotech*, **3**, 520–525, 2007.

Appendix A:
Expression for Electric Field and Potential Variation in the Semiconductor Space Charge under the Gate

Poisson's equation for the space charge region at the Si–SiO$_2$ interface of a MOS capacitor structure is described by the following equation:

$$\frac{d^2\phi(x)}{dx^2} = -\frac{\rho(x)}{\epsilon_{Si}} \tag{A.1}$$

where $\rho(x)$ represents the volume space charge density comprising electrons $n_p(x)$, holes $p_p(x)$, and ionized dopant atoms ($N_A^- = p_{p0} - n_{p0}$):

$$\rho(x) = q\left[(p_p(x) - p_{p0}) - (n_p(x) - n_{p0})\right] \tag{A.2}$$

Using Equations (A.1) and (A.2), we get:

$$\frac{d^2\phi(x)}{dx^2} = -\frac{q}{\epsilon_{Si}}\left[p_{p0}\left(e^{-\frac{\phi(x)}{\phi_t}} - 1\right) - n_{p0}\left(e^{\frac{\phi(x)}{\phi_t}} - 1\right)\right] \tag{A.3}$$

The LHS can be recast as:

$$\frac{d^2\phi(x)}{dx^2} = \frac{d}{dx}\frac{d\phi(x)}{dx} = \frac{d\phi(x)}{dx}\frac{d}{d\phi(x)}\left(\frac{d\phi(x)}{dx}\right) \tag{A.4}$$

Compact MOSFET Models for VLSI Design   A.B. Bhattacharyya
© 2009 John Wiley & Sons (Asia) Pte Ltd

Integrating both sides of Equation (A.3), and using the relation in Equation (A.4):

$$\int \frac{d\phi(x)}{dx} d\left(\frac{d\phi(x)}{dx}\right) = -\frac{q}{\epsilon_{Si}} \int \left[p_{p0}\left(e^{-\frac{\phi(x)}{\phi_t}} - 1\right) - n_{p0}\left(e^{\frac{\phi(x)}{\phi_t}} - 1\right) \right] d\phi(x) \quad \text{(A.5)}$$

Considering $\frac{d\phi(x)}{dx}$ as the variable, and using the boundary conditions $\frac{d\phi(x)}{dx} = 0$ at $\phi(x) = 0$ (at the space charge region–neutral substrate boundary), we get:

$$\frac{1}{2}\left(\frac{d\phi(x)}{dx}\right)^2 = -\frac{q}{\epsilon_{Si}} p_{p0}\left[\left(-\phi_t e^{\frac{-\phi(x)}{\phi_t}} + \phi_t - \phi(x)\right)\right.$$
$$\left. - \frac{n_{p0}}{p_{p0}}\left(\phi_t e^{\frac{\phi(x)}{\phi_t}} - \phi_t - \phi(x)\right)\right] \quad \text{(A.6)}$$

$$\left(\frac{d\phi(x)}{dx}\right)^2 = -\frac{2q}{\epsilon_{Si}} p_{p0}\left[\left(-\phi_t e^{\frac{-\phi(x)}{\phi_t}} + \phi_t - \phi(x)\right)\right.$$
$$\left. - \frac{n_{p0}}{p_{p0}}\left(\phi_t e^{\frac{\phi(x)}{\phi_t}} - \phi_t - \phi(x)\right)\right] \quad \text{(A.7)}$$

Substituting (i) $n_{p0}/p_{p0} = n_i e^{-\phi_f/\phi_t}/n_i e^{\phi_f/\phi_t} = e^{-2\phi_f/\phi_t}$, (ii) $p_{p0} = N_A$, and (iii) $F(x) = -d\phi/dx$ in Equation (A.7) and rearranging, we have the expression for the spatial dependence of the electric field in the space charge region:

$$F^2(x) = \frac{2q}{\epsilon_{Si}} N_A \left[\left(\phi_t e^{\frac{-\phi(x)}{\phi_t}} + \phi(x) - \phi_t\right) + e^{\frac{-2\phi_f}{\phi_t}}\left(\phi_t e^{\frac{\phi(x)}{\phi_t}} - \phi(x) - \phi_t\right)\right] \quad \text{(A.8)}$$

or,

$$F(x) = \pm\sqrt{\frac{2qN_A\phi_t}{\epsilon_{Si}}}\left[\left(e^{\frac{-\phi(x)}{\phi_t}} + \frac{\phi(x)}{\phi_t} - 1\right) + e^{-\frac{2\phi_f}{\phi_t}}\left(e^{\frac{\phi(x)}{\phi_t}} - \frac{\phi(x)}{\phi_t} - 1\right)\right]^{\frac{1}{2}} \quad \text{(A.9)}$$

Substituting $L_b = (\epsilon_{Si}\phi_t/2qN_A)^{1/2}$, where L_b is the Debye length for the extrinsic silicon, we have:

$$F(x) = \pm\frac{\phi_t}{L_b}\sqrt{f\left[\phi(x)\right]} \quad \text{(A.10)}$$

where $f\left[\phi(x)\right]$ represents the expression within the square brackets in Equation (A.9):

$$f\left[\phi(x)\right] = \left(e^{\frac{-\phi(x)}{\phi_t}} + \frac{\phi(x)}{\phi_t} - 1\right) + e^{-\frac{2\phi_f}{\phi_t}}\left(e^{\frac{\phi(x)}{\phi_t}} - \frac{\phi(x)}{\phi_t} - 1\right) \quad \text{(A.11)}$$

The sign on the RHS of Equation (A.10) is chosen appropriately based on the sign of the potential $\phi(x)$. As an illustration, assume that the electric field at the surface is directed in the positive x-direction. As $F = -\frac{d\phi}{dx}$, the potential gradient must be negative. The boundary condition dictates that the potential zero is located at the edge of the bulk neutral region in silicon which is positioned away from the surface $(x = 0)$, in the positive x-direction. Thus, $F(x)$ is positive only when the surface potential is positive. Similarly, for the surface field directed in the negative direction, the surface potential corresponds to a negative value. Thus, Equation (A.10) can be written as:

$$F(x) = sgn\,(\phi_s)\frac{\phi_t}{L_b}\left[\left(e^{\frac{-\phi(x)}{\phi_t}} + \frac{\phi(x)}{\phi_t} - 1\right) + e^{-\frac{2\phi_f}{\phi_t}}\left(e^{\frac{\phi(x)}{\phi_t}} - \frac{\phi(x)}{\phi_t} - 1\right)\right]^{\frac{1}{2}} \qquad (A.12)$$

At $x = 0$, $\phi(x) = \phi_s$ and $F(x) = F_{Si}(0)$, which gives:

$$F_{Si}(0) = sgn\,(\phi_s)\frac{\phi_t}{L_b}\left[\left(e^{\frac{-\phi_s}{\phi_t}} + \frac{\phi_s}{\phi_t} - 1\right) + e^{-\frac{2\phi_f}{\phi_t}}\left(e^{\frac{\phi_s}{\phi_t}} - \frac{\phi_s}{\phi_t} - 1\right)\right]^{\frac{1}{2}} \qquad (A.13)$$

where $sgn\,(\phi_s) = +1$ for $\phi_s \geq 0$, and $sgn\,(\phi_s) = -1$ for $\phi_s < 0$.

The surface electric field $F_{Si}(0)$ is related to the surface space charge density Q'_{sc} (cm^{-2}) by Gauss' law:

$$F_{Si}(0) = -\frac{Q'_{sc}}{\epsilon_{Si}} \qquad (A.14)$$

$F_{Si}(0)$, and field in the oxide F_{ox}, are related by

$$F_{ox} = \frac{\epsilon_{Si}}{\epsilon_{ox}}F_{Si}(0) \qquad (A.15)$$

due to continuity of displacement vector at the interface.

KVL across the MOS capacitor structure gives the *Input Voltage Equation* that relates the gate voltage to the surface potential ϕ_s:

$$V_{GB} = V_{ox} + \phi_s = F_{ox}T_{ox} + \phi_s = \frac{\epsilon_{Si}}{\epsilon_{ox}}F_{Si}(0)T_{ox} + \phi_s \qquad (A.16)$$

where V_{GB} is the input gate voltage across the capacitor, and T_{ox} is the gate oxide thickness.

For an *ideal* MOS structure with flat band condition, $V_{fb} = 0$, ϕ_s tracks the polarity of V_{GB}.

Appendix B:
Features of Select Compact MOSFET Models

The major features of a few currently popular compact models are given in Table B.1 [1].

Table B.1 Features of select compact MOSFET models

Feature	Model				
	BSIM3v3	BSIM4	EKV2.6	PSP	HiSIM
Approximate number of parameters	190	259	41	155	72
Model framework	Threshold voltage, V_T, based	Threshold voltage, V_T, based	Inversion Charge, Q_n, based	Surface potential, ϕ_s, based	Surface potential, ϕ_s, based
Approach to drain current modeling	Regional, piecewise for weak and strong inversion	Regional, piecewise for weak and strong inversion	Regional (normalized to specific current) with interpolation	Drift and diffusion, single equation for all regions	Drift and diffusion, single equation for all regions
Reference potential terminal	Source	Source	Bulk	Bulk	Bulk

(continued)

Compact MOSFET Models for VLSI Design   A.B. Bhattacharyya
© 2009 John Wiley & Sons (Asia) Pte Ltd

Table B.1   *(Continued)*

Feature	Model				
	BSIM3v3	BSIM4	EKV2.6	PSP	HiSIM
Non-quasistatic effect	Subcircuit simulation for large signal NQS effect	Subcircuit simulation for large signal NQS effect	Lumped circuit approximation for distributed channel for small signal NQS	Spline collocation method for solving differential equation	Numerical solution of quasistatic differential equation with transit delay
Lateral channel inhomogeneity	Step approximation for halo doping	Step approximation for halo doping	Arora's model for lateral channel doping profile	Lateral field gradient	Triangular doping profile and lateral field gradient
Short channel/DIBL effects	Quasi-2D model for solving Poisson's equation	Quasi-2D model for solving Poisson's equation	Charge sharing model with fitting parameters	Lateral field gradient; Voltage transformation	Lateral field gradient
Symmetry	No	No[a]	Yes	Yes	Yes

[a]BSIM5 satisfies symmetry requirements.

Reference

[1] G. Angelov, T. Takov and S. Ristic, MOSFET Models at the Edge of 100-nm Sizes, in *Proceedings of the 24th International Conference on Microelectronics (MIEL 2004)*, **1**, 295–296, 2004.

Appendix C:
PSP Two-point Collocation Method

In this appendix we shall discuss the two point collocation method ($N = 2$) given in Wang *et al.* [1]. The charge sheet model equation for the channel current of a MOS transistor and continuity equation give:

$$\frac{\partial q'_n}{\partial t} + \frac{\mu_{eff}}{L^2}\frac{\partial}{\partial y}\left[\left(\frac{q'_n}{\frac{\partial q'_n}{\partial \phi_s}} - \phi_t\right)\frac{\partial q'_n}{\partial y}\right] = 0 \tag{C.1}$$

where $q'_n = Q'_n/C'_{ox}$, and the coordinate y is normalized to the channel length L. The mobility is assumed to be constant along the channel with the value μ_{eff}. The non-quasistatic model considering the velocity saturation effect is discussed in Wang *et al.* [2]. Let the second term in the above equation be $G(q'_n)$. Thus:

$$G(q'_n) = \frac{\mu_{eff}}{L^2}\frac{\partial}{\partial y}\left[\left(\frac{q'_n}{\frac{\partial q'_n}{\partial \phi_s}} - \phi_t\right)\frac{\partial q'_n}{\partial y}\right] \tag{C.2}$$

$$G(q'_n) = \frac{\mu_{eff}}{L^2}\left[\left(\frac{q'_n}{\frac{dq'_n}{d\phi_s}} - \phi_t\right)\frac{\partial^2 q'_n}{\partial y^2} + \frac{\partial q'_n}{\partial y}\frac{\partial}{\partial y}\left(\frac{q'_n}{\frac{\partial q'_n}{\partial \phi_s}} - \phi_t\right)\right] \tag{C.3}$$

which can be written as:

$$G(q'_n) = \frac{\mu_{eff}}{L^2}\left[\psi(q'_n)\frac{\partial^2 q'_n}{\partial y^2} + \varphi(q'_n)\left(\frac{\partial q'_n}{\partial y}\right)^2\right] \tag{C.4}$$

Compact MOSFET Models for VLSI Design   A.B. Bhattacharyya
© 2009 John Wiley & Sons (Asia) Pte Ltd

where:

$$\psi(q'_n) = \left(\frac{q'_n}{\frac{\partial q'_n}{\partial \phi_s}} - \phi_t \right) \tag{C.5}$$

and:

$$\varphi(q'_n) = \frac{\partial q'_n}{\partial \phi_s} - q'_n \frac{\frac{\partial^2 q'_n}{\partial \phi_s^2}}{\left(\frac{\partial q'_n}{\partial \phi_s} \right)^3} \tag{C.6}$$

In Equations (C.5) and (C.6), $\partial q'_n / \partial \phi_s$ and $\partial^2 q'_n / \partial \phi_s^2$ can easily be obtained for $\phi_s > 3\phi_t$. The normalized bulk charge density q'_b and inversion charge density q'_n are expressed as:

$$q'_b = -\gamma \sqrt{\phi_s - \phi_t} \tag{C.7}$$

and:

$$q'_n = -(V_{GB} - V_{fb} - \phi_s - q'_b) \tag{C.8}$$

Using Equations (C.7) and (C.8), ϕ_s can be expressed in terms of q'_n as follows:

$$\phi_s(q'_n) = V_{GB} - V_{fb} + q'_n + \frac{\gamma^2}{2} - \gamma \left(\gamma^2/4 + V_{GB} - V_{fb} + q'_n - \phi_t \right)^{1/2} \tag{C.9}$$

Further, differentiating Equations C.7 and C.8 w.r.t. ϕ_s, and rearranging, we get:

$$\frac{\partial q'_n}{\partial \phi_s} = \frac{q'_b}{q'_b - \frac{\gamma^2}{2}} \tag{C.10}$$

and:

$$\frac{\partial^2 q'_n}{\partial \phi_s^2} = \frac{\gamma^4}{4q'_b} \tag{C.11}$$

More generic expressions for $\partial q'_n / \partial \phi_s$ and $\partial^2 q'_n / \partial \phi_s^2$ valid in all regions of operation are given in Wang *et al.* [2].

The original partial differential equation becomes:

$$\frac{\partial q'_n}{\partial t} + G(q'_n) = 0 \tag{C.12}$$

$$\frac{\partial q'_n}{\partial t} + \frac{\mu_{eff}}{L^2} \left[\psi(q'_n) \frac{\partial^2 q'_n}{\partial y^2} + \varphi(q'_n) \left(\frac{\partial q'_n}{\partial y} \right)^2 \right] = 0 \tag{C.13}$$

In the PSP model the above equation is solved using the spline collocation version of the weighted residual method.

Let $u(y, t)$ be an approximation to $q'_n(y, t)$. Divide the channel into three segments with y, ranging from $[0, 1/3]$, $[1/3, 2/3]$, and $[2/3, 1]$. Let the normalized inversion charge density be denoted by u_0, u_{1p}, u_{2p}, and u_1 at the points defining the endpoints of the ranges respectively. At the points u_{1p} and u_{2p} we can write:

$$\frac{\partial u_{1p}}{\partial t} + \frac{\mu_{eff}}{L^2} \left[\psi(u_{1p}) \left. \frac{\partial^2 u}{\partial y^2} \right|_{u=u_{1p}} + \varphi(u_{1p}) \left(\left. \frac{\partial u}{\partial y} \right|_{u=u_{1p}} \right)^2 \right] = 0 \qquad (C.14)$$

$$\frac{\partial u_{2p}}{\partial t} + \frac{\mu_{eff}}{L^2} \left[\psi(u_{2p}) \left. \frac{\partial^2 u}{\partial y^2} \right|_{u=u_{2p}} + \varphi(u_{2p}) \left(\left. \frac{\partial u}{\partial y} \right|_{u=u_{2p}} \right)^2 \right] = 0 \qquad (C.15)$$

The collocation points u_0, u_{1p}, u_{2p}, and u_1 are interpolated using a cubic spline polynomial to give a smoothly varying $u(y, t)$, with a continuity up to the third derivative. From the interpolation function (to be shown later), we can have $\partial u / \partial y$ and $\partial^2 u / \partial y^2$ at u_{1p} and u_{2p} in terms of u_0, u_{1p}, u_{2p}, and u_1. This converts the above two partial differential equations to ordinary differential equations as shown below:

$$\frac{du_{1p}}{dt} + \frac{\mu_{eff}}{L^2} \left[\psi(u_{1p}) f_1(u_0, u_{1p}, u_{2p}, u_1) + \varphi(u_{1p}) f_2(u_0, u_{1p}, u_{2p}, u_1) \right] = 0 \qquad (C.16)$$

$$\frac{du_{2p}}{dt} + \frac{\mu_{eff}}{L^2} \left[\psi(u_{2p}) f_3(u_0, u_{1p}, u_{2p}, u_1) + \varphi(u_{2p}) f_4(u_0, u_{1p}, u_{2p}, u_1) \right] = 0 \qquad (C.17)$$

The above set of ordinary differential equations, with appropriate initial conditions, can be solved using the SPICE inbuilt algorithm, or by the sub-circuit method, to give $u_{1p}(t)$ and $u_{2p}(t)$, and hence, $u(y, t)$ through interpolation. In the sub-circuit method each of the above ordinary differential equations is represented using a sub-circuit based on a voltage controlled current source.

The spline interpolation function for u is given below:

$$u = \begin{cases} u_0 + (3u_{1p} - 3u_0 + 3a_0)y + (3a_1 - 9a_0)y^2 - 9a_1 y^3 & 0 \leq y \leq \frac{1}{3} \\ 2u_{1p} - u_{2p} - 2a_2 + (3u_{2p} - 3u_{1p} + 9a_2 - 2a_3)y + (9a_3 - 9a_2)y^2 - 9a_3 y^3 & \frac{1}{3} \leq y \leq \frac{2}{3} \\ 3u_{2p} - 2u_1 - 6a_4 + (3u_1 - 3u_{2p} + 15a_4 - 6a_5)y + (15a_5 - 9a_4)y^2 - 9a_5 y^3 & \frac{2}{3} \leq y \leq 1 \end{cases}$$

$$(C.18)$$

The condition of continuity of u, its first, and second order derivatives at the collocation points u_{1p} and u_{2p}, together with the conditions that $u'' = 0$ at $y = 0$ and $y = 1$, give coefficients

a_1, a_2, a_3, a_4, and a_5 as shown below:

$$a_0 = \frac{1}{15}\left(-4u_0 + 9u_{1p} - 6u_{2p} + u_1\right) \tag{C.19a}$$

$$a_1 = a_2 = 3a_0 \tag{C.19b}$$

$$a_3 = u_0 - 3u_{1p} + 3u_{2p} - u_{1p} \tag{C.19c}$$

$$a_4 = \frac{1}{15}\left(4u_0 - 24u_{1p} + 36u_{2p} - 16u_1\right) \tag{C.19d}$$

$$a_5 = \frac{1}{5}\left(-u_0 + 6u_{1p} - 9u_{2p} + 4u_1\right) \tag{C.19e}$$

From Equations (C.18) and (C.19a–C.19e) we have:

$$f_1 = \frac{1}{5}(-7u_0 - 3u_{1p} + 12u_{2p} - 2u_{1p}) \tag{C.20}$$

$$f_2 = \frac{18}{5}(4u_0 - 9u_{1p} + 6u_{2p} - u_{1p}) \tag{C.21}$$

$$f_3 = \frac{1}{5}(2u_0 - 12u_{1p} + 3u_{2p} + 7u_{1p}) \tag{C.22}$$

$$f_4 = \frac{18}{5}(-u_0 + 6u_{1p} - 9u_{2p} + 4u_{1p}) \tag{C.23}$$

From Equation (C.18), and the solution of the set of ordinary differential equations, we get $u(y, t)$. The source and drain dynamic currents can be calculated from $u(y, t)$ following Ward–Dutton charge partitioning scheme:

$$I_S = -I_T + C_{ox}\frac{d}{dt}\int_0^1 (1 - y)u \, dy \tag{C.24}$$

and:

$$I_D = I_T - C_{ox}\frac{d}{dt}\int_0^1 yu \, dy \tag{C.25}$$

where $C_{ox} = WLC'_{ox}$, and I_T is the transport current as discussed in Chapter 4.

Substituting $u(y, t)$, given by Equation (C.18), in Equations (C.24) and (C.25), we have:

$$I_S = -I_T + C_{ox}\left(\frac{11}{91}\frac{du_0}{dt} + \frac{4}{15}\frac{du_{1p}}{dt} + \frac{1}{10}\frac{du_{2p}}{dt} + \frac{1}{90}\frac{du_1}{dt}\right) \tag{C.26}$$

and:

$$I_D = I_T + C_{ox} \left(\frac{1}{90} \frac{du_0}{dt} + \frac{1}{10} \frac{du_{1p}}{dt} + \frac{4}{15} \frac{du_{2p}}{dt} + \frac{11}{90} \frac{du_1}{dt} \right) \tag{C.27}$$

References

[1] H. Wang, T.L. Chen and G. Gildenblat, Quasi-static and Nonquasi-static Compact MOSFET Models Based on Symmetric Linearization of the Bulk and Inversion Charges, *IEEE Trans. Electron Dev.*, **50**(11), 2262–2272, 2003.
[2] H. Wang, X. Li, W. Wu, G. Gildenblat, R. van Langevelde, G.D. J Smit, A. J. Scholten and D. B. M. Klaassen, A Unified Nonquasi-Static MOSFET Model for Large-Signal and Small-Signal Simulations, *IEEE Trans. Electron Dev.*, **53**(9), 2035–2043, 2006.

Index